A Crude Look
at the Whole

A Decade Later: Insights into Real-World Complexity that Still Resonate

Exploring Complexity

ISSN: 2382-5901

Series Editors: Jan Wouter Vasbinder
Nanyang Technological University

Helena Hong Gao
Nanyang Technological University

Editorial Board Members:

W. Brian Arthur, *Santa Fe Institute*
Robert Axtell, *George Mason University*
John Steve Lansing, *Nanyang Technological University*
Stefan Thurner, *Medical University of Vienna*
Geoffrey B. West, *Santa Fe Institute*

For four centuries our sciences have progressed by looking at its objects of study in a reductionist manner. In contrast complexity science, that has been evolving during the last 30–40 years, seeks to look at its objects of study from the bottom up, seeing them as systems of interacting elements that form, change, and evolve over time. Complexity therefore is not so much a subject of research as a way of looking at systems. It is inherently interdisciplinary, meaning that it gets its problems from the real non-disciplinary world and its energy and ideas from all fields of science, at the same time affecting each of these fields.

The purpose of this series on complexity science is to provide insights in the development of the science and its applications, the contexts within which it evolved and evolves, the main players in the field and the influence it has on other sciences.

For the complete list of volumes in this series, please visit www.worldscientific.com/series/ec

Exploring Complexity – Volume 12

A Crude Look at the Whole

A Decade Later: Insights into Real-World Complexity that Still Resonate

Editors

Jan Wouter Vasbinder

Helena Hong Gao

NEW JERSEY · LONDON · SINGAPORE · BEIJING · SHANGHAI · HONG KONG · TAIPEI · CHENNAI · TOKYO

Published by

World Scientific Publishing Co. Pte. Ltd.
5 Toh Tuck Link, Singapore 596224
USA office: 27 Warren Street, Suite 401-402, Hackensack, NJ 07601
UK office: 57 Shelton Street, Covent Garden, London WC2H 9HE

Library of Congress Control Number: 2024947855

British Library Cataloguing-in-Publication Data
A catalogue record for this book is available from the British Library.

Reprinted 2025 (in paperback edition)
ISBN 978-981-98-0000-1 (pbk)

Exploring Complexity — Vol. 12
A CRUDE LOOK AT THE WHOLE
A Decade Later: Insights into Real-World Complexity that Still Resonate

ISBN 978-981-98-0212-8 (hardcover)
ISBN 978-981-98-0213-5 (ebook for institutions)
ISBN 978-981-98-0214-2 (ebook for individuals)

For any available supplementary material, please visit
https://www.worldscientific.com/worldscibooks/10.1142/14080#t=suppl

Desk Editor: Muhammad Ihsan Putra

Typeset by Stallion Press
Email: enquiries@stallionpress.com

Preface

Sir Isaac Newton famously said, *If I have seen further, it is by standing on the shoulders of giants*. It may have been modesty that made him say so, it may also have been the thrill he felt when contemplating what these giants made him see.

In March 2013, Para Limes organized the conference *A Crude look at the Whole*.

It turned out to be an extraordinary meeting. It *was* extraordinary *when it happened*, it *is* even more so *eleven years later*. During the conference, the speakers, all giants on their own turf, captured the excitement about what the new field of complexity science could mean for understanding our world and molded it in approaches to extract meaning from these budding insights. Now, eleven years later, the (videos of the) talks create a thrill, that may be similar to what Newton felt when he realized that giants allowed him to see what he saw. The thrill is so much stronger because, while eleven years ago the world looked very different from today's world, the views these giants offered at the conference bear more weight than ever.

Eleven years ago, all the global problems humanity is facing today were manifest already. But looking backward, they looked less grim. There was the threat of a pandemic, but there was always a threat of a pandemic. The changing climate posed a danger, but a change in climate always poses a danger, and this one seemed manageable still.

Eleven years ago, we could still have the illusion that we were in control. Now, we know that we cannot be in control (and we never could), because of the *complexity* of the natural, social, and artificial systems that we are part of. The giants of 2013 pointed to this complexity as the context within which human cultures and societies (and their problems) have developed in the past and they explored ways to deal with it in the present.

In the eleven years since then, the reasons for and the results of these explorations have largely been ignored. We, or rather our leaders,[1] still cherish the illusion that we can control the complexity of our systems through linear interventions, and ignore the disastrous consequences of acting on that illusion.

That is why the world looks grimmer now than it did eleven years ago.

The views offered at the conference *A Crude Look at the Whole* are even more relevant now than they were in 2013. When Helena Gao and I revisited the videos of the conference in 2023, we were struck by realization that, rather than moving toward dealing with complexity, we still reject it, and we do so more and more at our existential peril. But we can still make the turn and learn how to embrace complexity. A magnificent body of insights to do so presented itself through the video recordings[2] of the conference.

These recordings span a continuous spectrum of key aspects of complexity, from the broad view of the talk, *A Crude Look at the Whole*, by Murray Gell-Mann to the mechanism for steering complex adaptive systems, in John Holland's talk *Signals and Boundaries*. In between are explorative and enlightening talks on the following:

- the dynamic nature of ecosystems and the complexities involved in understanding and managing them effectively, by Simon Levin;
- the complex interplay between human activities and the Earth's systems, emphasizing the urgent need for a paradigm shift toward sustainable development within planetary boundaries, by Johan Rockström;

---

[1] As in government, politics, industry, academia, and institutions.

[2] The videos are accessible at the following: https://www.paralimes.org/past-events/conference-a-crude-look-at-the-whole/

- an examination of society through images, the impact of science and technology on societal change, and the exploration of the temporal dimension of complexity, by Helga Nowotny;
- governance, complexity, resilience, and strategic planning, by Peter Ho;
- the intricate relationship between bioenergy production, land use, and agricultural commodity markets, by Kristian Lindgren;
- the enigmatic phenomenon known as the Cambrian explosion, a pivotal event in the history of life on Earth that occurred roughly 550 to 600 million years ago, by Doug Erwin;
- the relationship between culture, cognition, and behavior, and how individuals from different cultural backgrounds perceive and react to their environment, by Ying Yi Hong;
- the significance of paleoclimate records, particularly focusing on the Asian monsoon system, and its impact on civilization, by Wang Xian Feng;
- the intricate complexities of the HIV pandemic and the interdisciplinary efforts required to combat it effectively, by Peter Sloot;
- the importance of schemas in reducing complexity and managing relationships, with implications for understanding global dynamics, predicting behavior, and making decisions, by Robert Axelrod.

While each of the recorded talks stood on its own, and none was informed by the other talks, together these talks exposed a view on a way of knowing that became largely hidden as modern science emerged. That is the "knowing" that we gain by looking at the whole, the whole being all the parts, the connections between the parts and the relationships and dynamic interactions between the parts along those connections. It is knowing how to deal with complexity at a human scale, in families, tribes, and small communities.[3]

---

[3] At a human scale, from families to small communities, people always knew how to deal with complex systems — finding food and shelter while maintaining a balance with their environment; managed through village elders, oral traditions, and living memories. People were part of a whole that they knew, had their place in, were dependent on, and could handle. Complexity did not get a name; it was *oneness*.

Our natural ways to handle complexity at that scale were replaced by the linear thinking that gained prominence through the advance of modern reductionist science. As this science spawned technologies that started to overhaul the ways we lived (through health care, communication, mobility, etc.), the human population and its communities grew in number and size, and the interactions between those communities intensified. Concurrently, complexity grew way beyond the human scale, and our natural ways of gaining knowledge of how to deal with it lost its human context.

> **A crude look at modern science**
>
> *Modern science began to emerge after Nicolaus Copernicus published "De revolutionibus orbium celestium." Through developments in mathematics, physics, astronomy, biology, and chemistry, it slowly transformed the views of society about nature. In the process, it created a belief that we can increase our understanding of the world we live in by searching for general laws and fundamental particles in laboratories that are totally isolated from that world. Within that isolation, we dissected complex problems into simpler problems that we thought we could handle. And we thought that by linearly adding the solutions to the simple problems, we could handle the complex problems. Only since the early seventies of the last century did we begin to see that those solutions do not work.*

As the human-induced global problems grow in complexity, humanity is in an urgent need to regain its sense of knowing how to deal with this complexity. Exploring ways of how to do that is what each of the speakers did at the conference *A Crude Look at the Whole*. It was such an extraordinary conference because each speaker was a giant in his or her field, and because the whole was a lot more than the sum of its parts.

In this book, we have tried to capture that whole while at the same time keeping the individual parts in view. We have done so by transcribing and editing the individual presentations, by adding a summary to all of them, and by indicating the relevance of each of the presentations to ongoing and further explorations.

We strongly recommend that you read the summaries first, the references to future explorations last, and the edited individual presentations in between.

Regarding the editing, we have opted for readability. Often that meant that we had to shorten the transcribed texts. We have tried to do so without changing the message or the context. If you want to get the full flavor of the talks and the discussions after each of the talks, we suggest that you listen to the original videos,[4] as you read the transcripts.

Unfortunately, two of the speakers have died.

John H. Holland[5] passed away in 2015 at the age of 86 years. In many ways, John shaped the field of complexity science. He was the father of Genetic Algorithms and Complex Adaptive Systems. The book *Signals and Boundaries* he presented at the 2013 conference was one of his last.

Murray Gell-Mann[6] passed away in 2019 at the age of 89 years. Murray was considered one of the most intelligent people on earth. He received a Nobel Prize for his discovery of the Quarks and inspired the conference *A Crude Look at the Whole*.

We have tried to capture the spirit of their talks by editing them as little as possible.

---

[4] The videos are accessible at the following: https://www.paralimes.org/past-events/conference-a-crude-look-at-the-whole/.

[5] See for an obituary: https://www.nytimes.com/2015/08/20/science/john-henry-holland-computerized-evolution-dies-at-86.html.

[6] See for an obituary: https://cerncourier.com/a/gell-manns-multi-dimensional-genius/#:~:text=Murray%20Gell%2DMann%20was%20one,Credit%3A%20WEF%2FM%20Wuertenberg.

Some Reflections on the Last Eleven Years and Before

Johan Rockström's talk (*Resilience for Human Development*) eleven years ago about the state of the global ecosystems and today's situation is a stark reminder that, while knowledge about the effects of human interventions in our ecosystems accumulates, our capacity to deal with the problems these effects cause stagnates. One major cause is that the complex adaptive nature of the ecosystems conflicts with the linear nature of human interventions.

Conversely, Peter Ho's talk (*A Crude Look at Governance and Complexity*), highlighting the challenges complexity poses to governance, is a stark reminder that during the past eleven years, very little headway has been made in finding effective ways to predict and deal with possible effects of human interventions in their ecosystems.

About change and complexity

Ecosystems are complex adaptive systems. Such systems change continuously. Human ecosystems are ecosystems where humans are major agents of change. Until about a hundred years ago, changes in such ecosystems took place over many generations. By the beginning of the twentieth century, such changes happened within only a few generations. In the last hundred years, enormous changes in human

Speed of change

World population: Between 1900 and 2020 (roughly six generations), the world population grew from 1.6 to 8.0 billion people. It took from AD 1100 to AD 1900 (roughly forty generations) for the world population to grow by the same factor (from 320 million to 1.6 billion).

Connectivity: In the last one hundred years, the mode of transportation changed from horses to cars to jet planes, and the mode of communication changed from letters to telephone to the internet. All this happened within a period of less than a hundred years. As a result of such changes, the world became a lot more complex, and that increase in complexity happened faster and faster.

ecosystems began to happen within a generation. Science, and in particular the technologies it gave birth to,[7] became the major driving force behind those changes. Technologies intervene, along linear paths, with complex adaptive systems, leading in an unpredictable way to unanticipated consequences. Often, technologies aim to simplify complex situations, but in practice, they lead to an increase in complexity that is more and more difficult to deal with. Climate change, plastic soup, and exponentially increasing inequality are examples.

About technology

None of the speakers specifically talked about technology. But, looking back eleven years, a most striking development has been

> **Our brain, digital technologies, and AI**
>
> *Our brain: Our sole tool to simplify that increasing complexity (our brains) did not change (at least not noticeably) and, as a result, the capability of humans to simplify their world in harmony with their ecology started to run out of sync with the accelerating rate of change in that ecosystem.*
>
> *Digital technologies: With digital technologies, men first increased the speed with which calculations could be done, and then started to use that speed in algorithms that could simulate certain patterns of behavior. That ever-increasing speed and those ever more sophisticated algorithms have now come to the point where we can entertain the illusion that we have improved or extended our brain and even that we can replace it by AI.*
>
> *Artificial Intelligence (AI) is a term that follows the notion put forward by René Descartes that you can separate body and mind. Natural intelligence emerges from a total integration of body and mind. If one separates the two, the mind part is not artificially intelligent, but dead. And it is only by enormous amounts of electrical energy that we can make a small function of it seem alive.*

the deepening and accelerating penetration of digital technologies in the networks of complex systems that make up our world. This penetration has enormous effects on the control and accessibility of information, the education of our children, the privacy of our lives, the inequality in the

---

[7]That includes medical technologies such as birth control.

world, the security of our life support systems, our capability to govern, and safety in the world in general. As our insights into the nature and dynamics of complex systems increased, we began to see that interventions in complex systems lead to unexpected consequences (for the systems being interfered with) that are often devastating.

Yet, that insight has (so far) not led leaders of society[8] to a shift in one of their dominant lines of thinking. In fact, there is a practically ubiquitous belief among these leaders (and to an unknown extent among the public) that technology will solve the problems humanity has created for itself, and that more technology will solve those problems faster. It is the false belief that more of the same will get us out of trouble.

That belief transforms into a line of thinking that we need to develop more technologies — that to develop those technologies we need more knowledge, that to generate more knowledge we need more science, that for more science we need more scientists,

An evolutionary solid line of belief-1

We cannot extend control of the ecosystems we are part of beyond the complexity that our brains have evolved to deal with (control). These brains have evolved as a complex adaptive system that "knows" how to deal with complexity at a human scale, from families to small communities.

Some 9–10 thousand years ago, with the onset of agriculture, the (collective) human brain started to interfere with its (complex adaptive) ecosystems through linear interventions. Initially, the effects of such interventions went unnoticed. But over time, as interventions accumulated, their effects led to an exponential growth of complexity that moved beyond our control. In our times, we will go through the last part of these exponential curves. More of the same will lead to inevitable disaster.

There is a way out: Rely on the complexity to which our brains have evolved to cope with the complexity of the systems around us. That means complexity at a human scale, in small communities, living in harmony with the ecosystems that we are part of.

---

[8] As in *government* (regional, national, and international), politics, industry, academia, and institutions.

and that to breed more scientists we need to invest more money in universities and research.

It is a singular line of thinking that will lead to disaster, as it already has led to climate change, loss of biodiversity, a plastic soup in the oceans, the usurpation of natural resources by a the few, and an increasing inequality around the globe.

About science[9]

Science is defined by the people working in it. It is also defined by the methods they use to find the true answer to a question. Science is not about raising the right questions. By "right questions" we mean questions to which the answer is relevant for mankind. Somehow "right" seems a wrong word to indicate what relevant questions are. At the same time, "relevant" seems to be the wrong word to indicate what the right questions are.

Yet, in some way, during the last 500 years in the West,[10] science evolved to become regarded as the only field of human activity that raises questions worth answering. Equally, the situation evolved such that the choice of questions raised by scientists being predicated on the assumption that they could be answered by rigorous scientific methods was taken for granted. In other words, because science has methods to find true answers to questions, it became accepted (by scientists and non-scientists) that the right questions to ask are those that can be answered using such methods.[11] To scientists, the only relevant questions are those that can be tackled by scientific methods. These questions mostly originate and are addressed within the boundaries of one discipline, where we can believe that control is real instead of an illusion. But most questions that occupy the minds of individual people and are relevant to mankind in general cannot be answered using scientific methods. They do not originate in scientific disciplines but in the real, messy, and complex world.

---

[9]This paragraph uses texts taken from Jan W. Vasbinder and Jonathan Y.H. Sim (2023), *Fit for Purpose? The Futures of Universities.* World Scientific Publishing.

[10]The West is defined as Western Europe and, in the last 150 years, the US and Canada.

[11] Karl Popper (1984), *Conjectures and Refutations.* London: Routledge & Kegan Paul.

That real, messy, and complex world is what Murray Gell-Mann addressed in his talk *A Crude Look at the Whole*.

Our messy world needs a <u>shift</u> in thinking

Heraclitus may have provided a great metaphor for the complexity of evolution by saying "No man ever steps in the same river twice. For it's not the same river and he's not the same man." In other words, everything changes all the time.

An evolutionary solid line of belief-2

In their book "A Darwinian Survival Guide," Daniel Brooks and Salvatore Agosta provide profound insight into the complexity of evolution. The following are some of their conclusions relevant to "A Crude look at the Whole."

- *The biosphere is an evolutionary commons with enormous potential for coping with change by changing.*
- *We need cooperating networks of low-density communities in climate-secure places.*
- *When a community outgrows its carrying capacity, add a new node to the network.*
- *Social institutions must be constrained by fiduciary principles.*

From reflecting on the eleven years of our immediate past, we have come to believe that what our messy and complex world needs is not more technology, but leaders who have the ability to see that everything changes all the time and feel the urge to understand the complexity of the situations they will have to deal with.

People who are trained to take a *crude look at the whole* and hone their feelings into what can be done through exploration and experimentation.

Contents

Chapter 8 A Complex World from a Virus's Point of View **287**
Peter M. A. Sloot

Chapter 9 Collective Phenomena, Collective Motion, and Collective Action in Ecological Systems **357**
Simon Levin

Chapter 10 Resilience for Human Development in the Anthropocene **421**
Johan Rockstrom

Chapter 11 Isotope Chemistry, Climate Change, and the Fate of the Chinese Dynasties: Implications for the Future of Asian Societies **485**
Wang Xian Feng

Chapter 1

A Crude Look at the Whole

Murray Gell-Mann

YouTube:　　https://youtu.be/UlPXQyFV3FY

Speaker:　　Murray Gell-Mann

Moderator:　Brian Arthur

Discussants:　Simon Levin, Bob Axelrod, Suman Bannerjee, Guaning Su, Jan Carlstedt-Duke, Sheila Ranas, Peter Sloot, Gregg Fisher, Peter Ho

1.1 Talk by Murray Gell-Mann

It's a pleasure to be able to discuss with this distinguished group the idea of a "crude look at the whole". I gave the talk with that title some years ago under the auspices of the Pardee Center for the Study of the Longer-Range Future at Boston University. There was a conference on the topic looking ahead and I was asked to give the keynote talk at the dinner and I should like to begin today by expanding on that title, which was a *Crude Look at the Whole*. Many of the points that I treat here were discussed in my book, *The Quark And The Jaguar*, which I'm glad to say is still mostly relevant, even 19 or 20 years after publication.

Over the years, I've organized some meetings devoted to aspects of the longer-range future, and I've seen how difficult it is to get most intellectuals to discuss future situations that are significantly different in some way from what we've already seen in the recent past or the present. As an example of reluctance to consider big changes, we might take the wide-spread failure to believe that the Soviet Menace came with an early expiration date. Usually, the intellectuals to whom I refer end up talking not about major departures from present conditions but rather about relatively small ones now. That's not a bad thing to do, but it's not what we mean by peering into the longer-range future. Despite my limited experience, I was able to share some ideas with the scholars at that meeting, and I'll share a few of them with you.

Until now, a great deal of research and teaching in the sciences and the humanities, especially in universities, has been confined to individual departments representing particular fields of knowledge, and while specialization and then subspecialization are inevitable and necessarily desirable, they need to be supplemented by research and teaching to transcend those sometimes narrow disciplinary boundaries. There are a number of institutions inside and outside of universities where such transdisciplinary activities are carried out. This one that we are at is an example; in particular, as most of you know, the Santa Fe Institute which I helped to found 20 years ago and where I now work is devoted entirely to such activities. There, the study of transdisciplinary problems by self-organized teams of people originally trained in many diverse specialties is the rule rather than the exception. Similarities in connections among topics in vastly different fields are recognized and exploited. The participants may have started as experts in subjects from physical sciences, life sciences, social and behavioral sciences, history, or the other humanities. They take part in discussions of all sorts of topics; each team does include, of course, at least one real expert in the subject matter, or particular subject matter under consideration. If nobody knows anything, you don't get very far.

I've often spoken in public about the need for such research and about institutional arrangements for getting it done, and what I have to say this morning may sound similar, but it concerns a very different topic, so one has to listen very closely to hear the difference. That topic has to do not with academic disciplines and pure research and teaching, but rather with subject matter in policy studies, especially broad policy studies concerned with the whole future of the human race and of the biosphere of our planet, including the other species with which we share that biosphere.

And here is the point of difference: in considering any very complex system, we naturally tend to break it up into more manageable subsystems or aspects defined in advance and study those more or less separately. For example, looking at the longer-term future, we might divide the various issues into such categories as, military and diplomatic issues, which some might call security and foreign affairs; political issues, including domestic politics in each country or region; ideological issues; environmental issues, including ones related to water, air, energy, and biological diversity; human health and wellness issues; family issues; demographic issues; economic issues, including the crucial socio-economic challenge of relieving extreme poverty; technological issues; scientific research issues; institutional or governance issues; democracy and human rights issues of transcendent importance, and so forth. Other items might be added, but the general idea should be clear.

In each of these categories, there are experts who have built their careers around the issues involved, and there are NGOs and, in many cases, government departments devoted to these. As a citizen, you might join one of the NGOs, such as an environmental organization, one promoting human rights, one working for arms control, one advocating for strong national defense, or one focusing on child health around the world, and so on. It's natural and common to try to break up the world's problems, as some people call it, to break them up in such a way. But, as was already hinted here, the difficulty is that understanding a nonlinear system, especially a complex one, by putting together descriptions of various parts or aspects will work only if those parts or aspects interact weakly so that the system as a whole is decomposable. That's not true of the world's problems. It's very far from being decomposable, and studying various aspects and trying to put those together into a complete picture is not possible.

So, in that sense, there's truth in the old adage that the whole is more than the sum of its parts, which sounds on the face of it to be wrong, but, it's not. Look at the list of categories we discussed. Can we really separate environmental issues from those involving population growth? I'll answer it — no! Can we consider these in isolation from technological change or from economic policy? No! Can we think about the attempts to alleviate extreme poverty without considering the unwise environmentally destructive projects that are sometimes carried out in the name of that worthy cause? Can we discuss issues of global governance without considering politics in the various countries and regions well and without treating the competition and conflict among different ideologies? Think if military and diplomatic policies fail, and mankind is plunged into a hugely destructive war, perhaps involving thermonuclear weapons, can our other objectives be attained? Of course not! Isn't economic growth threatened by the widespread prevalence of fatal or debilitating diseases? Can we separate questions about democracy and human rights? Again, no! And what if democratic processes bring to power elements of society hostile to human rights and to tolerance or ones that favor environmental destruction or aggressive war? All of that can happen. Think how Adolf Hitler took power legally before destroying the whole democratic political system in Germany, or think of Klement Gottwald in Czechoslovakia. Remember, some people make fun of naive idealism about elections with the slogan "one man, one vote, one time". So, while separate consideration of the various aspects of the world situation is necessary and desirable, it very badly needs to be supplemented by integrative thinking. That not only puts together the studies of various aspects but also takes into account the strong interactions among them.

We have to get rid somehow of the widespread idea that careful study of a problem in some narrow range of issues is the only kind of work to be taken seriously. It's really a widespread impression. Integrated thinking is relegated to a cocktail party conversation. Well, that really has to change. We face that situation in a great many places in our society, including academia and most bureaucracies.

Some of my remarks on this subject were quoted by Tom Friedman near the beginning of one of his books. I think it's the one called *The Lexus and the Olive Tree*. He came to a similar conclusion through his career working for the *New York Times*. He was assigned to cover first one set of issues, and then a different set, then a third set. Each time he was reassigned, he observed that what he was covering was intimately connected with what he had been studying earlier. No one had told him that it would be so. So, besides the work on the separate categories, we must have, in addition, the work of teams of brilliant thinkers, many of them specialists, devoted to considering the whole ball of wax, the whole thing. It can, of course, be argued that this is too big a job for any group of people, no matter how talented or erudite to do really well, and that's true. Of course, such an ambitious aim can be accomplished only crudely, and that's why I refer to it as taking a crude look at the whole. If we insist on perfection, the whole thing is doomed from the beginning.

Now, the chief of any organization, say a head of government, or a CEO, has to behave as if he or she is taking into account all the aspects of policy, including all the interactions among those aspects. It's not so easy, however, for the chief to take a crude look at the whole if everyone else in the organization is concerned only with a partial look. Even if some advisors to the chief are assigned to look at the big picture, it doesn't always work. Here's an example. Some years ago, the CEO of a gigantic corporation told me that he had a strategic planning staff to help him think about the future of the whole business, but then members of that staff suffered from three defects: One, they seemed largely disconnected from the rest of the company. Two, no one could understand what they said. Three, everyone else in the company seemed to hate them. I checked with several people high up in that company and got the same response. For an integrative effort to succeed, some kind of simplification is naturally required. Certain things have to be treated in a cursory fashion and others in more detail, but that process, what physical scientists would like to call coarse-graining, cannot be accomplished through the categories established in advance. It has to follow from the nature of the world system itself.

The necessary coarse-graining should rather be discovered than imposed. Think of the relation between weather and climate. It's a fairly common thing to consider, but I'm not sure that people really are aware of what they're saying when they talk about weather and climate. No clear results will follow from trying to examine the weather at each little place on earth and each short interval of time, while neglecting the strong interaction with other phenomena, but much can be learned from a study of weather suitably averaged over space and time and treated along with such things as, ocean currents, the nature and quantities of atmospheric pollutants, and the variations in solar radiation. Such a study of climate can be immensely rewarding, but it's essential to find the right coarse-graining of the weather and the other phenomena to obtain a useful definition of climate.

We didn't talk only about policy studies. Integrated thinking can be promoted by thinking about history and prehistory in ways similar to what has just been recommended for policy studies. In fact, they sort of go together and reflections on history can be very useful in considering what to do about policy. One thing, one trend that's encouraging is the increase in popularity of big history, history on a world scale. Instead of laughing at it as ridiculously, what should I say, ridiculously over-ambitious, one can try to show some sympathy for people who try to study big history. When Arnold Toynbee, just after the Second World War, published a study of history, many of his colleagues were not very impressed with the 12-volume book. He could, of course, be criticized for making a mistake here and there, and describing the rise and fall of some 21 civilizations. And he could be lampooned for implying that it was all leading up to the founding of the Anglican Church. That's perhaps going a bit far, but wouldn't it have been more constructive to help correct his errors large and small and try to build on his pioneering effort, rather than merely making fun of it? I don't mind making fun of it, but one should be able to do something else as well. I think, I'd like to think, that nowadays some critics might be more charitable.

Another encouraging trend is the increased tolerance these days of contingent or what-if history including counterfactual history, and you see now what you didn't earlier — very serious respected historians engaging in the study of counterfactual history. You know, what if Richard III had actually found a horse? That kind of thing. Alternative scenarios for the future are like a branching tree, with major branchings as the special points, where a chance has a huge effect or important transitions occur. One can try to estimate probabilities at the branchings. This happens, this happens, this happens, and so on. What are the possible situations and what are roughly their probabilities? Does thinking about such structures help us to identify occasions of great policy leverage with respect to the future? Well, I like to discuss many of the issues facing the world, under the rubric of sustainability, one of today's favorite catchwords. It's rarely defined in a careful or consistent way. So perhaps, I can be forgiven for attaching to it my own set of meanings. In the spirit of Humpty Dumpty. Remember, he used to pay off words on Saturday night for meaning what he wanted them to mean. They would all come crowding around on

Saturday night for their pay and that way he established control. Broadly conceived sustainability refers to quality that is not purchased mainly at the expense of the future. That is, quality of human life and quality of the environment, not purchased mainly at the expense of the future. I used to determine a much more inclusive way than most people, paying correspondingly on Saturday night. We were talking about the days of the six-day a week, only Sunday was off. But I use to determine a much more inclusive way than most people.

Sustainability is not restricted to environmental, demographic, and economic matters, but it refers also to political, military, diplomatic, social, institutional, or governance issues and ultimately sustainability depends on ideological issues and lifestyle choices. As used here, sustainability refers not only to sustainable population, economic activity, and ecological integrity but also to a considerable extent to such matters as sustainable peace and global security arrangements, sustainable preparedness for possible conflict, sustainable democracy and human rights, and sustainable communities and institutions. All of these are closely interlinked in the presence of destructive war. It's hardly possible to protect nature very effectively or to keep some important human social ties from dissolving. If huge and conspicuous inequalities are present, people will be reluctant to restrain quantitative economic growth in favor of equality of growth as would be required to achieve a measure of economic and environmental sustainability.

Let me say that again in slightly different words because it's not something that people emphasize very often. I claim that sustainability is intimately tied to the notion of restraining quantitative growth in favor of qualitative growth. We see that mostly in wealthy countries. A lot of resources are devoted to quality, and it's a different situation from that of countries with very large impoverished populations, where poverty cries out for quantitative growth. It's hard to recommend a future in which growth is all qualitative when so many people are poor, but as countries achieve a certain degree of economic success, they can better afford to divert resources from quantitative growth to qualitative growth. At the same time, great inequalities may provide the excuse for demagogues to exploit or revive ethnic or class hatreds, and provoke deadly conflict. In *The Quark And The Jaguar*, I suggested studies be undertaken on possible paths toward sustainability in this very general sense, during the course of the century 21st century, in the spirit of taking a crude look at the whole. This goes beyond trying to estimate the probabilities of various likely scenarios in the future. Rather, one would look for more sustainable futures, even if they seem to have low probabilities because we think of them as desirable. It's worth studying desirable situations, that to which we can't assign very high likelihood, but maybe somehow, we can make them happen. I like to employ a modified version of the schema introduced by my friend James Gustave Speth, who was president of the World Resources Institute when it was founded. He and I founded it together. Later, he was head of the United Nations Development Program. He was for a long time the Dean of the School of Forestry and Environmental Studies at Yale and the schema involves a set of interlink transitions that have to occur if the world is to switch over from present trends toward a more sustainable situation. Then I have long comments on these various transitions, which I don't think I should repeat here, unless people specially ask for them. But I can mention some of the names of these transitions — the demographic transition to a roughly stable human population worldwide and in each broad region; the technological transition to methods of supplying human needs and satisfying human desires with much lower environmental impact per person at a given level of conventional prosperity that depends, as indicated, on technology.

Decades ago, some of us pointed out the obvious fact that a measure of environmental impact, say in a given geographical area, can be usefully factored into three quantities multiplied together, so you get an identity. The three factors are population, conventionally measured prosperity per person, and the environmental impact per person per unit of conventional prosperity. So, the last factor is the one, as we just said, that depends particularly on technology. It's technological change that has permitted today's giant human population to exist at all, and while billions of people are desperately poor, quite a few others manage to live in reasonable comfort as a consequence of advances in science and technology, including medicine. The environmental costs have been huge, but nowhere near as great as they may be in the future if the human race does not exercise some foresight. Technology, if properly harnessed, can work to make the third factor, the one that depends on technology, as small as can be practically arranged given the laws of nature. How much the prosperity factor can be increased without damaging the environment even more depends to a considerable extent on how much is squandered on the first factor — mere numbers of people. Evidence of the beginning of the technological transition has started to show up in many places. Then, I have a long screed on a particular example which is fighting malaria, but I won't burden you with it today.

In fact, the economic situation, the economic transition, I mean, to a situation where humanity is not living so much on nature's capital, but mainly on the interest of that capital, that's very important. Let's live on nature's interests, not nature's capital. Imagine a visitor to a little New England village who notices an old man walking down the street, a man whom everyone shuns. Children run away at his approach; people cross the street to avoid meeting him. The visitor asked his host who is that, and why is he treated that way. The answer is, "Oh, that's Eustace Barnwell, he dipped into capital". But the human race is using up capital and not just dipping into it. We are mining fresh water in places like the Ogallala Aquifer in Nebraska. That's, by the way, why President Obama insisted on changing the route of the Keystone pipeline, if it is ever to be approved, because it was running right by the route, running right by the Ogallala Aquifer. So at least that is cleared up. We're wiping out gigantic forests at an almost unbelievable rate. We're depleting many of the world's fisheries. We're using up clean air by polluting it. The economic transition, I said this already, can be described as one in which growth in quality gradually replaces growth in quantity, while extreme poverty which cries out for quantitative growth is alleviated. By the way, analysts are now beginning to use realistic measures of well-being that depart radically from narrow economic measures, by including mental and physical health education and so forth. Happiness has become a serious subject of study by scientists, by philosophers, by medical practitioners. Many, many different kinds of people seriously study happiness. You remember that the former King of Bhutan talked about how his government was interested in gross national happiness, not some narrow economic measure. Some of us are thinking of sponsoring a conference here in a year or two on happiness looked at in the modern way. It may be that some holy new insights can emerge.

Well, I think I will leave the rest of these transitions to our discussion. What I wanted to do with this introductory talk was simply to get us into the general field. The rest of the transitions need to be thought about, but we don't have to enumerate them and list them carefully here. Well, what we should do though is to look around look, around the world, for some examples of people trying to study aspects of a "Crude Look at the Whole". What kind of progress is being made on that important task? Are there places? Are there schools of thought? Are there universities? Are there businesses? Are there arenas somewhere, where the crude look at the whole is being taken? Are there groups of people who are taking a crude look at the whole seriously, especially in connection with sustainability in the broad sense?

I think we have to look carefully for signs of this kind of thinking. It is too early for it to show itself in a dramatic way. We'll probably have to dig it out from among the various things that people are considering, but it might be worthwhile, it might have a really worthwhile effect to start to do that. Are there places where the "Crude Look at the Whole" is taken seriously enough so that people are making progress on it? There are a few examples of at least lip service to it. For example, people who think about what's called world systems analysis, such as Emmanuel Wallerstein, who spent some time at Yale, and another man named Richard Robbins. Their work does not impress me particularly, because it starts out from Marxist criticisms of capitalism and goes on to criticisms of Marxist criticisms, but it keeps its roots somehow in the soil of Marxist critique, and I don't find that particularly illuminating. But at least they're trying to do something that's integrative, and are there other examples? What about world modelers? I understand there's a whole class of people who are thought of as world modelers, but I'm not really familiar with what they do, but here in this distinguished gathering, I'm sure there are several distinguished people who can help us with this search to see what kinds of looks, crude looks at the whole, are being taken today that might indicate something about the practical courses in such thinking for the future. Anyway, either before or after a short break, let's see if we can find indications, examples, trends, thoughts, or anything that might indicate that we're getting closer on some front or several fronts to taking a "Crude Look at the Whole". Thank you.

1.2 Discussion

Brian Arthur: Murray said earlier I remember that he wanted some commentaries from one or two of the speakers first. Is that still true or do you want to sort of...?

Murray Gell-Mann: Whatever. You're the chairman. You should have a huge, heavy, lead gavel to suppress dissent.

Brian Arthur: Some are delighted to hear. I am glad, by the way, that Murray is happy about the happiness index...

Murray Gell-Mann: I think it's wonderful that people are beginning to look seriously at happiness.

Brian Arthur: Great. Let's start by having a comment by two or three people from the speakers. We can start with Simon Levin here.

Simon Levin: Simon Levin from Princeton University and the Santa Fe Institute. I have a comment that I'd like to turn into a question. First of all, Murray, I've had the pleasure of listening to you many times over the last 20 years or so, and I actually think this was the most inspirational. I really enjoyed your comments and it inspired me to ask you a question. You emphasized in various ways the importance of uncertainty and thinking about weather prediction. For example, in thinking about Eustace Barnwell and why he probably dipped into capital. Indeed, the idea that uncertainty might be a justification for using any of a non-renewable resource is a point that the ethicists, such as Peter Singer at Princeton, have used. In other words, if you have a non-renewable resource and you want to leave for future generations the same amount that we have today, then the argument would be that you can't use any of it — in order to preserve the rights of those future generations also not to use any of it, which obviously seems to be a contradiction. So, I asked Singer how you could justify using it, as there's some uncertainty as to whether the future will be here and that allows you to discount. So, you didn't mention discounting.

Murray Gell-Mann: No. In the book, I do mention it. I get furious at the idea of discount rates — discounting the future, discounting the future by gigantic factors.

Simon Levin: By gigantic factors. I am thinking more about uncertainty. You know, Stephen Jay Gould made the point that if the tape of evolution had been run again, it would have come out differently. And Francois Jacobi talks about issues of this sort, and indeed it's not just a quantitative difference, it's a qualitative difference as Brian Arthurs taught us about path dependence and economics. So, that implies to me that we have to accept, and I think I'm sure you'd agree with this, that there are limits to predictability and I'm interested in what you think that implies about the way we manage the future. It seems to suggest we have to have a trade-off between trying to determine what's the optimal future and trying to take a path which minimizes the chance of something really terrible. In other words, we have to trade-off optimality versus satisficing in some way as we think about the near term and distant future. So, I'd be interested in your views as to how we resolve this trade-off about the best versus the satisfactory.

Murray Gell-Mann: Well, I think you've eliminated the question already in what you just said. There is this question, though it always will be there, of balancing intergenerational equity and intragenerational equity. In other words, do we worry a huge amount about the existence of extremes of wealth and poverty, or do we worry more about preserving something for the future? Those need to be balanced somehow. And as you hinted, it's not so easy to make a decision for every particular analysis.

Simon Levin: So, one issue that you've hit upon here is what is sometimes in the literature called pro-sociality. How much do we care about others? How much do we care about our own children? But the same decisions have to be made even for our own lifetimes, even if we value them the same as we do ourselves. Our limited ability to make precise predictions about the future suggests the trade-off, the sort of trade-off you make in investments between high payoff and risk rating. And is there any framework which we ought to be thinking about that's different than what a portfolio manager would do?

Murray Gell-Mann: You know I'm not too familiar with what a portfolio manager would do. I don't contribute much business to a portfolio man, for lack of capital, yeah.

Bob Axelrod: Thank you. I just developed that — what portfolio managers try to do is have uncorrelated risk among the things that they invest in. So, that is a general principle that we should use to promote sustainability.

Murray Gell-Mann: Yeah, because you want things that are at least as much as possible independent upon one another piling up in the same direction.

Simon Levin: Decomposable was the term you used?

Murray Gell-Mann: Yes. That's different. Decomposable means that you can separate them, separate them out from one another.

Simon Levin: Well, decomposable implies a degree of modularity that is a way to buffer the system against risk. It was the lack of decomposability that in large part led to the economic financial crises of 5 years ago, that the system was too interconnected. So, the risks were not uncorrelated.

Murray Gell-Mann: That's true. That's quite true.

Brian Arthur: Murray, I'd like to get you talking about something a little broader. So far, the whole discussion has been about the benefit of human beings, but we're one species only among many, many species and you're someone who thinks a lot about many species.

Murray Gell-Mann: I certainly think that way.

Brian Arthur: Good, what do you have to say about the benefit of all sentient beings, if you want to put it that way?

Murray Gell-Mann: Well, I say, I make a big fuss about all of that. Saying, I don't know where our human tendency comes from to divide up into little groups that don't get along and sometimes get violent alone. Is it something that was at one time adaptive and evolved tens of thousands of years ago? Or is it completely dependent on social situations? Where does it come from? But, wherever it comes from, it would be desirable to reduce its effects. As I point out in *The Quark And The Jaguar*, almost any difference is sufficient to set groups of people fighting one another. I mentioned the example of Somalia, which has over the last 20-30 years frequently been in a bad situation. What makes it separate into these people that don't get along and fight with one another? It's not language; they all speak Somali. Religion? No, they're all Muslims. Sects? No, they're pretty much the same kind of Muslims, most of them. What do they have to fight about? Clan differences? Well, not even, it isn't even really clan differences. Sub-clan differences are responsible for a lot of the trouble. Any marker is good enough to set people fighting over it. Somehow, we are in urgent need of developing attitudes, such as Brian just mentioned, attitudes that enlarge the concept of who we are to include all kinds of human beings and also many non-human life forms, so that we have to pay some attention to the welfare of people in general and to some extent life in general. That kind of attitude can perhaps be cultivated over time. Maybe we can already see traces of it here and there.

Suman Bannerjee: I'm Suman Bannerjee from Business School of NTU. My question is about what I thought about your weather and climate comments; it was about the right chopping. It means we need to understand better by chopping things and looking at smaller parts, but if you make it too small, then we don't learn much about the whole. If it is too big, it's too complex to study. The problem is in a reward return system. You know, like a system that has ways to think it is coming from my teacher, and my teacher learned from his teacher. Moving away from the existing paradigm of thinking is always a challenge because the person who is peer reviewing here has been thoroughly trained, and they have a set way of thinking about, you know, the micro foundation of macro issues. I mean, I'm speaking generally here. So, I think one of the biggest challenges in transitioning to this new system, the peer review system, is the reviewers themselves. They are actually, in some sense, very competent people. There's no doubt about it. I'm surprised that you are, you know, 80 plus in age. You think like a fresh graduate, but there are very few like this. So, this is one of the biggest challenges, and I think you can make, I'll say, a lasting effect on these peers that, you know, you better change; otherwise, you will never encourage these low people. I mean, low means in the last hierarchy right now, to change their way of thinking about this whole.

Murray Gell-Mann: Yes. That's right. We have a whole array of institutions that are devoted in part to preserving how things are done and how thoughts are carried, and what thoughts are given prominence. We have departments. We have textbooks. We have curricula. We have entrance requirements, exit requirements. A whole set of conditions that make it very difficult to innovate. It's true, and the innovation itself is risky. You have to make sure that you know what you're doing. Sometimes, innovations can be sold, oversold, and actually be harmful. So, it's all very tricky, but I don't think we'll get anywhere if we don't try. So, trying is a big part of it.

Guaning Su: Thank you very much, Murray, for the wonderful talk. I'm Guaning Su, the President Emeritus at NTU. I want to bring the subject to what is your definition of sustainability, which I find very interesting. Quality does purchase not at the expense of the future. And one of the hot topics of discussion in Singapore these few weeks is on the population growth this week. We have a small island of 700 square kilometers. We're still trying to add to it by reclaiming land, but it is very limited, and recently the government put out a white paper regarding what is required in terms of population growth to continue the economic growth path that we had in the past, and there was a big uproar when the number 6.9 million was mentioned. We are right now at 5.3 million. So, I'm wondering whether there's something that academics can do to help model the situation and help government come up with thoughts about how to do this in a sustainable way because a sustainable growth for Singapore has to take into account that we're not going to grow too much inland area, we're not going to conquer anybody by military force, and so there is an expense for the future generations if we use up the capacity for population in this small island of ours. So, I'm quite interested to hear what your thought might be about academics contributing to this study.

Murray Gell-Mann: Well, that's a very important set of issues, and it's true, I've thought about them, but I don't have much wisdom to add to the stock of wisdom and the subject, I'm afraid. Singapore is in a very good position to make choices, to make sensible choices. Most places don't have the advantages of Singapore. Here, it's likely that some halfway-reasonable solution will be found. The tendency has been to concentrate on some particular thing. Either put up as a bad situation, something should be avoided, or put up as a good situation to be sought and defended. Either way, it's usually too strong an emphasis on the particular thing. One enormous advance has been to regard the position of women as critical to maintaining a fairly low fertility, not an outrageous runaway fertility. Education for women and liberation of women in general can make an enormous difference.

But there are other things involved too in the generation of production of children. The man has something to say. Sometimes, the man has too much to say. I have a story that illustrates how the role of the man can sometimes be under-emphasized. I was on a bird-watching trip to southern Africa. We explored two national parks, one in Botswana and one in Namibia. And we visited the park in Botswana, we had a wonderful local guide, a member of one of the 19 clans of Botswana, the Batawanna, this particular clan. And the guide that we had for mammals and birds was immensely knowledgeable. He was a nobleman of the Batawanna, but he grew up very poor and worked in the mines in South Africa, learning English and Famalago, and then he came to the national parks and was hired as a guide because he was so knowledgeable. He knew the habits of all the birds and the important mammals, and he knew their names in Sichuanah and English and Latin, and he knew where to find them, what time of day they would sing, and so on and so forth. It was quite impressive, and of course, he was a conservationist. His job is dependent on preserving the national parks and their flora and fauna, and he was a very important defender of biological diversity there in Botswana. I was there with Lydia, who was my girlfriend at the time, and Lydia wanted to make friends with various local people. She talked with the guide who was called "Vemootopei". "Mootopei" is a title, and she talked with him about children and wives and things. And he said, "Yes, I have seven children by three women who live on the outskirts of the park. I've agreed to pay the school fees for all seven of them". And she said, "Well that's a very good situation. Glad that you're doing that. How many children do you have?" He said, "Well, I told you. I'm supporting seven by three different women who live right here in the park". Lydia said, "Yes, but how many children do you have altogether?", and he said, "Well, if I had to guess". She was acquainted with physicists, so she knew about guessing. She said, "Well yeah, guess. Go ahead and guess". He said, "Well if I had to guess, I would say 50". That's the leading conservationist.

Jan Carlstedt-Duke: NTU and Karolinska Institutet. I'm glad you raised the question of women because I think I was feeling that this is definitely something that has been missed in a "Crude Look of the Whole", and I was actually quite surprised when I looked at the complexity community listed here. It's about 95 percent male and if you look around here as well, it's a very dominant male society and you've already identified several of the ways in which women are impinging very strongly on the future, not least the dramatic changes in fertility rates here in Singapore and Japan. Isn't this something that we need to more actively incorporate into our understanding of the driving forces of the future?

Murray Gell-Mann: Yes, of course. That's what I was saying to some extent. I think it's very important.

Jan Carlstedt-Duke: But also to include it among those working with complexity, some particular aspects of …

Murray Gell-Mann: The distribution of specialties is a very difficult question to embark on. That leads to lots of arguments and problems and troubles. And as you see, you've seen people thrown out of high positions because of their statements on each subject. Very, very tricky business, but as to the number of female scientists in various sub-disciplines, when we started the Santa Fe Institute 20 to 25 years ago, we thought a good deal about that. At that time, the question was, "Would brilliant young women in science want to go into newfangled fields, like what we now call complexity and so on?" They have enough trouble with discrimination against women, and we want to add discrimination against crackpot subjects at the edge of physical science. Probably, they are wise not to go into these things in great numbers just yet. But anyway, all that seems to fix itself. I don't see any shortage of brilliant young ladies in what you call complexity, and so on, it seems to me. They not only choose such subjects, but they do very well in them as well.

Sheila Ronis: Hi, I'm Sheila Ronid from Walsh College in Michigan and the project on national security reform in Washington. You asked for an example of a project that in fact is looking at the whole. We were tasked by Congress to assist in the redefinition of the national security system of the United States, and it was really shocking to those who became involved with the project that we in fact did take a true systems approach and our recommendations to Congress, many of which have generated into many other projects that have included my work as chair of the vision working group of that group, and we're actually prototyping a decision support entity. We're calling it a center at the moment that will support decision-making for the President of the United States. It's going to look at a tremendous number of different tool sets that will include many complex science-based approaches, and we're currently looking for partnerships and scientists who will assist us in the standing up of this center, but it's going to be very heavily systems oriented and very heavily complexity science oriented and we are incredibly excited about this project.

Brian Arthur: Sounds great. Peter?

Peter Sloot: I am Peter Sloot from the University of Amsterdam and NTU. Murray, I have a question regarding your remark on world modelers, and I think I consider myself to be one of those.

Brian Arthur: And good, we found one.

Peter Sloot: Now, now, one of the things that we always have to fight with is the following remark: It's about predictability, and so one of the things why you want a model is actually to get a feeling of how the future, what the future will look like, no matter what we talk about, the topic we talk about. We all know that, you know, lots of stuff that are not predictable. We know that the probability is very low and there are errors that are propagating if you like. I refer to the way you look away into the future. So, the question is, now, how do we defend ourselves in these terms? Because we find that many things that are being done now on predicting things are actually diverging so fast that we cannot really predict. It's only in the very near future we can really predict things, so that's the one step, you know. But then the thing that goes with that is the controllability. There's a certain arrogance in thinking that we can control the future once we know how, you know, to some extent to predict it. So, I'd like to have your opinion on this. I think this is a very fundamental aspect for, you know, any modelling, any future thinking, any decision support you're talking about. How can you… You know, there's a certain arrogance in that decision support idea. I'm doing the same thing there, so it's also my research field, but I think this is the thing we need to address more deeply and I would very much appreciate your thoughts.

Murray Gell-Mann: I don't have anything profound to say about it, but I certainly thought about these issues and written about them a little bit. It's hugely important to try even in the face of all these discouraging circumstances, to try to guess probabilities for various future scenarios and have to do it with scenarios and scenarios branch of course, if this happens, rather than that then you go into a whole chain of things, over here you go into a different chain and possibilities. In all cases, you're as crudely estimating probabilities for things, and your estimates are changed by new facts and new developments and you try to do the best you can to muddle through. You can compare notes with other organizations, for example, the Shell oil company, which tries to make predictions about the future. Many other outfits, many of them, are quite skilled actually, thinking about different alternatives and their consequences, and the possible chains of events that are unlocked by current developments. You can get somewhere with this. What I say in one of my essays on it is that the worst mistake is to take it very seriously. It's too risky to take it really seriously. You're guessing probabilities. You're not really calculating them enough in a very scientific way, and you'd better not place too much faith in the results. One of our visitors many years ago, I think he was from the RAND Corporation, he suggested that we think of these calculations as prostheses for the imagination, and I think that's the very best way to look at it. You try to equip your imagination with some relatively sophisticated tools for dealing with these chains of events of varying probability, looking for places where serious divergences occur with branchings, major branchings occur in the future. Those may also be places where action will make a huge difference. Look out for those perhaps, think about shifting sets of futures, depending on what happens, but all of it is tentative. All of it is regarded as prosthesis for the imagination. That's the best we can do, I think.

Brian Arthur: I think John Holland once called that flight simulators for the mind.

John Holland: A quick comment. I think it's helpful to distinguish steering from prediction. I don't know where my car is going to be tomorrow, but I have layers and, in traffic, I can steer a bit with a bit of prediction. If I'm going downtown, the prediction may because of a traffic jam or something, but steering seems to me to offer and then these kinds of proteases can be helpful.

Brian Arthur: Gregg Fisher at the back, if we can get a microphone way back up there? Please.

Gregg Fisher: Thank You Brian. As Brian will know, this turned into quite a Taoist conversation. Gregg Fisher, I run a little think-tank in London called Synthesis orientated around complex systems. Complexity theory helps me a great deal in making sense of problems, and many of the differences of social and environmental problems lead me to a conclusion — we need to act collectively. You see that kind of in multiple domains, we need to act collectively. But one of the main issues I have is the question of how do we do that, especially in the context of quite powerful political and economic ideologies, which are quite anti-state? Do you have any thoughts?

Murray Gell-Mann: Which are anti what?

Gregg Fisher: Anti-state. So, post-enlightenment kind of political and economics has evolved towards a quite liberal socially and economically, so therefore anti-state, anti-government.

Murray Gell-Mann: You mean these Tea Party people, for example?

Gregg Fisher: Yeah, exactly. Yeah, libertarian sort of thinking, but also social liberalism as well. Do you have any thoughts on, sort of, political ideology and, sort of, how complexity can help us actually act collectively together?

Murray Gell-Mann: Well, if these attitudes, political attitudes, were based on careful weighing of possibilities and careful concern for issues, that would be one thing. Another thing is people are influenced by, I don't know what it is, by raging hormones or something [laughter from audience]. People are influenced in ways that are not very rational. That's much harder to work with, much harder to do something about. I think one has to distinguish between the two. Attitudes that are arrived at through thought and consideration of real situations, and probably estimated probabilities for possible futures, and so on. That's one thing, and just gut reactions of various kinds are the other. And a lot of the political attitudes seem to be influenced more by the latter than the former. Isn't that true? This anti-state stuff, for example. They want to come and take away my anti-tank weapon. They'll have to take my anti-tank weapon; they'll have to pry my frozen dead fingers off my anti-tank weapon.

Brian Arthur: I have a comment on resilience. Murray talked a bit about policies and systems that are resilient, but if you think about flight simulators, real flight simulators, they're all about making people resilient, that is pilots resilient. You put them in many extreme circumstances artificially, and then you assure that beyond the bounds of most envelopes that they would face in an actual flight, they know how to respond and they're comfortable with a very wide range of responses. And when something extreme comes up in reality, they're not going to panic and they have a lot of resilience built in. But I'm wondering if I can entice Peter Ho one of whose jobs here in Singapore is to think about the future, strategic futures for Singapore, if I could entice you to make a comment on all of this being said here. Peter, I'm putting you on the spot here.

Peter Ho: Not really, if I comment on what Murray said, which I agree with, I'll be kind of cannibalizing from my talk on Wednesday. But I think it very much mirrors my own thinking about how we approach governance because increasingly, because of complexity, complexity generates a lot of problems for governments because the nature of the problems are what we call "we could" problems. They require the ability of governments to bring together people from many backgrounds, many experiences, and many insights to look at the problem in a holistic way.

1.3 Summary of Gell-Mann's Talk

The central theme of Professor Murray Gell-Mann's talk revolved around the necessity of integrating various disciplines and considering the interconnectedness of global issues to address the challenges facing humanity and the biosphere. Gell-Mann emphasized the limitations of conventional academic and policy approaches that tend to compartmentalize problems into separate categories without adequately acknowledging their complex interactions.

He highlighted the importance of transdisciplinary research and teaching in institutions like the Santa Fe Institute, where experts from diverse fields collaborate to tackle complex problems. Gell-Mann argued that understanding complex systems requires more than just studying individual components; it necessitates integrative thinking to grasp the interactions among different aspects.

Gell-Mann critiqued the prevalent tendency to focus on narrow aspects of problems, urging for a shift toward holistic perspectives. He emphasized the need for policymakers and leaders to consider the entire spectrum of issues and their interdependencies rather than compartmentalizing them.

The concept of sustainability was a recurring theme, encompassing not only environmental concerns but also economic, social, political, and ideological dimensions. Gell-Mann advocated for transitioning toward qualitative growth over quantitative growth, emphasizing the importance of technological advancements and equitable distribution of resources.

He discussed various transitions required for achieving sustainability, such as demographic stabilization, technological innovation, and economic restructuring. Gell-Mann stressed the significance of incorporating measures of well-being beyond economic indicators and highlighted the emerging field of happiness studies.

Throughout the talk, Gell-Mann called for a more comprehensive and integrated approach to addressing global challenges, encouraging scholars, policymakers, and institutions to embrace the concept of "A Crude Look at the Whole" to navigate the complexities of the modern world. He urged the audience to seek examples of integrative thinking and explore avenues for advancing this approach in academia, policy, and practice.

1.4 Relevance of the Talk to the Current Stage of Research

Professor Murray Gell-Mann's talk remains highly relevant to the current stage of research across multiple disciplines due to its emphasis on interdisciplinary collaboration, systems thinking, and holistic approaches to addressing complex global challenges. Here are some specific ways in which his ideas are pertinent to contemporary research:

Interdisciplinary Research: Gell-Mann highlighted the importance of breaking down disciplinary boundaries and fostering collaboration among experts from diverse fields. In today's research landscape, interdisciplinary approaches are increasingly valued and promoted as essential for tackling complex problems that require insights from multiple disciplines. Researchers are actively engaging in interdisciplinary collaborations to address issues such as climate change, public health, and sustainable development.

Complex Systems Science: Gell-Mann's emphasis on systems thinking aligns with the growing field of complex systems science, which seeks to understand the behavior of complex systems composed of interconnected elements. Contemporary research in this field explores the dynamics of complex systems ranging from ecosystems to economies to social networks, using approaches such as network theory, agent-based modeling, and nonlinear dynamics.

Sustainability Science: Gell-Mann's discussion of sustainability resonates with the interdisciplinary field of sustainability science, which aims to understand the interactions between human and natural systems and develop strategies for achieving long-term sustainability. Researchers in this field study issues such as climate change mitigation, biodiversity conservation, and sustainable resource management, often adopting holistic approaches that consider social, economic, and environmental dimensions.

Foresight and Scenario Planning: Gell-Mann's call for long-term thinking and foresight in policymaking is reflected in contemporary research on scenario planning, risk assessment, and futures studies. Researchers are developing methods and tools to anticipate and prepare for future challenges, such as technological disruptions, geopolitical shifts, and environmental crises. Scenario-based approaches are increasingly used to explore alternative futures and inform decision-making under uncertainty.

Quantitative vs. Qualitative Growth: Gell-Mann's distinction between quantitative and qualitative growth remains relevant in discussions about sustainable development and well-being. Contemporary research on alternative measures of progress, such as gross national happiness and sustainable development indicators, reflects a growing recognition of the limitations of GDP as a measure of societal well-being. Researchers are exploring ways to integrate qualitative dimensions, such as health, education, and social cohesion, into assessments of progress and prosperity.

Global Governance and Policy Studies: Gell-Mann's discussion of policy studies and global governance is pertinent to contemporary research on international relations, global governance, and public policy. Researchers are examining issues such as global governance reform, multilateral cooperation, and the role of non-state actors in addressing transnational challenges. Interdisciplinary research in this area seeks to inform policy-making processes and promote effective responses to global issues.

In summary, Professor Murray Gell-Mann's talk provides valuable insights and perspectives that continue to inform and inspire research across diverse fields, including complex systems science, sustainability science, foresight studies, and global governance. His emphasis on inter-disciplinary collaboration, systems thinking, and long-term foresight remains highly relevant to addressing the complex and interconnected challenges facing society and the planet.

Chapter 2

A Theory of Meaning: Or How A Schema Reduces Complexity

Robert Axelrod

YouTube: https://www.youtube.com/watch?v=RvVGVq7fkr8

Speaker: Robert Axelrod

Moderator: Ann Florini

Discussants: Chan Heng Chee, Simon Levin, Greg Fisher, Kai Xi, Suman Banerjee, Peter Sloot, Discussant A

2.1 Talk by Robert Axelrod

Thank you. It's a great honor to be here. This is new work, and I decided to present it first as far away from home as I could, which I've managed to do here today, to see what your reactions are. One way of looking at it is that I am taking up Murray's challenge from this morning of doing integrated work at some front, and the front that I have selected could be regarded as a study of coarse-graining. As Murray said, one of the major challenges in taking a crude look at the whole is how one does coarse-graining, and one can view, as I'll explain at the very end, how I'm talking about today as an effort to understand the tools and needs of coarse-graining. I've called it a theory of meaning, and I will give you a specific notion of that in a moment. Another way of looking at it is how a schema reduces complexity, so I will also explain what I mean exactly by a schema. But, the basic idea here is that I want to talk about how complexity can be reduced and how it is often managed. So, let's take an example of one schema called the Balance Schema, and it is, as President Bush once said, "You're either with us or against us." That's the kind of perspective that is represented in the Balance Schema.

Example: A Balanced Schema

1. Friends of friends are friends.
2. Friends of enemies are enemies.
3. Enemies of enemies are friends.

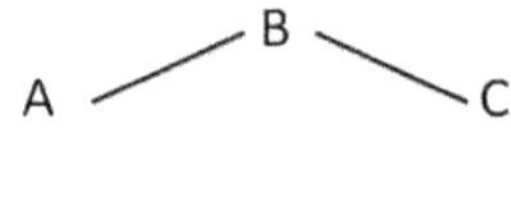

You know A likes B, B likes C, and D likes E.

Another way of putting it is that the whole world is divided up into two groups, or maybe just one group. For example, suppose you know that A likes B, B likes C, and D likes E. Then, you can't infer much else except that A likes C. But, suppose you know one more thing. Suppose you also know that A does not like D, and I'm using a dot and a dashed line for negative relationships or dislikes. Once you know that, then with a Balance Schema, you can infer all the other links. With five nodes or countries, or people, whatever the nodes represent for you, there are ten possible links. So, with these four, you can infer the rest. For example, once you know that A doesn't like D, then you know that E also doesn't like A because it's a friend of D. Those are all the links that you can infer. And so, with a Balance Schema, if you know a little bit, you could infer a lot more.

Example: A Balanced Schema

- You know: A likes B, B likes C and D likes E
- You learn that A dislikes D.

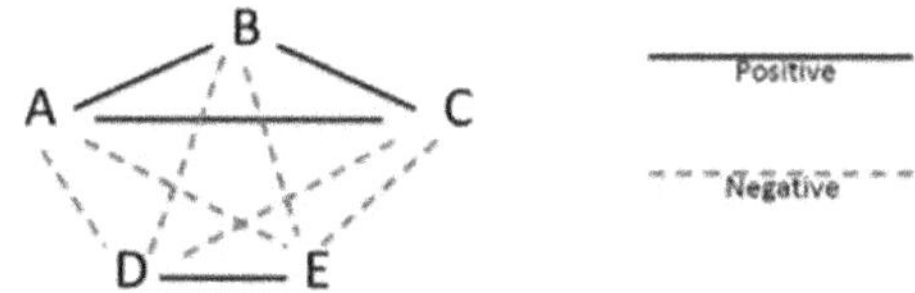

Now you can infer the remaining 6 links.

So, here's a definition of meaning: We could say that the meaning of some nodes is like this idea that A does not like D. What is the meaning of that? Well, given that you're using the Balance Schema, you're seeing the world in this way, and given your prior information, which are those three links that I started with, that A likes B, B likes C, and D likes E, if you know the news of this one further link, given the balance of those three links, there are six more things. And those are the other six that would make the complete graph.

So, the definition here of the meaning under the Schema of Balance is this new information given the old information. And that can be formally defined as the information available from knowing everything among these things minus what we know just from knowing X and what we've known from just knowing Y, that is, one of these examples where the whole is greater than the sum of the parts. So, knowing everything would be ten bits, and X was just 3 bits, which these three links you were told of. Y is this new piece of information, which is one more bit, and so the meaning of Y given X under the Balance Schema is 6 bits.

This is related to Dr. Gell-Mann and Lloyd's Effective Complexity, and if anybody asks about it, I'll refer you to Murray.

Here's another example of links between countries. I thought you'd be interested in how I see the Pacific Rim. It's not balanced. For example, South Korea really has a lot of problems with Japan, but they both like the United States, so just this triangle alone is not balanced. And there are others that are not balanced.

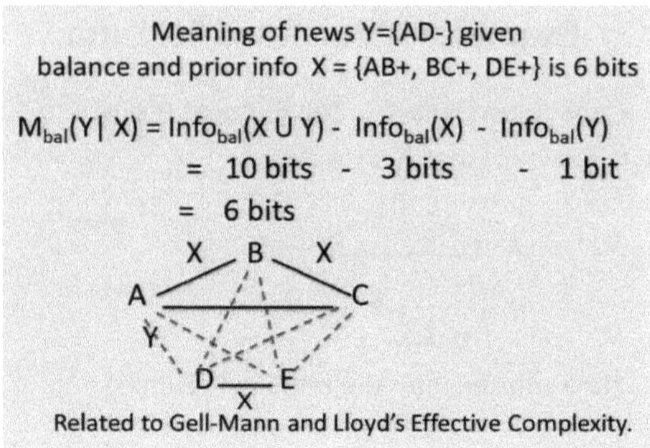

Another thing you could see from this is that there tends to be a hub-and-spoke pattern to a lot of this, which is very different from Europe, where, for example, Britain, France, Italy, Germany, the United States, and others all have a mutual alliance. But, in the Far East, it's more bilateral. I've put this one in red.

The relationship between China and the United States is basically positive now, but it's under a variety of stresses and will be under more stress in the future. And I personally think that this relationship, whether it stays positive or becomes negative, is the most important event or lack of change in the next, at least, twenty years. The world will heavily rely on whether this stays positive or not.

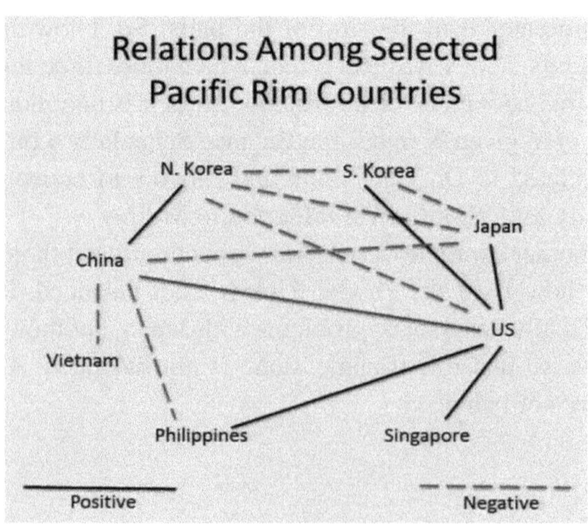

Okay, so some definitions. First, I want to describe a state space, which is an abstract term widely used in the physical sciences. Essentially, it means that every possible configuration is represented as a point in a space. In this case, the allowable configurations refer to the different ways that a graph can be, such as the one being shown. An example would be any graph, or what are commonly called networks nowadays, which is a set of nodes with a positive or negative link between each pair of nodes.

So, if that's the universe you're thinking about, then a schema would be any subset of that space. An example of the subset that I've already given you is the balanced graphs. Not all graphs are balanced, as we saw, but the balanced graphs would be a schema within this space of graphs with nodes, plus or minus links.

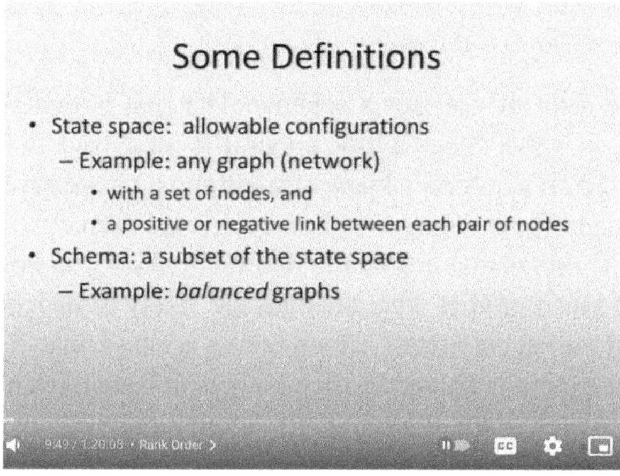

Here's another schema: Rank Order. You're all familiar with this. You can think of A being greater than B or B being greater than A, which could represent wealthier, richer, stronger, or anything like that. And of course, it's transitive, so if you know that A is stronger than B and B is stronger than C, then A is stronger than C.

One example that's simple enough is if you know A, B, and C, then the meaning of C being bigger than D is that A and B are also bigger than D. In fact, you could then summarize that with a simple chain. But, another example would be this: Suppose you know that another three things are ordered that way, but the meaning of the middle one is bigger than some new one (G), illustrated in the following graph, thus we can infer that the first one must be bigger than G. However, this doesn't tell you anything about whether the third one, F, is bigger or smaller than G. So, there is a different set of inferences and meanings from a rank order kind of schema than there is from a Balance Schema.

Another Schema: Rank Order

1. A > B or B > A
2. If A > B, and B > C then A > C.

Example 1: When A > B > C is known, meaning of
   C > D is A > D and B > D. A > B > C > D

Example 2: When D > E > F is known, meaning of
   E > G is D > G. But don't know if F > G.

$$D \longrightarrow E \longrightarrow F$$
$$\searrow G$$

There are a lot of uses for a schema. The first is that you can infer missing data, as we've talked a little about. I'll come back to each of these in sequence. So, if you don't know some things, as we saw before, you can infer some of the missing data. You could also correct errors if you've observed some things that are wrong. Now, I'll show you how.

You could also predict what changes are likely to happen in the system, and you can economize on memory in a sense that if you believe things are, for example, balanced, then you could keep track of things with a lot less memory than having to keep track of every possible relationship.

You could also evaluate source credibility. If some source tells you something that doesn't fit in with everything else you know, you might not trust other things they tell you.

You could also measure how much stress there is in a system and how good of a fit it is to that schema. And finally, if there are things that don't fit well, you could focus attention on that to kind of resolve the ambiguity that's implied by how some piece of information doesn't fit into the schema.

So, for example, starting with missing data, we saw that if you know just four of the links, namely, the three from X and the one from Y, you can infer the others. So, that's an example of how you could fill in missing data using this schema.

You could also do error correction. An interesting point is that any two balanced graphs, which remember our ways of dividing the nodes into one or two groups, have to have at least $n - 1$ links that are different. Why is that? Well, take this example: You have three of them in one group and five of them in another, and of course, within each group, there are positive links, and between groups there are negative links by definition of the Balance Schema.

Suppose that C moves from the first group to the second, so that would be another balanced graph, but it will require changing all of C's links because it used to like A and B and now it doesn't. It's in the opposite group, and it used to not like D, E, F, and G but now it'll have to like those. So, any one change, from one balanced graph to another with a single node changing sides, will change $n - 1$ links, namely, any one of them with all the others. As a matter of fact, you could prove that if more than one node changes sides, then there'll be at least seven and probably more links that are changed.

So, the point here is that these balanced graphs are not very similar to each other. They have to be at least $n - 1$ links apart, where n is the number of nodes. And that allows you to correct errors. So, suppose you have, the panicle A, and you saw that they were seven apart. That means you can correct one, two, or three links, because if you're one, two, or three links away from the balanced graph, you won't even be further than that away from any other balanced graph because they are seven apart. So, if you're less than halfway toward another balanced graph, then you have a uniquely close balanced graph. That means that you can correct up to half of the distance between graphs, which in this case was 7. So, you can correct any one, two, or three links that are in error when you were measuring the whole thing. For example, if you had 20 nodes, up to 9 errors could be corrected because they were 19 apart, and anything up to half of 19 could be corrected.

2. Error Correction

Two balanced graphs differ by at least n-1 links.
- Example: ABC versus DEFGH where n = 8.
- If C moves from ABC to DEFGH, C will change all 7 of its links.
- If more than one nodes changes sides, more than 7 links change.

So up to (n-1)/2 errors can be corrected.
- Example: An unbalanced graph of 8 nodes that is 1, 2, or 3 links away from a balanced graph is at least 4 links away from any other balanced graph.
- Thus 1, 2 or 3 errors can be corrected.
- For n=20, up to 9 errors can be corrected since 9 < (20-1)/2.

So, this gives you one value, one kind of thing besides filling missing data. If you have all the data but some of it is wrong, you can actually correct it provided there are not too many errors in it.

Another thing you can do is put in changes under the assumption that it's empirical and will only apply in certain cases, but if it's a case that an unbalanced graph tends to move to the closest balanced graph (I'll give you some examples to show you why that might be) with that assumption, you can actually predict where they'll go. For example, if you start with this one where there are two on one side and five on the other, if C likes A but not B, then C has got a problem because it likes one of the people in the left group but not the other. But, let's suppose it also likes one of the people in the right group but not the others. Well, that's going to be an unbalanced graph.

There's a prediction about what will happen here, which is the closest balanced graph, that is, C will come to like A because on the left side of the A–B group, there's only one problem that C has, namely, what is its relationship with B. Since we assumed that it likes A but not B, and so if it comes to like B, it solves this problem of being in the left group, but then it also has to change its mind about D, which it likes but it's now in the other group. So, just a few changes, two changes in this case, will make a balanced graph where C would prefer to join the left side rather than the right. So, in this sense, links can be predicted if they're going to change toward the nearest balanced graph. You could make predictions about how it's going to change.

3. Predict Changes

- Empirical claim: An unbalanced graph is attracted to one of the closest balanced graphs. (If n is even, this is unique.)
  - Example:
    - AB vs. DEFGH, but
    - C likes A but not B, and likes D but not E, F, G and H.
    - Prediction: C will come like A, and dislike D, making the balanced graph ABC vs. DEFGH.
- So changes in links can be predicted.

So, here's another example from a couple of years ago, when we had the uprising in Syria, before Syria was siding with Iran, with the Syrian Regime of Assad. But, Israel opposed the Syrian Regime and so it sided with the Syrian Opposition, one that became dominant, and of course the Syrian Regime and the Syrian Opposition didn't like each other, and so you did have, in this case, when the uprising took place in 2011, a balanced situation where Turkey, Iran, and the Assad Regime were in one group, and Israel and the Syrian Opposition were in another, and there was mutual dislike across the two groups.

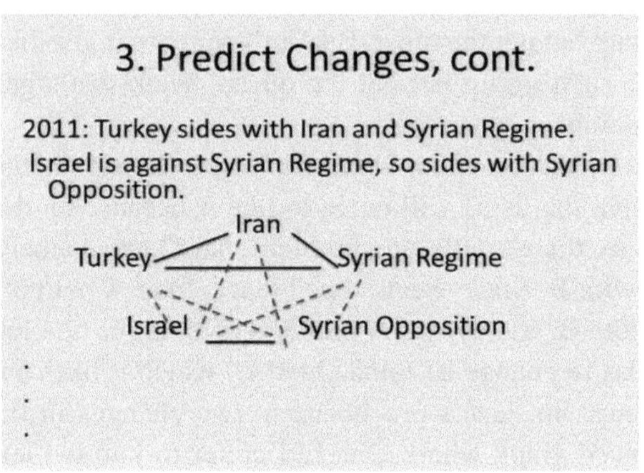

But, the next year, things changed, and the changes are given in the bold links. So, Turkey came to oppose the Syrian Regime because there were lots of conflicts at the border. The Turks came to give refuge to the Opposition. So, what tended to happen was that Turkey became hostile to the Regime and favorable to the Opposition. Now, that's also not balanced. For example, Turkey and Syrian Opposition and Israel, Turkey, the Syrian Opposition being favored by both Turkey and Israel might suggest that Turkey and Israel would come to like each other and that Turkey and Iran would come to dislike each other. If that happened, it would be balanced again with those two changes.

I wrote this slide a week ago and the very next day, I read in *Reuters* that despite the widespread belief that the ties between Israel and Turkey

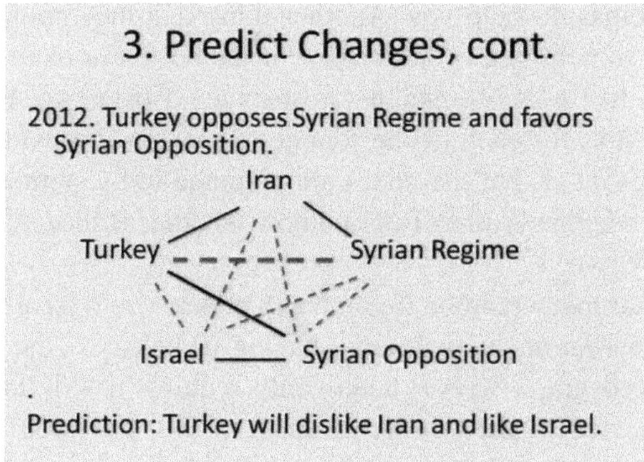

are virtually non-existent, in fact, they've been talking to each other, and maybe something will come out of that.

So, that is a change to the closest balanced graph as represented by this, where Turkey and Israel would not only talk to each other but something would come out of it. And of course, there are big barriers because Israel killed 9 Turkish citizens when a ship from Turkey tried to supply Gaza and Turkey is asking for an apology but maybe something will work out in that regard, in which case Turkey would probably then break with Iran because Israel is an enemy of Iran. That would give this a new possibility. So, there's an example of a prediction and as Murray says, "Don't take it too seriously."

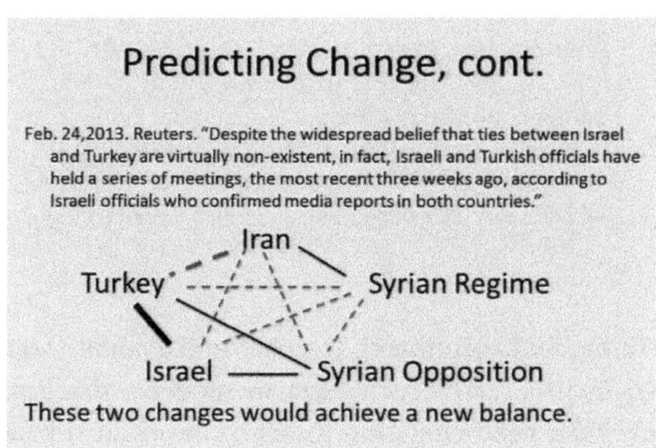

But, schemas do help you. Another thing that they could do besides prediction is to help you economize your memory. For example, a graph with n nodes has a lot of links, it's n times $n - 1$ over two. How did you know that? Well, for each of the n nodes, there's a link with the others. That's n times $n - 1$, but the nodes are assumed to be symmetric, and so if you know whether A likes B, you know whether B likes A. That's why you divide by two.

So, an example would be if you had 8 nodes, you'd have 28 links. We gave you an earlier one with 5 nodes having 10 links. So, the point here is that a balanced graph with n nodes only requires $n - 1$ links provided they're connected in the correct orientation. So, provided they're connected, you can infer all the rest with $n - 1$ links.

Okay, so what is the efficiency of this? Well, for example, with 8nodes, you only need seven links, whereas there are actually 28 altogether, so 27 are connected. All the nodes will be sufficient to predict all 28. And, for example, if you have a bigger number like 20, then well you need 20 links to infer everything and 190 if you don't have a balanced schema. So, by the Balance Schema, if it's reliable, will save you 90% of your memory, and that's very efficient.

4. Economize on Memory

- A graph of n nodes has n(n-1)/2 links.
    - Example: 8 nodes have 8*7/2 = 28 links.
- A balanced graph of n nodes requires only n-1 links to infer the rest (provided the n-1 links connect all the nodes).
    - Example: 8 nodes need 7 links, e.g. all of A's likes and dislikes. This compared to 28 links in general.
    - Example: 20 nodes need 19 links if balanced, but 190 if not. This saves 90% on memory.

Another thing, as I mentioned, is you could evaluate source credibility. You could use the error correction method we discussed before to identify which links need changing to get to the nearest balanced graph. Then, you could say the people that provided the information about those links might have given you misleading information, wrong information, or badly measured things. By discounting what they tell you about other things in the future, you might be able to prove your accuracy by knowing which sources to believe and which to doubt.

5. Evaluate Source Credibility

- Use the error correction method to identify links that need changing to get to the nearest balanced graph.
- Then reduce the credibility of the source(s) that provided the "erroneous" information.
- This can improve accuracy in the future.

Another one is you could measure the stress in the system. So, we could define stress in a graph as the distance to the nearest schema. If it's zero distance, there's no stress. You already have a balanced graph. If you're one link away, then there isn't much stress, and if they are several links away, then there is more. So, you could define stress as just the number of links you need to get to the nearest balanced graph. And then you could use stress to evaluate how good a fit is, say, a balanced graph, to given sets of information like the Pacific Rim example I showed you.

Suppose you have a graph where there are three groups of friends, all of whom like each other within the group and they dislike the other two groups. Now, that's not a balanced graph because it fails on the idea that an enemy of an enemy is a friend. If there are only two groups, the enemy of an enemy is a friend, but if there are more than two, an enemy in one group might be an enemy of a second group and not be a friend of yours. So, if you use this three-group schema, you could actually have less stress. You might be able to account for the data with fewer errors if it fits a three-group schema than if it fits a balanced schema. So, you could actually measure the stress in one kind of schema versus another for the same data. That can lead you to a prediction for a given schema going back to balance, that the corresponding change will be one that minimizes stress.

That reduces the stress from where you are to the closest place where there's no stress.

You've heard of the maxim politics makes for strange bedfellows? And that's true, but we have an addition to that idea that it minimizes the strangeness of bedfellows. That is to say, it actually minimizes the stress and brings together, for example, in a political coalition, the parties that are most similar to each other. So, even though there'll be some stress left because not everybody likes everybody completely because they have different positions, it'll still minimize the stress.

6. Measure Stress

- Def. Stress in a graph is the distance to the nearest schema (e.g. the nearest balanced graph).
- Stress can be used to evaluate goodness of fit of a graph to a schema.
  - Example: Let G be the graph that has three groups of friends all of whom dislike those in other two groups. Then G is not balanced, but it does fit a 3-group schema.
- Prediction (reformulated from before): For a given schema, links will change to minimize stress.
  - Axelrod's Maxim: "Politics minimizes the strangeness of bedfellows."

And finally, another final use I mentioned of using a schema is that it helps you focus attention when there's ambiguity. So, if the graph is nearly equidistant between two balanced graphs, you can identify which links would need to change to send it one way or the other way. That would mean that if you wanted to see which way it's likely to go, you might study just those few links that would distinguish the critical experiment whether it's going to go one way or the other way if they're equally or nearly equally distant from two balanced graphs. That is a major advantage because you don't have to study and closely observe everything. It'll suggest which places and which kinds of information you need to resolve the ambiguity.

Now, that brings me to the interesting topic of beliefs and reality. I would say there are three reasons why the world might fit a Balance Schema. The first one is that the changes in the world tend to move toward

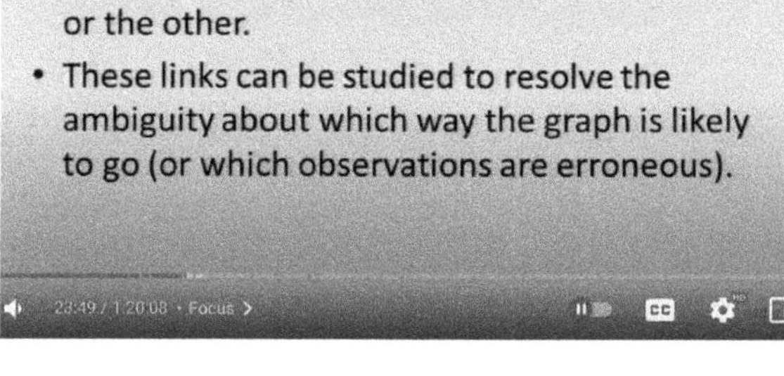

balance. For example, we tend to see things as friends and enemies, and if you think there are only two groups, then you'll think that the world would actually move that way because those that don't belong will join the coalition that they're most similar to or that they would have the least conflict with. So, under certain notions of balance and under certain situations, the world might actually move toward a balanced schema, and in general, it might move toward a particular schema that is appropriate to the situation you're in. I've been using balance for friends and enemies. But, one major issue about schemas and relations, in reality, is that the world tends to move that way, but the second is that people tend to believe it because it's cognitively efficient as well as because it would tend to make good predictions. So, if you have certain friends and certain enemies and there's somebody else that you find is a friend of one of your enemies, then you could make all kinds of predictions about them and you might not trust them, for example, whereas if they're friends of a friend, you might trust them. And because of that, we get to number three: that if you see things this way, there actually could be a self-fulfilling prophecy because if you only know that somebody is a friend of an enemy, and then you treat him or her as an enemy, which would tend to be self-confirming, but if you know that they're friends of a friend and you treat them as a friend, that too can be self-confirming. And so, belief in a schema can be self-confirming because the first idea is that the world moves that way and the second idea is that it's cognitively efficient for people to take cognitive shortcuts to think of it that way. So those are the three reasons about the relationship between beliefs and reality about why schemas would tend to be valid.

> **Beliefs and Reality Can Interact**
> **To Sustain a Schema**
>
> Three reasons to expect the world fits a
> balanced schema:
>
> 1. Changes in the world move toward balance (stress reduction).
> 2. A balanced schema is cognitively efficient.
> 3. Belief in a schema can be self-fulfilling, because reasons 1 and 2 can reinforce each other.

Now, there are various schemas. I've emphasized the balance one and I told you about the rank order one. Another example would be a multi-dimensional one, for example, a two-dimensional (2D) mileage chart. If we had a mileage chart between cities then that has a certain structure, for example, the triangle inequality must hold if it's on a flat part of the earth. And you could have a higher dimension to account for data, such as the scheme we had on the crude look at the whole, which was a 2-D view of the similarity relationships among disciplines.

Another example, of course, is trees. A tree structure is like an organization chart: one at the top and then several subordinates and each subordinate has subordinates. So, a typical hierarchical organization is represented as a tree, and a tree of life is a biological idea, of course, where there's a kingdom, phylum, class, order, genus, and species, and those are layers of the hierarchical structure. And hierarchical structures sometimes have, what John Holland and others have emphasized, default hierarchies which are things that every level inherits — the properties of the level above it. So, dogs are an example of mammals. They inherit the properties of mammals such as they give live birth. So, tree structures are another kind of schema, and there are many others that you can imagine. These schemas differ on which situations in the real world they apply to. Obviously, trees don't apply to the same thing that balance clusters do.

So, you need to pay attention to what kind of situation you're dealing with to pick the appropriate schema. But, one interesting thing is a difference in robustness, and robustness can be thought of as the basins of attraction. What I mean by that is the idea that we talked about prediction that if you're several [stages] away from a balanced schema, that balanced one, if it's the closest, attracts the world to itself, so there's an area around it. Likewise graphs that are similar to the balance graph are attracted to it, and so the size of the basin of attraction is just the number of unbalanced graphs that will move to a given balanced graph. And the size of that basin of attraction, I'll show you now the geometric view of that in a moment, gives you properties of different schemas.

Various Schemas

- M clusters (e.g., balance is 2-cluster)
- Rank Order
- Multidimensional (e.g., 2 dim mileage chart)
- Trees (e.g. organizational chart; tree of life)
    - Default hierarchies: e.g., dogs have properties of mammals
- Etc.

Schemas differ on their robustness (the size of basins of attraction).

Now, how do you choose between schemas? Well, here's one very simple example: Suppose that the cost of choosing a particular schema is the number of incorrect beliefs that you'll get if you use it, and of course, that's what you want to minimize. And suppose you are a Bayesian and you have priors, so you think it's probably a balanced schema but maybe it's a three-cluster schema. So, if those are your priors, then for any given graph that you observe, any full set of relationships, you can actually choose whether they balance — you can actually calculate, if it's balanced, how many things have to change, if it's three-cluster, how many have to change. And since you have those priors, you could actually calculate what minimizes the expected cost. And so, this gives you one way of thinking about optimal beliefs — over schemas.

Choosing Between Schemas

- Suppose cost = # incorrect beliefs
- Suppose there are priors over schema.
  - Example: 80% balance, 20% 3-cluster
- Then for a given (complete) observed graph, one could choose the schema and (nearest) graph that minimizes expected cost.
- This gives "optimal beliefs."

So, here's a more complicated example: dealing with 17 countries. I'll show you five at the beginning, and then we'll deal with all 17. So, the question is, who will wind up with whom in World War Two in Europe. Before the war in the late thirties, the basic situation was that Britain disliked Communism and the Soviet Union, and it also disliked Germany and the Nazis. Germany disliked the Soviets and it disliked Britain. The underlying nodes are the major actors, and I'm giving you a couple of others — Poland and Hungary. Germany was aligned with Hungary, and Britain had a defense treaty with Poland, so was aligned with Poland. You can see just from the heavy-duty lines that this is an unbalanced schema. For example, Britain is against two of the other major powers, and each of them is against Britain and the other major power, so that's unbalanced.

What happened first was that in 1939, Hitler and Stalin got together; the Germans and the Soviet Union had a pact, a non-aggression pact, which divided up much of Eastern Europe between them. Then, you had this situation where that negative relationship changed to positive. Now, that's balanced, as the Soviet Union, Germany, and Hungary are on one side, and Britain and Poland on the other. However, that didn't stick because, well, let's see, this one shows that it's still balanced even though these two countries are aligned, they knew that they still didn't like each other, and so this was kind of a very fragile link, and they both knew it.

And then, in 1941, of course, Germany invaded the Soviet Union, which made that link negative; and now you have another balanced graph. Now, going back, they asked Churchill, "Say you've been anti-Communist, anti-Bolshevik your whole life, and how come you can align with the Soviet Union now?" And he said, "Well, if Hitler invaded Hell, I would then make at least a favorable reference to the Devil in the House of Commons." So, this is his notion: Well yes, I could have a positive link with Hell, which is the Soviet Union, which he always said was Hell, but he's going to make a favorable reference, so he's going to treat them positively just because it was Germany that got involved. And so, Britain did change its mind about dealing with the Soviet Union. This too is balanced, and this is what the rest of the war in Europe was organized around. So, let's be more ambitious and actually try to predict the countries taking sides. Now, we don't want to use alliance structures because that is just rephrasing the question, so I'm going to use the properties of the actors and I'll show you how I do that.

Now, first of all, there were 17 countries that were involved and now that we know there were 17, there'd be 136 links involved, and there are actually a lot of balanced configurations. Well, how many balanced possibilities are there? Let's take Britain. If you think about Britain and if you divide all the other 16 into friends and enemies, that'll be a balanced schema. If it is also the case that the friends of Britain like each other and similarly the enemies of the British, and its friends, like each other, then any way of dividing up the other 16 will give you a unique balanced arrangement. And that's 2 to the power of 17 minus 1, which is 65,000, so there are a lot of possibilities to get balanced graphs out of these many nodes. But, now, in order to help us make a prediction, let's take into account that countries don't just like or dislike; they have propensities for liking each other that are positive if they actually have a lot of reasons why they prefer to work together or negative if they have reasons to work against each other. Now, how are we going to estimate that? Well, one way I chose to do it is that I took five characteristics of a pair of countries and I used those to decide whether they had a propensity to work together.

So, one is whether they have a similarity in religion, another is whether they have a similarity in government type, another is if they've fought a war against each other, whether they have ethnic conflict relating to across that border, and whether there is a border issue. Now, most countries don't have any borders, so they don't have issues, but some of them, of course, do. So, if you take these five things and you don't know better, you just weight them equally, and that's how I got the propensity of any two countries to want to work together. For example, the more similar they are and the fewer problems they've had in the past, the more likely they want to align today.

And then, how about a country's stress? Now, we could talk about stress at the individual country level. So, if you arrange things into some tentative arrangement of two sides, then you could take for a given country, let's say Britain, the weighted propensities of the countries that are on its own side in this tentative arrangement and subtract the propensities of the countries on the other side. So, ideally, you'd only have positive numbers in this first factor and negative numbers in the second factor; then, you'd really be happy. And you want to weight those because Britain cares more about its relations with, let's say Germany, than it does with Luxembourg, and so obviously you want to weight the bigger countries more. And there's a standard way in international relations research to do that using national capability which includes the factors of the demographic — the population, the industry, and the military, standing military.

And so, when you take these parallelized propensities, for any given configuration, you could say that if it's balanced it will have a certain amount of stress. For example, it might be balanced in a sense that you've grouped people into friends and enemies, but it might be that some of the friends don't like each other so it would still have stress. So, the premise or the prediction would be that a country would change sides if it's in some tentative arrangement — it'll change sides to reduce its stress. And this can be regarded as moving from one balanced graph to another if presuming it only changes sides, it changes all its linkages to become positive with the side it's now on, which they might not like. So, if you start with a random alignment and then look at a country changing sides until no one wants to change anymore, then you get to a local minimum, and that'll be the prediction that this approach will make.

Predicting Alignment

- $N = 17$ countries involved. There are $n(n-1)/2 = 136$ links.
  There are $2^{(17-1)} = 65{,}536$ balanced configurations.

- But now links not binary. A link is the propensity of two countries to get along with each other. Propensities are estimated by:
  - Similarity of religion, Similarity of government type, Previous war, Ethnic conflict, and Border issue

- A country's stress in a given configuration =
  (weighted propensities with countries on its own side)
  - (weighted propensities with countries on the other side).
  Weight of a country is its national capability (demographic, industrial, military)

- This gives a configuration space: points in this space are balanced graphs.

- Premise: A country will change sides if it would reduce its stress. This can be regarded as movement in the configuration space.

- Start with a random alignment, and let countries change sides until no one wants to change any more. Prediction: local minima.

And here is what you get, remember that there are 35,000 possibilities, and actually you only get two local minimums; in the first one is the vertical division where you have these countries against those countries, and that's what actually happened: Britain and France for example, were on the same side as the Soviet Union against Germany and a bunch of its allies. However, there's also another local minimum, which is if you get there, nobody wants to change interestingly enough, and that's this alignment — Britain is with Germany against, so this is everybody against the Soviet Union and its friends, and that actually is also local minima if you get there it won't change.

And one way to visualize that is with an energy landscape or stress, where there is a configuration space.

Now, remember I'm representing this as two dimensions as that's all I can draw and all I can think about, maybe I can think about 3 or maybe 4, but it's really a 17-dimensional hypercube. That's what this is down here, but we can imagine this as just each one of these points here is a different balance graph. So, if you start with a given random arrangement of countries, and then give them a chance to switch, they'll switch to one that lowers their stress, and because everything is symmetric, it'll lower the group stress, and then it turns out that with these propensities based on the

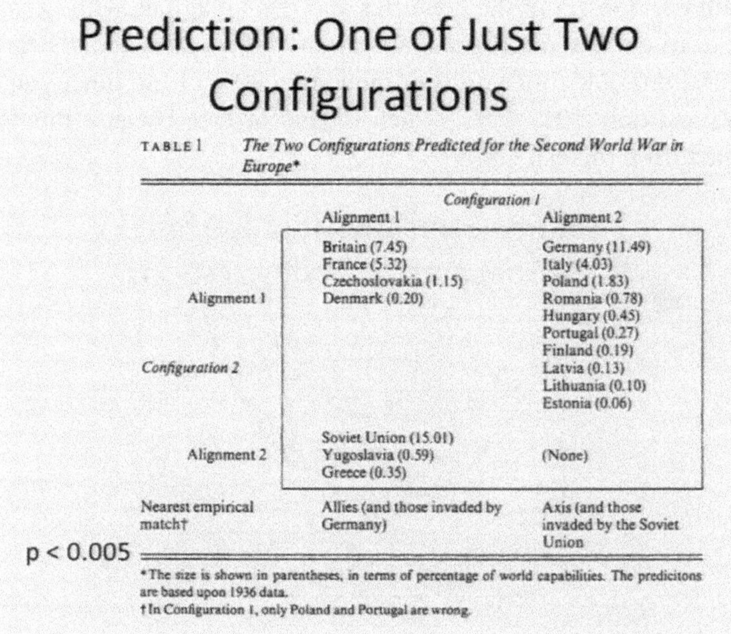

TABLE 1	The Two Configurations Predicted for the Second World War in Europe*	
	Configuration 1	
	Alignment 1	Alignment 2
Alignment 1	Britain (7.45) France (5.32) Czechoslovakia (1.15) Denmark (0.20)	Germany (11.49) Italy (4.03) Poland (1.83) Romania (0.78) Hungary (0.45) Portugal (0.27) Finland (0.19) Latvia (0.13) Lithuania (0.10) Estonia (0.06)
Configuration 2		
Alignment 2	Soviet Union (15.01) Yugoslavia (0.59) Greece (0.35)	(None)
Nearest empirical match†	Allies (and those invaded by Germany)	Axis (and those invaded by the Soviet Union

p < 0.005

*The size is shown in parentheses, in terms of percentage of world capabilities. The predicitons are based upon 1936 data.

†In Configuration 1, only Poland and Portugal are wrong.

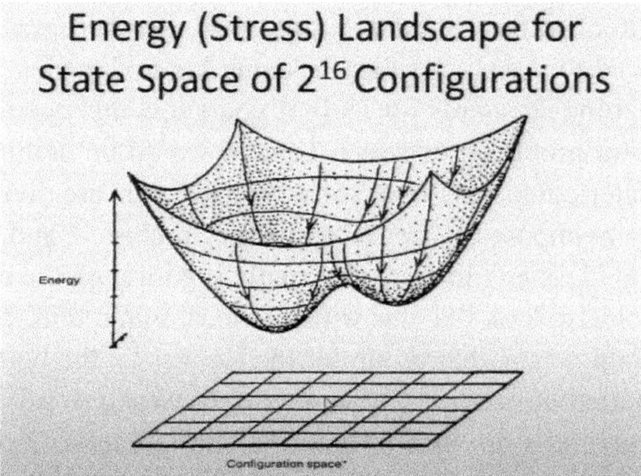

empirical measurements I told you about, there are two local minimums. And what are they? Well, this one is the second configuration: Soviet Union versus the others. And this one is the Axis versus the Allies, which is what actually happened. Remember, I didn't use any alliances; I just used information about the countries and their relations with each other. And it turns out that the basin of attraction of this one is about 90% of the configurations, which if you start randomly, 90% of the time, you'll end up here and only 10% there, which is kind of nice because this is what happened. But, this isn't silly.

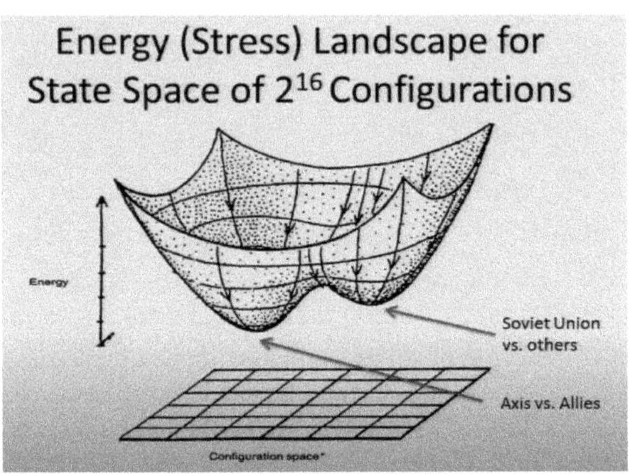

I presented this particular work which was done with Scott Bennett to a business school audience and they suggested that there was another setting that looked to them a lot like the same kind of model, which is the problem of setting standards for Unix if you know this operating system. And there were problems of which version would be dominant and the Versik computer companies both software and hardware divided up in an alliance to try to impose the standard that they preferred, and we use basically the same ideas and in fact we actually got through the different alliances that look also look like this with two local minimums, and here they are, these group versus that group, or the top versus the bottom, and the first configuration only one of these nine were wrong.

Incidentally, let's go back to this first configuration: 2 out of the 17 were wrong, but the probability that you would get that by chance is less than 1 in 100. It is, in fact, 1 in 200, and so this is a statistically significant prediction and this same thing actually happened here, that is, this is also statistically significant. So, it's kind of interesting that the same basic

Setting Standards for Unix

TABLE 2 — *The Two Configurations Predicted for Unix Alliances**

	Configuration 1	
	Alignment 1	Alignment 2
Alliance 1	Sun (28.9, S) Prime (1.0, S)	DEC (20.0, G) HP (11.5, G)
Configuration 2		
Alliance 2	AT&T (28.5, G) IBM (3.8, G)	Apollo (21.2, S) Intergraph (4.4, S) SGI (4.4, S)
Nearest empirical match†	Unix International	Open Software Foundation

*Size is shown in parentheses, along with whether the firm was a computer generalist (G) or technical workstation specialist (S). All firms had a design orientation.
†In Configuration 1, only the IBM prediction is wrong.

approach worked for two completely different setups — two completely different settings — countries and computer companies.

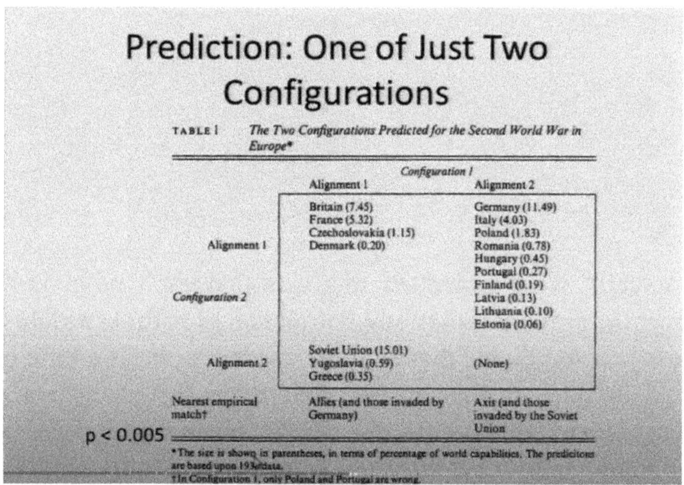

Prediction: One of Just Two Configurations

TABLE 1 — *The Two Configurations Predicted for the Second World War in Europe**

	Configuration 1	
	Alignment 1	Alignment 2
Alignment 1	Britain (7.45) France (5.32) Czechoslovakia (1.15) Denmark (0.20)	Germany (11.49) Italy (4.03) Poland (1.83) Romania (0.78) Hungary (0.45) Portugal (0.27) Finland (0.19) Latvia (0.13) Lithuania (0.10) Estonia (0.06)
Configuration 2		
Alignment 2	Soviet Union (15.01) Yugoslavia (0.59) Greece (0.35)	(None)
Nearest empirical match†	Allies (and those invaded by Germany)	Axis (and those invaded by the Soviet Union)

$p < 0.005$

*The size is shown in parentheses, in terms of percentage of world capabilities. The predicitons are based upon 193?data.
†In Configuration 1, only Poland and Portugal are wrong.

Now, here's another kind of application: we've talked about graphs with nodes and links, so here's another one which is curve fitting. So, suppose you see data points like that, and you want to understand it and simplify it to make sense of it. Well, you might say, "Well, a straight line looks pretty good, with some noise of course." But now, suppose that this was housing construction, and let's say these were just typical years and not these actual years when there was a huge dip right here.

But, let's say this was some period when things were more or less normal, and knowing that it's something that might be growing at an exponential rate, like a few percent a year, now you might say, well, this may not be a straight line, maybe it's more of an exponential though you can't really tell for sure, but it would change which kind of curve that you would prefer to fit, either normally you'd want to fit a straight line if you could, but if it's something that represents growth over time, you might fit in exponential; and those are in effect different kinds of schema.

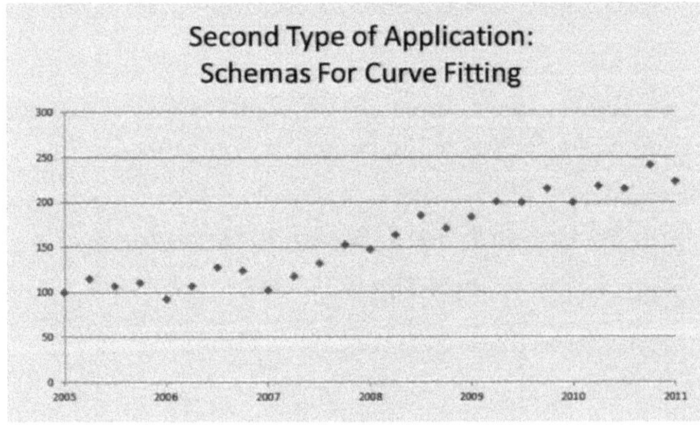

So, actually what generated this was that each point is a certain growth over the previous point, say, 3% annual growth — and this is quarterly data — so you divide that by four, so every point is a little bit multiplied by over one — for growth. And then, there's also a seasonal component which both boosts the second quarter and decrements the fourth, and now once I told you that, you could probably see "Well actually maybe it is seasonal," which you probably couldn't see before. But, once you start thinking that maybe it's something like agriculture or housing that might have a seasonal component, then you could probably actually see that those have a seasonal component, and, of course, there's also some randomness in this. So, knowing what it's about also helps you decide what kind of schema you want to use to fit it.

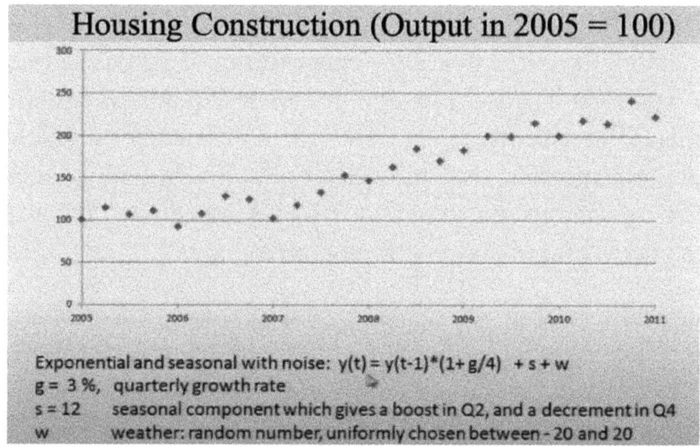

It's an interesting relationship between which schema you use — there's a trade-off between achieving better goodness of fit and greater efficiency in terms of cognitive ease. So, to return to the first set of examples, if you use a balanced schema to understand alignments of countries, for example, it might not be a very good fit. You could achieve a better fit by positing three or four clusters, but then you lose some efficiency. It requires more parameters and you have to remember more things if you're fitting things to these more complicated clusters. Thus, there's a trade-off between goodness of fit and cognitive efficiency.

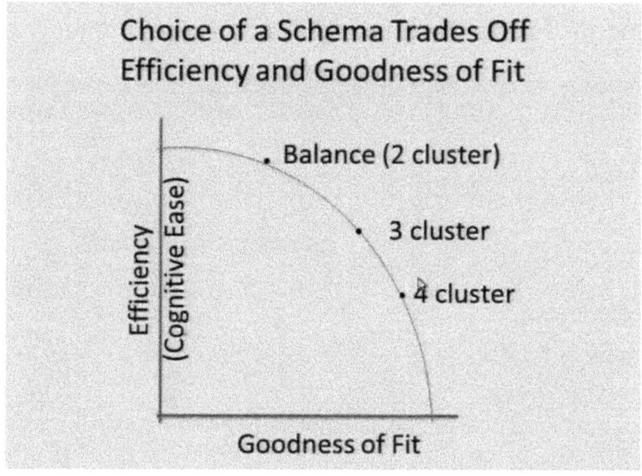

The same thing happened in the last example I showed you, where you're selecting the curve to use for curve fitting. If you use a linear equation, you won't get as good a fit as if you used a quadratic or third-degree equation, and likewise, there's a trade-off in efficiency — this has fewer parameters to remember than the earlier one. So, in both of these cases, applications of schema, the choice of which schema you pick, is an effective choice between goodness of fit and efficiency.

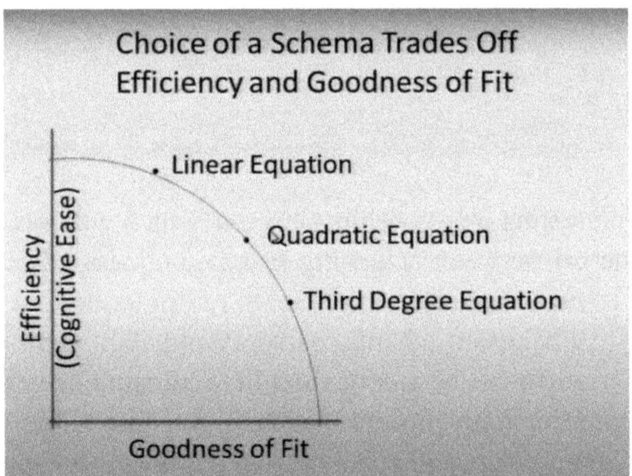

And if you had a course in Algebra you can do better because, for example, quadratic equations will have the same goodness of fit, but once you have some proficiency with it, it's easier to understand and use.

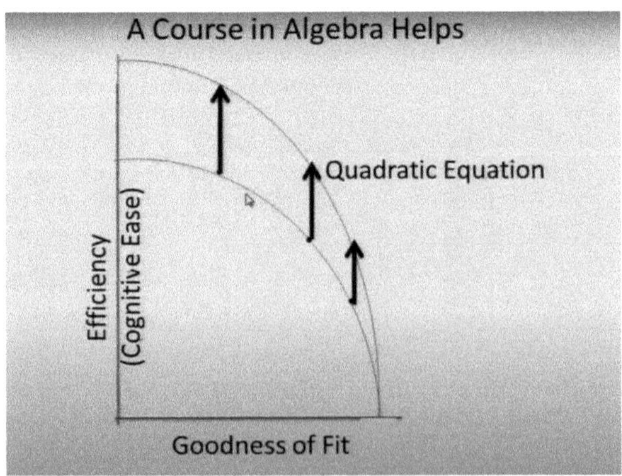

So, just to review so far, we had various things you can do with a schema that help you reduce complexity, and you can use it in various ways to *correct* errors, missing data, prediction, and so on.

> ### Review.
> ### Uses of a Schema Reduces Complexity:
>
> 1. Infer Missing Data
> 2. Correct Errors
> 3. Predict Changes
> 4. Economize on Memory
> 5. Evaluate Source Credibility
> 6. Measure stress and goodness of fit
> 7. Focus attention to resolve ambiguity

So, here's a third application. We saw the graphs for alignments and balanced schema, and we saw curve fitting. Here's another one, which is interpreting pictures. This is a typical problem that artificial intelligence has: If you give it a picture that a human would understand, how can a computer come to understand what is going on? This is a famous example where it's clearly a face, but it's a face which is a very common schema. We're ready to interpret ambiguous things such as faces, but in this case, there are two different interpretations of this one drawing in the middle. You can either see it as an old woman looking down to the left or as a young woman looking over her shoulder. Now, can you see both?

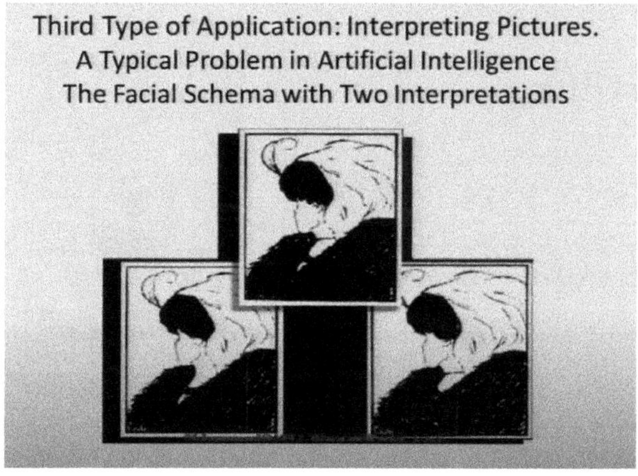

How many can see the older woman and how many can see a young woman? And how many can see both? And how many can see both at the same time? No, that's really hard to do. You might understand both, and you might be able to switch between them, but it's really hard to see them both at the same time. And if you had trouble seeing both of them, this is a slight distortion that emphasizes the older lady, and this is a slight distortion that emphasizes the younger one. It's like moving from a graph between two different balanced schemas — you can move one way or another to resolve the ambiguity. But, there's a case where the schema of a face can be used in two different situations.

Okay, so we have one more after this. The summary is that you can use schemas to reduce complexity and the sense of information. I want to be clear that I'm using complexity in the sense of information efficiency, and reducing complexity means that you can do better with this schema. So, we saw how this could work for networks of relations, which balance m-clusters or rank order, and we saw how it could work for curve fitting and interpreting pictures.

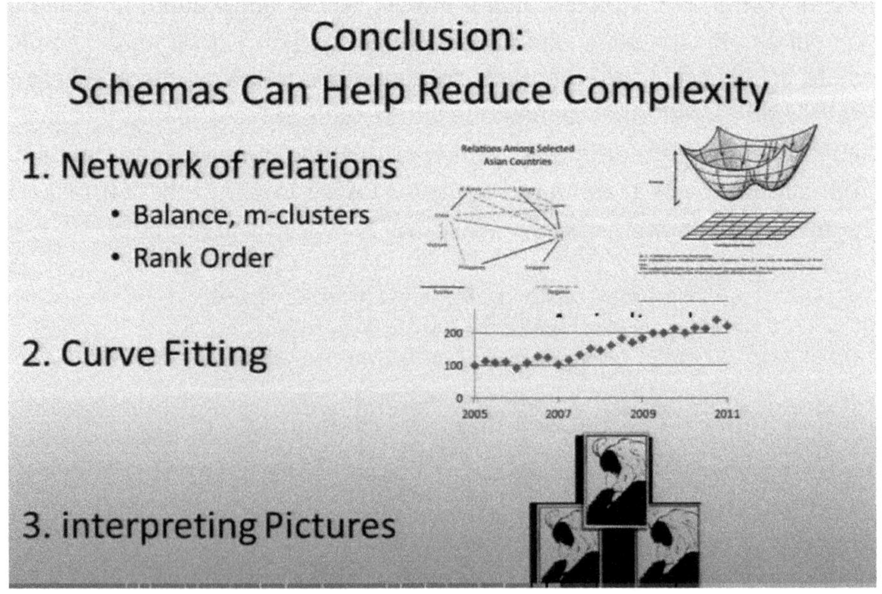

Now, I want to show you how Murray inspired me. Let's see how this all relates to what he was talking about, which means I want to go to a slide I just did over the coffee break.

Integrative on One Front:
Coarse Graining

Poli Sci:	**Alignments and Conflict**
Economics:	**Cost/Benefit of a Schema**
Psych:	**Mental Models, Perception**
Anthro:	**Ingroup/Outgroup (Balance)**
Philosophy:	**Meaning as Inferable Knowledge**
Statistics:	**Goodness of Fit, Choice of a Fcn**
Physics:	**Stress, Energy, Lyopenov Fcn**
Game Th:	**Deception**
A.I. :	**Perception, Meaning from Big Data**
Complexity:	**How to Cope with Complexity**

So, Murray recommended doing integrative work on at least one front to get started. I'm not as ambitious as to do all the fronts, but I'll do one. This talk could be thought of as dealing with one of these integrative issues, which is how you do coarse-graining. Finding the most important features of taking "a crude look at the whole" is the answer. And this approach toward using schemas, to simplify the information and to organize it, is definitely integrative. For example, in political science, you could certainly use it to study alignments and conflict as I showed you. You could think of it as relevant to economics because I did a cost–benefit analysis of choosing schemas, a kind of cost–benefit analysis at the heart of economics. It's also about psychology because it dealt with mental models and perception, and anthropology as Murray says we should try to minimize in-group and out-group thinking. But, in fact, this helps us understand at least some of the ideas about why it's cognitively efficient to organize things that way even though it has unfortunate political consequences. So, it speaks to anthropology about in-group and out-group problems.

Philosophy: Well, this is a theory that suggests that meaning is one of the major philosophical concepts. One kind of meaning at least is what is inferable knowledge that you get if you know some things, you know what schema to apply, and you get some news. What's the meaning of that news? Well, the meaning of that news can be regarded as the additional knowledge that you get by virtue of having that schema in your prior knowledge. Statistics, of course, is about goodness of fit and about the choice of which function to use to represent data. In physics, we actually used here the concepts of stress and energy, and the landscape with the two local minimums is actually the Lyapunov function which can identify a notion of returning to an equilibrium state — another physics idea.

Now, Game Theory. If you know what schema the other one is using, you could actually use this to deceive them because if you could set it up so it's similar to one of the things that is close to one of their well-formed beliefs, then they could round off the errors that they might observe and decide that the evidence fits a certain schema that you want them to have. And so this is a game theory problem where deception can be used with the schema concept.

Artificial intelligence is also dealing with perception, of course, but also the meaning of big data. A major problem is when you have this huge amount of data like all the Twitter links for something. How do you find meaning in that? How do you find the patterns in that? Well, one way to at least help is to have some schemas to start with and then use these techniques to find the closest example of that particular schema that would fit as well as possible. Now, of course, there's always going to be stress in that fit, but we can talk about how to reduce that stress with different schemas.

And finally, complexity theory is one where you think about how to cope with complexity and how to reduce the complexity that we have. So, it is integrative, but it's not integrative across everything; but, it does, I think, speak to the important question that you want to deal with when you take a crude look at the whole, which is how to do coarse-graining. Thank you.

2.2 Discussion

Chan Heng Chee: Thank you very much, Professor Axelrod. I'm from the Singapore University of Technology and Design, and I'm with the Lee

Kuan Yew Centre for Innovative Cities. I told you earlier I'm a political scientist. I find your schema interesting, but I wonder how applicable or usable it is because I think it does not take into consideration some of the more interesting developments of late. For instance, you have friends and enemies, but maybe the nature of relationships between states these days can fall into the category of neither friend nor enemy, and countries don't want to choose. It is also the end of the Cold War, so to some extent, it's a bit of the end of ideology. And even if it is a civilizational conflict, it may not end up in friends or enemies. Also, I think economics now plays a very major role in determining the directions of relationships, so when you just take it simply as friends or enemies, it is a bit more complicated. Take, for instance, the relationship between the United States, Australia, and China. Australia is an ally of the United States but Australia has a very deep economic relationship with China. Australia really doesn't want to have to choose you know, and I ask, "Can an ally have as its largest trading partner a country that your ally has some problems with?" But even the United States today says that it really is not an enemy. China is not an enemy. It's just a competitor. So, these are more complicating considerations which should be new nuances in how you apply these schemas. I can also think of, you know, the sets of relationships between India, the United States, and Iran and Japan, Iran, and the United States because economics comes in and so when you try to apply sanctions and get people to follow the sanctions, the NDAA is really troublesome. Thank you.

Robert Axelrod: I agree with all of your comments about the real world. I'd like to say some of those are pretty easy to incorporate into the formulation that I have proposed and others are not. For example, if there's an economic neutral gain from trade, that would be one reason why two countries might want to have good relations and overlook some problems that they might otherwise have. That would be built into the parallelized propensity as one of the factors that would determine what they want in order to maintain good relations. So, I think you could build economics in the way that you suggested in a straightforward way. Second, there are going to be situations that are neither clear friends nor clear enemies, which is why I moved to this propensity idea where there's a degree of willingness to cooperate versus taking a zero-sum orientation, and you could actually take that into account. So, for simplicity, I drew guided lines or solid lines, but I also showed you how you can think of those as real numbers, plus or minus. Now, I grant you that even that is an over-simplification that you could put these various factors into a single scale, but on the other hand it might be useful as a crude look at the whole to think about if you do put these various factors. Then, you could actually see some of the problems that you're pointing out, which would reflect this stress in the system, so that if the United States has a kind of a mixed relation with China, to take your first example, but Australia wants to have good relations with both the United States and China, to the extent that the United States and China are competitive and start saying nasty things to each other or mobilizing their military, their navies against each other, to that extent, it's going to be unbalanced and Australia is going to have some stress in maintaining a very cooperative relationship with both sides if they're not getting along with each other. That's exactly what I wanted to show. So, I think that some of these issues can be dealt with as measurement problems and some are not just reflections of the stress that is in the system. And for still others, this kind of simplified scheme won't work at all, but that's what is to be expected from a crude look at anything in particular, such as how to do coarse-graining.

Ann Florini: My initial reaction was exactly the same as Heng Chee's, and then I started thinking about the scope to which the balanced schemas would apply. The situations you were talking about were situations of declarations of war, war and peace, crisis decision-making where this kind of simplification down to friend and enemy does happen. The literature shows that very clearly. But, then, I also started thinking about the partisan polarization happening now in the United States where people seem to be grouping themselves in exactly the way that you're describing in the absence of that kind of crisis pressure of decision-making. So, I think I want to give more thought to how widely this applies.

Simon Levin: So, when you talked at the beginning about the example of Iran, Syria, Turkey, and Israel, or any of the other examples, you deliberately simplified by ignoring the fact that they're all parts of bigger networks involving the United States. And so the first question is this: If you have at the broad scale a network that's unbalanced, or that's got high degrees of stress in your second approach to the problem, what can you say about the sub-graphs? For example, what's the maximum number of balanced sub-graphs you can have of a degree k in a larger graph that's unstable? And secondly, can you say anything about how the stress or the unbalanced nature will be resolved? Will it be the sub-network that would cause the greatest change in the direction you want? Is there some hierarchy theory that allows you to go from the larger networks down to the components?

Robert Axelrod: Very good questions. The first one about how many ways you can have sub-graphs that are balanced, that you can actually do mathematically. I couldn't do it in my head, but you can actually say if there are 17 nodes how many triangles could be balanced without the whole thing being balanced. That's a knowable question. On the first one about the resolutions, suppose you had different clusters of balances but then the various subsystems are interacting with each other, how can you predict what's going to happen? And I think you could. The way you would expect it to happen is that some of those sub-clusters would merge with each other and until if you want to assume it's going to be balanced in the end, then obviously you're going to have so many sub-clusters. The most likely thing that will happen is that the fine of two sides will consist of some of the sub-clusters on one group and some of the sub-clusters on the other. You then have the emerging property that the sub-clusters are going to emerge according to the, probably some aggregate, propensity between those sub-groups. And so, the two sub-groups that really hated each other would more or less likely work together than two sub-groups that were indifferent to each other. And so, I think that you can use this to study the emergence from sub-graphs to a unified graph if you took the larger system into account. Now, this is meant to be a study of how one can do coarse-graining, but it's not really that it claims to take into account to be a full theory of conflict for example. But, we did see that you can make some predictions that work not only for countries but for companies, and so it probably does have some ability. And so, I think you can study emergence.

Simon Levin: Presumably these transients are data for making this test, but I'm wondering also if there's a link here with Duverger's Theorem.

Robert Axelrod: So, you want to state it for us?

Simon Levin: I mean the theorem which I thought was a political science theorem.

Robert Axelrod: It is. I can state it too, but I…

Simon Levin: Oh, within a political system like the US, you won't long to have more than two political parties because…

Robert Axelrod: Right.

Simon Levin: It seems to me that your approach to resolving the balanced graph thing would give you this as a prediction.

Robert Axelrod: Well, I don't think so, and I'll tell you why. I think the Duverger Theorem comes from the idea that the American electoral system says that whichever party gets a plurality in the vote like if it's 40–30–30, then the 40 becomes the president, and therefore there's no point in supporting the weakest of the three parties, and therefore you'll tend to move to only a two-party system. That's a function of the way you count the votes. In many other democracies, in fact most of them, there's proportional representation. And what you do then is you give the number of seats in Parliament in proportion to the votes they got, and then in Parliament, there's bargaining over which group will get together to form the Cabinet, and that I think does correspond to this idea that the Cabinet, that the coalition that gets together, will minimize the strangeness of the parties working together. And therefore when you get the ins and the outs in that case, the reason I don't think it's permanent is because in the next election, there could be a different distribution and a different movement in the ideological space, and so Duverger is really about the consequences of having a plurality electoral system; most democracies have the proportional which could keep the diversity at the party level, but then after each election there'll be bargaining which will in fact bring some together, and they might not like each other, but what the prediction would be is that the ones that get together were the ones that least dislike each other.

Greg Fisher: Greg Fisher from the think tank Synthesis. I saw one of the charts had a balanced graph with Germany and the Soviet Union having a friendly relationship, but then the Germany invaded the Soviet Union. Can you explain in your sort of theoretical framework how that happened?

Robert Axelrod: Well, we could do it at two levels, of course. I could do it sort of as a political scientist and historically, which should be the short version, of course. That is, it was convenient for the two of them to get together in a temporary way by dividing up Eastern Europe while both of them were arming themselves to the teeth, and they both expected that there would eventually be a war. The interpretation in the graphical sense would be that they actually still had very hostile propensities, but for the time being, they both had an advantage in working together against Britain. And so, they did that on a temporary basis and, of course, not surprisingly, it broke down. The more surprising thing would probably be that Stalin was quite surprised when it broke down, but he wasn't surprised that it would break down. He knew that there were underlying propensities to have a showdown between the Communist and Nazi governments. He was only surprised that it happened when it did.

Greg Fisher: Forgive me if I've misinterpreted you, but I thought the idea was that if there's a balanced graph, there wouldn't be a change.

Robert Axelrod: Well, yes. That would be a falsification of the theory right there. But, I would say that it's not a dramatic falsification because if you measure the propensities, what was a positive relationship was about as weak a positive relationship as you can get, and so when it breaks, it wouldn't be very surprising.

Kai Xi: Good afternoon, I'm Kai Xi from Nanyang Technological University. I have to point out two possible limitations of this theory: The first one is that the scheme actually predicts how the relationship between the different countries will change after some variable happens. For example, the relationship between these two countries gets sour. So, is there any way we can incorporate these changes in the variables, also into our scheme, so we don't just make some *post-hoc* predictions but rather can predict, next time, what kind of variables or how the variables will change? The other question is this: Is there any way we can incorporate time into this scheme? Because the scheme seems to presume that the relationship between the countries will always tend toward a peaceful relationship, be the premise that all the countries can become friends but they cannot all be enemies, or the propensity to go toward lower stress. They all predict that the countries actually want to become peaceful. But the countries won't become peaceful immediately. There may be a time period where there is a lot of conflict, but in the long run, maybe the countries will tend toward peace. So, if we predict that countries will always tend toward peace at any point of time, it might result in some, or we might encounter some problems. So, is there any way we can incorporate the time scope into the scheme as well? Thank you.

Robert Axelrod: Thanks for those questions. The first one is, if there's a change in one of the variables, can you predict the change at the system level? And the answer is yes, and to take a specific example that's based on what I showed you, that was based on 1936 data about military capacities and other things. At that time, as you might expect, Poland was hostile to both Germany and the Soviet Union. It was caught between them not only geographically but also politically and was hostile to both. However, in 1936, the Soviet Union was so much stronger than Germany, which was rearming quickly, but by 1936, the capacity of the Soviet Union looked bigger. And so, Poland, deemed negative to both, would have been predicted to be more hostile to the Soviet Union because it was the bigger, more important one as far as it was concerned. However, by 1938, after Germany had continued its very fast mobilization, against the Soviet Union's slower industrialization and mobilization, Germany became the stronger one and Poland would be predicted to be anti-German. Or another way of putting it is that Germany would invade first rather than the Soviet Union. So, when a variable changes like, in this case, the weights on the size of the country affects the weights on their military capacity, as well as, in that case, you're also taking into account industrial capacity and demographics, but the military capacity is what changes fastest. So, when that changes, you can make a different prediction about how the countries will align based on their propensities, which are weighted by that. So, that's an example of how if one of the parameters, in this case, the relative strength of the Soviet Union and Germany, changes, you can make a different prediction about, say, how Poland will react. So, you can do that with any of the variables. Your other question was about predicting change, and certainly, you can do that and you could be wrong and you wanted any good theory that has got to be falsifiable, and so it's going to be wrong sometimes. And the cruder it is and the more ambitious it is in what it's about, the more it's going to be wrong and so one doesn't have to be too embarrassed by that. But it's much, much, much better than chance, so that's good. Now that's what we in the social sciences use as our standard.

And then you asked if there is a possibility of moving toward peace between all countries. Let me just be clear that while I usually talk about balances having divided into two groups, also everybody in one group, or everybody's friends with everybody, is also consistent with balance. If everybody is friends with everybody else, then nobody has to change and it is balanced. So, a balanced graph could either be actually two clusters or one. And so, if the relationships become positive, for example, if economic mutual gains from trade dominate concerns about competitive security, then one can imagine in the distant future that most important relationships between countries, and maybe almost all of them, would be positive in some areas like scientific cooperation. That's not so far from what it is. And so, the possibility of parallelized propensities being positive and therefore having one large group is consistent with the theory if the parallelized propensities lead you in that direction.

Suman Banerjee: I don't think I'm adding anything new but building on the question already asked to you. I take you from the real world and ask you that, you know, we know something called Arrow's impossibility theorem, and in some sense you are having these pairwise things with a single issue. But where friendship–enmity can be boiled down to a single scaled variable, we can easily rank them and think about stable coalitions and things like this. But in a multi-issue world, where I consider one country as a friend, as a friend point and one as an enemy point, like she raised the group contraction lemma and the field expansion lemma kind of preclude any aggregation and any conclusion about stable coalitions. You know, which is, in some sense, arrow's impossibility theorem in a very simple, my understanding, language. So, really possible, I mean at points, one single issue dominates others; it becomes so important like the Syrian crisis right now. You know probably dictating their neighbors, but we have seen like, you know, strange coalitions like the United States and Saudi Arabia and things like this, which is again probably a single point-driven thing. But, in a multi-issue world, one I'm friend to somebody and one I'm enemy to somebody. Well, it's quite common, and then probably this kind of aggregation, this kind of coalition formation might be difficult to say. I'm just trying to get a clarification from you.

Robert Axelrod: I'm not sure if I follow what the question was.

Suman Banerjee: The question is in a multi-issue in a world like where I like A, B, C, D…

Robert Axelrod: Right, oh, okay.

Suman Banerjee: I put one slant A-B-C-Ds, it might…

Robert Axelrod: Yeah. Well, if there's one issue, for example, in domestic politics, there's often an important dimension of Left–Right, and then parties in a parliamentary system will be arrayed on that dimension. One would expect that the coalition that can achieve a majority would be a connected coalition in that space and have the minimum range in order to achieve the majority. So, this same kind of thinking can get you the idea of what it takes. Now, if there are multiple dimensions, there may be multiple ways if you scatter the parties in two dimensions. Some might be international and some domestically oriented, for example, as well as Left–Right, then you can also imagine which collection of parties will have the least conflict of interest between them, which is to say are closest to each other in a way that can you get the least stress. Well, put it this way, the prediction you would get is that the coalition that forms is the one with the least stress and still has a majority, so the parties are basically close together but enough of them, so that they do have a working majority. And so, that's what you would expect in that kind of system. Occasionally, you get the interesting idea regarding the outs who might typically be on the Left and Right extreme, then this is balanced in the sense that they usually don't like each other at all, and they usually are even more opposed to each other than they are to the middle coalition that might have formed, and so that expectation will not necessarily be a balanced schema, but probably be three groups — the Left, Center, and Right — and you most typically would get the group that excludes both extremes, with some above the extremes, to form a group that has a majority. You then have three. You tend to imagine three clusters, and very Left parties that are left out tend to not be too different and various Right parties left out tend not to be too different. If you're doing an electoral coalition situation, you have three clusters that would probably be the best interpretation.

Discussant A: I mean, in the context of your framework, you're suggesting that the link between two nodes is classified as either friendly or hostile. Is there a way of introducing a metric of intensity and making that metric time-dependent? Because then you could examine the fragility of existing relationships as time evolves.

Robert Axelrod: Well, I've already provided one metric in the example from the European system, where there were five different aspects that contributed to the measurement of whether two countries would have a positive or negative propensity. These aspects included similar religion and government, border conflict, ethnic conflict, and previous wars. So, those are just five examples. Putting those together gives you a metric for how each pair of countries is positively or negatively related to each other. And as I mentioned, that metric is time-dependent. In other words, for example, if the countries were mobilizing throughout the thirties at different rates, then you could conduct that kind of analysis, say, once each year, which I did. Then you could see how the fact that Germany was mobilizing faster than anybody throughout the late thirties changed the propensities somewhat, especially because countries that disliked others would worry more about working with friends against Germany than they would against the Soviet Union if Germany was growing faster. So, you can definitely incorporate both a metric and time dependency.

Peter Sloot: What I'd like to do is to actually go from a crude look at the whole to a very precise look at the small part, so actually fine-graining rather than coarse-graining. There have been a lot of studies on social networks, right? Tremendous amounts of studies where we have data, much more than we have on countries actually. And there have been a lot of studies on how opinions spread through social societies and how one individual, depending on how you look at that individual, might, you know, be part of at least ten or twenty different sub-nets.

Robert Axelrod: Right.

Peter Sloot: So, people can be part of a balanced and a non-balanced net at the same time, if you'd like. I don't understand anymore, you know, how I can use your theory to understand, for instance, the percolation of opinions. So, that's point one. Let's say having two being balanced and non-balanced at the same time, and that then is dynamically changing. And the second thing is what we actually discovered ourselves, which is called stochastic resonance in these networks, where there's noise in these networks that will actually move you from one minimum to another minimum, and you cannot even determine where you're going to go. So, my question is this: Does it still hold for these systems? Because it should hold for these systems, when you actually take a crude look then, so when you zoom out...

Robert Axelrod: I didn't have time, but I'll use it to just give you a couple of minutes on the time scales of the dynamics, which is your second part. So, this is an egg carton, which is meant to illustrate the idea that there are many basins of attraction shown through the graph with two, and this is a lot. Now, again it's multi-dimensional space, not just two. But, suppose the system is where the blue line is, so a very short time scale once you measured very quickly after that, not much would change and so short time scale you'd expect to be just what you'd observed. But a little later, you'd expect it to be moving toward the nearest balanced graph, and if you allow more time, it might get to the bottom of that local basin, but this is the point that you are making that over a longer time scale with enough noise and you couldn't have noise as shaking the system. It might shake off into a different basin, and then it might be there, or later it might be here.

Time Scales of Graph Dynamics

Assume links change at random, with bias toward neares balanced graph.

- 1. Very short time scale: at observation
- 2. Short time scale: move toward nearest balanced graph (within the basin of attraction).
- 3 Medium time scale: at or very near the nearest balanced graph, i.e. the local minimum.
- 4. Long time scale: at or very near *any* balanced graph: a slow random walk among neighboring balanced graphs.

So, over a very long time scale, you can't predict where it will be, except that it'll probably be near one of the local optimums. And that's one reason why long predictions, when there's noise, you know, become less and less feasible. But, it is an emergent property of this kind of thinking. It's not built into it that the longer time scales have the more diffused predictability. But, it's more specific than that because it doesn't say it'll be anywhere. It will say it will be close to an attractor.

2.3 Summary of Robert Axelrod's Talk

Robert Axelrod's talk explores the concept of schemas, focusing on their role in reducing complexity and managing relationships, with implications for understanding global dynamics, predicting behavior, and making decisions. Axelrod begins by framing his discussion within the challenge of coarse-graining, which involves simplifying complex systems to make them more manageable. He introduces the idea of schemas as tools for reducing complexity and managing relationships.

Axelrod illustrates the concept of schemas using examples such as the Balance Schema, which categorizes relationships based on alliances and enmities. He discusses how schemas like the Balance Schema help simplify complex networks of relationships by categorizing actors as friends or enemies and establishing rules for interactions between them.

Using examples such as the relationships between countries in the Pacific Rim and the evolving dynamics between China and the United States, Axelrod demonstrates how different schemas can be applied to analyze real-world situations. He emphasizes the importance of understanding the underlying dynamics of relationships and how they influence global events.

Axelrod introduces the notion of state space, which represents all possible configurations of a system, and explains how schemas can be seen as subsets of this space. He describes how schemas help infer missing data, correct errors, predict changes, and economize on memory by reducing the amount of information that needs to be stored and processed.

Axelrod discusses the use of schemas in measuring stress within a system and evaluating the fit of data to a particular schema. He explains how schemas can help identify areas of ambiguity and focus attention on resolving inconsistencies.

Axelrod explores the relationship between beliefs and reality, arguing that schemas can influence perceptions and behavior, leading to self-fulfilling prophecies. He illustrates this with historical examples such as the alignment of countries during World War II and the adoption of standards in the computer industry.

Finally, Axelrod discusses the relevance of his work to complexity theory and its implications for understanding and coping with complexity in various domains. He suggests that schemas offer a way to navigate complexity by providing a framework for coarse-graining and reducing the dimensionality of complex systems.

In summary, Robert Axelrod's talk introduces the concept of schemas as a tool for reducing complexity and managing relationships in complex systems. He demonstrates how schemas can be applied to analyze real-world situations, predict behavior, and make decisions, with implications for understanding global dynamics, navigating complexity, and shaping outcomes.

2.4 Relevance of the Talk to the Current Stage of Research

Robert Axelrod's talk on schemas and their role in reducing complexity and managing relationships remains relevant to the current stage of research in several ways:

Network Science and Complex Systems: With the increasing availability of data on complex networks such as social, economic, and biological systems, there is a growing interest in understanding the underlying structures and dynamics. Axelrod's discussion on schemas provides insights into how complex systems can be simplified and analyzed using principles of network science.

Predictive Modeling: In fields such as political science, international relations, and economics, there is a constant need for predictive models that can anticipate changes and outcomes. Axelrod's approach to using schemas for predicting behavior and changes in relationships offers a framework for developing such models and understanding the underlying mechanisms.

Cognitive Science and Decision-Making: Understanding how individuals and groups perceive and interpret information is essential in fields such as psychology, cognitive science, and behavioral economics. Axelrod's exploration of how schemas influence beliefs and behavior provides valuable insights into decision-making processes and cognitive biases.

Conflict Resolution and Diplomacy: In contexts involving conflict resolution, diplomacy, and negotiation, Axelrod's discussion on how schemas can be used to manage relationships and reduce stress within a system offers practical implications. By understanding the underlying dynamics and applying schema-based analysis, policymakers and diplomats can make informed decisions and develop strategies for conflict resolution.

Machine Learning and Artificial Intelligence: In the field of artificial intelligence and machine learning, there is a growing interest in developing algorithms that can understand and navigate complex environments. Axelrod's talk provides a conceptual framework for developing intelligent systems capable of managing relationships, predicting behavior, and making decisions in complex, dynamic environments.

Overall, Axelrod's insights into the role of schemas in managing complexity and relationships have broad implications across various fields of research, including network science, predictive modeling, cognitive science, conflict resolution, and artificial intelligence. By understanding and applying these principles, researchers can gain deeper insights into complex systems and develop more effective strategies for decision-making and problem-solving.

https://doi.org/10.1142/9789819802135_0003

Chapter 3

Curiosity, Innovation, and Complexity

Helga Nowotny

YouTube: https://www.youtube.com/watch?v=tVqe0_z6nhY

Speaker: Helga Nowotny

Moderator: Jan Vasbinder

Discussants: Sulfikar Amir, Sheila Ronis, John Richardson, Caroline Pluss, Rhett Gayle, Discussants A, B, C, D, E, and F

3.1 Talk by Helga Nowotny

Thank you very much. My talk has three parts — the first part I take, literally, a look at society. I know there are various ways of looking: You can look at figures, you can look at graphs, you can look at the way people think, you can look at patterns, but I decided to look at images. It's my first part. My second part, I want to take us back to the notion of change in society, and here, no surprise, it's science, modern science, and technology that have really changed the way our societies work. And in the very last part, I want to come back to something that I think is there. It's there but not very well explored, namely, the tempered dimensions of complexity. So, let me start with my literal first question: Where is society? This is a society, a eusocial insect society, and as you know, there are many species of wasps, of bees that differ in the way they organize themselves, although there are certain characteristics they share, and of course, there are ants and other important eusocial species. And if you talk to researchers working on this, it is fascinating to discover not just the division of labor that we all know about, non-producing workers and the queen, but actually, they're interested because they found out the insects in this society know who the next queen will be. And this is a very interesting, fascinating question: How do they know? So, it's an interesting topic, and it's a kind of society.

This is perhaps somewhat closer to us — groups of chimpanzees — and here we have already some affinities with the way the group conflict is just as endemic in them as there is cooperation, and also recent work has tended to emphasize a notion that has been taboo 20 years ago, namely, that they also have something called culture. Now, this was a very controversial topic, but now, more and more people agree there is a way of transmitting acquired knowledge to the young — to the next generation — which is one dimension of what culture is all about.

And therefore the question is this: Where is human society, and what do we understand by it? Balance of conflict and cooperation. Robert Axelrod's talk was very much about conflict and we heard about cooperation in the very first talk — altruism that goes beyond kin selection, which is one way how humanity has learnt to cooperate going beyond genetically related individuals. And while we know what the unit of measurement is in biology and biological evolution, namely, the reproductive success to have as many offspring as possible in your life, the unit of measurement

in cultural evolution, in my view, is knowledge production and innovation. Now, what does human society look like? Are we talking about a good society — an old topic of philosophers — an ideal society, a utopian society that was very much *en vogue* at the beginning before the Renaissance — people looking for ways to change the society, they projected it into the future where utopia does not have the place as yet. But there were also failed societies and sustainable societies. This is an ideal society — Plato's Academy. And as you remember, he proposed that the philosopher should be the ruler, something that in real life has not worked out very well.

There are societies that have gone bad and this is a vision by Hieronymus Bosch of the *Last Judgement, Hell* — the failed societies.

Hieronymus Bosch 1550–1560, Last Judgement Helll

This is the very well-known book by Acemoglu and Robinson on the importance of institutions in why certain societies succeed and others do not.

Why Nations Fail. The origins of power, prosperity and poverty. Daron Acemoglu & James Robinson, 2012

Jared Diamond and many others looked at unsustainable societies that have disappeared from the earth literally.

One thing that human societies certainly have is the notion of underlying power. I'm not going to give you a lecture on what political scientists say about power and how it differs from authority and how it differs from status, but power is a notion and power can be institutionalized either through coercion or through persuasion and you have everything in

between. But something that is very important to say about power is that it needs to be legitimized. In other words, the people who are governed, the people who are ruled have to have some way of accepting that power is there, and once this is accepted, and *The Leviathan* was one way in which Thomas Hobbes tried to explain why people accept absolute government by the absolute monarchs of the day as embodying literally the

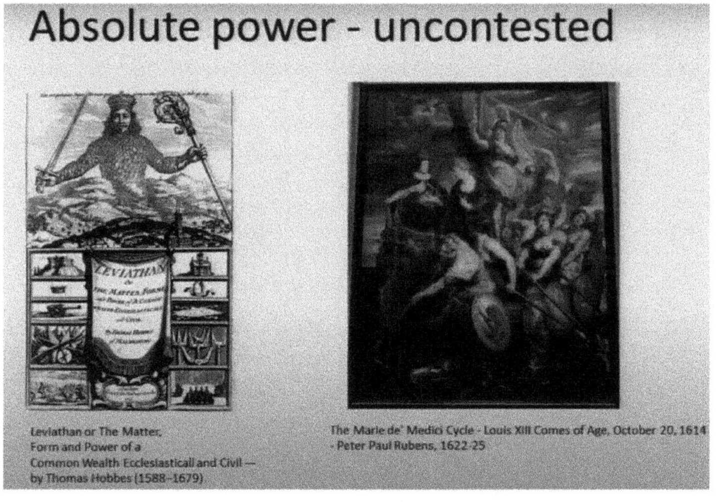

Leviathan or The Matter,
Form and Power of a
Common Wealth Ecclesiasticall and Civil —
by Thomas Hobbes (1588–1679)

The Marie de' Medici Cycle - Louis XIII Comes of Age, October 20, 1614
- Peter Paul Rubens, 1622-25

leviathan embodies the will of the people. You have other imageries here sailing together again in an uncontested way.

But founded in history was the contestation of power by various means — the fall of the Bastille and we see here workers striking. We see other angry people speaking up against the power that ruled them.

Now, there are of course the component parts — what makes up society, and here we see a collection of individuals representing visually component parts of various societies. You see Augustus on the left representing the power of his state — the Roman Empire; you see Saint Aquinas for the Middle Ages, you see Lao Tzu, you see individuals here in East Asia, and you see on the very right side, you see the utopian image of the genetic individual of the future. This is the way life scientists look at individuals as component parts and the way we are all unique but nevertheless related to each other.

There is yet another way of looking for representations of component parts. These are groups, social groups, and what is interesting here is that none of you have any difficulty in contextualizing what you see. By looking at this group on the left side, you see that this is a well off noble family somewhere in Europe, you see that it's a related group, and on the right side you see a politician addressing the people to rally behind her, so we have no difficulty making inferences from such group pictures, and you could go on and on, show different kinds of pictures, we have no difficulties in interpreting them in a wider context.

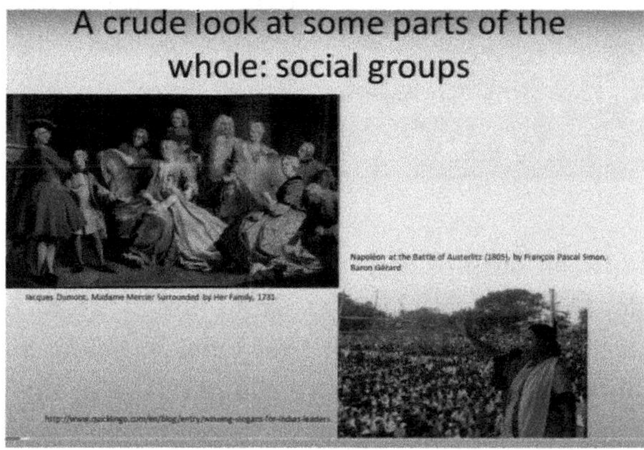

And today's societies are composed of consumers. This has become a very important, although not the most important, feature of the way we interpret and see societies today: voters certainly, producers. And the question is this: are these producers more than just producing material goods? Are they also producers of knowledge? Is there something that can enable them to make them innovators? Is there something out there that can make them participants in what is these days called citizen science?

Nowit has become a truism that it's not the individual components but the relationship between individuals, whether they are in groups or in whatever arrangement — social, political-economic, or knowledge arrangements we find them. But we also see that there are shifts going on in this globalized world, and the sense of the whole is also changing.

Now, to summarize briefly this first part of trying to find how we represent, how we actually look at society in terms of visual imagery: We end up, or I end up, by saying there is a sense of un-wholeness. It's very difficult to imagine society as a whole. We imagine parts. We can imagine

certain kinds of relationships, but it's very, very difficult to imagine society as a whole. If you close your eyes and think of what comes to your mind now in terms of society, you will get some conscious and some very unconscious images cropping up in your mind. Asking this question, of course, is highly context-dependent. It depends on the time when it seemed different, historical episodes, the place where we are, and who selects what for which purpose. This also means that the media of representations differ, whether we look at a statue of ancient Roman times or

whether we look at a photograph of today. It makes a difference also in terms of the imagery that it evokes in our minds; and there are various media here that we have used, and of course, advertisement has become one of the most interesting, not just predominant but most interesting ways of influencing us today in terms of representing this un-wholeness.

The un-whole representation can also take on a life of its own. You see, on the left, a flag is being burnt. Now, you can say, you know, why do people get so enraged if their flag is being burnt? Precisely because it is a symbol. It stands for something else, and this is why people can get tremendously enraged and angry if their flag is being burnt, while the others, of course, in an aggressive way, choose this way of proceeding. So, this un-wholeness, and especially if we look at particular parts of it, we see hierarchies, we see segmentation, we see fragmentation. So, between these various parts, there is a void, or maybe I should say in-betweenness — the layers, the groups, the individual components — and this in-betweenness is filled out because the human mind abhors a vacuum. We fill it out with what Yaron Ezrahi and others called collective imaginaries. Collective imaginaries are images that people have in their minds in terms of what they think exists and should exist. It's a very vague and fluid notion, and yet it is very, very important to perform these imageries in order to have these voids functioning in terms of connecting the various parts.

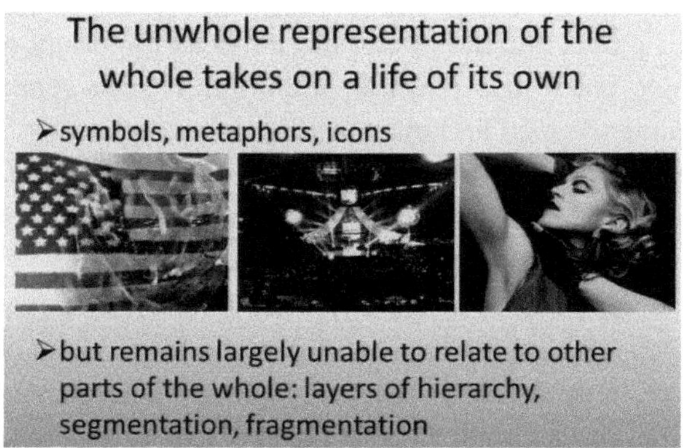

Let me now turn to the second part of my presentation. Here, I wanted to take a crude look — a crude and very simplified look, using images again, across time and space. Due to time constraints, I had to eliminate details, and I apologize for that. But let me just say, delving into the history of maps is a fascinating topic because it shows that all early maps have always had the collective "we" at the center. So, people were looking at the world from their we-centered perspective, and although now, since Mercator, we have a standardized version of what the world looks like, there are interesting tendencies to change the way maps are made and represented, namely, in terms of the inequalities that exist on this earth. It's a fascinating topic, but as I said, I don't have the time to go into it. So, I want to concentrate here on a crude look across time.

What is interesting here is that we know change goes on continuously. It goes on sometimes with ruptures. We have our own way of describing epochs and important events. If you go to Europe and ask people about the significance of 1969, you get a certain set of answers which we share with the US. If you go and ask about the importance of '89, you get some different answers because '89 — the fall of Communism — meant the liberation of one part of Europe, making it possible to have the European Union in the way it exists now, while in this part of the world (Singapore), '89, while on a cognitive, intellectual level, you know it happened, it does not have this very deep meaning in terms of changing biography or the fate of a country or continent.

So, how do we represent these social dynamics across time? One of the oldest ways is through an organic metaphor, as you mentioned metaphors. So, it is a tree, and the tree of life plays an important role in evolutionary biology, but it also plays an important role in genealogy — the way people have mapped and visualized changes in terms of their family history: Where do I come from? With whom am I related? It contributes to giving you a sense of where you have come from and where you are now — your sense of identity.

But of course, we have also seen, in terms of modern economic volatilities, especially in recent years, an increase in these kinds of representations of change in the stock market and related phenomena.

But let me go back to what I believe has been one of the most important events in the history of humanity, which happened to take place in Europe — the institutionalization of modern science. Historians have asked this question: Why did it happen in Europe? It's typically a

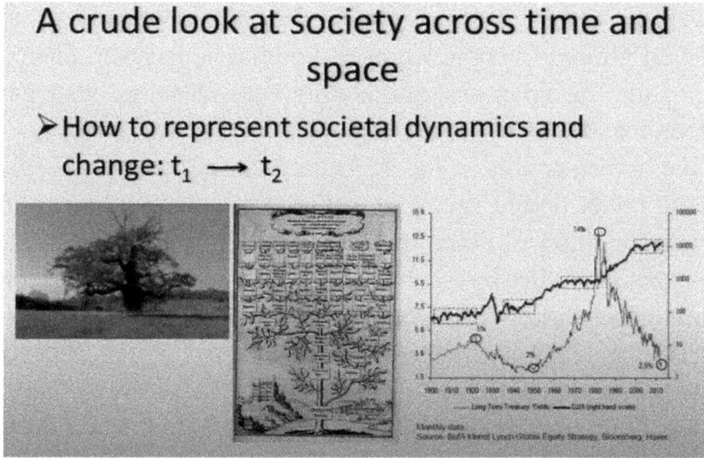

hypothetical question because it did happen in Europe and not anywhere else. But we know that institutions did play an important role, and we also know that individuals played an important role. In the beginnings of modern science, there were two very important premises on which it was based. One was to claim (and this was a claim in the beginning) that the laws — and of course, "law" is a metaphor taken from the social world, the world of politics — that Nature's laws are above the will of the rulers of the day. They're above whatever the political authority, typically the monarch, says, or what the religious authorities, the church in those days, say; the laws of Nature are not subject to human arbitrariness. This was an important premise, going on to explore these laws of nature. Sometimes, I think that this very easy division — that Science is about facts and Society is about values — overlooks one very important fact, namely, that the free inquiry of Science is based on a value, namely, free inquiry. So, you already have a mixture there.

The second important premise is linked to the Baconian program. I singled it out not because it was better than others but because Francis Bacon articulated his vision in a very coherent way. There were many other individuals throughout Europe who formed part of what we would now call a social movement, and historians speak about the Scientific Revolution, but it was really a social movement going on across quite a length of time involving small groups. Social movements always start with a small group of people who see a common purpose. The common purpose of the Baconian program and its equivalents in other parts was that through systematic inquiry into the way Nature works, you can change the human condition for the betterment of society. So, this was a very strong belief tied to very pragmatic goals, namely, how to improve the human condition to make life better for everyone and, at the same time, systematic inquiry by means that would yield answers. Bacon spoke about imitating nature, and in one of his utopian pieces called *The House of Salomon*, which was a secret society devoted to the betterment of humanity, he sent out the members of this society everywhere in the world to pick up knowledge and skills wherever they could find it, and bring it back. If you read it today, you're just struck by the visionary words that he had; many of the technological applications that we take for granted today were already in Bacon's dream, if you want to call it that way. And this systematic inquiry of Nature needs to use methods, particularly experimental methods, that would allow you to ask Nature the right kind of questions. If you ask her the right questions, you get some answers, but never all the answers you are looking for. You have to ask again, and this is the continuous process that we call research — asking good questions of nature, getting some answers, improving on that, and moving on. So, the image that you see here again is a telling one; it shows the sense of openness in exploration, and the two pillars have specific meanings. They were the Pillars of Hercules that were the limits in the Ancient World beyond which humans could not trespass, and here you have the ships moving in and out of the harbor, symbolizing this opening of the world.

Later on, technology, of course, played an important role, even if the emphasis was on the systematic inquiry of Nature. Technology, in the beginning, was trial and error, but later on, it became science-based technology, and this is something that we now take for granted — that science and technology mutually influenced each other, worked with each other, and led to something that Joel Mokyr, the economic historian, calls the Industrial Enlightenment.

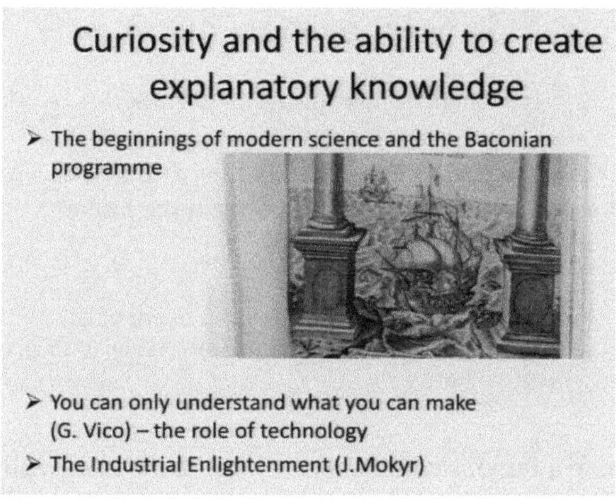

Now, curiosity played an important role in all this. Curiosity was viewed by the Catholic Church as something to be frowned upon, something to be constrained, because you would never know where it might lead you. And this is precisely one dimension of curiosity. If you follow your curiosity, you do not know where you will end up, you do not know where it will lead you. And this is also one of the reasons why all societies at any time do something to tame this curiosity. Curiosity comes in two variants: One is artistic curiosity and the other is scientific curiosity. When I say "tame," I mean either directing it in certain directions, like wanting impact, economic growth, etc., or forbidding certain activities by saying they are too risky, etc. But curiosity is an important driving force that cannot go completely unrestrained.

So, this then brings me to the role played by innovation and what research, science, and innovation do. I would come back to prediction. Woody Allen famously said, "It's difficult to predict, especially the future." The enormous way research has changed our lives means that we are actually at the very beginning of further changes. It is a fat-tailed probability that research will yield much, much more that will continue to transform our society. The thing is, in many ways, this imagination, this collective imaginary, as we see it in art, film, visual arts, and so on, is part of the cultivating, seeding ground in which other innovations take root.

Curiosity, innovation, complexity

Change and coping with change

➢ Past-present-future: continuity and ruptures
➢ The difficulty to predict (especially the future)
➢ Power laws

➢ Artistic representations open the realm of the imagination and imaginaries: film, visual arts, theater, literature

Now, I have a famous countryman by the name of Joseph Schumpeter who wrote a book in 1911 — so it is more than a hundred years ago now — which was frowned upon by the economists of his day, although later he went on to become a famous and later infamous finance minister, and so on. But Schumpeter was one of the first economists to really look at what happens in innovation. He made it clear that innovation creates economic expansion; it does not just follow it. He made it clear that it's about the individual entrepreneur, which for him was an individual that never only followed the profit motive. There was something extra, something that, you know, the entrepreneur wanted to change, and the difficulties in innovation were there for this heroic entrepreneur to overcome. He looked very closely at the necessary finance capital that was needed for innovation in his days, etc., but again, he gave a lot of emphasis to the individual acts of the entrepreneur. And he realized very early on that innovation is a double-bladed sword in the sense of creative destruction. It opens new avenues, new jobs, new professions, new ways of doing things, but this also means that others are disappearing.

Innovation is also inherently uncertain, just as research — not development — is a process with uncertain outcomes. You do not know yet what you will find, and therefore it means that you have to explore and exploit the opportunities you have for newer improved products. Innovation is a matching process of the opportunities that are there and what you can offer. It's also very difficult to predict accurately, as Brian Arthur and

> ## Innovation, according to Joseph Schumpeter
>
> ➤ Industry-specific innovation does not follow, but CREATES expansion
> ➤ The human element of LEADERSHIP: to overcome these…difficulties is the function characteristic of the entrepreneur
> ➤ All companies react adaptively to change, but creative responses come only from INNOVATIVE ACTS by entrepreneurs
> ➤ Innovation is a DOUBLE-BLADED SWORD

others know very well. Even if you know the beginnings of it, there are lock-in situations and so on. What is most difficult — this is one of the reasons why so many technological predictions fail — is not that they cannot precisely see where we are and what the future technological improvements can be, but what they cannot predict because it's so difficult is the uses to which innovation will be put.

> ## The obsolescence of the linear model of innovation
>
> ➤ The production of scientific and technological knowledge remains essential
> ➤ But: no linear translation of knowledge into working artefacts responding to, influencing and anticipating market demand
> ➤ User's requirements and latent needs
> ➤ Co-invention of users and uses

Therefore, I'm just here bringing to your attention that normally we speak about two different kinds of innovation: the incremental innovation that goes on all the time — firms, organizations have to continue to innovate and they do so in order to reproduce success. James March, the organizational sociologist, has looked into this question: Why do successful firms innovate at all? They have it all; they want to reproduce success, so for them it is counterproductive to try to innovate, and his answer is, first of all, there's the over-self-confidence of the leaders of the firms, so they think they know it all, and the other is that in replicating success, like in replicating DNA, errors creep in, so you have something similar to mutations. This is how the new gets even into the most successful firms.

And then, of course, we know that the innovation process varies greatly depending on the sector. If you speak about the pharmaceutical sector, it's a different way of innovating compared to the IT sector due to regulations, etc.

**Two strands of innovation:
(1) incremental innovation**

➢ Occurs all the time in firms and organizations
➢ Organizations attempt to reproduce success
➢ Due to overconfidence and errors in
   replicating: the new enters through the
   interstices (J. March)
➢ Heterogeneity in innovation processes,
   contingent on sector, firm, technology field

But then, there are radical innovations, and all radical innovations are science-based. This is one of the reasons why societies that want to make a difference have to continue to invest in what Abraham Flexner famously called the "uselessness" — "the usefulness of useless knowledge" — of what is seemingly useless knowledge because you do not see the uses as yet to which it will be put. Nevertheless, it will turn out to be useful, but it is impossible to predict whether this will happen in five years or in thirty years. There are many examples: GPS and other devices that we use. Without Einstein and the like-minded people who preceded him, we would not have these kinds of radical innovations. And radical innovations also come together to induce these kinds of radical transformations that we see.

(2) Radical innovation

➢Paradigm change of how economy works,
  major transformation of societal life

➢All radical innovations are science-based

➢long-lead time from fundamental research to
  transversal innovation, inherent uncertainties

Now, I would therefore conclude this part on innovation by saying it's an emergent property of interactivity in a time-compressed present, and this is what I call elaborative practices of innovators. They have to take form; they must become visible; they have to find an organizational space into which they fit. We know about the inertia of large systems. You can have a bright idea; you can even have a bright technical artifact, yet if the system is not ready for it because you have to change too much in it, it will take a longer time until you can transform the entire system. So, there are these inertia and other ways that hinder, and this is part of the double-edged sword of innovation of which Schumpeter spoke.

Innovation as emergent property of interactivity in a time-compressed present

➢product, process, work evolves at rate in
  which elaborative practices take form, become
  visible
➢elaborative practices based on
  indistinguishable individual and collective
  contributions
➢multiple loops of mutual interventions
  continue to evolve & modify processes of
  creation

Now, let me turn to the third part of my presentation, namely, the temporal dimension of complexity. In a previous section, I attempted to take you through change in society: The way we imagine it, the way we visualize it, and also what has made the difference, namely, scientific curiosity, systematic inquiry into nature aided by technology and innovation. And now, we come to the way, and this is something that I have been interested in for a long time, namely, how we construct time. Social time is a construct because it's based on us being biological beings, so we have an inbuilt arrow of time in our life that goes from birth to death. This, again, is based on the physical concept of time, so if you want, there's a hierarchy of time, but we have some leeway in the way we perceive time, how we construct time, and how we interact with and through time. Now, the most familiar, of course, is the concept of linear time — past, present, and future. The image that you see here is how a successful gambler looks at the world. Successful gamblers who are willing to take risks succeed for the most part, not always, but for the most part in taking the risks.

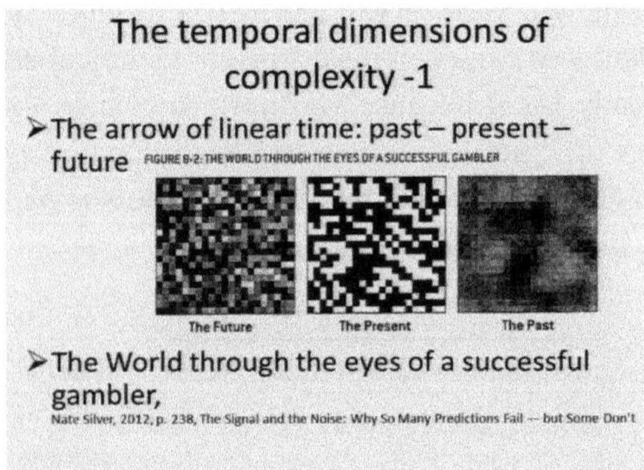

So, it's an interesting question: How important is the past to them, and their experience, of course, which gamblers also depend on? To what extent are they Bayesians, or how do they use Bayesian concepts? But perhaps a more interesting question is how a gambler looks at the future. And here you see that they give the greatest attention to the present. The past is there, but it's in a more blurred vision. The future is less blurred, but it's the present that matters to them.

Cyclical time is a concept that many societies have been familiar with, the eternal going back to the origin. Or you have in some civilizations the idea of time as coming from a golden age where everything was wonderful and, now, we are on a sort of anthropic path of degradation, and then the world starts anew from hopefully a golden age again. So, there are all these notions of cyclical time; of course, we are also familiar with cyclical time because we speak of business cycles, we speak of life cycles, we speak of the life cycle of products, so it's built into our sense of time as well.

This here I found quite interesting because it shows that our perception of what we can predict also depends on the historical context. If you start from a 50% evenness — what is predictable, what is unpredictable — you see a drop in predictability in the 30s, which was, of course, the economic depression, the war years, where indeed life had become in general much more unpredictable, the outlook on the future very, very blurred. Then you see a rise again in the 70s, and now we are slowly declining again in our perception or you can also say now our confidence in being able to predict.

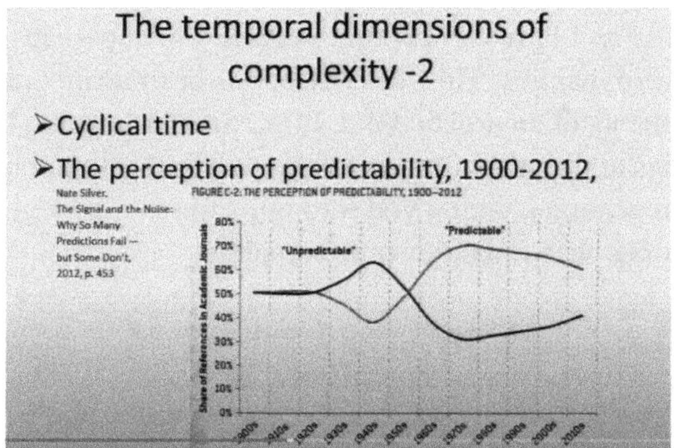

These are other forms of representing business cycles in particular, and again, it's interesting: on the left side, the spiral, an old geometric form that allows for a different sense of going up or going down depending on how you look at it, versus a linear form of prediction.

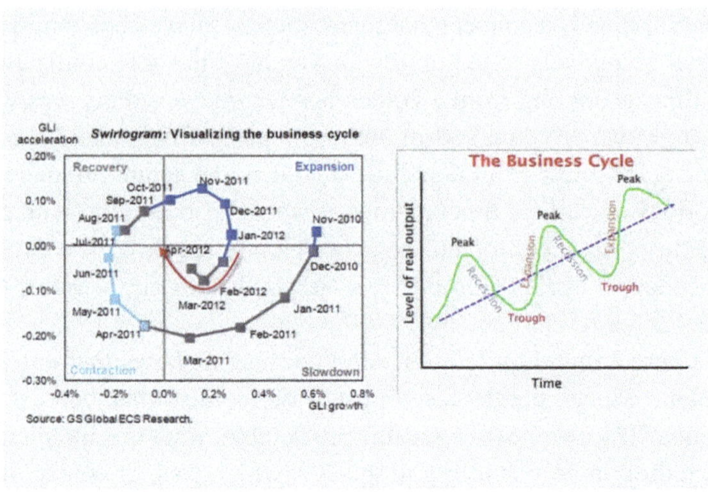

And so, this is my question: Whether with science and technology, curiosity, and innovation having really opened up the present era of enormous potential, and if you look into the things that life scientists speak about in particular, it's an enormous range of potential that is being opened up to us, whether we are not also entering a new time regime, that's the question I have. A new time regime which perhaps one can call nonlinear time dynamics. This, of course, has been again prefigured with trials, experiments of an artistic kind, film, visual arts going back in time, showing things at the same time or taking you in the emotional responses that films can generate, taking you through various time regimes, theatre, music, of course, that embodies time somehow.

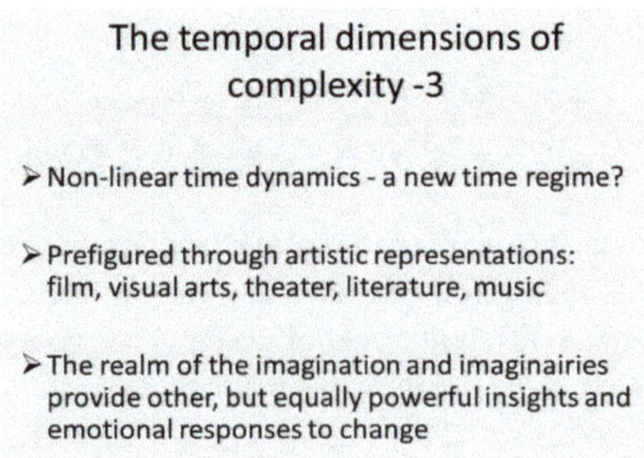

So, all these give you again one of these cultivating grounds out of which perhaps a new time regime is about to emerge. And if indeed there is such a nonlinear time dynamics in the making, it means not that it's replacing linear time. As I said, as biological beings, we have the arrow of time in our genes, we have it in our body, we know it, we feel it, we live it. Likewise, the physical world we know, cyclical time, we recognize the cycles that are built into this linear time concept, but nonlinear time may add another dimension as yet. By this nonlinear time dynamics, I mean the time sense that is generated by the creative processes that involve individuals and collectivities, knowledge, and technology in novel ways, in the way they interact.

Non-linear time dynamics in the making

A plural time regime: linear time, cyclical time, non-linear time dynamics

➢ the creative process involves individuals & collectivities, knowledge & technology
➢ approximates simultaneity

Very often, when people say, "everything is going so fast now," they draw attention to this speed, the acceleration: Look at where we were 10 years ago, 5 years ago, and where we are now. And I tend to think there may be something to acceleration, but it is more a question of approximating things happening at the same time, and our ways of trying to cope with things happening almost at the same time. Now, certainly, things happen at the same time.

This is a representation of what happens on the Internet every 60 seconds. Now, 60 seconds is almost over while I speak, and look at the numbers here: 168 emails have been sent while I began this sentence, and so on and so on.

And of course, by the interaction that is generated in this non-virtual space here, you can imagine data generating more data, big data generating more big data, and it is this what I mean by this nonlinear

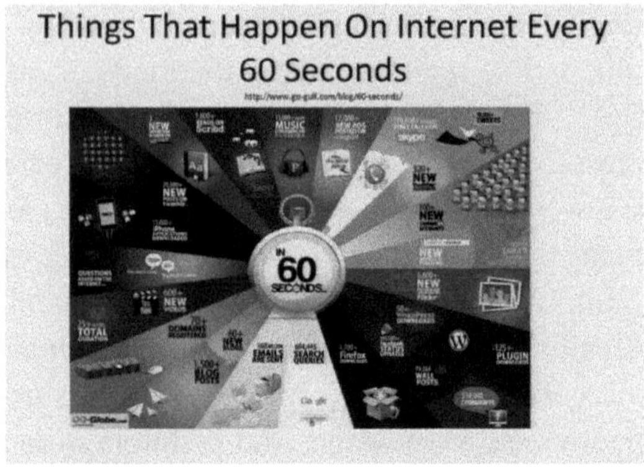

time dynamics. Now, there is one thing to be said about the creative process, namely, that underlying curiosity and innovation, there is human creativity — the creative process connecting what was not connected before, finding new forms, new contents, and new interactions. All these are always inherently uncertain. You do not know success, and by success, I don't mean a big notion of success. I see success as being one of two things: either you have a predefined goal that you want to reach, after which you have successfully reached this goal and you can move on to your next goal, or you simply have the sense of having achieved something — it works. That's my second definition of success: It works. It can be a small thing, it can be medium-sized, or a big thing — it works.

And the way social systems, that are high-performance selection systems like science, operate is evident when observing all the young people here in the audience. You are accustomed to going through exams, participating in competitive events all the time, applying for grants, and striving for various achievements. You are part of high-performance selection systems, and these systems operate to reduce uncertainty. Of course, you are integral to the system. We observe similar dynamics in sports and other fields, all aimed at coping with and reducing the uncertainty of success.

Now, returning to this nonlinear time dynamics, as I said, the approximation of timelessness I think is more important than just the sense of speeding up. They're co-temporalities, meaning there are different time regimes coexisting next to each other, and creativity and innovation are important emergent properties of this interactivity in what I call a time-compressed present. It's not so much about the future; it's about the time-compressed present.

Non-linear time dynamics

➢ Approximation of timelessness: everything happens – almost – at once
➢ Not so much acceleration, but simultaneity and co-temporalities
➢ Creativity and innovation as emergent property of interactivity in a time-compressed present

Now, creativity and work have something to do with each other. When I said one definition of success for me is "it works," I want to draw your attention to what work can also mean. "It works" — when you go back in human history, the first tools that we found outside of caves where our ancestors lived and the like, and the first tools according to the paleontologists studying this, were an extension of the human hand, so technology as the extension of the human hand. We have long since moved on to technology as the extension of the human mind, and we have found new interfaces between technology as an extension of the human hand, the human mind, and of course the technological artifact.

> ### Creativity and the changing nature of work
>
> ➤ Changing nature of work: individual and collective contributions become indistinguishable
> ➤ But - attribution of credit and of rewards still matter

I don't know how many of you have read an article that made it to the front page of the *Financial Times* on Friday, and this was science news on page 1 of the *Financial Times*. This was about an experiment conducted with rats in the US and Brazil. Now, there was a Brazilian who was working in the US and he had set up with his former lab an experiment with implanted electrodes in the minds of rats. The rats were rewarded for doing certain tasks, responding to visual signs; but only one part of the rats had the visual signs, the others did not have the visual cues but they were connected through electrodes with the mice that could do it. In 70% of the cases, the mice that had no visual cues were able to mimic the results that were obtained with the other ones. Now, mice are rewarded with food. It's very simple, but humans want food and something else. Humans also want recognition, and it is therefore very important to start to rethink work and what this kind of technology that connects hands, minds, and technological artifacts does to the way we attribute credit and how we attribute reward. It's becoming much more difficult to say this is the individual contribution because it's not just workers and the queen. We are much more egalitarian in terms of our contributions. Nevertheless, we have to find ways of rewarding, which means also motivating individuals and collectivities in order to interact and to be creative.

And the more we put the emphasis on technological innovation, I think the more we have to start thinking seriously also about social innovation. Sometimes, I call it the "dark energy of society." You know, physicists and astronomers tell us there is this huge percentage of unknown dark energy in the universe. Sometimes, I feel social innovation is the dark energy of society. We don't quite know what it is, but we know it is very, very important. Embedding technological innovation requires social organization. It necessitates innovative social arrangements to appropriate and utilize it effectively.

> **Social innovation – the dark energy of society?**
>
> ➢ Multiple loops of mutual interaction and interventions continue to evolve and modify processes of creation
>
> ➢ Creativity, scientific knowledge and technical know-how keep social innovation going

As I conclude my presentation, I offer three perspectives on the societal whole or lack thereof, depending on one's viewpoint.

The importance of the connections, whether from parts to the whole or vice versa, is crucial in this complex, multi-agency interaction. This interaction is not solely mediated through technologies but is also fostered by them, as technology becomes integral to the creative process itself. Rather than a linear progression from past to present to future, we observe numerous feedback loops and interactions in this extended, time-compressed present.

So, I come to the end of my presentation with three concluding looks at society as a whole, or as fragmented, depending on how you look at it. The importance of connections from parts of or for the whole to this very complex, multi-agency interaction between parts and whatever interact, not only mediated through technologies but also fostered through it. Technology becomes part of the creative process itself. Instead of a causal arrow and linear time concept going from the past to the present to the future, we see many feedback loops and interactions within this extended, time-compressed present.

Second conclusion: We know this is not evenly distributed. There are some places, and Nanyang has become one of these places, where a

A concluding look at the societal whole -1

- only connect -

- from parts of/for/the whole to multiagency interaction between parts (human and technological)
- from causal agency past-present-future to mutual manifold feed-back loops in extended present

creative environment has been created. Creating a creative environment implies that you cannot foresee everything. You must leave room for things to emerge. We know how important in any research process and part of my present function as President of the European Research Council, I speak to a lot of researchers. Part of it is that they tell you how important serendipity is. Serendipity in the research process simply means you find something that you have not been looking for but you recognize its importance, and so you have to be prepared and yet if you're over-prepared, the chances are big that you will miss it, but you have to be able to see it. Serendipity plays an enormous role. The more we open up toward technology being part of the way we interact, the more we have to think we can also optimize serendipity to occur. And then, of course, the diversity of encounters, and all this is part of what makes a creative environment creative.

A concluding look at the societal whole -2

- creating creative environments -

- allowing for randomness and contingency, cultivating serendipity, density and diversity of encounters
- reduction of uncertainty through competition in hi performance selective systems

And, my last, third look is at the relationship between evolution and complexity. In biological terms, we know that there have been major transitions, but there is nothing in biology that tells us or that follows that evolution should move in one direction because evolution is non-directed; it's non-teleological. So, there's nothing in evolution that tells us that there should be increased complexity; nevertheless, it happens. And I think cultural evolution, which I have been speaking about — science, research, technology, and innovation — demonstrates this increase in complexity, evident in phenomena such as nonlinear temporal dynamics.

So, can curiosity, creativity, and innovation hold the societal whole together?

And there are three related questions with which I will leave you: What kind of society do we want to live in? How to organize this kind of society in which we want to live effectively? And how can we keep this kind of society open for further evolution and the future? Thank you very much.

3.2 Discussion

Sulfikar Amir: My name is Sulfikar. I am from Sociology at NTU. I truly enjoyed your talk. It's really inspiring, but I have one question related to the notion of wholeness that we have been discussing in this conference. As we understand, social science researchers, particularly sociologists and economists, are divided into two big groups in the way they perceive society. The first group looks at the micro-level, the interaction between individuals, while the other group focuses on the macro-level, attempting to understand society's behavior from a systemic perspective. Over the past decades, perspectives or concepts offered by thinkers like Giddens' Structuration or Bourdieu's Habitus have aimed to bridge these two levels — micro and macro.

I think social scientists also have a valuable opportunity when natural scientists provide us with tools like network analysis. Network analysis has emerged as a method to connect these two levels, providing a clearer understanding of how society operates. This approach becomes increasingly practical as we observe that many institutions and individuals are interconnected through networks, a characteristic of globalized society. However, the proliferation of networks everywhere leaves me puzzled.

My question to you is this: To what extent, in your view, do structures remain relevant in our society? If structures are still relevant, what are the epistemological and socio-political implications that we should consider in understanding the complexity of society?

Helga Nowotny: Okay, this was a long question, but I will have to give you a shorter answer. I think one of the reasons why social networks have also become very attractive for sociologists, for instance, is because they offer new kinds of data that did not exist before. The methods open to social scientists, sociologists in particular, were traditional ones where you could conduct micro-level observation, survey analysis, or various analyses between the micro-level, mid-range level, and macro-level. Of course, there were always people who were modeling, as Bob showed us this morning. But with the data now becoming available, such as who contacts whom, whether through *Facebook* or telephone calls, etc., there's a new way to look at networks as structures.

Now, structures are not the same as what you refer to as "structuration." I think the next challenge is indeed to look beyond it because we will see what methods in the social sciences have always been telling you: You need to have the right kind of method for the questions you ask. There is no universal method in the social sciences. You have to be very careful in choosing the method that fits your question. Depending on your research question, you can use network data. For others, if you ask, for instance, after we celebrated how *Facebook* lead masses to Tahir Square and saw this as a way of predicting the future. But the future did not unfold this way, and therefore, you have to go back and say, while it's a great mobilizing force for bringing people to Tahir Square, this does not solve the problems that the people in Tahir Square wanted to address.

Sheila Ronis: Sheila Ronis from Walsh College. I think it's interesting that your view of creativity assumes a lot of freedom. One of the problems that we face, at least in the United States, and I believe many other nations also face, is that the way we have structured our doctoral education is very unidimensional. You end up learning more and more about less and less, and frankly, it hasn't been helpful in solving the large complex problems that require an interdisciplinary approach.

In the United States, one line of research we've been pursuing involves examining the methods by which faculty members are promoted and tenured, for example, and the narrow scope of journals in which they can publish. I'd like to ask this: Have Europeans found some solutions to this issue? I believe European scientists are a little more open to systems approaches and interdisciplinary approaches, although I know it is a struggle everywhere. We've traced some of these challenges to our higher education accreditation bodies and their criteria for accrediting institutions. However, the political reality is that effecting change in this area is enormous. Yet, I believe it would greatly enhance the creativity and innovation of society as a whole if we could at least liberalize the way we educate people. Any thoughts on this?

Helga Nowotny: Yes, I think you're right in identifying the underlying need for creativity to thrive with freedom and autonomy. It requires individuals who can leverage this freedom effectively. Within academia, there are numerous anecdotes illustrating narrow-mindedness and other challenges. However, unless academia acknowledges and addresses these issues, it risks becoming obsolete.

I recently spoke with someone in the cybersecurity field who mentioned that the US Pentagon is actively seeking to fill many positions. When I inquired about the source of these recruits, he informed me that they were turning to the hacking community, bypassing traditional academia. According to him, the desired talent isn't readily found within academic circles, so they're exploring alternative avenues.

Additionally, with the emergence of online courses and the opening of new educational spaces, academia faces significant pressure to adapt or face creative destruction. This presents a pivotal moment in the history of education, where academia must either evolve or risk becoming a footnote in the narrative of how we educate our people.

Discussant A: Thank you. I enjoyed the lecture very much. The comment you made toward the end about the need for recognition of creative workers resonated with me. I work at a large software company. We have about 4,000 creative software engineers working in teams, solving complex software problems creatively. We've actually studied and researched their motivations and behaviors, and we find that recognition matters even more than money, although we do compensate them well. It's a kind of minimum requirement, but the truth of the matter is that recognition from their peers and from above is actually one of the most important rewards from their point of view, and a failure to provide that is highly demotivating.

Helga Nowotny: Yes, thank you for the confirmation.

Discussant B: Hello. I actually have a question about when you talked about the different views of time. You mentioned linear time, cyclical time, and nonlinear time dynamics, right?

Helga Nowotny: Yes.

Discussant B: So, my question is this: With nonlinear time dynamics, it's a new conceptual way to view time, right? Are there any applications of this conception of time to, let's say, optimize society, or could you provide a particular example, such as innovation, to bring about optimized innovation into society using such a conception of time?

Helga Nowotny: This is difficult to answer because one would have to look at very specific cases. If you're talking about "optimizing," incentives play a role, but once you have a predefined goal or target, you provide freedom in how to achieve it while keeping the target in mind. Now, you can compress the time, but if you compress it too much, you won't achieve the desired results, so finding the balance of slack time is crucial. Creative time cannot be planned. However, if you only allocate creative time without any time constraints, you won't achieve the goal either. It's a delicate balance that you have to establish to reach the desired outcome.

Discussant C: A comment on the rewards system, particularly in a large software company, where recognition from peers often proves to be even more important than financial rewards. I can't help but wonder if this principle could be applied to the financial community as well [laughs]. Perhaps it could aid us in addressing the issue of bonuses.

Helga Nowotny: Well [laughs], you know, wearing my other hat, I'm actually the president of the most important European funding agency for fundamental research, which we call "Frontier Research." Each year, we allocate funds in euros. I'm not sure about the equivalent in Singaporean dollars, but this year we've allocated 1.7 billion euros for fundamental research without any predefined theme. This is crucial; it's a truly bottom-up approach, focusing on individual researchers rather than groups. While they may have a team, we assess them as individuals. Our evaluation covers all areas of science and scholarship, including the humanities.

Now, when it comes to evaluation, each community has slightly different criteria, and our panelists are adept at handling this diversity. Interestingly, the only panel that consistently approaches evaluations differently is the economists. I'm not sure why, but unlike other panels where consensus is sought through discussion and deliberation, economists seem to enjoy voting. They engage in clever games, attempting to anticipate their colleagues' reactions. While they may find it enjoyable, they certainly stand out from the rest.

John Richardson: John Richardson from the Lee Kuan Yew School. You haven't said much about failure, and as a funder and leader of research, I know I have students who are conducting research on entrepreneurship. We know that failure is one of the most important ingredients of creativity because if you aren't prepared to fail, you can't be creative. However, institutionalizing failure as a funder, university president, corporate leader, or government leader is quite difficult. So, could you talk a little bit about how, personally, as a research leader, you create opportunities for failure and how you counsel others to do the same?

Helga Nowotny: Well, as a research funder, the European Research Council encourages applicants who have reached a certain threshold to carefully review the comments provided by the panel members and to resubmit their proposals. Over the years, we have observed a relatively constant reapplication success rate of 40%, which is an interesting empirical figure given that there were no comparable data before our establishment. For those who almost made it but were not successful due to resource constraints or lack of competitiveness, we have a success rate of only 12%. Thus, this group is encouraged to submit again.

Regarding innovation, permit me to also comment on the difference between the US and Europe. While in the US, failure often means "get up and try again," in Europe, there is a perception that Europeans tend to give up after failure. This difference is partly attributed to the availability of venture capital, which is much greater in the US than in Europe. In the

US, failing entrepreneurs have better opportunities to rebound, whereas in Europe, failure often signals the end of the road.

In terms of research careers, particularly in Continental Europe, there is a tendency to select individuals for full professor positions relatively late in their careers. Consequently, many researchers are deemed too old to secure suitable positions. This contrasts with the tenure track system in the US, which allows for more mobility and opportunities for individuals who may not succeed at prestigious institutions like Harvard.

As a research leader with a research group, one of the most challenging aspects is assessing whether an individual has the potential to succeed. This requires relying on personal judgment. Key considerations include the individual's perseverance, inner drive, curiosity, and willingness to persist despite the likelihood of failure. If these qualities are lacking, it may be better to advise the individual to pursue a different path outside of science.

Caroline Pluss: Caroline Pluss from Sociology. It ties in very nicely what you said because I'm far from convinced that creativity comes out of having space. I think a lot of people are creative because they're under a lot of constraints, leading them to be creative out of necessity. Consider immigrant entrepreneurs, for example. There's a disproportionate number of successful entrepreneurs among them, and their success often stems from being forced into new environments and having to navigate numerous contradictions.

For me, creativity involves synthesizing disparate elements. If you're too comfortable, you may lack the drive to do so. So, in a way, creativity cannot be told, if it comes out of necessity. I think a lot of people or painters have a lot of creativity. It does not come out of tranquility. I didn't quite understand what you meant by "time-compressed present." You mean the time–space compression? Or what is "time-compressed present"?

Helga Nowotny: Okay, so "time-compressed present" means we are putting many things together in the same unit of time. This is a way of speaking, of course, compressing time by multitasking essentially. Since we have many more options available, we want to do many more things.

This, in comparison to a generation or two generations before, which had far fewer things. So, this is the answer to your first point. The answer to your last point, on constraints and creativity, you know, my favorite researcher, my ideal researcher, is what I would call a "competent rebel." Competence must be there, but at the same time, you need a rebellious spirit in the sense of not accepting what the older generation tells you, but you have to be competent enough to ask the right kind of questions to challenge the authority that is there. So, it's both the spirit of rebellion, but you also need the competence to do it, and this is yet another form of creativity that has many different incarnations.

Discussant D: Yes. Just a clarification. You made two statements. One was that there's no linear translation I think, or linear model. Right?

Helga Nowotny: In innovation, yup.

Discussant D: In innovation, right. Then, you later said that all radical innovations are science-based. That sounds like a linear model statement a little bit. So…

Helga Nowotny: Yeah, no, it's not [what I] meant. The linear model of innovation is something that exists in the literature of people who work on innovation, and it dates back to the time after the Second World War when something like this did exist. You were going from basic research to application, to market, and this is what Vannevar Bush outlined in his manifesto *Science, the Endless Frontier*. Now, since then, things have changed a lot, and we know from empirical studies and theoretical ideas looking at how innovation actually occurs, as well as many case studies putting things together, that this linear model makes no sense.

You also have many areas where you cannot draw a distinction anymore between fundamental research and applied research; it simply doesn't make sense. This is particularly the case in the life sciences, but not only. In the life sciences, for example, one hour you might be conducting fundamental research and the next hour you might be working on something applied. This is due to the fact that the cell has become a laboratory. In many labs today, you need research technologies that help you ask your next question, but you also need them to answer your question. So, the linear model of innovation in the technical sense is obsolete.

Now, radical innovation just means you can trace back the innovations that really transform whole sectors or society as a whole. If you look at the way we use computers today, for instance, it's an ongoing long process. You cannot point to a single day and say that everything changed. These sorts of ruptures do not occur.

For instance, George Dyson wrote a wonderful book *Turing's Cathedral*, which came out last year. It's about the Institute of Advanced Study in Princeton. He shows that Turing's ideas were crucial, but at the same time, without von Neumann and without the war effort, we would not be where we are today. This is what I mean by "science-based," and it's through these radical transformations and innovations that we see significant progress.

Discussant E: Going back to the concept of radical innovation, part of my curiosity lies in the distinction between incremental and subsequent radical innovations. In essence, any real innovation is radical by nature. For instance, with quantum innovation, we don't talk about something that starts as incremental and then becomes radical — it's inherently radical from the start.

However, the missing link here is replicative innovation. Too often, we see many instances of replicative innovation that do not lead to quantum leaps in progress. So, my point here, in wanting to ask you, is about the difference between disruptive innovation and your concept of radical innovation, and how you define disruptive innovation compared to your own definition. Christensen doesn't use the term "radical" because, in any innovation, the term "radical" inherently implies a significant departure from the status quo. Thank you.

Helga Nowotny: Well, as with any distinctions, you can use them analytically or to address specific questions. Regarding the distinction between incremental and what you referred to as "replicative," it's essentially replication but with a mutation. When I mentioned replicating success earlier, it implies replicating with some variation because innovation occurs in an environment that is constantly changing, necessitating adaptation.

So, while I prefer the term "incremental," it does involve aspects of replication. Perhaps we can settle on that.

Now, your other question is more challenging for me to answer because it also depends on the time frame you consider. A radical innovation can have immediate and destructive consequences. For example, if academia doesn't change and becomes obsolete, many universities that rely on outdated teaching models could face radical destruction. However, if you look at it over a longer time frame, you'll see that such changes have occurred throughout history. The next generation adapts, and there are shifts toward interdisciplinary studies, for instance. These adaptations help mitigate the destructiveness. I think Schumpeter's insight was essentially correct — that every creation also entails something being displaced, but this doesn't mean that nothing new will emerge afterward.

Discussant F: You posed this question: "What kind of society do we want"? I believe nobody has yet put forward any ideas, and I'd like to introduce an interesting new concept from a recent book by Philip Korinsky, which unfortunately doesn't translate very well into English. It's called *Les Traité* in French. This concept suggests that every member of society not only has responsibilities but also certain rights. A clear example of this is vaccination: When people agree to be vaccinated, it's not just for their own protection but also for the well-being of the community and to prevent the spread of diseases. If we extend this concept further, a good society could be one where individuals embrace this mutual commitment to responsibility and also have corresponding rights. How can these age-old concepts become integral to shaping the future world?

Helga Nowotny: Well, I can't provide a definitive answer to the question I posed earlier about the kind of society we want to live in. It's a question for all of us to discuss and work toward. However, you're absolutely correct that a society comprised solely of self-centered individuals is akin to what we see in game theory. You have the free-riders, but at a certain point, a society with only free-riders is unsustainable and unlikely to materialize. Instead, we encounter tipping points where individuals may either cooperate or defect. In game theoretical terms, this is an example of defectors and cooperators, and what we need is the right blend of both.

Rhett Gayle: Hi, great talk. I'm wondering… We've talked a lot in various talks about trans-interdisciplinary research studies and so I was wondering if any research is being done on the kinds of skills and intellectual background that would make people successful in doing that kind of work as opposed to just going, "Okay, we have these specialists and now we're going to bring them together." Is there any research being done on, sort of, what kinds of people, what kinds of training might actually, sort of, start people on that road rather than, sort of, later down — am I making sense yet? No?

Helga Nowotny: [Laughs] I think what we… No. I cannot answer in any precise way, but I think what is clear is that our present university system, with a focus on disciplines, to some extent, I would defend it because a discipline teaches you what is a good research question. Unless you know how to pose a good research question, you will not get very far. But at the same time, you have to enable it. It's like speaking in more than one language. You have to become multidisciplinary, at least to understand what is happening in another discipline. But how do you do this? There are various suggestions out there. For instance, teaching in universities could involve a disciplinary course accompanied by a seminar that brings in another discipline, allowing for a sort of meta-interdisciplinary approach on top of that. So, there are various kinds of these experiments going on.

Rhett Gayle: Okay, thanks.

3.3 Summary of the Talk by Helga Nowotny

Helga Nowotny's talk covers three main parts: an examination of society through images, the impact of science and technology on societal change, and the exploration of the temporal dimension of complexity. Each part is summarized as follows:

Examining Society Through Images: Nowotny begins her talk by contemplating the concept of society and how it is understood and represented. She uses visual imagery to explore different aspects of society, from historical representations to contemporary depictions. She discusses the complexity of societal structures, the interplay between power and legitimacy, and the role of collective imaginaries in filling the voids between different societal components. Nowotny emphasizes the challenge of imagining society as a whole and highlights the contextual nature of societal representations, influenced by historical, cultural, and media factors.

Impact of Science and Technology on Societal Change: Moving on, Nowotny discusses the transformative impact of science and technology on societal dynamics. She traces the historical development of modern science, emphasizing key premises such as the autonomy of natural laws from human authority and the Baconian program's focus on systematic inquiry for the betterment of humanity. Nowotny highlights the intertwined relationship between science, technology, and innovation, illustrating how they have shaped the Industrial Enlightenment and driven economic expansion through creative destruction. She also explores the role of curiosity and uncertainty in driving innovation, emphasizing the importance of embracing uncertainty as inherent to the innovation process.

Temporal Dimension of Complexity: In the final part of her talk, Nowotny delves into the temporal dimension of complexity, exploring different conceptions of time in societal contexts. She discusses linear, cyclical, and nonlinear understandings of time, drawing attention to how societal perceptions of predictability and uncertainty have evolved over time. Nowotny suggests that contemporary society may be entering a new time regime characterized by nonlinear time dynamics, where multiple temporalities coexist and interact in a time-compressed present. She highlights the role of creativity, innovation, and serendipity in navigating this complexity and emphasizes the need to cultivate a society that is open to further evolution and future possibilities.

In conclusion, Helga Nowotny's talk offers a multifaceted exploration of society, science, and time, inviting reflection on the challenges and opportunities presented by complexity and uncertainty in contemporary societal dynamics.

3.4 Relevance of the Talk to the Current Stage of Research

Helga Nowotny's talk touches upon several themes that are highly relevant to the current stage of research across various disciplines:

Understanding Society: Nowotny's exploration of society through images and visual representations resonates with current efforts in social science research to employ interdisciplinary methods, including visual analysis, to study complex social phenomena. In an era of big data and multimedia communication, researchers are increasingly turning to visual data to gain insights into societal structures, interactions, and representations.

Impact of Science and Technology: The discussion of the transformative impact of science and technology on societal change is particularly relevant in the context of ongoing debates surrounding the ethical, social, and political implications of technological advancements such as artificial intelligence, biotechnology, and digital communication. Researchers across fields are grappling with questions about the responsible development and deployment of emerging technologies and their effects on individuals, communities, and societies.

Temporal Dynamics: Nowotny's examination of different conceptions of time and the emergence of nonlinear time dynamics speaks to current discussions in fields such as complexity science, systems theory, and futures studies. As researchers seek to understand and model complex systems, including social systems, they are increasingly recognizing the importance of temporal dynamics and the need to develop new conceptual frameworks and analytical tools to capture the multi-dimensional nature of time in complex adaptive systems.

Innovation and Creativity: Nowotny's insights into the role of curiosity, creativity, and innovation in societal dynamics resonate with ongoing research in fields such as innovation studies, entrepreneurship, and organizational behavior. Researchers are exploring how creativity emerges, how innovation processes unfold, and how organizations and societies can foster environments that nurture creativity and innovation. Additionally, there is growing interest in understanding the interplay between technological innovation and social innovation, as well as the implications for governance, policy, and sustainability.

Future Directions: Finally, Nowotny's reflections on the future of society and the need to envision and organize societies effectively in the face of uncertainty are highly relevant to current discussions about global challenges such as climate change, inequality, and political instability. Researchers are exploring alternative visions of the future, experimenting with new forms of governance and social organization, and engaging in participatory processes to co-create desirable futures.

In summary, Helga Nowotny's talk offers valuable insights and provocations that resonate with current research agendas across a wide range of disciplines, from social science and humanities to natural and computational sciences. Her exploration of society, science, and time provides a rich foundation for interdisciplinary dialogue and collaboration aimed at addressing pressing societal challenges and shaping more inclusive, sustainable, and resilient futures.

Chapter 4

Land Use, Economy, and Complexity

Kristian Lindgren

YouTube: https://www.youtube.com/watch?v=iiTDKILQgp8

Speaker: Kristian Lindgren

Moderator: Kevin Xiao

Discussants: Brian Arthur, Peter Schwartz, Robert Axelrod, Greg Fisher, Discussants A, B, C, D, E, and F

4.1 Talk by Kristian Lindgren

Kristian Lindgren: Thank you. First, I'd like to thank John for this kind invitation, and actually also for providing me with the title. It was first actually "Energy Economy and Complexity," and I chose to switch from energy to land use because that is the topic where we are focusing right now. But I would also like to start by confessing, like Peter Sloot did, that I sort of belong to the global world modelers in the sense, taking a very crude look at the whole, but using fairly simple and rough models, making what-if scenarios for the future. But I try to be careful about what conclusions can be drawn from such crude looks at the whole world in the future, and in the talk today, I will try to problematize this type of what-if scenario, which is usually based on various types of modeling approaches, and problematize that by looking at the other end of the scale where I have been engaged, which is in the complexity area on the micro-level, so to say, where game theory, agent-based modeling, and sort of microeconomics assumptions are playing a more important role. So, what I will try to do today is to first give sort of a polished background starting with energy and climates and looking toward the area of land use, and to show, via a sequence of models, how the typical modeling approaches in land use can be problematic. I've chosen to go through a number of, sort of, simple economic models at different levels of description to illustrate how, when you go down to more microscopic descriptions of the system, many new features may appear. So, I must also say that this is ongoing research in our group, so I've done this together with a group of four Ph.D. students, and I'm very glad that I got the opportunity to present this work. It sort of excites us the most right now.

But the energy background is the following: Over the past hundred years, we have witnessed a more than tenfold increase in the supply of primary energy, primarily being fossil fuel energy use. The effect of that has been an increase in atmospheric carbon dioxide from below 300 ppm to passing 400 ppm within one or two years, I would expect.

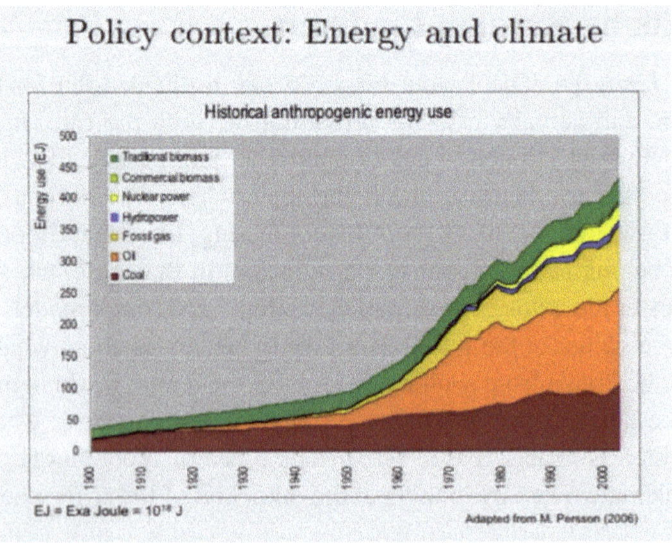

For the future, if we want to meet the reduction requirements and avoid the risks of having an increase in global mean temperature that stays below plus 2 degrees from pre-industrial levels, we would have to see a decline in the use of, or at least in net emissions of, carbon dioxide. So, the use of coal, oil, and gas should move in that direction. But at the same time, we have economic growth that will continue to demand more of the energy services that primary energy supply can support. Part of this increase in energy supply can, of course, be mitigated through energy efficiency and possible lifestyle changes, but there will certainly remain a gap that needs to be filled by new technologies — technologies that will have their different pros and cons.

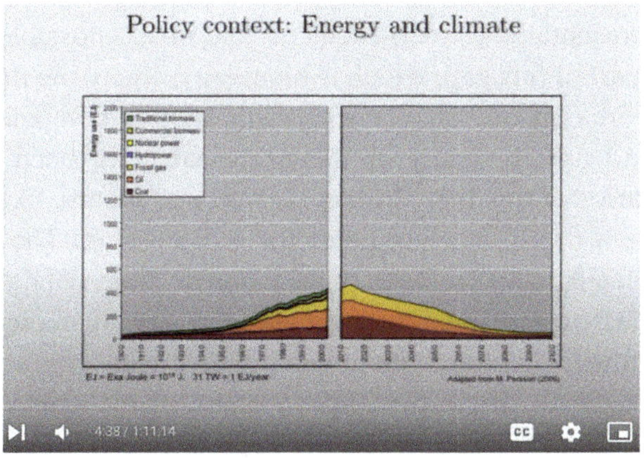

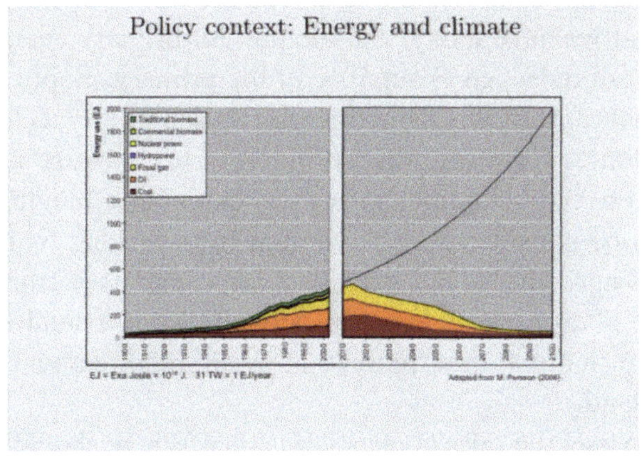

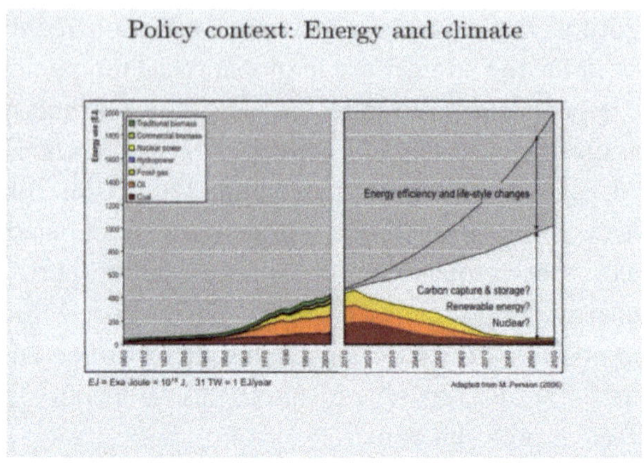

In this presentation, I will focus on one of the possible candidates for filling a part of this gap, which is bioenergy, where the main problem is that an increased use of bioenergy will start to compete with other forms of land use. In many of them, I mean, I've been involved in energy scenarios of this type, I think that a rough picture like the one we see might be a useful starting point for a discussion. The problem is, I think, when you start to make these pictures more detailed, then you might get a false impression that you are more certain about what to say about the future. So, a rough picture could give you some sort of framework as a basis for discussion, but in many of the scenarios that are made using various types of modeling approaches, they suggest that the bioenergy supply in the future might increase from the order of 40 to 50 exajoules that we have today, and the present primary energy supply is around 500 exajoules, so about 10% of the primary supply comes from bioenergy, but most of it is from traditional use of biofuels and only less than 10 exajoules from commercial biofuel use. In many scenarios, the order of 100 to 300 exajoules is suggested, about a hundred exajoules of those can be considered to be coming from residue fuel flows from forestry and agriculture, but usually a large part is assumed to come from the use of crops from agricultural land, and if a hundred exajoules is to take part in this future figure, that would require about 500 million hectares of land.

So, of course, the critical question is this: Where should that land be placed?

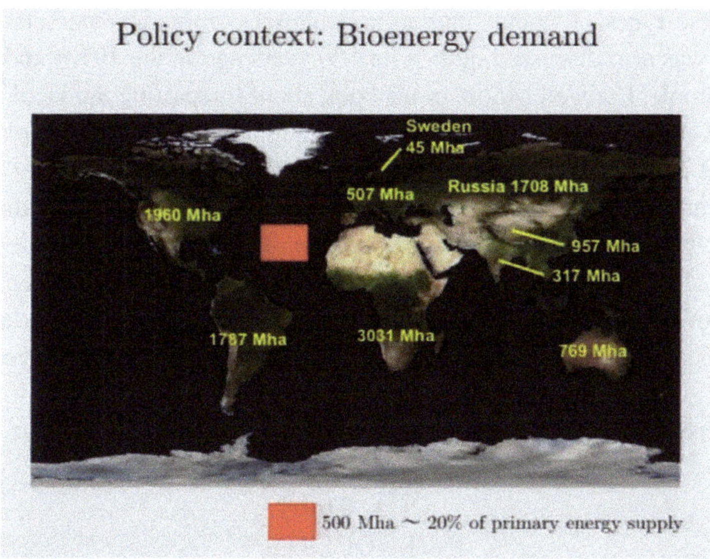

And this problem, this issue, has been accentuated in the last five years. If we look at the price dynamics in markets for different agricultural commodities, we have seen high fluctuations, at least two price peaks during the past five years. Some of the causes that might be behind part of this increase in food prices have been suggested to come from increased competition for land between the production of crops for food purposes and for bioenergy.

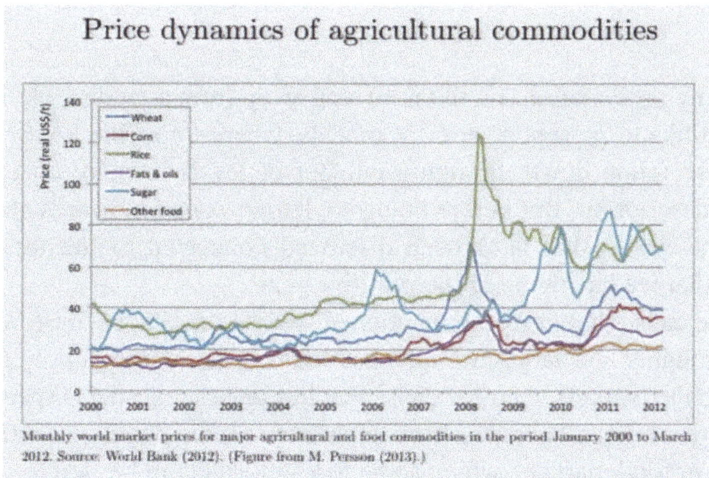

Monthly world market prices for major agricultural and food commodities in the period January 2000 to March 2012. Source: World Bank (2012). (Figure from M. Persson (2013).)

These types of fluctuations in agricultural commodity markets are not new. It was also discussed quite a lot 100 years ago in the 1910s and 1920s. For example, Ludwell Moore in his book about forecasting the yield and the price of cotton starts that chapter with this quotation that I'd like to read: "An eminent economist has recently told us that economists no longer talk so confidently as they once did of forecasting social phenomena, and that, confronted with the complexity of social relations, 'the sober-minded investigator will be slow in laying too much stress on single causes, slow in generalization, slowest of all in prediction'" (Moore, 1917). I think that, in a sense, supports a crude look at the whole perspective, and it's almost 100 years old.

> **FORECASTING THE YIELD AND THE PRICE OF COTTON**
>
> **CHAPTER I**
> **INTRODUCTION**
>
> As eminent economist has recently told us that economists no longer talk so confidently as they once did of forecasting social phenomena, and that, confronted with the complexity of social relations, "the sober-minded investigator will be slow in laying too much stress on single causes, slow in generalization, slowest of all in prediction."
>
> (Moore, 1917)

Typed from presentation slide

In my presentation, I will take a step away from a crude look because I would like to present a series of models, where I go from a very aggregate description down through a number of levels to more of a microscopic description. But before doing so, let me say a few words about the different causes that have been discussed connected to the food price spikes that we have seen in the past five years.

One cause that has been mentioned several times is because weather, both droughts and large rainfalls, has led to a decreased supply of food crops to the markets. Another cause that has been suggested is speculation on future markets. There has been an inflow of money trading contracts or futures in food markets, which some researchers claim has led to some of the price spikes. There is also an increased demand from various growing economies, including diet changes. Especially China has been mentioned as one country that has increased its import of food, and its national policies on import and export. For example, India in 2007 put a ban on the export of various types of rice, which really led to a decrease of rice traded in the international market. That probably was a major cause of one of the price spikes. But the final point that I would like to discuss more today deals with land competition due to increased bioenergy demand.

Food price spikes — possible causes

- Weather related decrease in suppy
- Speculation in future markets
- Increased demand in growing economies (diet changes)
- National policies on export or import
- Land competition due to increased bioenergy demand

The reason why I'd like to go into this question a little bit more in detail is that a colleague of mine made a systematic review in the literature of what kind of modeling approaches have been tried to make a quantitative assessment of how an increased demand of bioenergy affects food prices. What he found was the following: Out of 126 papers that have been published since the year 2000, 117 were based on equilibrium economic modeling approaches and 9 were statistical. None of the models explicitly talk about dynamical features. If we remember the volatile character of the price curves for agricultural commodities, and also bearing in mind that these studies actually are talking about responses within a 10-year period, one can question what kind of answers these types of models really give.

The effect of bioenergy demand on agricultural commodity prices

- A systematic review (M. Persson, 2013)

Methodology	No. of studies
Modeling	117
partial equilibrium (PE)	86
computable general equilibrium (CGE)	31
Statistical	9
Total	126

So, that is sort of the background for our work, where we have learned some economics from constructing a series of models, starting with actually two types of equilibrium models, then going to dynamic models and also agent-based models. In order to be able to compare the models across these levels of hierarchies, I will base the models on the same microscopic foundation. So, these are the four or possibly five levels that we are considering.

There are two types of equilibrium models that I will describe. The top one is usually used when you do partial equilibrium modeling. That is when you focus on, in this case, the agricultural sector, and you don't look at the effect in other sectors in society. I will briefly discuss how that can be transformed into a land rent equilibrium problem, but then we will leave the equilibrium description and talk about price and quantity dynamics, which is an old topic in economics, connecting back to Ludwell Moore's work actually, and finally reach agent decisions based on profit maximization, which will illustrate new features that may appear when you really model things at the agent level. The very bottom part connects to my earlier research in game theory that is strongly inspired by Axelrod's work in the eighties, and also by Brian Arthur's ideas of avoiding equilibrium models. I will not have time to talk about that, but it's still an area that interests me quite a lot.

Model hierarchy

- Maximisation of consumer and producer surplus (PE)
- Land-rent equilibrium
- Price and quantity dynamics
- Agent decisions based on profit maximisation
- Basis for decisions — rationality?

So, I will describe those four levels of models in this hierarchy, and I will start by describing the macroeconomic basis that we are using. So, at the very bottom, we assume that there is an agent that holds a parcel of land. The five giga hectares of agricultural land is assumed to be ordered in a way from the best quality to the left to the worst quality to the right. This is a relative quality measure from one to zero.

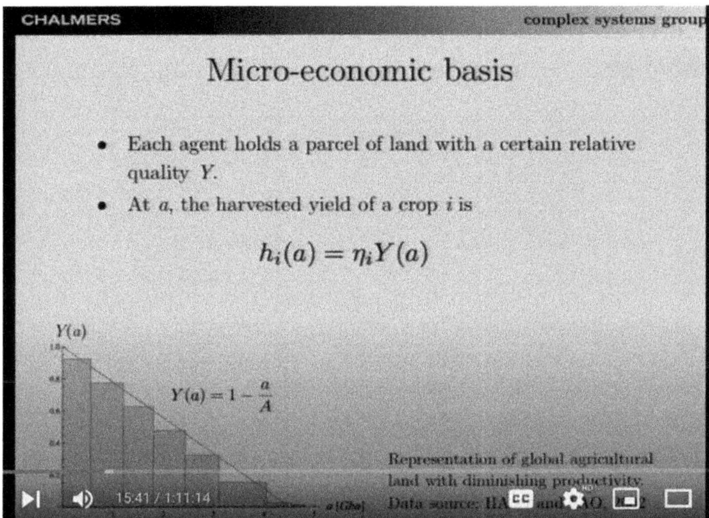

That relative quality measure is multiplied by the crop-specific yield constant, which will give a certain harvest at a certain point, denoted by a, along this line, and for the specific crop i.

So, each agent is situated somewhere along this line, receiving this harvest. An agent also has a cost for producing the harvest. If the agent chooses the crop i, and given that cost, the agent also finds the profit function, considering that there is a certain price that the agent expects to be paid when the crop is harvested and sold on the market. So, it's the income minus the cost, of course.

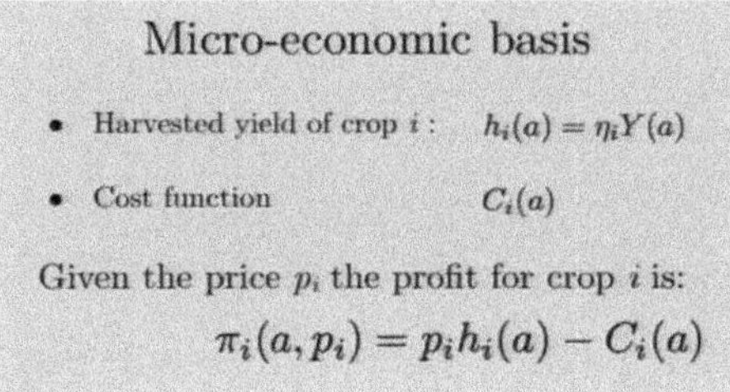

The agents are profit-maximizing, so we are assuming what is usually done in such ideal situations: The agent chooses the crop that maximizes the profit, and the total production of a certain crop i, denoted by qi, is obtained by integrating overall land qualities and the different harvest yields. Here, we select the optimal crop at the different positions along this quality line.

Micro-economic basis

- Harvested yield of crop i : $\quad h_i(a) = \eta_i Y(a)$
- Cost function $\quad C_i(a)$
- Profit function $\quad \pi_i(a, p_i) = p_i h_i(a) - C_i(a)$
- Optimal crop $\quad i^* = i^*(a, p) = \arg\max_i \pi_i(a, p_i)$

Total production q_i of crop i

$$q_i = \int_0^A da\, h_i(a)\, \delta_{i\, i^*(a,p)}$$

And then finally, the crucial point is that the produced quantity determines the price. I mean, this is a critical point: We assume that when the crops are harvested, they are put on the market, and that will set the price according to the demand function, and we typically use an elastic demand function. This is certainly critical because this price was not known to the agent when the agent decided which crop to sow.

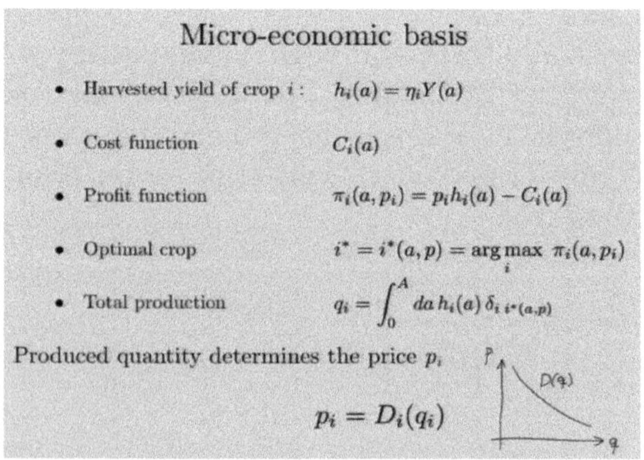

So, there is this delay between the decision of what to sow and when the price is revealed. This is not a problem in equilibrium models because equilibrium models assume this price to be the same, and they just find this situation, so to say. I will, in a few minutes, show how it works, but when we move down to lower levels in this hierarchy, we will see that this delay in information plays a crucial role.

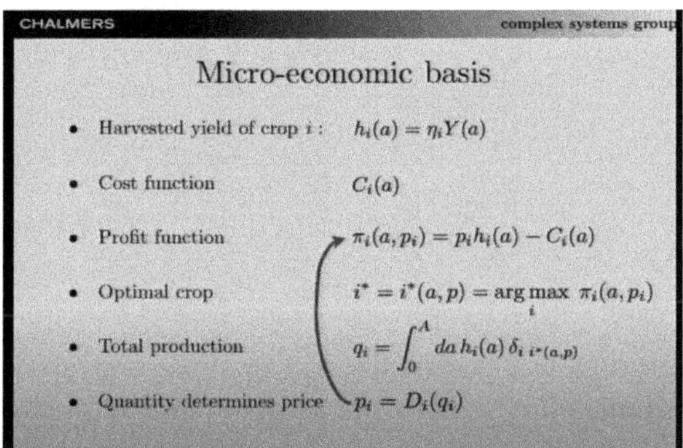

So, now I will describe this. This may be elementary for many of you, but I'd like to go through these steps anyway. This approach of maximizing consumer and producer surplus is the standard approach in all project liberal models used to assess price responses to increased bioenergy demand. The idea is simply the following: We have the demand curve I showed before. We have a supply or a cost function which we construct from the individual agents. We build this up. So, the supply function models the cost of production as a function of quantity. If there is a certain aggregate quantity Q produced, that will lead to a certain price, and the consumer surplus is simply the value that is represented by this area. This is sort of an economic value that all those consumers buy the commodity that they experience. Similarly, the area below the price line constitutes the producer surplus or the profit that the producers will gain. The idea with surplus maximization is to maximize this area. That sounds reasonable. We will then maximize the total area and everybody would be happy. So, this looks fine. For a single-commodity situation, this is quite simple.

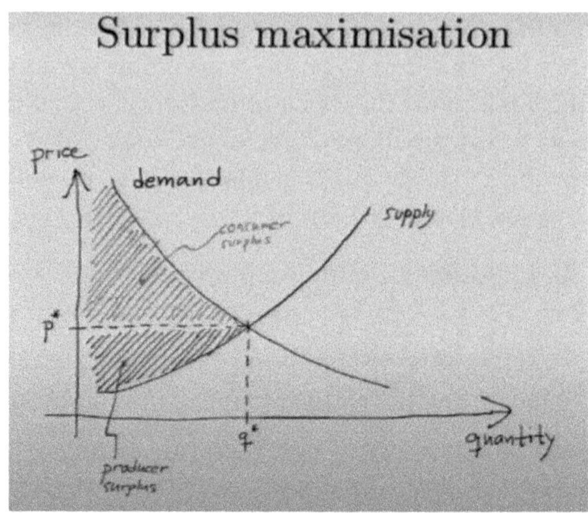

But if we go to two commodities, like if we consider two crops, it wouldn't be simple.

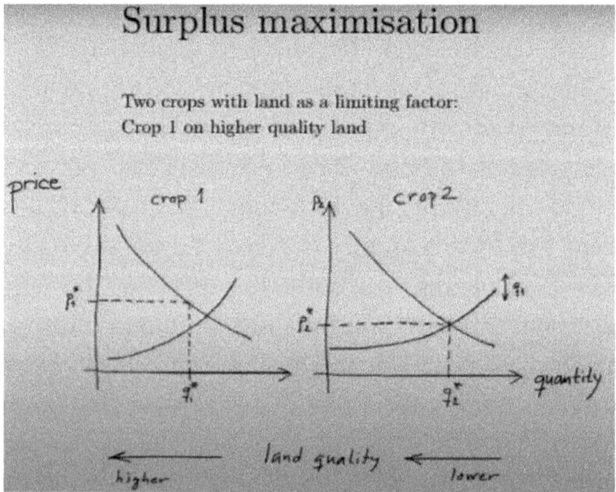

Let's assume that crop one is placed on higher-quality land and crop two is pushed to land of lower quality, then it means that the cost function or the supply function of the second crop will depend on how much we produce of the first crop. So, in equilibrium, this situation may look like this:

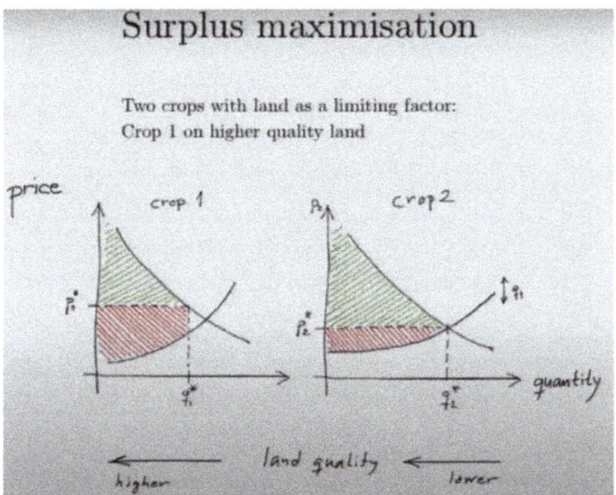

The maximum total area may actually then lead to the fact that we will not reach this crossing line because land is a limited factor of production. So, this enters as a scarcity price for land, and in fact, the distance between the supply curve and the price level constitutes the land rent. The profit that the farmer can get now, assuming that the farmer owns the land, would correspond to the land rent in this situation.

This brings us to the other formulation of the maximization problem, which is the land rent equilibrium situation, where the main point is that the land rent of the best-quality land that produces the second crop needs to be the same as the land rent for the worst land that produces the first crop. If there were a difference there, farmers would switch to one or the other crop. So, it means that we can put these two figures into one graph, where we replace quantity on the horizontal axis with land area instead. We still keep the price on the vertical axis, so we see that up to point *a1*, the best-quality land, we will produce crop one, and after that, crop two would be produced until this point, and at this point, the value of land will be zero.

And the land rent equilibrium problem is then such that we get to continue it in land rent between the first crop at the very right side and the second crop at the very left side, so we get a graph like this one:

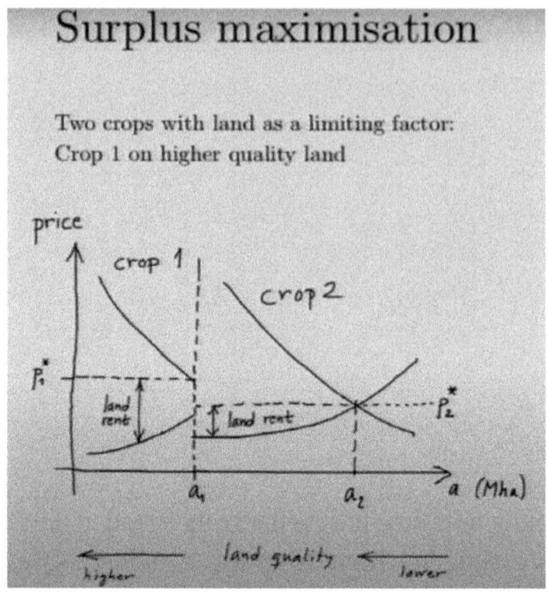

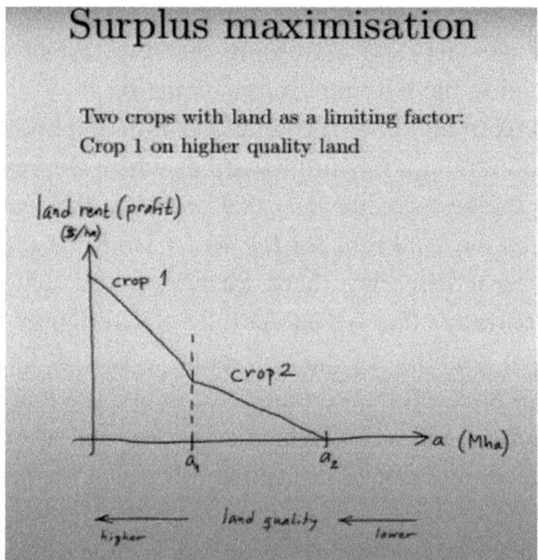

So, of those land rent curves or the profits that the farmer would gain from producing crop one and crop two, the curves for crop one would continue down like this and for crop two would go up like that, just indicating that in the interval *0–a1*, the crop 1 will be the most profitable one, and in area *a1–a2*, crop 2 would be the most profitable one.

So, this is just an illustration of the process that finds such an equilibrium. In this case, we consider three generic crops. To the very left, we have intensively produced crops *IP*. In the middle, we have bioenergy crops, and to the right, we have extensively produced crops which could be passed to land or fodder for animals. What we see to the right here are the lines or the price levels. We have different scales here. That's why we don't have equality in the profit made there. The production of bioenergy is larger, which will make continuity in inland grain curves still to hold.

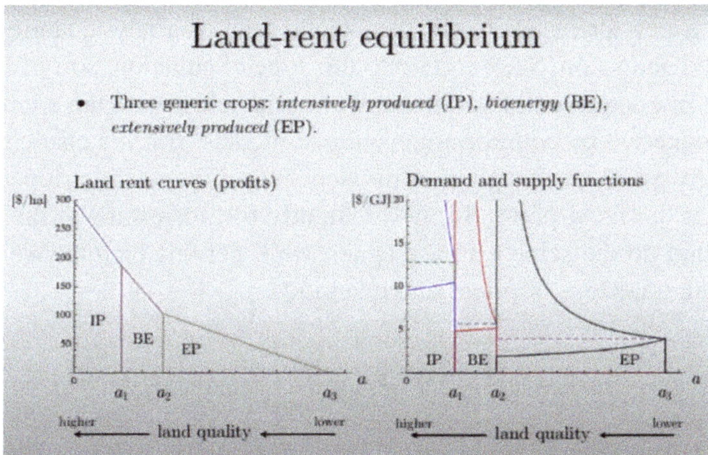

So, this would correspond to a situation whereby only two crops are present, and bioenergy is excluded. If we add bioenergy, that will be typically placed in between depending on what type of bioenergy crop we are considering; the immediate very visible result is that we will get a much higher land rent resulting from this. Now, this is depicting a very large introduction of bioenergy, in the order of a hundred exajoules, that I discussed that is used in many scenarios.

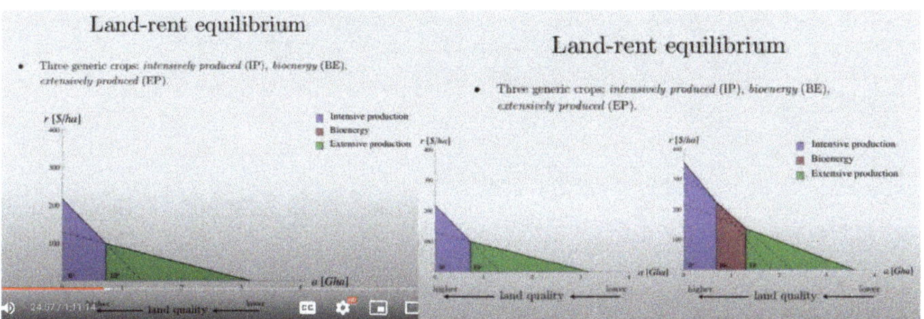

So, next, I'd like to move away from those equilibrium models. I mean, in those equilibrium models, we're not asking the question about the stability of the equilibrium. I mean, the equilibrium is assumed to hold. What I would like to do is to think about it in a very simple way. This was done some 80 years ago, actually. Think about the stability of the equilibrium in this type of picture, assuming that at some point in time a certain quantity QT is produced, which will set the market price P of T. So, this is a very simplistic view, assuming now, that will lead to agents choosing to produce this amount of the crop in total, and that would then lead to a new price level, and that would lead to a new quantity being produced and so on. So, we get into this type of situation, sort of indicating that in a sequence of iterations, we would actually hit the equilibrium point suggested by equilibrium economic models. But it's also clear that whether we will hit this point or not depends on the relative slopes of the curves at the fixed point. So, if we slightly tilt, for example, the supply curve, and do the same exercise again, we'll get this picture. We have a diverging trajectory of prices and quantities.

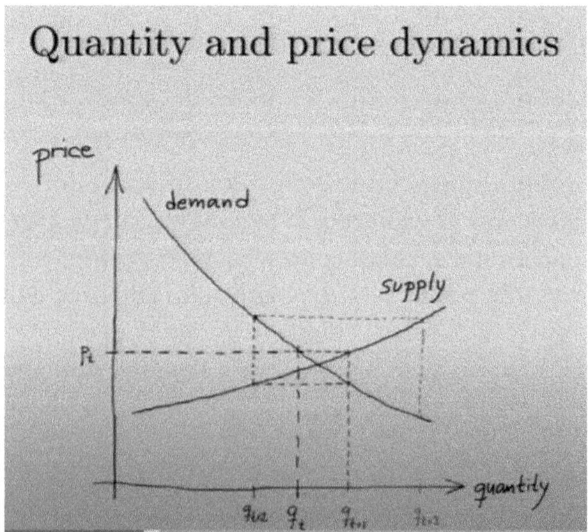

Because of the picture we get here, this was termed the Cobweb Theory in the 1930s. It followed from Ludwell Moore's and many other researchers' investigations of price and quantity dynamics in the agricultural markets. Ezekiel was one researcher involved in this, talking about the apparent inconsistency between the persistence of these observed cycles and the tendency toward an equilibrium posited by economic theory. Tinbergen drew those diagrams in his paper from 1930, for example, and Richie Schultz and others were other economists involved in this work.

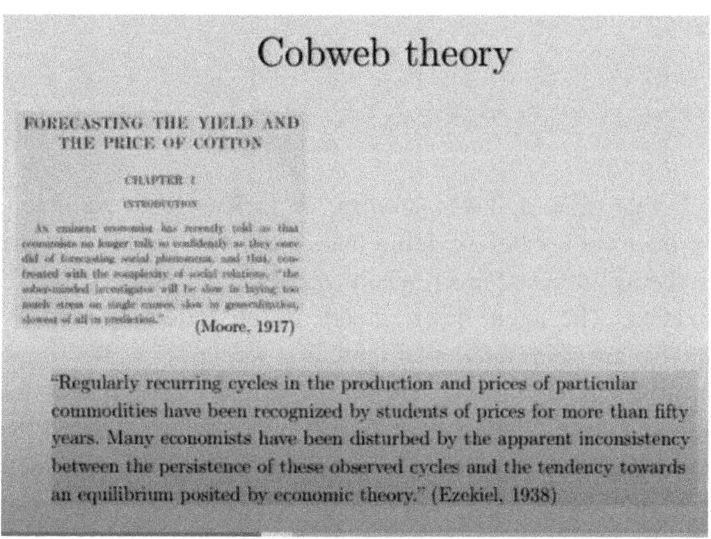

What the Cobweb Theory really dealt with was the analysis of the stability of this point. So, the stability comes from the ratio of the slopes, and if the absolute value of the ratio of those slopes is not less than 1, then this fixed point is unstable. This type of model class was further developed and explored to a large extent in a series of papers in the 1990s by Holmes, Brock, and Arifovic, who investigated chaotic behavior, periodic cycles, and also various types of learning mechanisms that could affect the stability of the dynamics.

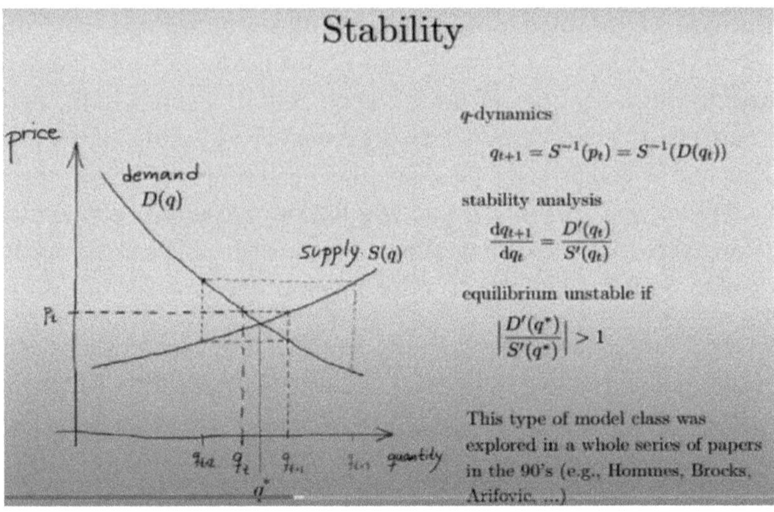

$$q_{t+1} = S^{-1}(p_t) = S^{-1}(D(q_t))$$

$$\frac{\mathrm{d}q_{t+1}}{\mathrm{d}q_t} = \frac{D'(q_t)}{S'(q_t)}$$

$$\left|\frac{D'(q^*)}{S'(q^*)}\right| > 1$$

So, I will show a few examples, or at least one example, of this dynamic later on, but before doing that, I'd like to go down to the fourth level, to the agent-based level, where agent decisions are based on profit maximization. The agent-based model is the following one: We have N agents that are distributed over land, and we will now, for simplicity in a sense, consider an infinite number of agents there. It will make the description of the modeling a little bit simpler. We considered two types of agents actually. An agent is characterized by what kind of method the agent uses in order to determine how the agent finds or guesses the price at harvest. So, that is the e part. The i part is the crop that the agent is currently growing, and each agent is sitting on a certain parcel of land at position a that determines the certain quality that the agent has on his or her land. So, at each a, we have an infinite number of agents in this simple model. There is a mix of which crop they grow, and there are also two different ways in which they guess or estimate the price at harvest time. We assume that there are discrete time steps. One time step is the complete season, and we also make this nice if you want the system to find an equilibrium: We only allow a certain fraction γ of the agents to reconsider their choice of crop. Having γ being very low means that we will slowly change the crop types. The system may then reach an equilibrium. We will see that this parameter may be critical.

Agent-based model

- N agents are distributed over land
- In the limit of $N \to \infty$, we consider densities of agents, $x_i^{(e)}(a)$ located at land parcel a with land quality $Y(a)$.
- Agents may differ in:
  - which crop i they presently produce, and
  - how they estimate (e) the price at harvest time.
- Discrete time steps, one step is a complete season.
- At each time step, a fraction γ of the agents reconsider their choice of crop.

The two types of agents that we consider, one being naive agents or zero-intelligent agents, will guess that the price at harvest time will be the same as the prices now. The other agent type is rational agents. We again now have two equilibrium models by assuming that the rational agents actually know exactly what the price will be at harvest, so they know actually two things. They know how many agents are zero intelligent and they know what the zero-intelligent agents will do, so they perfectly know what the price will be because they also know that all the other rational agents will do the same as they do. So, the rational agents are in a sense both in a good and in a bad position. They are in a good position because they know exactly what to do, but they have no choice either. The zero-intelligent agents act just as they want to without thinking about what happens. The rational agents will have to act in a way that will actually bring them to a temporary equilibrium, but they are not very rational because they are not looking forward in time and more than one time step. So, if all agents were rational, that would directly bring the system to the equilibrium point, but when there is a mix of naive and rational agents, that might not happen.

Now, in this first setting, I am assuming that the distribution between naive and rational agents is fixed. There is a certain fraction of the land owned by one of the agents, and the rest belonging to the other agent. There is no dynamics between the agents or for agents to switch between naive and rational types. The only dynamic that is entering here is the choice of the crop. This can be summarized in these equations: The zero-intelligent agents occupying crop I at the next time step will be equal to one minus gamma, those who don't change, plus a fraction gamma change term. The optimal crop is given by I^*. The optimum crop is then determined by, or depends on, the land quality a and the price that they see. The naive agents assume the price to be what the price is now. The similar equation for the rational agents would look like this, where the agents switch to the optimal crop *I** given the future price *PT* + 1 that is assumed to be known then for the rational agents.

Two types of agents

- *Naive* (or zero intelligent) agents:
  - expect the present price to hold at harvest time.
- *Rational* (perfect 1-step prediction) agents:
  - know what the price will be at harvest.
- Each land parcel a is characterised by densities of agents growing crop i:
  - naive agents, $x_i^{(0)}$, and rational agents, $x_i^{(R)}$.

- Agents switch crops depending on what they consider the optimal choice

Naive: Switch now to what is optimal if prices remain the same.

$$x_i^{(0)}(a, t+1) = (1 - \gamma)x_i^{(0)}(a, t) + \gamma \rho \, \delta_{i\,i^*(a,p(t))}$$

$$x_i^{(R)}(a, t+1) = (1 - \gamma)x_i^{(R)}(a, t) + \gamma \rho \, \delta_{i\,i^*(a,p(t+1))}$$

Rational: Switch now to what is optimal at harvest ($t+1$).

This is now an illustration of the dynamics. Here, we see the land quality going from high to low quality. Here, we have the extensively produced crops. The green is the intensively produced crops and the red is the bioenergy crops. What we see here is a situation where 30% of the agents are the rational ones, and we see that even though they know what is optimal for them to do, that is not a sufficient number in order to stabilize the system. It should be noted that after the season, what the zero-intelligent agents are doing is that they mimic the actions of the rational agents, so there is a delay in what the naive agents are doing, but that sort of delay is enough in order to destabilize or to keep the system out of equilibrium.

We can see what happens if we ramp up the fraction of rational agents. Now, it's 38%. We see some kind of stabilization, okay, so these are the price levels for intensively produced crops, bioenergy crops, and extensively produced crops, and now when we pass 50% of the agents being rational, the system looks more or less stable. Please interrupt if you would like to ask a question.

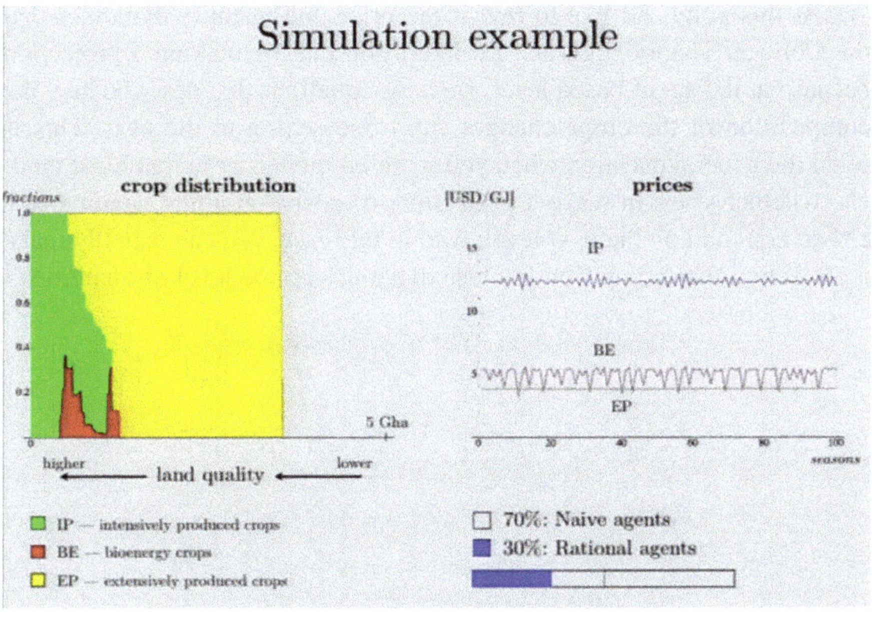
Simulation example
crop distribution
fractions
prices
[USD/GJ]
IP
BE
EP
5 Gha
higher
lower
land quality
seasons
IP — intensively produced crops
BE — bioenergy crops
EP — extensively produced crops
70%: Naive agents
30%: Rational agents

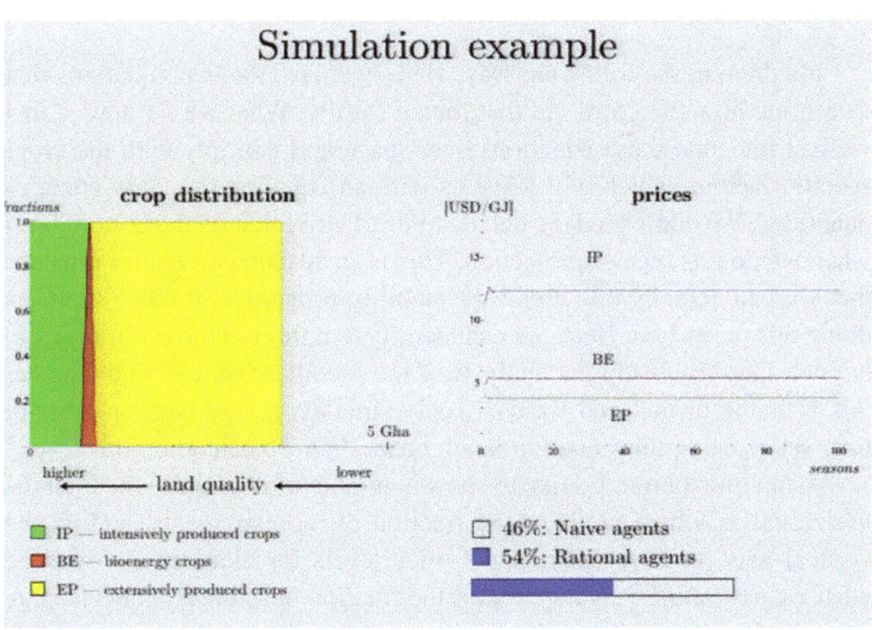
Simulation example
crop distribution
fractions
prices
[USD/GJ]
IP
BE
EP
5 Gha
higher
lower
land quality
seasons
IP — intensively produced crops
BE — bioenergy crops
EP — extensively produced crops
46%: Naive agents
54%: Rational agents

At this point, I'd like to revisit the price and quantity dynamics that the Cobweb Theory suggests. I'd like to do that by making a projection because at the agent-based level, we have equations that describe how the composition of the crops changes from one season to the next. This is often the general question when you work on models or hierarchical models: whether you can make a projection, or coarse-graining, leading to a closed equation on the next level. And in this case, you can actually make a projection that brings you to close dynamics on the level of quantities.

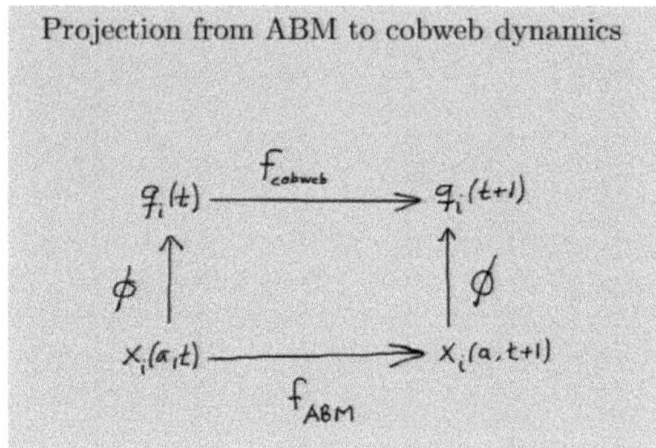

It's done in the following way: Here, we have the two equations that determine how the crops are distributed locally. What we do now is that we just integrate these equations over space and multiply with the crop-specific factors, which will leave us with an equation that only contains quantities. We don't need the details of the distribution on the bottom level when we do this type of projection. This is an advantage because it means that we can actually talk about the stability properties of this system by doing this projection. Here, we can identify that this is a three-dimensional iterated map. By using the analysis of the Jacobian, we can find whether this is stable or not, and we can easily simulate a very large number of time series using this equation much faster than we can do on this level.

As an illustration, I'd like to show a number of runs then where on the horizontal axis here we have the fraction of rational agents, and on the vertical axis we have the realized price levels for bioenergy, so we see what we saw at the very end: that if the fraction of rational agents is large enough, the system stabilizes at a certain level, but when we go down to smaller fractions of rational agents, we see a period-doubling sequence.

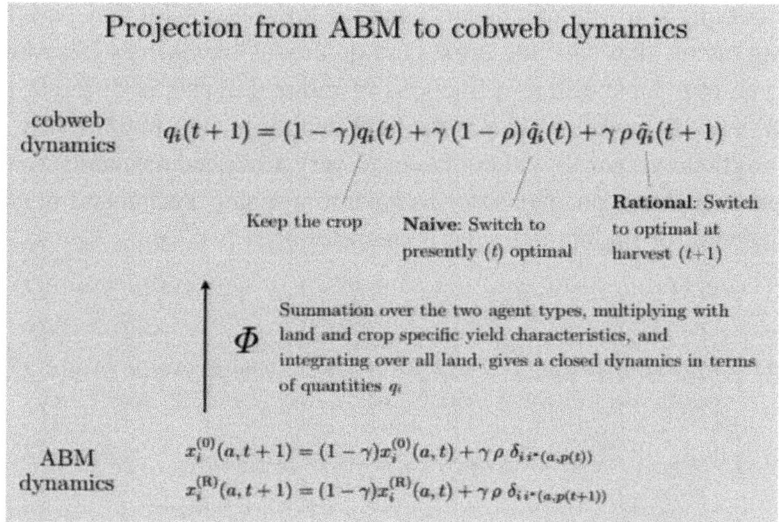

$$q_i(t+1) = (1-\gamma)q_i(t) + \gamma(1-\rho)\hat{q}_i(t) + \gamma\rho\hat{q}_i(t+1)$$

$$x_i^{(0)}(a,t+1) = (1-\gamma)x_i^{(0)}(a,t) + \gamma\rho\,\delta_{i\,i^*(a,p(t))}$$

$$x_i^{(R)}(a,t+1) = (1-\gamma)x_i^{(R)}(a,t) + \gamma\rho\,\delta_{i\,i^*(a,p(t+1))}$$

This clearly shows that the system is chaotic when the number of rational agents is too small.

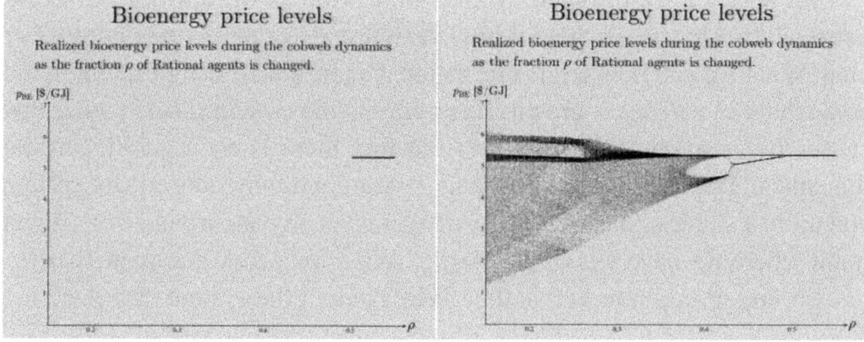

I would like to finally talk a little bit about what we are currently working on, which is adaptive agents. Then we go back to the level of agent-based modeling. We relax the constraint that at each land plot there is a fixed fraction between the rational and the zero-intelligent agents. We assume that there is some cost CR for an agent to use the rational expectation mechanism. After each season, the fraction of agents may reconsider what type of mechanism they should use in order to make a guess for the price after harvest time. They do this change depending on their comparison with other agents and see how successful they have been among agents that have the same land quality as themselves. This means that you may have different fractions of rational agents in different parts of the land diagrams, so that if the land quality is very high or very low, you might do very well without having a very advanced mechanism, while at certain points you may want to have a stronger mechanism or better mechanism for finding out what is the price after harvest.

Adaptive agents

Can the system dynamics be less unstable if **agents adapt their mechanism forming expectations** based on performance?

- Cost c_R to use the rational expectations mechanism.

- A certain fraction of agents (here 20%) switch to the most successful mechanism (measured by accumulated profit).

- Comparison with agents at the same land quality.

So, here we see the same type of simulation as we saw before. At the bottom left, what we see here is how the rational agents are distributed over land. What we can now tell is that rational agents are mostly present in the area where bioenergy is grown. Bioenergy is the crop in this situation that causes the instability. But we also see that there is an instability in the dynamics. The question is this: Why is that, and why doesn't the system end up in a stable situation, where all agents locally are around this critical point where we have lots of bioenergy? Why don't they dominate there?

So, my question to you is this: Why doesn't the system stabilize?

Discussant A: Two types of agents: sophisticated agent and naive. Now, the sophisticated agent adapts in a particular way, and you've got a naive agent who is going to obviously not adapt in that particular way or do they have the same adaptive mechanism?

Kristian Lindgren: Yeah. They can. Also, the zero-intelligent agent can switch to being rational, but there is a small cost to be paid for being rational.

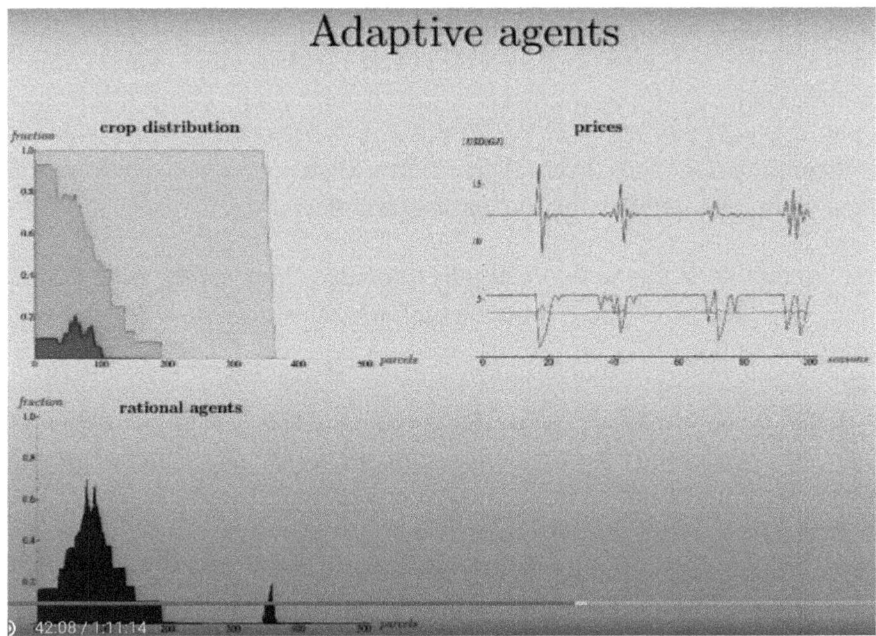

Discussant A: And is that determined endogenously or do you specify this cost?

Kristian Lindgren: This cost is specified and I should say that it's very low in this situation. I keep the cost very small actually. I'm trying to have you think about what might cause this destabilization. What happens is that we get very close to the equilibrium situation, and then you don't have to pay the cost for being rational. You can actually be a zero-intelligent agent, so what we see is that at those points we get more and more free-riders. They don't need to pay the cost because we're still sitting in the equilibrium, so then we can forget about how we came there. Then, my question is what a simple mechanism could be to stabilize the system.

Discussant A: So, what you're saying is that as you get close to equilibrium there are more and more free-drivers, and that in some sense less information is being aggregated as you get closer to equilibrium, and that when that happens, you go back into some instability, and then again people start adapting. You can get back again to equilibrium. Less information gets aggregated and then that cycle continues and then we don't see stability.

Kristian Lindgren: Yes, that's right exactly. So, my next question to you is this: Could we stabilize this system in a simple way?

Discussant A: It has to be externally provided because, the way you're modeling the agents, there is no internal mechanism which will aggregate that information.

Kristian Lindgren: That's right. Either we would have to introduce another agent or we could possibly modify how we model the system. Let me show two simulations here. The top one is the one I showed before. The bottom one is slightly modified, and only slightly I would say. Actually, it's modified in a way to be more realistic. So, my question is this: What have I done? Insurance is a good point. Actually, we haven't done that, but that is part of our next step. That would also help. But that is not what I did.

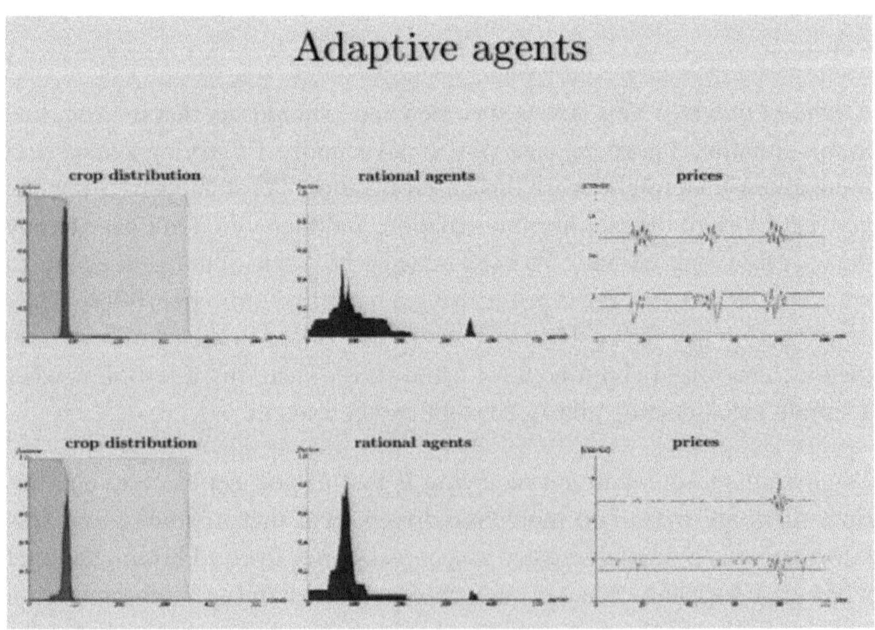

Discussant B: Add some noise?

Kristian Lindgren: Yes, exactly. There is noise added, a very small level of noise, 0.3%, added independently to the three different crops. That noise keeps the system slightly off equilibrium and makes it possible for the rational agents to stay in the system.

Suman Banerjee: There's a paper by Grossman and Stiglitz on the impossibility of an efficient informational efficient market. It is the general political economy of 1983, which basically handles this in the stock market environment. It shows this noise can actually give some extra profit to this, and have this informed or rationalized in the way you're saying. It basically is a part of equilibrium. It is this free-riding versus this totally basically revealing prices, versus totally noisy prices. It has to be in between. It can be totally noisy prices or it cannot be totally efficient prices basically. Yes. The JP paper in 1983.

Kristian Lindgren: Very good. Yes. If you would increase the noise much more, then the system would be very unstable. If you decreased the noise, you wouldn't be helped by the noise. You could also make other types of modifications. Here, we have introduced a third agent which is a trend-following agent. It is without noise actually. The trend-following agent works in concert with the rational agents, and it leads to some degree of stabilization, but remember that in all of those cases, where we, in a sense, get pretty close to an equilibrium situation, we still have to assume that we have these ideally rational agents that have a tremendous amount of information about the future prices.

In our current work, we are now exploring a more open set of possible agents, but that would not include this sort of very rational agents that have a tremendous amount of information about the future price.

I think we can move on to a summary of what we have done. One of the key points of this presentation was to show you this hierarchy of four different model types that are all based on the same or similar microeconomic situation. It is clear that we do get new behavior appearing at the third and fourth levels, at the bottom level. One key finding, I think, is that we have this ideal projection from agent-based dynamics to quantity or cobweb dynamics, which provides us with a tool for analyzing an

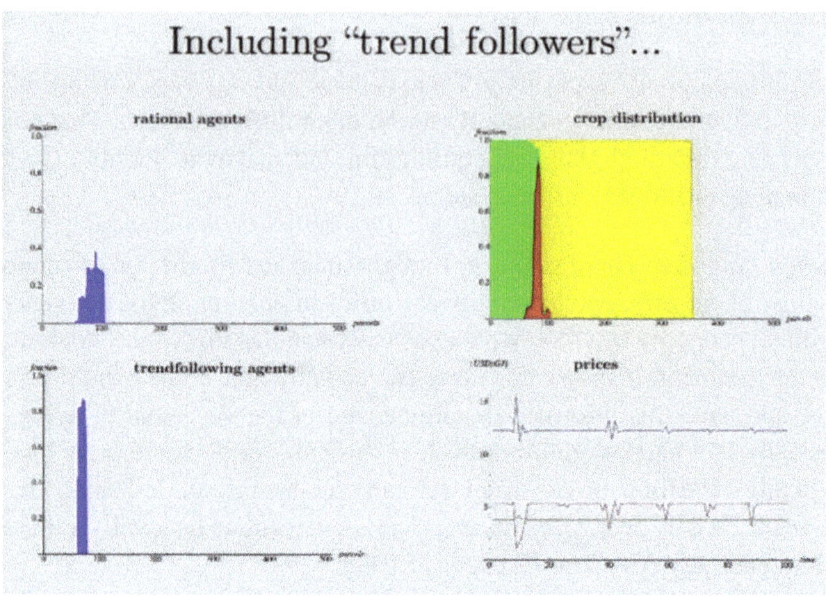

agent-based model when such a projection can be done. Then, we introduced adaptive or evolutionary behavior, which will not allow us to make this type of projection any longer. The analysis of this type of system then requires that we perform the analysis on the level of the adaptive agents.

Summary

- Hierarchy of models based on same micro-economic agent description
    - Maximisation of consumer and producer surplus
    - Land-rent equilibirium
    - Price and quantity dynamics
    - Agent decisions based on profit maximisation

- Projection from <u>ABM dynamics</u> to <u>quantity ("cobweb") dynamics</u>
    - Allows for stability analysis of ABM (in the limit of ∞ many agents).

- The competition on land between different crops leads to unstable dynamic behaviour: periodic and chaotic oscillations

- With an evolutionary or adaptive agent dynamics the analysis has to be done on the micro level, and this work is in progress...

I thank you for your attention.

4.2 Discussion

Brian Arthur: Thank you. One of these things to do is to match the price and output data with real statistics over time. Many crops in many parts of the world have been measured. There's lots of data as price data, land use data, and so on. The second statistical thing that can be done is to look at how farmers do make their decisions statistically from one year to the next. So, my question is this: Have you thought of matching this qualitative behavior and the agent-based learning with actual real market data?

Kristin Lindgren: Yeah. We have thought about that. I know that there is a fairly recent paper dealing with time series of food commodities and comparing statistical features of the price series with what you find in reality. That was the model where they specifically looked at storing crops in between seasons. We intend to extend our model also to that type of mechanism and also to explore whether we could capture some of the statistical features of real-time series. When it comes to decisions on the level of agents, we haven't thought about how to do that, but it's certainly of interest to us to go in that direction. The global land project came out with the report less than two weeks ago, where they called for the development of models for making assessments of land use, where you include the decision-making on the level of the farmer in those models. Let me also then go back to the reason for doing this exercise. That is all as you saw in this systematic overview. All of the models that have been used in making assessments are equilibrium models. There are not yet really any good models that are based on agents' decisions that make assessments on the global level. There are lots of models where you model local situations, farms, and so on, where you do agent-based modeling, but not where you aggregate those to the global level, and so on. The global level is where we have a dominance of equilibrium modeling, which is sort of strange, I would say.

Peter Schwartz: In Singapore, obviously the question of land use is a very critical issue. Agriculture is not the competitive force here. It's people's choices about how they choose to live and so on. Do you think the models that you're talking about here would apply to those situations, where some people view their home as simply a home, others and economic activities where the price increased matters lead to selling behavior? Which then may put pressure on others to sell, and so on? Are there dynamics that would be similar that would allow similar kinds of predictive models to be developed for real estate forecasting and decisions about land use planning for Singapore?

Kristin Lindgren: I can very well think of this delay between when you take a decision and whether it's a decision to build a house or to make an investment. You don't know the future price, so [there is a] delay between the price you know, when you take your decision and when you want to sell something. I know that this has been investigated also for real estate markets, but I don't know any details of those models though. But I would expect that not exactly this type of model, but models with the same idea should be applicable to that type of problem, I think.

Discussant C: I assume you go into the dynamic models because you're interested in these peaks and spikes and so forth, and to learn ways to reduce those. It just seems to me unrealistic to have any perfectly future-aware agents, so it seems that's a simplification that has to be eliminated if you're going to get practical. The second question is, in many cases at least that I'm aware of, there are also subsidies acting, which may well reduce uncertainty in parts of these, particularly in the energy side. Have you any intention of going in the direction of putting subsidies in and reducing those unreal agents?

Kristin Lindgren: Yeah. Let me just think for a few seconds. I mean, first, I agree that these agents are unrealistically well-informed. That is just a starting point, actually. To make these models, in a sense, more realistic, we need to move away from those well-informed agents. Rather, we would like to have some kind of adaptive or evolutionary mechanism that selects what type of agents should be in the population, so to say. When it comes to subsidies, we haven't thought about that explicitly. We have thought about introducing various types of mechanisms that could possibly lead to stabilization, but the primary goal of this investigation was to point out a number of mechanisms within this class of economic models that could lead to stabilization. Then, we know that there are a variety of different mechanisms that could be added to achieve stabilization. But that could also lead to the system locking into two equilibria, different types of equilibria. So, the system could very easily be more dependent on history, I would say. We haven't explicitly thought about subsidies in this respect.

Greg Fisher: Greg Fisher from the think tank Synthesis. If the prices you're observing are traded in the financial markets now, I'm guessing that's what you're looking at. I'm wondering to what extent you're examining how prices have formed in financial markets. Because I worked for the Bank of England for nine years and in a hedge fund for three years, I've gained a lot of experience, particularly sitting on dealing desks. David Tuckett at UCL wrote about *Minding the Markets*, discussing narrative formation, emergence, and perpetuation brilliantly. I think to understand how it really works in the financial markets, I'd recommend anyone read that book. It's excellent. I'm wondering if you've captured that at all, or if you want to capture that, because some of those narratives can be completely detached from the fundamentals you're talking about. Not always, but they often become detached, and price formation can kind of go all over the place, which has nothing to do with fundamentals.

Kristin Lindgren: Yeah, when it comes to price formation here, it's extremely simplified, of course. We're assuming that there is a demand function that determines the price. That's a high simplification. One parameter that we have to modify is that we assume that not all of what is produced after a season is immediately put on the market; instead, it's stored, and then a fraction of the storage is released onto the market. So, there is a parameter that you can use in order to dampen the effect of large and small produced quantities. That's a simple way we have introduced to modify the demand function. But I will look at the reference you mentioned.

Discussant A: The model you have doesn't include any financial factors. You're looking at a real supply-and-demand model with no financial factors at all. For example, in real estate markets, historically you see real estate booms when there's a huge increase in credit. Real estate purchases are typically highly leveraged, with buyers putting down a 10 or 20% down payment. When borrowing is easy and low cost, as we've seen during the current run-up in prices, not just in the US but also in Europe, the dynamic shifts. The point being made is that once you get the dynamic going with cheap credit and the expectation that you can borrow, flip, and make a profit, it takes on a life of its own, unrelated to fundamentals. So, in the model you have, credit and leverage play no role.

Kristin Lindgren: That's not introduced at this point.

Discussant A: At this point, we're looking at a different set of models when trying to gauge the build-up of real estate prices. In fact, if you draw a graph — I've done this during my time at the IMF — if you plot the data for European countries like Spain, the UK, France, Germany, and Ireland, you'll see a clear correlation between the increase in credit and the rise in prices in those countries. So, if leverage isn't available, you won't see a surge in real estate prices. Financial factors play a fundamental role historically.

Kristin Lindgren: Yes, and those are factors that we plan to introduce. The work we've done so far has been inspired by the Santa Fe artificial stock market model, but at this point, we've kept the model very simple. Introducing those types of mechanisms into this model is important, I think.

Discussant D: So, what is that? Maybe this also hasn't been considered and maybe it's not worth consideration. But 500 million hectares of land is an awful lot of land. I'm just curious about the angle of water, and where that water would come from and how that's factored in.

Kristin Lindgren: I'm not working in that area. I'm just sort of referring to that as a scenario that is often suggested. What we want to do here is to illustrate what that would mean in terms of land competition, but water is an extremely important issue in that context, but it's not part of our research.

Robert Axelrod: If you go back to your original motivation about preventing climate change, doesn't biomass create problems in the way that fossil fuels do, and therefore they aren't really a contributor to solving of the problem?

Kristin Lindgren: Well no. I see your point. There are indeed various biomass production methods that can have a net-zero or close to net-zero effect on atmospheric carbon dioxide.

Robert Axelrod: I didn't know that.

Discussant E: I have one question: Do you use any special geometric distribution for the land resource in your model?

Kristin Lindgren: For the land resource or the land quality?

Discussant E: The land resource. I mean in your model, do you just put the land into your model or do you follow some special geometric distribution?

Kristin Lindgren: Ah okay. There is no geography in this model.

Discussant F: My question pertains to whether this model could be adjusted for real estate land use. In the case of real estate, having a roof over one's head is a highly emotional issue. It evokes strong emotions. Therefore, a government that seeks election would respond to that in a definitive way, as seen in Singapore, for example. To address concerns, the government has intervened seven times successively. In your model, is there room to account for such governmental interventions, or is it beyond the scope of modeling? Thank you.

Kristin Lindgren: It is beyond this model. The purpose of this sequence of models was to show what kind of various types of dynamic features appear when we leave the equilibrium modeling area. I mean this is a very highly conceptual, very simple model. We do intend to include a variety of different mechanisms to make it possibly more realistic, but to me, it's hard to say whether it can be transformed into something that is useful in that sense. We consider this as a conceptual model that is useful as a demonstration of the problems that we have with the equilibrium models that are being used to make assessments.

Suman Banerjee: Just a clarification on the question that was asked about the rational expectation model agents. I think you don't need fully rational agents. What you need are better-informed agents. So, if you have naive agents and if you have a set of agents who are receiving a noisy signal, I think you can achieve all the results that you are getting with the noisy signals. The precision of the noisy signal has to be better than zero. Basically, they have to be better informed but not necessarily with rational expectations. Obviously, you see these pronounced cycles if differences between the naive and the rationals are very high, and there is a cost to becoming rational, but the sufficient condition is that they are better informed rather than rational. That's what Stiglitz and Grossman had in that paper. Thank you.

4.3 Summary of the Talk

Kristian Lindgren's presentation focuses on the intricate relationship between bioenergy production, land use, and agricultural commodity markets. He begins by highlighting the challenge posed by increased bioenergy demand, which competes with other forms of land use. Lindgren emphasizes the complexity of modeling future scenarios, cautioning against overly detailed predictions that may create a false sense of certainty. Despite this complexity, he suggests that rough conceptualizations can provide a useful basis for discussion.

Lindgren discusses various modeling approaches used to assess the impact of increased bioenergy demand on food prices, noting that most studies rely on equilibrium economic modeling, which may overlook dynamic features and fail to capture the volatile nature of agricultural commodity markets. He introduces a hierarchical framework comprising four levels of models, each building upon the previous one: equilibrium models, price and quantity dynamics models, agent-based models, and adaptive agent models.

At the macroeconomic level, Lindgren describes equilibrium models that analyze the interaction between supply and demand for different crops, considering factors such as land quality and crop-specific yields. These models aim to maximize consumer and producer surplus but may overlook the dynamic nature of market interactions.

Moving beyond equilibrium models, Lindgren explores price and quantity dynamics, illustrating the Cobweb Theory, which describes the cyclic nature of agricultural markets. He discusses the stability of equilibrium points and the role of factors such as relative slopes in determining market dynamics.

Lindgren then delves into agent-based models, where individual agents make decisions based on profit maximization. He distinguishes between naive agents, who assume current prices will persist, and rational agents, who anticipate future prices. He demonstrates how the composition of agents and their decision-making processes influence market stability and dynamics.

Finally, Lindgren introduces adaptive agent models, which allow agents to adapt their decision-making strategies over time. He discusses the distribution of rational and naive agents across different land qualities and explores the impact of introducing trend-following agents on market stability.

In summary, Lindgren's presentation provides insights into the complexity of modeling bioenergy production and its effects on agricultural markets. By examining different modeling approaches and their implications, he highlights the need for nuanced analysis to understand the dynamic interplay between bioenergy demand, land use, and food prices. His hierarchical framework offers a systematic way to explore these complex interactions and uncover new insights into market dynamics.

4.4 Relevance of the Talk to the Current Stage of Research

Kristian Lindgren's talk on bioenergy production, land use, and agricultural markets remains highly relevant to the current stage of research for several reasons:

Bioenergy Production and Sustainability: As the world continues to seek alternative sources of energy to mitigate climate change and reduce dependence on fossil fuels, bioenergy remains a prominent option. Lindgren's discussion of the challenges posed by increased bioenergy demand and its competition with other land uses highlights the ongoing importance of understanding the sustainability implications of bioenergy production.

Complexity of Modeling Future Scenarios: Lindgren's caution against overly detailed predictions and emphasis on the need for conceptual frameworks for discussion resonate with contemporary research practices. In an era of complex systems and uncertainty, researchers are increasingly recognizing the importance of acknowledging and navigating uncertainty in their models and analyses.

Modeling Approaches and Methodological Innovation: Lindgren's presentation of hierarchical modeling approaches, from equilibrium models to adaptive agent models, reflects ongoing efforts in research to develop and refine methodologies for understanding complex systems. In particular, the incorporation of agent-based models and adaptive strategies aligns with current trends in interdisciplinary research, where scholars seek to integrate insights from economics, ecology, and other fields.

Dynamic Nature of Agricultural Markets: The discussion of price and quantity dynamics, including Cobweb Theory, remains relevant in understanding the behavior of agricultural markets, which are influenced by factors such as weather, speculation, and government policies. Lindgren's exploration of stability and instability in market dynamics provides valuable insights into the challenges of predicting and managing agricultural commodity prices.

Sustainability and Land Use Planning: Lindgren's focus on the critical question of where bioenergy crops should be cultivated underscores the ongoing importance of sustainable land use planning. As global demand for food, fiber, and fuel continues to grow, researchers and policymakers must grapple with questions of land allocation, environmental sustainability, and socioeconomic equity.

Overall, Lindgren's talk offers a rich and multifaceted exploration of the complex interactions between bioenergy production, land use, and agricultural markets. Its relevance to current research lies in its methodological insights, theoretical frameworks, and practical implications for addressing pressing global challenges related to energy, food security, and environmental sustainability.

Chapter 5

Signals and Boundaries — Gated-Urn Models for a Crude Look at the Whole

John Holland

YouTube: https://www.youtube.com/watch?v=8BppamGhulA

Speaker: John Holland

Moderator: Helena H. Gao

Discussants: Martin Riser, Peter Sloot, Peter Breton, Greg Fischer, Sheila Ronis, Nikolay Berezhnoy, Martin Reiser, Peter Breton, Greg Fischer, Brian Arthur, Maarten Boasson, Robert Axelrod, Song Ji Young, Alain Wouters, Discussants A, B, C, and D

5.1 Talk by John Holland

Thank you, Helena, for an impossible introduction. And thank you for pointing out one of the advantages of old age, that you were young at the right time.

What I'd like to do is relate this idea of signals and boundaries to some of the questions that we've been talking about and especially some of the things that Murray raised yesterday and some of the other questions.

And I wanted to start off with a couple of questions that have been posed when I've been here at previous times. Many of you will recognize the iconic casino downtown, and I wanted to look at two things that have been done recently that, in my opinion, do involve this notion of boundaries. One is that behind the casino, they took a saltwater marsh and by preparing the soil around it in the right way, it now supplies fresh water to Singapore. And the other thing is many of you have experienced the automatic traffic system which, while not exactly solving the traffic problem, makes things much better in Singapore than most large cities. So, one of the points I'll try to make is that once you could put the appropriate boundaries in place, and notice that each of these has a filtering operation, then many of the results you desire are automatic from that point onward.

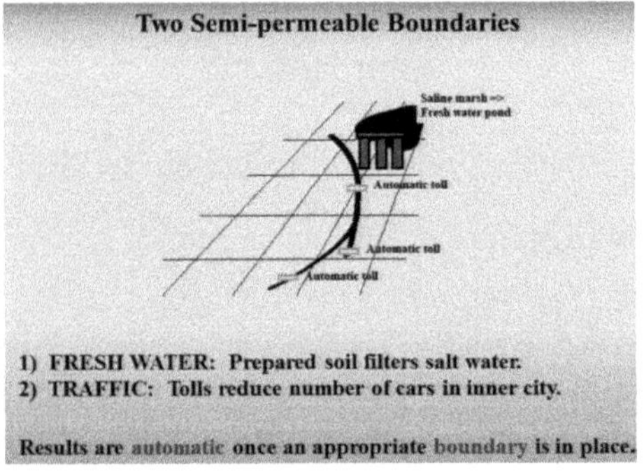

When I was in high school, biology texts sketched a biological cell as a thing full of a jelly called cytoplasm with no structure at all. And now, and this is a very crude sketch but shows you just how much structure there is in the cell and it's very crowded with boundaries of all kinds. One of the points I will make is that boundaries give rise to specialists. And this is an important thing in many different ways.

Now, some subset of you have seen this diagram way too many times but here is a diagram of a complex adaptive system where you have multiple agents interacting and adapting or learning, and because the way the agents interact is nonlinear, that is, they work typically by conditional if-then rules, the aggregate behavior is not, as Murray and others have pointed out, it is not the simple sum of the behavior of the individuals involved. Another thing that's important in these systems is that the influence is both top-down and bottom-up. It's not one or the other.

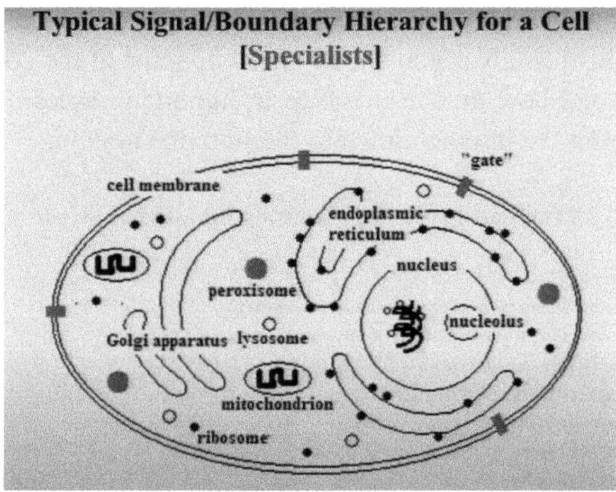

So, if I look at something like the stock market, the Dow Jones average in the United States influences the individual agents; if the market average is rising, some agents will buy, some will sell, so there's this top-down influence, and yet the market average is an average of their net summed behavior. So, these two things interact in important ways and all of these interactions are mediated by boundaries. And I'm going to use the term semipermeable boundary to emphasize the filtering aspect of these boundaries — that they let some things through but not everything.

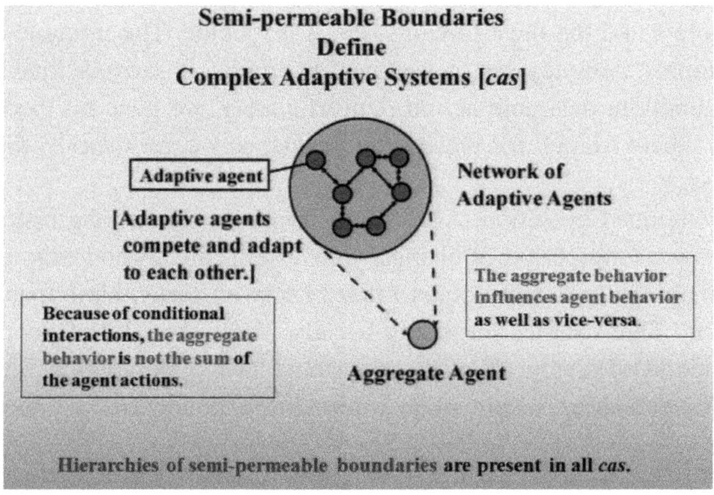

And just to give you some examples, and I won't spend a lot of time on this, but this just gives you a kind of list of some of the complex adaptive systems and how they're involved in important issues of the day — everything from the internet down to the immune system.

Problems Involving Complex Adaptive Systems

Encouraging innovation in dynamic economies.
Controlling the internet (e.g. controlling viruses and spam).
Predicting changes in global trade.
Understanding markets.
Providing for sustainable human growth.
Preserving ecosystems.
Strengthening the immune system.

Each problem involves many interacting agents (components) that learn or adapt.
Semi-permeable boundaries set limits on the interactions.

In each of these problems, you have many interacting agents and they all change over time. They do learn or adapt. And the semipermeable boundaries constrain the interactions. And the questions of course are these: How and why? And is this a good procedure?

Just so I can give you some examples of the things we don't know, all the complex adaptive systems we've looked at involve something I'll call lever points and the most obvious one is a vaccine. The immune system is extremely complex but a vaccine is simple; it gives a long-lasting directed action, desirable action. Unfortunately, we have no theory that tells us where to look for these lever points, so we get them by God and by guess.

Open-ended evolution: All of these systems have a long history and they show a long period of change. Ecosystems are a good example. We have no computer models, at least that I know of, that exhibit open-ended evolution. That's rather surprising because there are a lot of people that have tried. And finally, all of these systems have a hierarchical organization, of boundaries within boundaries within boundaries — biological cells, like the one I just drew an example. And again, there's no theory or general model that tells us what are the mechanisms that cause these boundaries to form. How do they form?

> **(Three Examples)**
>
> 1) All *cas* exhibit lever points – **points where a** simple intervention causes a lasting, directed effect.
>   **Example: vaccines.**
>   **There is no theory that tells us where or** how to look for lever points.
>
> 2) Open-ended evolution **is typical of** *cas* – **an initially simple system** exhibits increasing diversity of interaction and signaling.
>   **Example: ecosystems.**
>   **There are no models that exhibit** open-ended evolution.
>
> 3) All *cas* **have a** hierarchical organization **of boundaries enclosing boundaries.**

And just to point out again some examples of the relations between boundaries and what I'll call mechanisms, which are what typically generate the boundaries and the effects, I've just put up a few things and again most of these are familiar. Let me talk about the last one, which deals with language. Where we have communities, we have utterances as the mechanisms of communication and they coordinate the activity of the community. Now, just to give an example of this, which always amuses me, Chimpanzees have a set of utterances that they use in various circumstances and one circumstance is when there is another tribe of chimpanzees invading their territory. And they have a warning call. Now, chimpanzees are a male-driven society or tribe. The males do everything — they get the choice of all the food and the alpha male drives the group in various ways. So, when these foreign groups of chimpanzees come into the area, the alpha male and all the other males take up shouting this warning cry that there are invaders. Occasionally, and I love this, occasionally, one of the females in the group will pick up the idea of shouting that there is this group coming in when there is no group; the males rush off to tackle the group, the female rushes and grabs all the food and goes the opposite way. That's just a lovely example. They say that you can't lie unless you have language, but in that case, I think we have to face the fact that the chimpanzees have language, or at least the females do.

Some Effects of Semi-permeable Boundaries in *cas*

Boundary	Mechanism ['gate']	Effect
Markets & Manufacturing	"buyers" & "sellers"	increased efficiency
Ecological niche	species interactions (sight, sound, etc.)	recirculation of resources
Cell nucleus (protein-encased chromosome)	conditional gene transcription	chromosome acts like a computer program
Communities (tribes) [national boundary]	utterances [visa]	coordinated activity [" them" and "us", selective immigration]

Okay, so I will discuss this notion of a boundary that lets some things through and some not in terms of the notion that there is a condition in the boundary that says which kinds of identifiers I require on some chemical or reactant or signal in order to let it through the boundary. So, in this case, we have the thing that's coming in biology that would be typically called wide-end and we have the receptor in the membrane and this thing is looking for a particular site and it says, I don't care what else is there. So, it's just picking out a particular site on the signal. This could be effectively an address. And it says if it has the right address, it can go through.

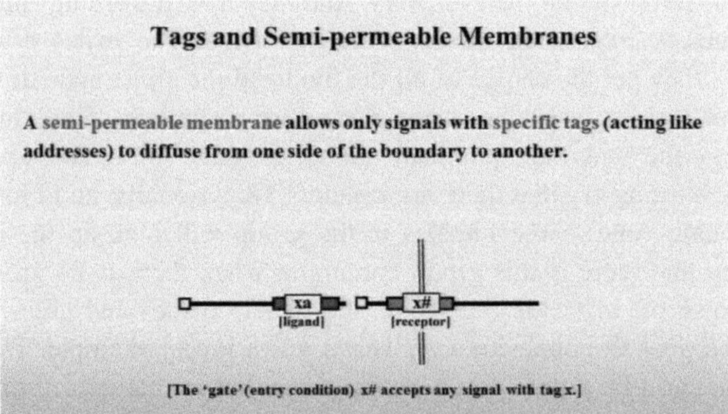

These things I'm going to call tags to have a general name for them and here are some examples in standard areas where we can talk about, say, in the case of the cell, the protein signals and in the case of food webs and so on. In each of these cases, it's rather surprising that we have this notion of tags, that there is a small, we could call it, active site that determines what happens — not the whole thing just a small part.

Tags Mediate Signal/Boundary Interactions

In a typical *cas* there are spatially distributed sets of interactions between signals and boundaries. Tags, small parts of the signals, typically direct the interactions.

Signal-processing system	Tag
protein signalling in a cell	ligands, active sites
Food webs in an ecosystem	images, pheromones
Situated language acquisition	intonation, gesture
Trade in economies.	'headers' on buy and sell orders.
Central nervous system	synapse values in neural network

And one thing that was pointed out and emphasized yesterday that I take as very important is that in order to understand these organizations much as in evolutionary biology or anyplace else, we have to know something of how it came about — how this organization rolls. And you'll notice that a lot of the things that we call now big data are really just snapshots. They're the way the system looks right now, and the diagram that we had at the very start which showed this complex network is a snapshot. It doesn't tell us much about how the thing changes, it tells how it looks right now. And to understand that diagram, I'm going to make the claim that you've got to know more than that. You've got to know how that diagram came to be, how it evolved.

The Temporal Dimension is Critical

Organization arises from the co-evolution of signals and boundaries.
Darwin's tree of life.

A 'Snapshot' (such as the current network of agent interactions) is inadequate
(Music is meaningless if presented as a series of snapshots.)

It is important to discover the mechanisms of co-evolution.

In complex adaptive systems, we have this notion that we should look for mechanisms that cause the signals and the boundaries to co-evolve. They work back and forth with each other. Now, there's a nice idea Christian Mortensen said — in language, that not only do we adapt to language but language adapts to us. And his experiment is a very pretty thing. He simply takes various sounds and lets them go around a group. Many of you have played this whispering game where you sit around a table and you whisper something to the person next to you and that person whispers on till it goes around the table and almost always it comes out to be something quite different from what started out. He does this with phonemes, with individual utterance sounds, and some of them survive much better in this circle than others. So, in this sense, his idea is that the sounds that we tend to use have adapted to our nervous system. So, it goes both ways. I find this a very pretty idea and very suggestive for other areas in complex adaptive systems.

Another thing that we talked about yesterday and that to me is important is that we're not trying to optimize, we're trying to improve in many cases, we're trying to steer the system. We're not trying to somehow decide, "Oh this is the best." And it's easily shown that even for very fundamental parts of a biological cell, one fundamental part is the Krebs cycle which is in every aerobic organism. The Krebs cycle you would think would be highly refined. Now, it is pretty efficient but in fact a modern biochemist can go in there and design a version of that cycle that's more efficient. Okay. So, it's adequate to the cause, more than adequate, but not optimized. And this is important to keep in mind I think when we talk about complex adaptive systems.

Another thing is that most of us, in particular, in my own training, we like the idea of matrices, fixed points, eigenvalues, eigenvectors and so on, and basins of attraction, but in these systems, it's relatively easy to show that the basins of attraction change over time as the system evolves. So, even there, we have to be very careful. Now, yesterday, it was mentioned, and I like this idea, that metaphors can play an important role in science and especially in these complex systems where it's difficult to make connections sometimes. And the thing that I like about this is that if you read the collected papers of James Clark Maxwell, you will find that he's one of the very few scientists who've written very carefully about how his ideas came about. And metaphor was a major technique that he used, and I have just given one example, that he went from the notion of tow lines familiar in the canals in Britain at the time to the notion of lines of force. And he goes on in this way, talking about vortices and so on in a way that's very nice, and if you want to find out about how one of the world's major scientist thought and came up with these incredibly elegant set of equations, this is a good place to read.

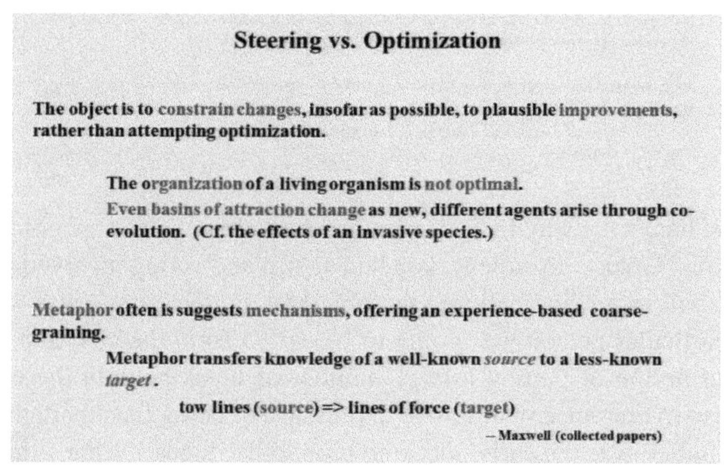

Now, we mentioned tags, these particular sites, and I will claim that tags provide a way of steering signals through these membranes in the system. And so, if I can change the tags, I can change the network of interactions. And so, we're going to talk a bit about that. If I put a new tag in a system like a biological cell, then in effect I have suggested a new routing through the membranes in the cell. Now, that may or may not be any good, but the nice thing about these complex adaptive systems is they involve many agents, so they're highly parallel. So, I can divert a small part of the system much as, say, an ant, and in that nest, one ant can go wandering off and may find a new source or may not. But the hive or the group or the nest as a whole continues to work pretty efficiently. We can find things like this in complex adaptive systems. So, my suggestion is this: I can afford to try new tags because the system is highly parallel.

Tags Provide Steering

By controlling entry and exit, tags steer signals and resources **through boundary hierarchies.**

    Examples:

        (i) **the** active sites **on catalysts and** gates **on semi-permeable biological membranes concentrate particular reactants;**

        (ii) **the title 'chief financial officer' confers a range of powers for controlling a firm's budget.**

New tags, **by generating new pathways, amount to** hypotheses for ways to steer **signals and resources through the agent's** boundary hierarchy.

In systems with many agents, new tags can be tried without seriously disrupting the system.

    New tags only disrupt a small part of the flow, unless they become widespread through adaptation.

And here's the way I'm going to go about this. We're now going back to classical Greece. In Athens, you had an urn and voting involved putting a black ball or a white ball into the urn. And a typical issue was whether some particular person was going to be exiled from the city, and it is in fact that notion of putting a large number of black balls in the urn that gives rise to our current still-alive expression of being blackballed. Okay, so this notion was the early notion in probability theory. What's the probability that if I reach into this urn at random that I'm going to get a black ball? And so, in the late Middle Ages, I can't remember which century, but people started using this notion of balls and urns as a way of getting at the notion of probability. The proportion of the balls of each color in the urn was the probability, or the concentration, and these are interchangeable. So, we're going to use this idea but we're going to modify the urns so they have dates. The urn will only let certain colors of balls go in. So, if I am in classical Greece and I don't want to be exiled, I forbid anybody to put black balls in the urn. Now, if I think of semipermeable membranes, this means that certain proteins can go through the membrane and some cannot. And we're just going to imitate this with the urns, where, in fact, then the concentration of each color of ball is the concentration of each different kind of reactant, say, in the biological cell. It's a simple idea.

And then in order to get the effect of multiple agents, we're going to put these urns in a geometry and we will allow diffusion to take place by simply reaching into one urn, picking out a ball, and putting it in some nearby urn. Now, notice if I have these gates, I can reach into one urn and say pick a black ball and then find that it won't be allowed in the adjacent urn, in which case it just stays where it started. Okay, now I'm going to make the point as we go along that with this very simple model, it's possible to do some real mathematics. And more to the point, and I won't spend time on this, but you can use the theory of Markov processes to good advantage here.

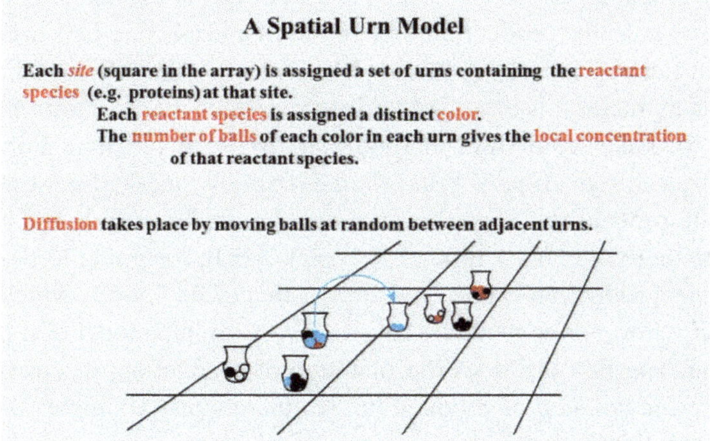

My agent says that every time I add an equation to a book, I cut the audience by half. I'm now giving permission to at least half of you to fall asleep. What I'm going to look at is a very simple example. It actually does the calculations in front of you. This is easy arithmetic. I'm going to look at two urns where there are two irreversible reactions. And in the first urn, I have A plus B yields a new thing Y, and then Y can go through the gate and come into the gate of an adjacent urn where Y plus Y produces C. Okay, so that's all there is to it.

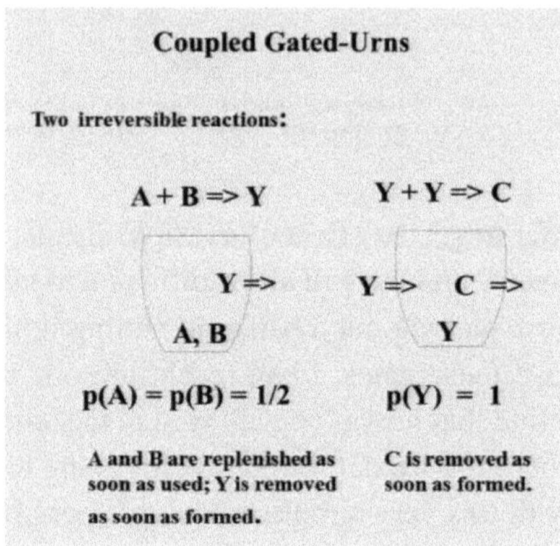

We use now something that I'm sure all of you have seen many times, elementary chemistry, where the proportion of A and B in the first urn is one-half, the production of Y in the first urn is one-half times one-half, but I can either take A first or B first, so BA and AB, and that's the reason why factor 2 is there. And this is a mistake that still after 84 years I tend to make from time to time and forget to put that 2 in there. Anyhow, so, production of Y in urn one is simply one-half per time step, that's the probability. Then, because urn 2 contains only Y, the production rate there is 1. It's 1 times 1. If I pick a ball, it's going to be Y. And so, the net production is one-half times 1 that of the 2 urns, which is just one-half. Now, notice what we have is a very simple model of a production line. The first urn does the first step of production, the second urn does the second step of production. So, there's just a simple two-stage production line.

Reactions in Coupled Gated-Urns

[Reactions result from random collisions – the "billiard ball" mechanics of elementary chemistry]

Reactions in the coupled urns:
$$p(a) = p(b) = \tfrac{1}{2}$$
Production of Y in urn 1: $2p(A)p(B) = \tfrac{1}{2}$
Production of C in urn 2: $P(Y)P(Y) = 1$

Net production of C per time-step: $2p(A)p(B)\,P(Y)P(Y) = 1/2$

I want to recall for you something that most of you have encountered one way or another, one of Adam Smith's major examples was talking about, at his time, how they had just, in effect, produced the first production line. It used to be that blacksmiths would produce a straight pin just for pinning things and that straight pin was a highly expensive item; only if you were in the upper-middle class could you afford to buy a straight pin. At the time that Smith was writing, specialists took over — one blacksmith would draw the wire, another would clip it, another would sharpen it, and another would add the head of the pin. This increased the production rate of straight pins by a factor of 10 and made them a common item available to most people. So, this business of going from the general craftsman to a production line of specialists made a huge difference.

I'm going to try to illustrate that point now. So, now we're going to take a single urn. This is a craftsman blacksmith but does everything. So, all of the reactions, both the A plus B yields Y and the Y plus Y yields C, take place in a single urn. Now, at this point, you notice that they're in equal amounts in their urn, so one-third and one-third and one-third, they're just in equal proportions. And so, we do the same calculation, it comes, one-third times one-third times two is two-ninths, and then, we do the calculation for Y. Now, notice Y is only a third of the population, so the rate is much slowed, one-third times one-third and that yields on-ninth. To be fair, we use two urns — two blacksmiths in the specialist line — so we're going to compare this to two that are working side by side. What's the difference between the two specialists and two working side by side? We see, lo and behold, a factor of 10. With this very simple model, we get this huge speed up by putting some specialists in there.

Comparison of Throughput in a Pair of Coupled Urns
with
Throughput in a Pair of Independent Urns

Throughput (net production of C) in a single urn:
$$p(A) = p(B) = p(Y) = 1/3$$
Production of Y: $2p(A)p(B) = 2/9$
Production of C: $P(Y)P(Y) = (1/3)(1/3) = 1/9$
Production of C per time-step in a single urn: $2p(A)p(B)P(Y)P(Y) = 2/81 \sim 1/40$

Production of C per time-step in <u>two</u> uncoupled urns: $2/40 \sim 1/20$

[from previous slide]
Production of C per time-step in a pair of coupled urns: $1/2$

Speedup (provided by coupling) $(1/2)/(1/20) = 10.$

[Cf. Adam Smith's 'pin factory']

Now, my claim is that I can do some much more sophisticated things with multiple urns and hierarchies of urns where I can still do reasonable calculations and still find out something about what's going on in these complex adaptive systems.

And one of the things that we would like to look at is how these things co-evolve over time to produce specialists. How is it that we can change the tags and urns where we had independent urns and suddenly the urns become coupled? Because if they become coupled, then by just ordinary biology, Darwinian Fitness, this increased efficiency is going to favor them in any set of interactions that they're involved in.

And at least some of you know that, at this point, I began shouting about recombination in contrast to mutation. In most even in current biology textbooks, for even undergraduates, it is usually said that mutation drives evolution. I'm just going to say, not so. Mutation takes place about one in every million to ten million times in a given place in the chromosome. Everything in the DNA is there to prevent mutation, to make it a stable chemical. Okay. What about recombination? How frequent is something called a crossover? Why is it that you look partly like one parent and partly like the other? Every one of us is a combination of some genes from one parent and some genes from the other. This is recombination. If you look at two successive generations, nobody in the second generation is identical to anybody in the previous generation. This is all because of crossover or recombination. And, of course, humans have known about this for a long time. I mean, they have used it. When we breed horses or cattle or something like that, you try to combine those elements, those animals that have good characteristics to get animals with combinations of those characteristics, and this has been known for a very long time. My claim is that evolution is driven primarily by the crossover and recombination that occur in every individual, in every generation, rather than mutation, which is extremely rare. Now, there's a good question here which I'm not going to answer for you right off: If I have this recombination and no individual in the next generation is like any individual in the previous generation, we only get one Einstein, we only get one Murray Gell-Mann; what information is being preserved from one generation to the next? Since we're not preserving details, what's being preserved?

Co-evolution in Gated-Urn Models

(1) A population of tagged urns generates a directed interaction network.

(2) Changes in the tags used by the urns' entry/exit conditions change the network of interactions.

(3) Recombination of entry/exit conditions (using, say, a genetic algorithm) modifies the network by recombining parts of extant tags to generate new conditions.
New urns can establish new niches, or enlarge existing niches.

I'm going to talk now in terms of recombination, of building blocks. These pieces, and you can think of them as tags or parts of tags that are being recombined to yield new tags. And one thing we know from genomics is what they call their "motifs," are building blocks that occur in many different functional combinations.

Alright, and this is a slide that's awfully familiar to some of you. This is simply to make this point about the role of building blocks.

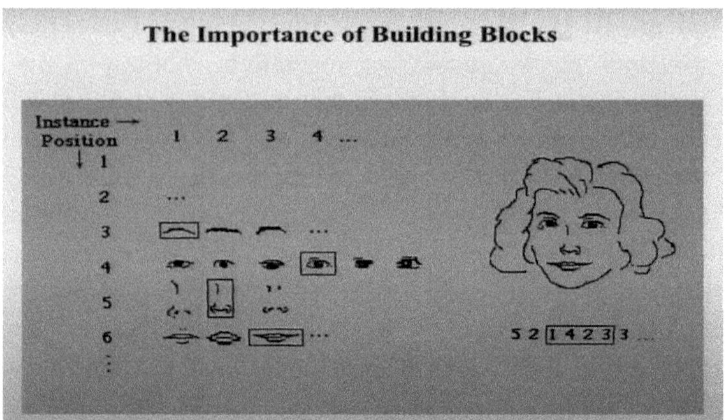

What I'm going to do is take the human face and I'm going to divide it into ten different features — hairstyle, shape of the forehead, eyebrow shape, eye shape, and so on. Ten different features are layered. And then I'm going to give ten alternatives for each feature. So, in here, we have ten different eye shapes, okay, ten different eyebrow shapes, and so on. This face was created by picking one building block for each feature. This style of hair, this style of eyebrow, and so on.

Okay, now the question that I love to ask my classes, really two questions: How many building blocks are there and how many faces can I make from them? First off, how many building blocks? I hear some people saying 100. There are 10 sacks with 10 blocks which comes out to be a hundred building blocks. How many faces? I can pick one of ten from the first sack, one of ten from the, … ten times ten times ten…, ten to the tenth power. 10 billion faces, all different, all legitimate faces from one hundred building blocks. They claim that's what politicians use to recognize all the people they know. And by the way, this is something you can test, as to whether you have the right building blocks: Unless you've actually dealt with chimpanzees in a zoo, they're going to look pretty much alike to you. You're going to have a hard time distinguishing their faces because they have different building blocks by and large for their faces from those that we have and you just haven't had any practice in those building blocks. So, building blocks are important and when we come to talk about building blocks for tags in this business of getting through these urns with different kinds, I would claim that recombinations of the right kinds of building blocks are signals, and the membranes or boundaries co-evolve, recombining those building blocks. Let me make new gates for the urns and new signals to go through those gates.

So, another point that fits with this is that when we look at these semi-permeable boundaries, they tend, as we saw in that simple example, to increase the concentration of selected signals or selected reactants. By increasing the concentration, you increase the rate at which they interact or communicate with each other. So, the net effect is all these semiperme-able boundaries in the cell acts much like catalysts: They speed up reac-tions of the selected reactants and that's an important effect.

So, quickly then, what we've been saying is that when we look at these reactions, they are controlled by tags. And when we look at that, then we can have conditions for these gates on the urns, these semiperme-able membranes with their gates, that let us sort the various kinds of reac-tants. And then, by increasing the concentration of particular reactants, you increase the rate of combination or reaction within that membrane. So, this is the point that I've been sort of making. Actually, if there's some questions at this point, I'm quite willing to take an interruption. Okay, most of you fell asleep when I gave you permission [audience laughing].

> # Summarizing: Reactions, Conditions, & Tags
>
> 1) **A reaction requires a given** combination of tags **(active sites) for each reactant involved.**
>
> 2) **An entry/exit** condition **, using 'don't care' (#) symbols, specifies the particula** combination of tags **required for balls (reactants and signals)** to enter or exit a gated urn.
>
> 3) **Accordingly, when an urn admits balls with** tags required for a given reaction, **the concentrations of the reactants will usually be substantially increased,** increasing the reaction rate.
>
> **Semi-permeable boundaries act much like catalysts.**

Now, I'm not going to go through these, but these are some of the more mathematical things that we can do to study these things. We can, for particular purposes, use difference equations. We can use the notion, the urn model that we talked about, we can use this kind of billiard ball mechanics, Markov processes, and finally we can look at rule-based signal processing. These are all more formal techniques. I'm not going to go through them, it's just to point out that there are ways to handle this other than trying to just do simple models.

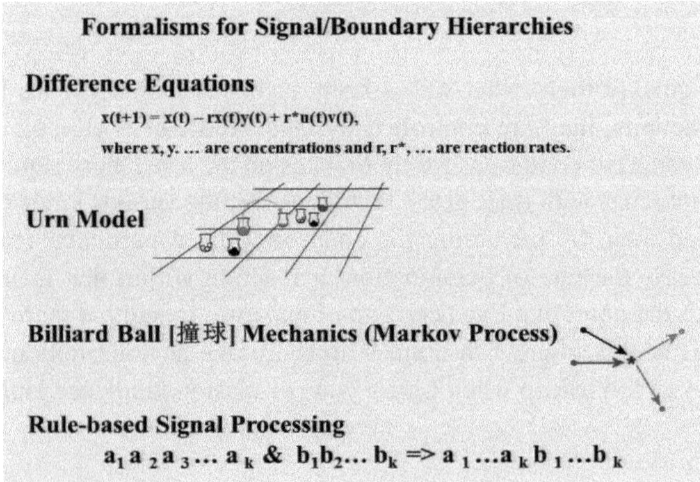

$$a_1 a_2 a_3 \ldots a_k \ \& \ b_1 b_2 \ldots b_k \Rightarrow a_1 \ldots a_k b_1 \ldots b_k$$

And if I went in the right direction, I would like to do this in terms of something that's very hard to implement and I find that an institution I dearly love, for instance, is losing the capacity to do this and this is what I call exploratory modeling, where I have some notion of mechanisms or pieces that I want to explore. I want to see how these mechanisms work together. I'm not trying to imitate data or forecast or do anything. I just want to explore the possibility of these mechanisms.

> **Quick Summary**
>
> 1) **The co-evolution of signals and tags can provide the hierarchical organization typical of *cas*.**
>
> 2) **Exploratory gated-urn models allow us to explore these possibilities.**
>
> [The emergence of language from pre-primate abilities provides a good test of these ideas.]

What can they do? You'll find that in most places it is very hard to fund exploratory research. In the United States, in particular, the National Science Foundation and most of the other funding agencies like the National Institutes of Health want to see what they call deliverables. And that usually means a 3-year horizon. And a 3-year horizon is way too short for big problems.

Discussant A: What happens when in the signaling you have conscious agents like including societies and you can anticipate what the consequence will be and can differentiate different possibilities and make twists across those boundaries?

John Holland: Yes, yes. This is very important and what we find is, for instance, that in these systems, we build what I could call internal models. And so we can look ahead, we can do anticipation, but notice that this doesn't simply require consciousness. *E. coli* anticipates food by swimming up a food gradient. So, anticipation is essential. It plays a big role in this and I will just refer to Helena to answer all such questions [audience laughing]. But this is, for instance, part of the transition from warning calls to the ability to communicate in a forward-looking sense. So, when a tribal leader can say look over there, beyond the hill, beyond the light of sight, this is an important advance and the kind of stuff that I and Helena are working on. We're trying to see what these levels are as you progress from warning calls all the way up to the kind of things that we expect of a child even at, you know, a year and a half. Other questions?

Sheila Ronis: A little earlier in the lecture, you talked about improving or steering systems to make them better. And I guess my question is this: What criteria do you use? Are they normative judgments? You know, I mean, how do you know?

John Holland: Whether it's an improvement?

Sheila Ronis: Yes.

John Holland: Yeah, yeah. Now, I could refer to Peter Ho. Risk assessment, you know, if I think of steering, I mostly think of, I have some knobs that I can twist. This could be, you know, like the control panel, a panel on an airplane. And this has a lot to do. I don't know, in general, when I start out what an optimal flight path is going to be from the US to Singapore. But what I do along the way is some steering in terms of knobs that I can twist, but I might not be doing something optimally, but I can usually measure something, you know, and this is the thing we see on the panel in front of us. You know, what's the headwind, how much fuel am I using, and so on. In general, then, steering is a bit easier because I look and use local criteria rather than global criteria. So, that's part of the answer, but this is a complicated problem. But I do advocate steering. Yes, in fact, the people who commissioned me to write this book for Oxford, that was their first question: Can you distinguish complicated from complex?

Here are a couple of hypotheses that I would like to find out more about. And one of these is that local concentrations induced by feedback and recycling provide opportunities for the formation and adaptive radiation of agents. In other words, these things that the urns allow, for instance, recycling, give the agents opportunities that they wouldn't have otherwise. Now, I mentioned an example that is very urn-like. In a tropical rainforest, we have plants that are cup-like called bromeliads. Now, interestingly, in a rainforest, they grow on the poorest soil in the world because the heavy rain leeches all the nutrients out of the soil into the nearest creek. So, the soil is very poor and if you look at a tropical rainforest, it seems like everything is directed to keeping these nutrients from reaching the ground — all kinds of beasts. The bromeliad is a good example; it catches the water. Frogs, various kinds of insects, and so on breed in that pool of water; the waste products feed the bromeliad, and this whole recycling plays a big role. And so, what we would like to know in more general ways, and you know we talk about recycling all the time. If we have a production process, we'd like to recycle this and you can actually do in these urn models some calculations that show that it does make a difference and so on. That is something that we need to know a lot more about.

Agent formation: We find that in this system we get more and more when we look at them. Observe the real systems — we see more and more kinds of agents. And some of these specializations and adaptations are really just totally remarkable. Darwin had studied this; he wrote a very nice book. If you are interested in orchids, it is one of the best books on orchids that you can read. And in that book on orchids, he discussed an orchid called the Comet orchid. The Comet orchid had a nectar tube [about this long]. He knew from his studies that moths pollinated and fed on that kind of orchid. But that would mean that this moth would have to have a tongue [this long]. People say he never made any predictions, but Darwin actually said there must be such a moth. About ten years ago, for the first time, they filmed that moth in Madagascar. So, it's just, you know, it's a lovely thing that kind of specialization: If either that orchid or that moth disappeared, then both of them would die out. What allows them to sustain themselves in this highly specialized condition? The other thing that we noticed is in these systems unless we have a big meteor coming down to destroy all the dinosaurs, these systems tend to increase in complexity over time — more specialists, as specialists make possible still more specialists and so on. These three things I would like to know a lot more about.

> **Hypotheses to be Tested**
>
> Local concentrations of resources induced by feedback and recycling provide opportunities for the formation and adaptive radiation of agents.
>
> This process of agent formation leads to increasing diversity of agents and progressively larger amounts of resource "tied up" in agents.
>
> Under 'tranquil' conditions, increasing agent specialization should be observed.

We claim that these gated-urn models give us at least one method to explore some of these things. And I hope that some of these kinds of ideas can be applied to this whole question of language acquisition which is still a very open question.

Here are two big problems. One is that there is no comprehensive theory of signal modulation and that is to me a very important problem, whether we're talking about governments or whether we're talking about ecosystems. And this is a problem that is not going to be solved rapidly. This is going to take extended effort over a long period of time. And co-evolution in networks is another thing and here we have lots of good theories about networks in certain forms. So, we have the small world models, we have scale-free networks, and so on, but notice again these are snapshots. These are not how these things develop over time. Again, I think this is a very difficult problem; it's going to take quite a bit of time.

> **Two BIG Problems**
>
> 1) There is no comprehensive theory of the signal-modulated behavior in niches and autonomous systems.
>
>   For example, a theory of mind requires internal models for anticipation of future actions of other agents within the physical world.
>
> 2) There is no over-arching theory of the co-evolution of agent interaction networks and the flow of resources and signals over those networks.
>
>   For example, an increasingly diverse range of carbon exchanges and recycling produces the complex hierarchies of an ecosystem or a biological cell.

Finally, this is the final slide. I claim that even if we're going to take a crude look at the whole, it can't be at one level. It has to be layered in some way or other. And in order to really get this crude look, we've got to look into all of those systems that we know about that are complex adaptive systems that have hierarchies; they have urns within urns within urns. And in order to even look at this crudely, we have to take into account that we're going to miss something critical otherwise. I'm ready for questions.

5.2 Discussion

Discussant B: [inaudible] from School of Biological Sciences NTU. In one of your first slides, you mentioned that there is no program for open-ended evolution. And I would like your thoughts on the program of memes from the book of Richard Dawkins, *The Blind Watch Maker*. So, is that not a good example?

John Holland: Your question raises something that's very important. Yeah, Richard did essentially mutation recall, and the problem there for open-ended evolution is Richard — although he did a few experiments along these lines on human preference for some of these things — did not have a fitness function. The result is that I can get unlimited variation by using a random number generator. The question that's critical is this: What are the selection criteria and then how do these things interact as they're selected? And that's where I don't know of a model. I know of some interesting attempts but they all flattened out way too soon.

Discussant B: Thank you.

Discussant C: So, thank you very much for your talk. I thought it was very interesting how you very elegantly segmented things into those, it was really interesting. So, this is a question that doesn't touch upon your main themes. I made the mistake of claiming that biological systems were optimal to an entire panel of biologists.

John Holland: Uh oh.

Discussant C: And I wanted to clarify because I'd be interested in knowing what your opinion on this is. So, I come from the perspective of optimal control. So, I see everything as being this optimal control problem. And so, the claim that I would make is that rather than saying that something is not optimal, because you need some sort of element of repeatability, I'd be interested in saying that maybe it's just not the same sense of optimal, so maybe it's not energy-conserving, maybe it's not something else, but maybe it's something where it follows some guide, some, like, objective, and that allows us to do or have some element of repeatability. I don't know if you would agree with that statement, or if that fits within what you had in mind when you were saying that.

John Holland: It is within, but it's a statement I would take with great caution. I actually work with a group of people at the Chinese Academy of Math and System Sciences where adaptive control is a major feature of what their interest is. So, adaptive control theory is something that I like and it's fun to work with. Let me try to give an example this way. Take the Krebs cycle. The Krebs cycle is well defined; it's eight enzymes. You could look at it and you could ask what the throughput of the system is. And could I tweak those enzymes a little better to make, you know, what I'd call the plant, to make the plant produce more of the critical chemicals? And the answer is, you can, but there's more to it than that. As I said, the Krebs cycle is amazing. You find it in carrots as well as elephants. It's everywhere, it's ubiquitous. Now, this is back to this notion of building blocks. The thing that, so to speak this is my conjecture, cost the system to stop before fine-tuning this any further was that it has attachments that let it attach to other reactions and it was better to fix this and the attachments around it so that it could be used as a building block one level higher. It was better to have a standard building block than to keep fine-tuning this and not have a standard building block. So, the claim is that this is what happened evolutionarily. Now, if you were careful, I think by putting in the layers, you might come up with a new definition of optimum that would encompass this, but I don't know of one as yet.

Martin Riser: I'm Martin Riser from NTU School of Art Design and Media. I did not understand your point about mutation or driving evolution. Recombination would account for the variability of different individuals in a gene pool, but for a gene pool to evolve into a new species, as I understand the theory, requires mutation. So, why do you say mutation does not account for evolution?

John Holland: Okay, let me first make a couple of points along that line. We now know, although relatively recently we did not, that there's a free exchange of genes between bacteria. Horizontal transfer, it's called. It is massive, it's massive. And the genes that move around actually give rise to different capacities, different kinds of bacteria. So, even if I fix the pool of genes and allow this exchange, I'm going to get new species at the level of bacteria. There's another side point here that's worth noting. We tend, how should I say this, to look down on bacteria, but in fact, 90%, by weight, of the bio-carbon flow in our world is through bacteria and we know so little about bacteria. We take a cubic meter of soil and most of the bacteria can't be cultivated by normal circumstances. We know so little about bacteria that in effect we're like a CEO in a corporation that is trying to budget things not knowing 90% of the budget. It's really incredible that we don't pay more attention to that kind of thing. Now, going back to your point. One could ask quite reasonably this question: Can I get new species by recombination alone? My claim is yes indeed. And these variants that we get, originally, gradually, if they get isolated, and these are many things, but if I have these barriers, these semipermeable membranes, then I can get separate kinds of mixing of genes just in different boundaries. So, my claim would be that mutation is critical but only for replacing genes that get lost. If I don't have the gene I can't recombine it. So, I can afford a very low rate of mutation to get these new genes, but most of evolution and most of the fitness changes come from recombination. That's my claim.

Martin Riser: Then, I have a second question. Can you give some examples of the application of this very interesting urn model? Somewhere it was applied to a problem successfully.

John Holland: It's so new that, at least, to me, the closest I've come is like the calculation of the coupled urn versus the pair of uncoupled urns. I think that's a fruitful direction to move and I would try to do some simulations along that line as well with more complicated versions.

Martin Riser: Well, let's form an institute.

Peter Sloot: Before we form an institute, I have two more questions. John here I am.

John Holland: Oh Peter, hey…

Peter Sloot: So first, the reaction to this thing about the mutation. You're right about mentioning the jumping genes. They seem to be much more important than we ever thought. In addition to that is the epigenetic effect, where we can actually get, kind of, Lamarckian evolution going on within, you know, one or two generations, so that's really remarkable. That was a remark. I have actually a question coming back to this optimization and control if you like. Personally, I don't believe in that, and I want to have your opinion on it because of the following thought: Most biological systems operate completely out of equilibrium, and if there were some kind of optimum and some disturbances come along, the whole system would collapse. You would have a cascade of errors propagating through the system, and that's what we see in metabolic regulating systems there, where, you know, you can actually disturb them and they somehow find another way to operate. And that would be your urn model, which could really help us start reasoning about these things. We have never done that. We've never really thought about if you have your urn model as a production chain or whatever you want to call it, and then we add noise to it, how would that say the signal transmission still survives that noise because we know noise is critical? It's not just an annoyance, but it should be there. It might be there to actually drive the operation of this, of the system. There might be a co-evolution, if you like, of noise and order at the same time. And we never were able to study these things but maybe this urn model could give us some, you know, some way of thinking to deal with that; so what's your opinion on this?

John Holland: I would hope so very much and I think as part of the reason you could see how you might imitate some of this. I mentioned Morton Christensen's whispering around the circle kind of thing, and I think you could imitate that thing with a bunch of urns and selective transmission. One thing that goes along with that is that I find perhaps part of the reason tags are so prevalent is because they're shorter and therefore more resistant to noise and also more easy to recover if they get a little bit distorted. So, I don't know. That's again a conjecture that one might try.

Peter Sloot: Maybe this is to get us thinking — we try to think where we send noise in through a network, a metabolic network or whatever you have, and then look how the signal transport can actually be enhanced by the noise through some kind of stochastic resonance that you get. And it seems to be very crucial, so maybe we can take this offline and discuss this. I would love to do that with your urn model.

John Holland: Yeah, yeah, sounds like fun. Yeah.

Peter Breton: Hi, Peter Breton. I was wondering if you can elaborate more on how we can interpret the boundaries and the signals in the social system. And also whether it has to be a physical system or if it could be a non-physical system.

John Holland: That's the latter, first, certainly not physical. And my being exposed to language, I think about dialects. In Southeast China, there are 200 so-called dialects, but in Europe, they would be languages. People that are 50 miles apart along the same river will not understand their dialects. So language in that case is the boundary, and it filters. it's not as simple as that, but there's this "us and them" kind of thing again. I'm going to trust the people more who speak my dialect than the people who don't. And you know, there was a time in the United States when we had all these Native American tribes that spoke a great range of dialects. It's interesting that if they were pushed, then there was an almost universal sign language that Native Americans used to communicate, and this was for certain kinds of trade. But then when it came down to the critical issues as to whether we're going to war on that tribe over there, that was dialectical. So, boundaries, you know, they in effect need not be physical, they could affect some things and not others, and all we have to do is look at the Congress of the United States to see how this works.

Greg Fischer: Greg Fischer. John, can we go back to the discussion about anticipation? I'm thinking specifically about the problem of reflexivity. Okay. So, if we're interacting, I'm anticipating you and you're anticipating me, and you get a kind of regret. Would you say, it's my view, that reflexivity, that problem, is a greater, more significant problem in human systems because our imagination allows us to create this sort of reflexive problem? Would you agree with that? That reflexivity is a greater problem or a greater issue in human systems than it is in natural systems?

John Holland: I'll start by trying to be a little more, you know… which we don't really have time for, but defining what reflexivity is — let me use another set of terms to try to get at what I think is a very similar issue. If I look at the central nervous system, I have what typically is called feed-forward through several synapses from, you know, eyes, three sets of synapses there, lateral geniculate, and then finally to the optic lobe. But then, once I get beyond there, I get reflexivity or recirculation as spikes come back in circles and so on, and this was the best, still the best book, biased, in psychology that tries to relate macro behavior to the central nervous system, in my opinion, is the organization of behavior, written, you know, more than half a century ago. But very, very carefully thought out. Well, alright. Now, if I'm talking about this, one thing that humans are very good at — they're not the only ones, but they are good at — is this business of being able to run things ahead internally. I use the term autonomy. That means that that section of your nervous system is only modulated by the input. It's not determined by it. And so, I do these things called look ahead and, you know, it shows up especially if I'm trying to play a game or something, but it's all the time. Well, what we would like to know more of, and maybe the urn models could help even in this, is to know how do these autonomous subsystems arise. And we do know things, which is also part of reflexivity. While this incoming information is mostly feed-forward, there is stuff that comes down that fine-tunes it and says, "Well, I'm looking for this kind of thing." I think that's an important part of language. And some of these experiments in language are incredible, where you take a sentence, you keep the first letter and the last letter of a word, and scramble everything in between, and you'll find that you can almost always read the sentence.

Brian Arthur: Brian Arthur. John, brilliant as usual.

John Holland: He says that and then he hits me.

Brian Arthur: I'm really taken by this and I think that trying to push you a little bit — you stopped short of saying that what you're trying to do is put a finger on or talk about a major mechanism of evolution that hasn't been talked about that much and that's the evolution of biological forms via modules and a hierarchy of modules. And I'm wondering if I can get you to stick your neck out and make that claim. You know that's an idea that goes way back as well, and we've even heard Simon talk about them in the sixties. Another comment is that there is another system that works by the evolution of modules and hierarchies, and that is the evolution of technology itself. For example, the laser printer invented in 1972–1973 was put together by getting a laser to write images on a Xerox drum and using a computer to direct the laser to do that. So, you've got modules getting strung together all the time, and getting combined and recombined, making ever new technologies. But would you stick your neck out and claim that you're pushing evolution a little farther, please?

John Holland: Yeah, restate the very first part of that Brian, so I answer it more directly.

Brian Arthur: A lot of the work on evolution and neo-Darwinism has really been, you know, a gradual accumulation of small changes and then bifurcation into species and so on. But what people haven't talked about anywhere near as much is evolution via modules, building blocks, and hierarchies of modules. Now, this seems to be what you're doing.

John Holland: Yeah, and this broad range of problems is typically called morphogenesis — how do I start from a single fertilized cell and build a multicellular organism where every cell in that organism has the same chromosome, the same building plan? Interestingly enough, one of the most significant papers written in this area was by Turing, the same man who invented computers on morphogenesis. It's a very simple idea: You have two chemicals, one of which presumably yields one color and the other which presumably yields the other. One of the chemicals is auto-catalytic in the sense that it speeds up the formation of itself and the other chemical, and diffuses rapidly. The other is an inhibitor that stops these actions and diffuses slowly. You start with a homogeneous system and you run it, and you could actually do, in this case, differential equations, and you wind up with something with spots. Then, if you change the shape, it winds up with stripes instead of spots and so on. It's a nice thing but nobody really followed up on that paper. I would hope that by putting urns within urns, you could in fact watch the formation of a multicellular, multi-urn thing with hierarchy. But, you know, that's conjecture. I mean, the format is there. The question is, what would happen, whether it would be interesting.

Maarten Boasson: Maarten Boasson from the Netherlands. Thank you, John, for formalizing these ideas that have been floating around. I would like to take this one little step further maybe. Assume that the filters allow for errors and let through signals that almost fit the filter. If that ability of false signals to get through your boundaries is not too big, it will probably not disturb your system too much; but do you have any idea how far you can take that and at what level the system will break down?

John Holland: I'm afraid I've got a quick answer: I don't. I think that's one of those things that I would like to do, and it would be very productive with an exploratory model, to get some initial idea of what to look for. This happens in biological cells often.

Maarten Boasson: Even in technical systems?

John Holland: Yeah, and even in music.

Maarten Boasson: You'll hear it tonight here.

Martin Riser: There is a discussion in this context of complex adaptive systems, of the question of reductionism or the method of reductionism. And I would very much like your point of view. And if I look at the urn model, it is a reductionist model, in small particles, I can understand, which in interaction have emergence. But some people say reductionism must be given up in order to understand a complex adaptive system. Please comment on this question.

John Holland: This question is tied up with the notion of emergence and whether emergence is something that's holistic or whether I can show it as something coming from beneath. Now, again, I attribute this to Murray. Murray says that if you go three orders of magnitude in any direction, you have a new science. So, quartz, to nucleons, to atoms, each time you've got about three orders of magnitude and you have a new science. Now, the question is this: What's the relation of the new science to the old science? And in my terms, this is very much a matter where the laws at the lowest level constrain what can happen at higher levels — I don't dare violate them; I'll be unhappy, and so will those scientists if I violate laws at a lower level. But I still have room. So, if I talk about, say, a water molecule and then I talk about fluid flow, I use the Navier–Stokes equations for fluid flow. That wouldn't make any sense for individual molecules any more than it would make sense to call an individual water molecule wet. Wetness is an emergent aggregate property. The properties of H_2O constrain what can happen in the fluid, but they don't determine it. So, my view of reductionism is that you have these layers, you have room for new laws at new levels, and that reduction isn't in the same sense that I can't violate laws at lower, more fundamental levels, but I can still have Navier–Stokes equations, and I can still have, even with physical things, this notion of emergence, that wetness is an aggregate property. It doesn't make any sense to call H_2O wet any more than it would to call it blue. And that's the kind of thing — that's the way I tend to look at it.

Robert Axelrod: John, do you think that…

John Holland: When your friends attack you, you really are in a problem.

Robert Axelrod: When you assume that a question will be an attack you've got a problem.

John Holland: Yeah, yeah.

Robert Axelrod: Actually, my question was, what more do you think you would have to add to your urn models in order to have open-ended evolution?

John Holland: One thing I would have to discuss, and I didn't talk about it much here but I've actually looked at some of that — it's even in the book — is how to have hierarchies of urns. Urns within urns within urns. I think in fact...

Robert Axelrod: And you get that already?

John Holland: Yeah, yeah, I can. I have to be careful when I do it if I'm still going to use Markov mathematics because urns are not balls. So, if I put an urn within the urn, I've got to figure out how to do that. But there is a way to do that. Okay. So, I would like to keep that possibility for analysis. So, my claim would be, Bob, that you really have to get to this question or morphogenesis to get open-ended evolution. You really have to do something about having multi-cells, and that gives you agents made of agents — that kind of notion. Now, this is even true of individual cells. You know, the Margulis theory was that free-living organisms got involved, actually engulfed by early bacteria — they weren't quite that — and so the mitochondrion has its own DNA. It has the remnants of what it was as a free-living organism but it in fact was... you got this extra structure, this extra boundary by simply taking it in and adapting it to the local.

Robert Axelrod: But can't you get boundaries within boundaries of the system you've already got?

John Holland: Yeah.

Robert Axelrod: And therefore, my question again is, what else would you need to get open-ended evolution? You say you need hierarchies that you've got already. What more do you need?

John Holland: No. I'm at the stage where I believe I can do it. Yeah, but that's where the exploratory model comes in. I'm optimistic but great. You know me well enough to know I'm always optimistic.

Song Ji Young: Song Ji Young from Singapore Management University. I'm a political scientist myself, but with a mathematical background, so you can tell me all about the mathematical equations; I can process some of your ideas. Let me give you an example. I'm working on irregular migration, trying to use complexity theory, and I think your recent book, *Signals and Boundaries*, has become sort of my Bible for applying some of the very fundamental concepts listed here. So, it's also quite similar to some of the previous questions raised about the agent in the system and the multi-layered system within. My problem arises when the system becomes an agent. For example, in migration, you have a state system, but individuals working on different levels, and I can probably use some of the concepts of a semipermeable membrane to understand this human migration political system. There is regular migration, but irregular migration such as human trafficking, smuggling, undocumented labor migrants, refugees, and stateless people — they move through these small, tiny bits of semipermeable membranes, but the state system doesn't allow these people's irregular movement. When the state system becomes an agent, trying to, as you call it, "steer," "manipulate," or, as New Johnson says, "tweak" the system to narrow down the semipermeable membrane so that these irregular migrations don't happen, it's also relevant to the previous discussion about the open-ended system. Some systems are open, but not every system is as open as the others. And there's a political question: When a system becomes an agent and tries to manipulate or tweak the system, there's a question about the normative justification for this tweaking behavior or the system becoming the agent. That's my first question; do you have any comments on that?

The second question is about the whole idea of complexity systems using evolutionary biology. The evolutionary idea, Darwinism, let's say, it only works for the survival of the fittest. But what about the survival of the unfit? Because we are human beings, we are different from molecules in movement. We have consciousness, purposes, and intentions. Why do we move the way we move? So, that's my second, more normative question for complexity theory.

John Holland: Okay. Let me talk about the first system, and, well, quickly I'll tell a joke. Most of you who've been around young kids reach a point where they say, you know, "What holds the earth up?" and many fathers, in order to short-circuit, or mothers to short-circuit, this question will say, "Well, it's this giant called Atlas that holds the earth up," and then of course, the child, not being dumb about it, says, "Yeah, but where does he stand?" And let's say the father, since I've done this, says, "Oh well, he stands on the back of… the elephant stands on the back of the turtle," and the child, of course, says, "Yeah, but what holds the turtle up?" And the father, if he's smart, says, "It's turtles all the way down." If we try that kind of thing about agents, when a government becomes the agent, the government has certain ideas. We see this, one place where we see this right now is between China and the U.S. There is what Bob would call tit-for-tat. Each one makes the visa harder and harder to get, charges bigger fees, and so forth and so on. Okay, so this — other things get involved [rather] than simply filtering who can come in and out. These are statements of one kind and another too. I think all these things make this very complex. You know, and we're having this big argument in the United States. We have this semipermeable boundary with Mexico. The migrant laborers, as they're often called, are important to the economy, but there's also this political issue. So, I find if we're going to study this kind of thing, we need a fairly rich kind of model, and particularly this means we tend to be awfully cavalier about what we say when we say something is fit. We need a much more multi-dimensional notion of fitness, you know. If not, we wind up with the kinds of things like eugenics and so on, where you try to get rid of the unfit, whatever that means. You know, so I think we have to be much more astute, and here perhaps is where big data can help — if we had much more sampling and data about these things, about what actually happens, who gets through the boundary, and what they do after they get through, what happens to their children and so on. You know we're about to see in China the effects of the one-child policy. But we also know enough about that — that's going to be about a 20-year span and then this thing is going to level out again where there won't be too few young people supporting the older people. But it's very hard to take those into account. And without being overpraised-worthy by the host, I think Singapore does better about this than most countries in the world.

Discussant D: Hi John. I used to be a biophysicist, and I happened to study morphogenesis by simulating the coupling of Turing's reaction-diffusion equation with a lattice-based cellular automata model. Essentially, the reaction-diffusion described what the tags would do, and once the tags are picked up by the cellular automata, it means the urns would allow the tag to go in. This system exhibits very interesting dynamics. Also, I have a question: Did you say that the generation of new tags will drive the evolution of the network?

John Holland: Did I say — repeat it again.

Discussant D: The generation of new tags would indeed drive the evolution of the system. I'm just trying to think of examples in developmental biology where many developmental signals are conserved. So, should I expect a more rapid evolution of the tags through recombination, or is there a lot of conservation there? I have a question here. Also, by evolving from a single-urn model to a chain of urns, of course, you increase the efficiency of the system, but you also get the threat of the risk of breaking down the chain where urns don't communicate with each other. Additionally, you're eliminating the possibility of new reactions, right? Because life originated from a soup where molecules randomly found each other and so on. So, that made me think of the evolution of universities. You know, we have more and more departments, and every department is more specialized in one thing. That's why I think Santa Fe is a great place.

John Holland: One thing I didn't speak much about, but it has to be in the urn model if you're doing it right, in my opinion, is that urns have to be able to reproduce themselves. That is, I have to be able to go from an urn with a certain set of tags to another replica of that. Now, we have a population, and now we get back to this notion. If the urn, or a set of them, has to collect the resources necessary to make the definitions of the gates that define the urn, then I have an implicit definition of fitness rather than explicit, which I much prefer. An urn is only going to show up in this system if, in fact, it can make replications by collecting appropriate resources. Now, the same thing goes for a production line, okay? So, a cascade is what it would be called. Now, in all of these, one of the things that I look at as a guide is slime mold. Slime mold, through most of its lifetime, looks — it depends on what kind of medium it's in — either like an amoeba, a single-celled organism, or else it has flagella if there's a lot of water and it swims around. But it spends its whole lifetime that way until water gets in short supply, and then it sends out a signal, and a whole bunch of these, literally thousands of them, gather into an aggregate, so it's a kind of very simple multi-celled organism, but that's not the end of it. What's really interesting is once they aggregate, then some of those cells get signals that cause them to specialize, and they send up a tall stock, and on the end of the stock they put a bud, and on that bud come spores. The only thing that survives of all those thousands of cells is the spores. And they sit there, sometimes for years, until water comes along again. Now, if you're going to look at this kind of thing, you've got to be very careful again about what you define as fitness. What's the fitness of these individual cells? And, of course, the same thing goes when we talk about social organisms like ants. Is the fitness all in the queen? Well, that doesn't make any sense. And so, my claim is, in order to study these, I've got to first of all talk about what it means for an urn to reproduce itself, and I just avoided that here because that's a whole topic in itself. And then, once you do that, and if you close this, you require resources that are available to do the reproduction, then you get away from this business of explicit fitness, and you get this implicit fitness, and this really goes back to this question of government agents, where, in the long run, if the government can't keep its population going one way or another, then that government is going to fall. And we've seen this, you know, over and over again, even in a great empire like Rome. It gets to the point where they can no longer sustain the amount of input that's required. That's not the only reason, but it was a critical reason.

Alain Wouters: Yes. Hello. I'm Alain Wouters from Whole Systems, and I'd like to offer some perspectives that may inform the question of what else is needed to model, but also inform the question about optimality or steering in which direction evolution goes. And that's the following, and I'd like to apply it to the ecosystems, but also to organizational development. So, my answer would be to consider what else would be needed:

A relevant representation of context. When we're modeling a system, what I think is needed is a relevant representation of context. I say this for the following reason: I think in the fitness function that we look for in a system, there are two components. One is related to existence, and that's outward-directed, and the other one is related to what I would call performance, and that has to do with the use of resources. The first one is really about connectivity. When a system comes into existence, it will remain there if it can connect sufficiently to other elements in its surroundings. That's why the ecosystems remain as ecosystems. It could link in particular to other biochemical processes or chemical processes. For a company, it's the same thing. You know, a company comes into existence, and if it can link to customers and suppliers and whatever, it can actually be a viable system, existent. It is not yet optimal, you know, that's when management comes in and we start cutting costs and, you know, getting the maximum bang for the buck. So, my thing is the fitness function has these two components: One is about existence, so connectivity, and the other one is about performance, so leverage — that's the notion of leverage. And underlying that for me is a thermodynamic principle in that both actually increase free Gibbs energy. When a system emerges, a new entity comes into being; when that can connect to its environment, you get increased degrees of freedom, so you're basically increasing entropy. So, whereas at one level, we think, you know, why would the structure emerge, why isn't it going to maximal chaos, why isn't — you know because that's what we think with entropy, but because you have a new entity coming into existence, you actually...[Recording abruptly ended].

5.3 Summary of the Talk

John Holland's talk addresses the concept of boundaries in complex adaptive systems, drawing examples from various fields such as biology, economics, and language. The following is a detailed summary of the key points discussed:

Introduction to Boundaries: Holland starts by discussing two real-world examples where boundaries play a crucial role: a saltwater marsh converted to provide fresh water to Singapore and the automated traffic system in Singapore. He emphasizes that once appropriate boundaries are established, desired outcomes often follow automatically.

Importance of Boundaries in Biology: Holland recalls how early biology texts depicted cells as simple blobs, contrasting it with the reality of cells being highly structured with numerous boundaries. Boundaries within cells give rise to specialization, a crucial aspect of cellular function.

Complex Adaptive Systems: Holland presents a diagram of a complex adaptive system, highlighting the nonlinear interactions between multiple agents. He notes that the influence in such systems is both top-down (e.g., market averages influencing individual behavior) and bottom-up. Boundaries in these systems act as semipermeable filters, constraining interactions while allowing some elements to pass through.

Examples of Complex Adaptive Systems: Holland lists various systems, from the internet to the immune system, where many interacting agents adapt over time within boundaries. The dynamics of these systems are constrained by semipermeable boundaries, raising questions about their effectiveness and implications.

Challenges and Unknowns: Despite advances, there are still many unknowns in understanding complex adaptive systems. Holland discusses the lack of theories guiding the identification of leverage points (e.g., vaccines in the immune system) and the formation of hierarchical boundaries within systems like ecosystems and biological cells.

Language and Boundaries: Holland provides an amusing example from chimpanzee behavior to illustrate how boundaries and communication (language) shape social interactions. He introduces the concept of "tags" as mechanisms that allow certain elements to pass through boundaries based on specific identifiers.

Mechanisms and Boundaries: Holland discusses the relationship between boundaries and mechanisms, highlighting how mechanisms generate boundaries and their effects. He uses examples from various domains, such as biological cells and food webs, to illustrate the role of active sites and tags in determining interactions.

Recombination and Evolution: Holland argues that evolution is primarily driven by recombination rather than mutation, emphasizing the importance of understanding how building blocks (tags) are recombined over generations. He explores how recombination leads to the formation of specialists and adaptations in complex systems.

Mathematical Modeling and Exploratory Research: Holland discusses various mathematical techniques, including difference equations and Markov processes, for studying complex systems. He advocates for exploratory modeling to understand mechanisms and interactions, lamenting the lack of funding for such research.

Future Challenges and Perspectives: Holland highlights the lack of comprehensive theories for signal modulation and the co-evolution of networks as significant challenges.

He concludes by emphasizing the importance of considering hierarchical structures in complex systems to gain a comprehensive understanding.

Overall, Holland's talk underscores the fundamental role of boundaries in shaping the behavior and evolution of complex adaptive systems, offering insights into how mechanisms, interactions, and adaptation occur within these systems.

5.4 Relevance of the Talk to the Current Stage of Research

John Holland's talk remains relevant to the current stage of research in several ways:

Complex Adaptive Systems (CASs): CASs continue to be a major focus of research across various disciplines, including biology, economics, sociology, and computer science. Understanding how components interact and adapt within boundaries is crucial for tackling complex real-world problems.

Interdisciplinary Approaches: The interdisciplinary nature of Holland's talk reflects the ongoing trend in research, where collaboration across different fields is increasingly valued. The concept of boundaries and specialization applies not only to biological cells but also to social networks, economic systems, and technological infrastructures.

Mathematical Modeling: Holland's discussion on mathematical modeling techniques, such as difference equations and Markov processes, remains relevant as researchers continue to develop and refine models to study complex systems. These models help elucidate the dynamics of interactions and provide insights into emergent behaviors.

Exploratory Research: The importance of exploratory research, as advocated by Holland, is still recognized in the scientific community. Exploring mechanisms and interactions without the constraint of predefined hypotheses can lead to new discoveries and insights, particularly in areas where understanding is limited.

Challenges and Unknowns: Despite advancements, many challenges and unknowns persist in understanding complex systems. The need for comprehensive theories of signal modulation, co-evolution of networks, and the role of boundaries in shaping system behavior remain areas for further investigation.

Practical Applications: Holland's examples of real-world applications, such as the automated traffic system in Singapore and the conversion of a saltwater marsh into a freshwater supply, highlight the practical relevance of understanding complex systems. Researchers continue to seek solutions to societal challenges by leveraging principles derived from complex adaptive systems theory.

In summary, Holland's talk provides foundational concepts and perspectives that continue to inform research on complex adaptive systems, interdisciplinary collaboration, mathematical modeling, exploratory research, and practical applications in addressing real-world problems.

Chapter 6

Innovation and Diversity

Douglas Erwin

YouTube: https://youtu.be/5_gklD5788Q

Speaker: Douglas Erwin

Moderator: Su Haibin

Discussants: Brian Arthur, Robert Axelrod, Discussants A, B, C, D, E, F, G, and H

6.1 Talk by Douglas Erwin

Douglas Erwin: Thank you. It's a pleasure to be here. I'd like to thank Jan and the complexity group for the invitation to come to Singapore. For an evolutionary biologist, this part of the world is the second center of evolutionary thought, of course, because Alfred Russell Wallace also came up with the idea of natural selection not that far from here. I want to take a look today at innovation in a very broad sense, in the sense meant by this meeting, and I want to look at a question about whether or not the factors that drive innovation in biological, cultural, and technological systems are similar. It's obviously similar. These obvious similarities are at a very broad level in the sense that variation, inheritance, selection, and drift are current in all these systems. There are some people who call this approach to looking at things universal Darwinism, although that's also a term that's loaded with a lot of baggage. But in essence what we want to do is understand the processes that drive innovation, with respect to the talk yesterday. This is what I'm going to be focusing on, not adaptability, but what was termed yesterday as radical innovation. The hope is that understanding some of these systems may shed light on the processes underlying others, and ultimately the goal is to build models of what drives innovation in any of these systems.

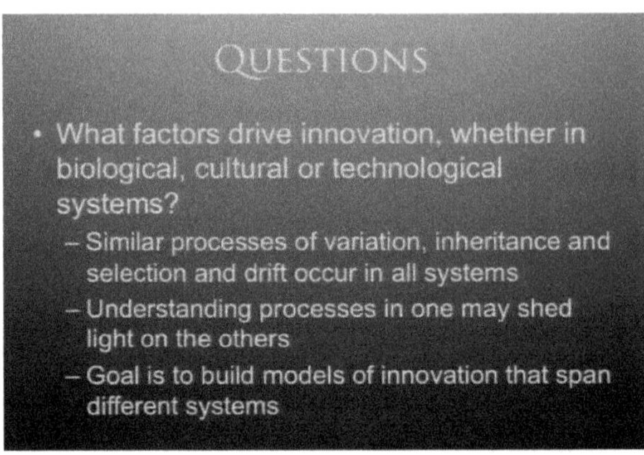

I'm going to begin by talking about what I'm not going to be talking about, what I really don't mean. Many of us, when we think of innovation, often think of adaptive radiation. This slide shows work done by a whole series of scientists, including most recently Peter and Rosemary Grant at Princeton University, looking at the species dynamics of finches in the Galapagos Islands on the other side of the Pacific.

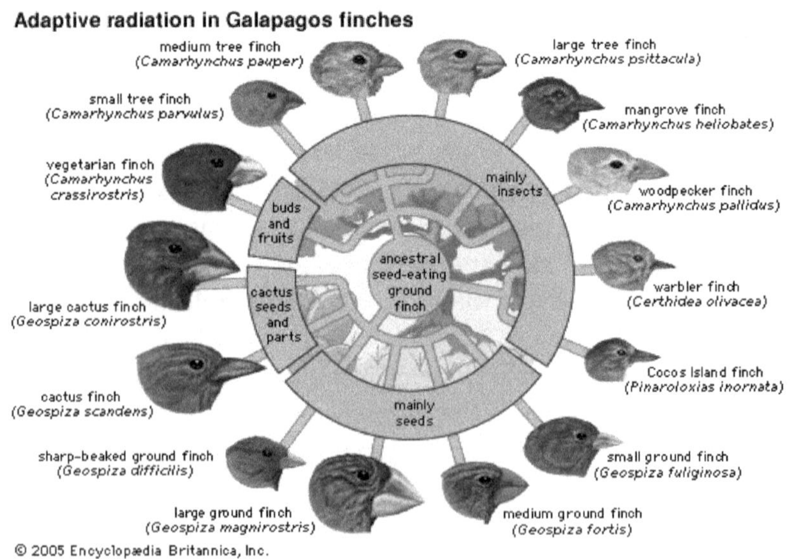

In this case, you have the diversification of a whole lot of different bill forms from a single ancestral form that was presumably flown in from Ecuador many hundreds of thousands or several millions of years ago, and on the Galapagos Islands, they've diversified into a variety of different species. This diversification is often taken as a basic model for evolutionary innovation. One of the points that I want to make today, in talking about a very different kind of innovation, is that this is merely one type of diversification and innovation, and shouldn't straitjacket our thinking about the breadth of changes that happen in evolution. Another example of such an adaptive innovation is the Hawaiian silverswords.

This is a diverse group of plants in the Hawaiian Islands, also in the Pacific. They look a lot like yucca. The silverswords have all come from one incredibly undistinguished shrub that lives in coastal California. They got to Hawaii and diversified into things that looked like yucca, little button plants, shrubs, trees, and a whole host of other things. So, this is one indication of the power of morphological change that can happen within adaptive radiations. And these are certainly an important component of evolution.

In technological systems, we see similar sorts of things. This is the presumably adaptive diversification of nails that serve some function. This is the diversity of Crest toothpaste that you can find in your nearest drug store. It's less clear to me whether there's actually an adaptive function to all these different toothpastes, but there's certainly a purpose to get shelf space and people's money for things they don't actually need.

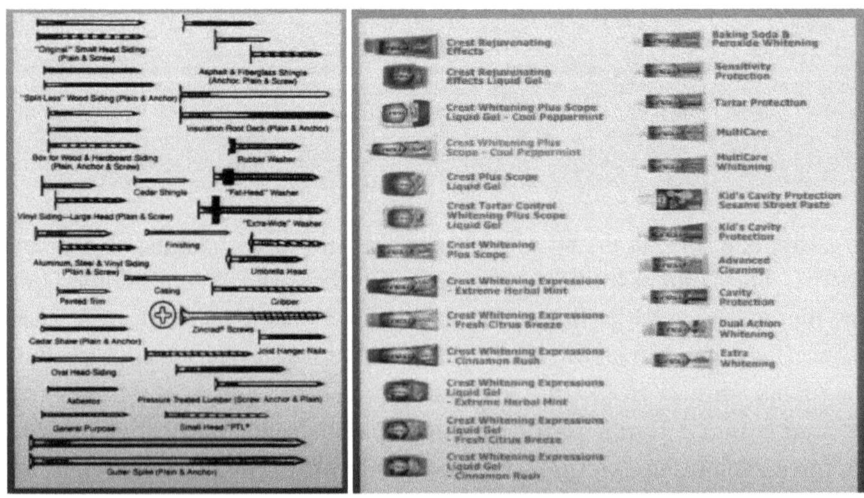

What I want to focus on for most of this morning's talk is an event that began a little over 550 to 600 million years ago called the Cambrian explosion. This is a diversity of fossils and reconstructions representing a few of the taxa that first appeared during this explosive radiation of animals. You see some sponges up here, some very early echinoderms. This is one of your earliest relatives, the cephalochordate named Pikaia, as well as a number of bizarre and unusual organisms, including the coolest animal that ever evolved in the history of the planet, in my humble opinion, which is Opabinia, which I'll talk about a little bit more at the end. That's a really neat animal, by the way. We'll go back. It is about three or four inches long. It's got five eyes on stalks on its head. It's got this great proboscis that comes out in front of it. This is the actual fossil from the Smithsonian, with a claw on the end of it. And these lateral flaps. It's like something out of the bar scene in the first Star Wars movie, before they started going downhill.

So, what does the Cambrian radiation represent? It represents the construction of a design space that occurred, as far as we can tell, at least in geological times, relatively rapidly. And what we want to know is how that design space is constructed. What components are reflected in the physical environment, which I won't talk about much today? How much is reflected in changes in genes and developmental interactions, and how much is reflected in ecological processes? These are the same questions that I'll come back to at the end of the talk in a slightly different vein that we have to think about in talking about cultural and technological innovation as well.

This is a depiction of the appearance of major groups of organisms during this period of time. Let me take you through the slide here. Along this axis, we have time from 580 million years ago up to 440 million years ago, encompassing two intervals of time known as the Ediacaran down

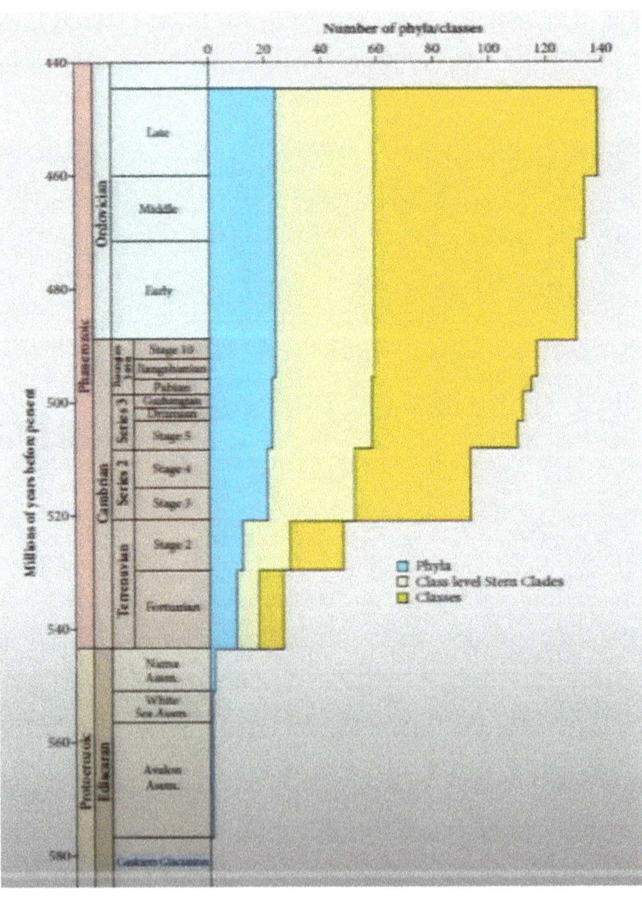

here and the Cambrian, and a bit of the Ordovician, which you can forget about. The main point I want to make here is that in looking at the biggest groups of animals that are called phyla, as well as the components of them in the Linnean system, classes, many of these groups first appear at very low levels between about 542 million years ago and 520 million years ago, and then there's a burst of both phyla and classes in what's called the Cambrian stage three here, and basically by the end of the Cambrian, all the excitement is over.

We have all the major groups of animals, including vertebrates. The sole exception is the group of animals called the Bryozoans. Now, how many of you have ever heard of Bryozoans? Amazingly, of course, Murray heard of them. Murray and Simon. Okay. So, four of you have heard of Bryozoans. You don't actually need to know much about Bryozoans, so we can skip past those. Our knowledge of this interval of time is greatly aided by the fact that we have a series of exquisite fossil deposits in many different parts of the world. These fossil deposits preserve many of the soft parts of the anatomy that usually aren't preserved in fossils. These things include, up in the Yangtze gorges in South China, fossilized embryos from the Doushantuo formation. We have series of deposits of

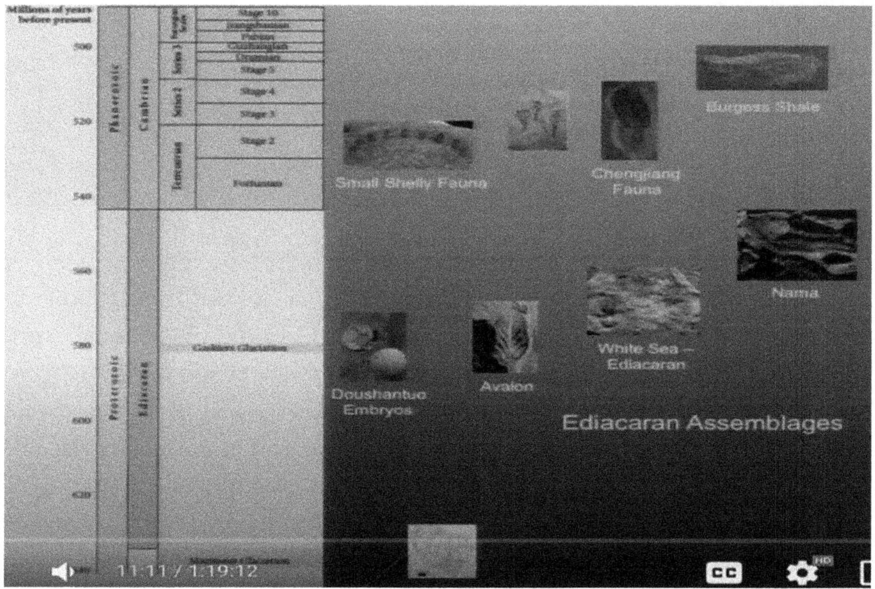

what are called Ediacaran fossils from Kangaroo Island in Australia, the Avalon Peninsula in Newfoundland, the White Sea in Russia, the Ediacara Hills in South Australia, in Namibia, and many other places in the world. That gives us our first view of these animal fossils.

I'll show you some photos of those in a minute, and then in the Cambrian itself, we have another set of exquisite fossils, and I'll show you a number of things from the Burgess Shale, which is up in the western Rockies in British Columbia, Canada.

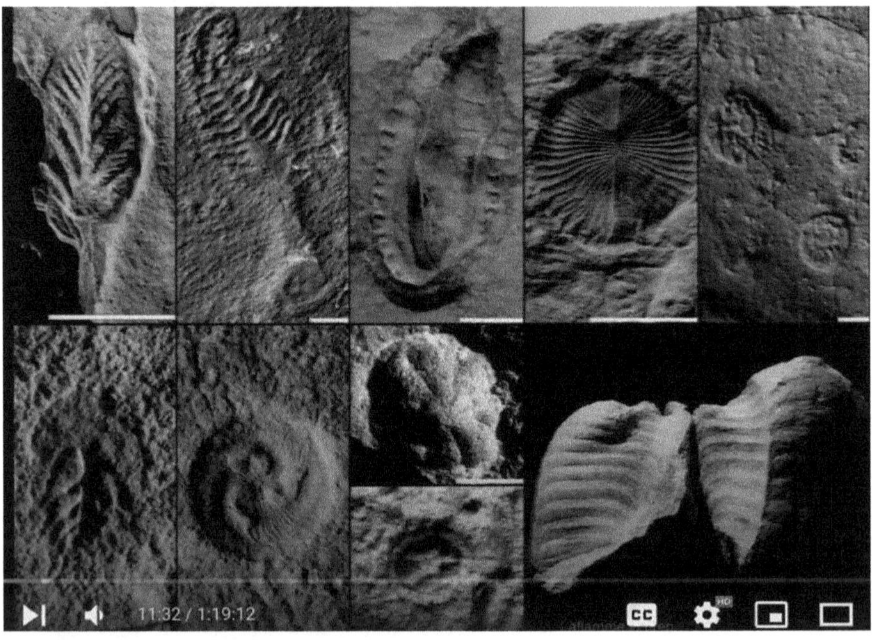

This Ediacaran biota that predates the Cambrian explosion itself is composed of a variety of fronds and discs. This is a little frond, about an inch long, and some discs as you can see here. Some of these are very small, only a couple of millimeters across, with the exception of this form up here called Kimberella. None of these things have appendages. None of them have eyes. None of them have any sign of a gut. They're very simple, frond-like or disc-like organisms. The evolutionary affinities of most of these things are unclear. This form here from the White Sea in Russia is called Kimberella. It actually turns out to be a very primitive relative of a mollusk, so we have a diverse set of fossils before the Cambrian explosion, but they're very different from what we see 542 million years later.

This is a shot of the Burgess Shale quarry up in British Columbia, Canada. It's a shot from about 20 years ago because if we were up there now, this glacier would have retreated even further. Sadly, 520 million years ago, this was along the coast of North America in very tropical waters. It's a deep-sea deposit where the shale has been brought down from shallower waters, and we have exquisite preservation of shale in these deposits.

This is a slab of trilobites and other forms that were collected by Charles Walcott beginning in 1909. You can see a variety of things here.

I just want to point out that some of these things, for example, trilobites here and the annelids include impressions of legs along the sides, there's a bit of an antenna coming up off the top, and you can see the same thing over here.

If we look at some of the other fossils, this is one of the more recently described forms known as Hurdia. Gretchen Daley did her Ph.D. thesis on this in Uppsala in 2009. This animal is about ten inches long — body of the animal too with a tail — and hasa large head shield covering the mouth and then some appendages hanging down below it.

This is a very early relative of arthropods. It's related to Anomalocaris. This is a top view of an Anomalocaris — the tail down here, these lateral flaps on either side, and it's got these two giant appendages. Hurdia basically is an Anomalocaris with a head shield stuck on the front end. Until 2009, this was believed to be three in three different parts, and only assembled during Daley's Ph.D. thesis. This is a large predator. These things in China got up to be about 2 meters long.

Trilobites were, of course, quite common.

This is Aysheaia. It's an Onychophoran worm, about three inches long. You can see the body of the animal here. This is the head, with some slight appendages, pairs of appendages with small claws on the end of

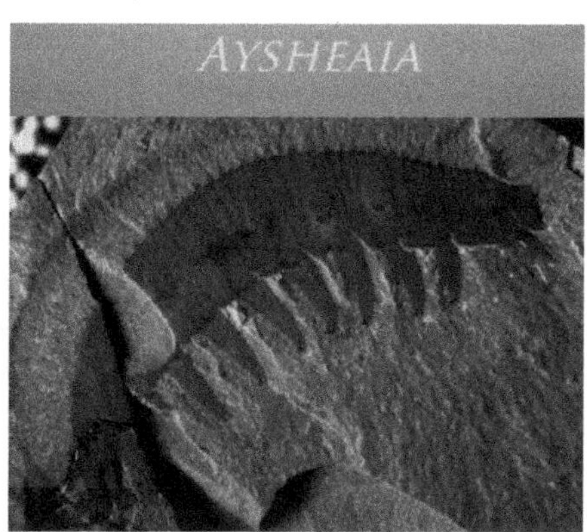

them, leading people to think they may have crawled up sponges. Onychophorans or lobopods are very common in Cambrian deposits.

This is another form reconstructed from fossils in China. The body of the animal is down here. You can see these spines coming up off it, and the legs are mostly buried in this specimen down here, with a big bulbous head at one end.

This allows us to make reconstructions like this, with the big Anomalocaris peeking over the arms of the sponge in the corner, along with Kerygmachela and a variety of other things.

The main point is that we see a microdiversity of design forms, although not really that many species, within what, to a geologist, is a very short amount of time. We have a group of things down here in the early part of the Cambrian called the small shelly fossils; these are mostly tubes and plates and things. But then, major forms that add to this diversity have been described, all appearing very rapidly in some of these exquisite fossil deposits.

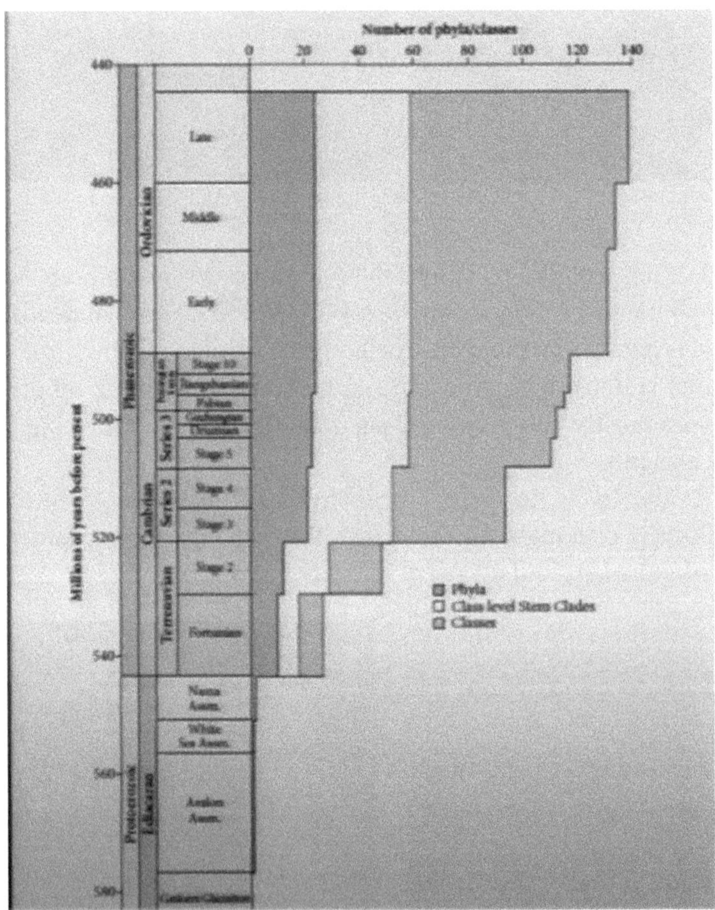

One of the other ways of looking at the morphological diversity that we see in the Cambrian is to measure quantitatively what we call disparity. Disparity assesses form rather than simply the number of taxa, such as phyla, classes, families, and so on. When we ignore taxa and simply look at form, the pattern we get is that the morphological space defined by arthropods, Priapulid worms, or many of these other groups is defined

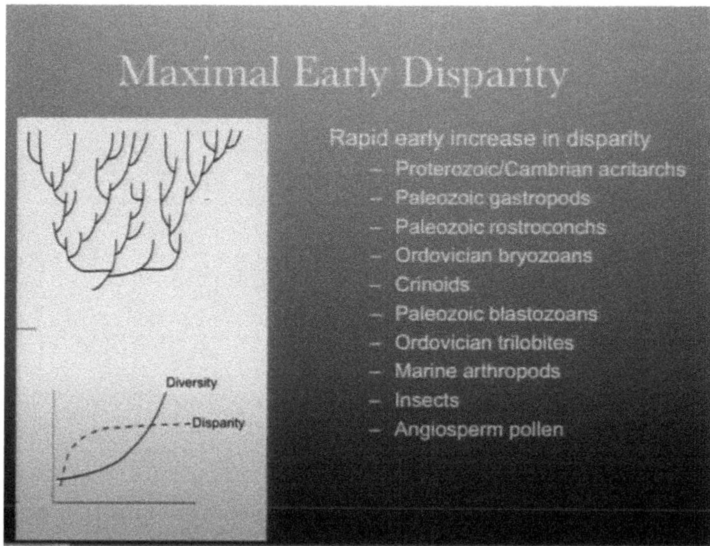

very early in their history. So, this state of form is defined early and then it's filled in. The space doesn't grow throughout time. This is the claim that Stephen Gould made in 1989 in his book *Wonderful Life*, and later his student Michael Foote and several other generations of students at the University of Chicago turned Steve's ruminations into science in a quantitative way. A whole host of these groups that I've shown here have now been studied, all indicating this very rapid increase in disparity, only later in diversity, shown here on the dashed line, with time going this way. Disparity goes up rapidly, and only later does diversity, the number of taxa, climb as the space is filled in. So, that gives us one indication about the pattern of innovation at the morphological scale that's happening during the Cambrian radiation.

Let me now move to ask what the role of genomic and developmental changes is in structuring this. Of course, one hypothesis is that what's driving this is the changes to the genomic and developmental system. If that's true, then we have an explanation for what's happened. To set the scene for this, let me briefly take you through a figure from a paper in 2010 by Eric Alm's group from MIT. This paper in *Nature* looked at the birth–death loss transfer of genes, similar to what we were discussing in John Holland's talk just a few minutes ago. If you look on the left here, this is a scale through time from the beginning of the oldest evidence for

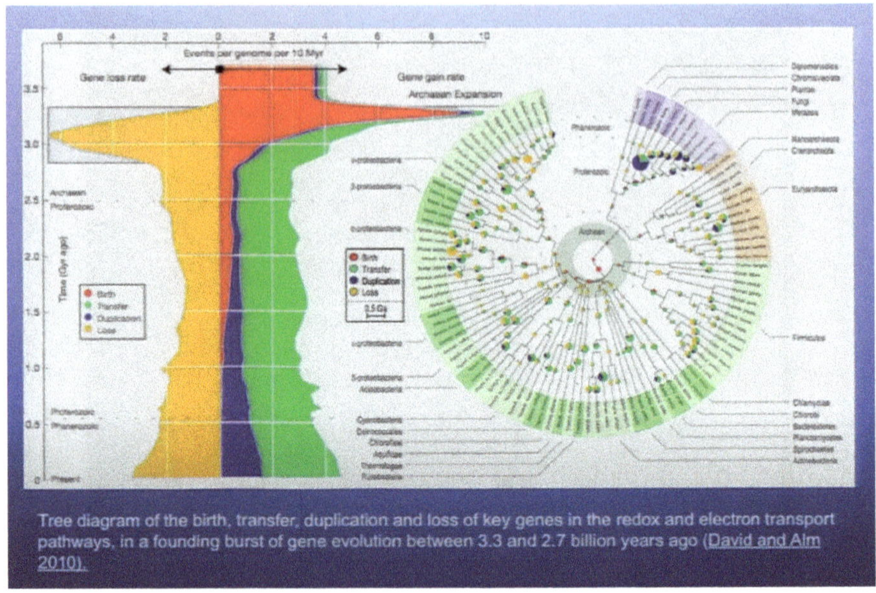

Tree diagram of the birth, transfer, duplication and loss of key genes in the redox and electron transport pathways, in a founding burst of gene evolution between 3.3 and 2.7 billion years ago (David and Alm 2010).

life, about three and a half billion years ago now. The data here are from living organisms, so they're projecting back to the origin of life, and there are ways of doing that. As we come down here to the present, the Cambrian radiation that I'm talking about is about that dotted line there, just past half a billion years ago. This figure shows gene loss on this side, gene gain on this side, and the kinds of changes you can see in these different colors. So, red primarily concentrated at the base here represents the birth of new genes. Green represents horizontal gene transfer events. Blue indicates duplications of genes, and loss is represented by this mustard color. This analysis focuses only on redox and electron transport genes. You can see there's a very different pattern of what happens at different points in time, but then one of the more interesting things happens. If you look at this evolutionary tree from the figure again, the center takes us back around three and a half billion years ago. Some billion years later we reach the small green ring. Time is going outward, where the outer ring is today. The main thing I want to draw your attention to is that within this clade here, that includes plants, fungi, and animals, most of the changes, the bulk of them, are gene duplication events. And that's true not only of the genes studied by Eric Alm's group but also a series of other genes as well.

Let me focus on our closest relatives. The closest groups to animals are something called choanoflagellates, and the genome of choanoflagellates, specifically the Monosiga genome, has been sequenced by Nicole King at Berkeley and her colleagues. Additionally, we have whole genomes of many other organisms, including this fun Amphimedon, a Placozoan called Trichoplax, the sea anemone Nemostella, and flies or Drosophila. One thing I want to draw your attention to here is that although there are significant differences in the size of the genome of these different animals and differences in the number of genes, basically

GENOMIC COMPLEXITY

	Monosiga	Amphimedon	Trichoplax	Nematostella	Drosophila
genome size (Mb)	41.6	167	98	450	180
# genes	9,100	?	11,514	18,000	14,601
# cell types	1	12	4	20	50
# T.F.'s	?	57	35	min. 87	min. 87
# T.F. families	5	6?	9	10	10
microRNA	0	8	0	40	152

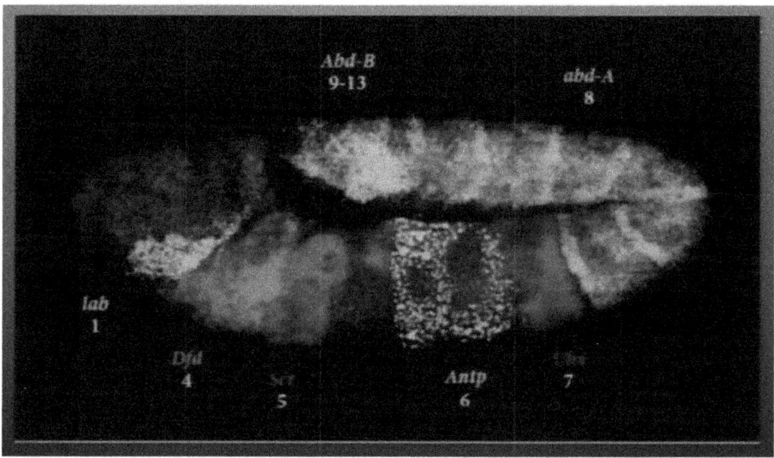

once you get to about 18,000 genes, you can build whatever you want. It's not the number of genes that produces this morphological complexity; rather, it's the interactions between those genes, and one index of that is the number of transcription factors and also microRNAs, which I won't discuss very much, but they are in these lower couple of rows.

Let me focus on one aspect of these transcription factors: the hox genes. This is a multi-colored in situ hybridization of a developing Drosophila embryo. This is the sort of thing that developmental biologists love to do, and it shows, in different colors, the expression of different genes in the developing embryo, from the head end around to the tail, indicating the production of different components. From the front to the back of the developing embryo, we have labial and DfD at the anterior end, and then a series of other genes controlling body plan regionalization as we move toward the posterior end.

This schematic illustrates a more simplified representation of a generalized insect on the left, showing a series of hox genes. The interesting aspect about the hox genes, both in arthropods and across animals, is that they are often lined up together on a chromosome. Therefore, a chromosome contains a series of hox genes that have arisen through duplication events. Different genes have then specialized to control the formation of different parts of the embryo. For instance, the gene in green controls the

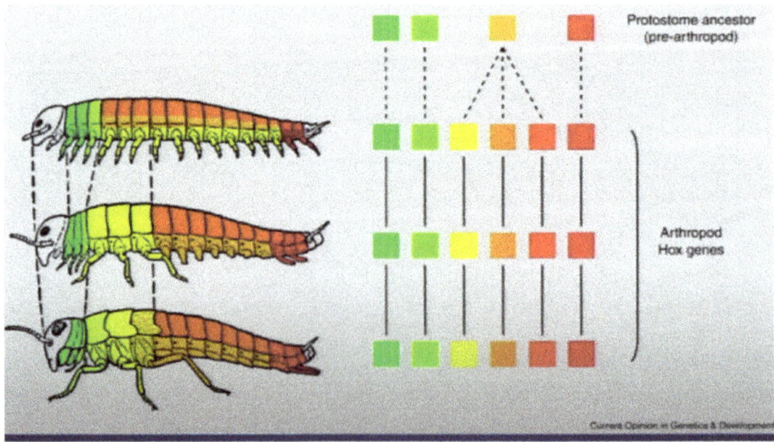

anterior end, and as the colors progress toward warmer tones, they represent regions toward the posterior part of the body. This pattern of diversification of these developmental genes, known as transcription factors, happens in much of the body plan formation, including appendages and other aspects of the body plan controlled by the duplication and divergence of these transcription factors.

By studying these in different organisms, developmental biologists can actually build a tree, which is shown here, illustrating the duplication and growth of these hox gene components.

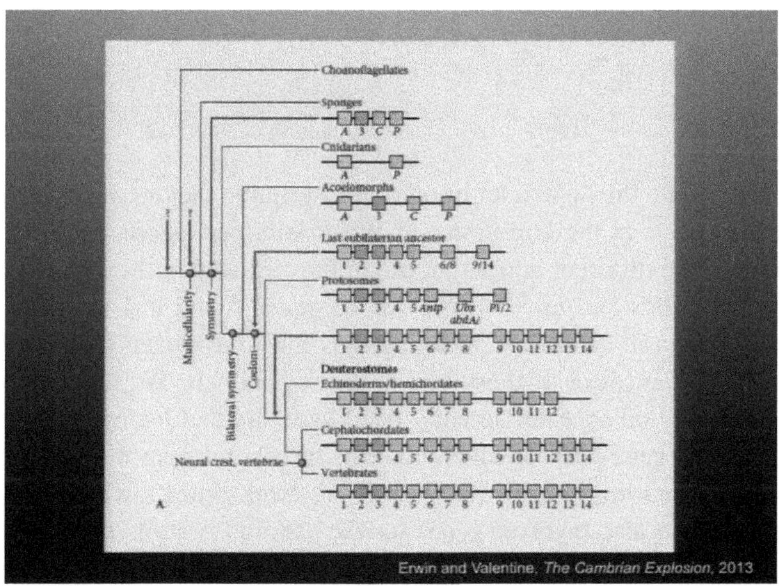

What you see on the slide, with sponges at the top, is an increasing number of these transcription factors. The different classes of genes, the sort of subfamilies if you will, are shown by the different colors. So, the anterior genes are in baby blue. These genes don't undergo much duplication. The next set of genes undergoes a little duplication, but then it's the two posterior classes of genes found in Bilaterian animals that have undergone many duplication events and control the body plan formation seen in arthropods.

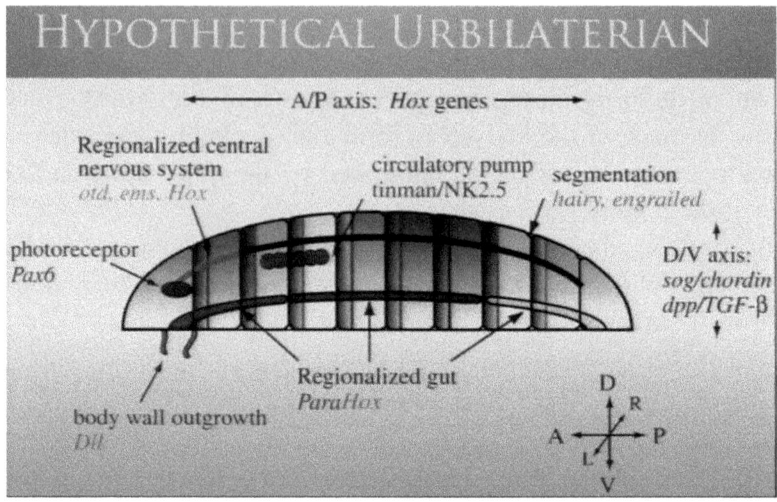

You can do this with a lot of other transcription factors. In 2001, Sean Carroll, who is at the University of Wisconsin, put together this sort of cartoon of the different genes that might have existed in the last common ancestor of flies and mice. These are all genes found in both mice and chickens, as well as in flies, meaning they must have been shared by the last common ancestor of those two groups of animals. We'll get to when that last common ancestor actually lived later, but that includes genes for things called gene pack 6, which controls the photoreceptor distalis, controlling outgrowths of the body wall — Sean wasn't willing to say appendages. It also involves genes for the nervous system, even the heart with a wonderful gene name "Tinman," segmentation, and various body axes. Now, I don't actually believe that this organism ever existed, for reasons that I will get to in a moment, but this shows you that through this comparative developmental approach, one can begin to understand the developmental aspects of organisms that lived half a billion years ago or more.

Let me now address the question of when these things existed. The point that I'm going to get to shortly is that while I don't doubt that all these genes were present, they likely weren't performing the functions that Sean has assigned to them in this cartoon. This skepticism arises from considering when this organism existed. What I'm going to try to convince you of is that the origin of these major groups of animals and the development of this toolkit happened around 150 million years before the fossils we see in the Cambrian explosion. The reason for this gap, what we call a macroevolutionary lag, is not due to a poor fossil record. In fact, the fossil record of that time interval is quite good. Many of us researchers have spent a lot of time walking up and down rocks in China, Africa, North America, and other places looking for animal fossils. The absence of animal fossils from that period isn't because we can't find them; rather, it's because they hadn't evolved yet. This absence tells us something important about the nature of this innovation.

This is a molecular clock analysis we published at the end of 2011 in *Science*. For those of you who are interested in the intricacies of molecular clock analysis, you can refer to the paper. Essentially, a molecular clock enables us to utilize sequences from living taxa, in this instance, seven different housekeeping genes and 118 living taxa. We sequence the genes and employ a series of algorithms, along with calibration points in the fossil record, to project backward and determine when these different lineages branched through time. This is the result we obtained.

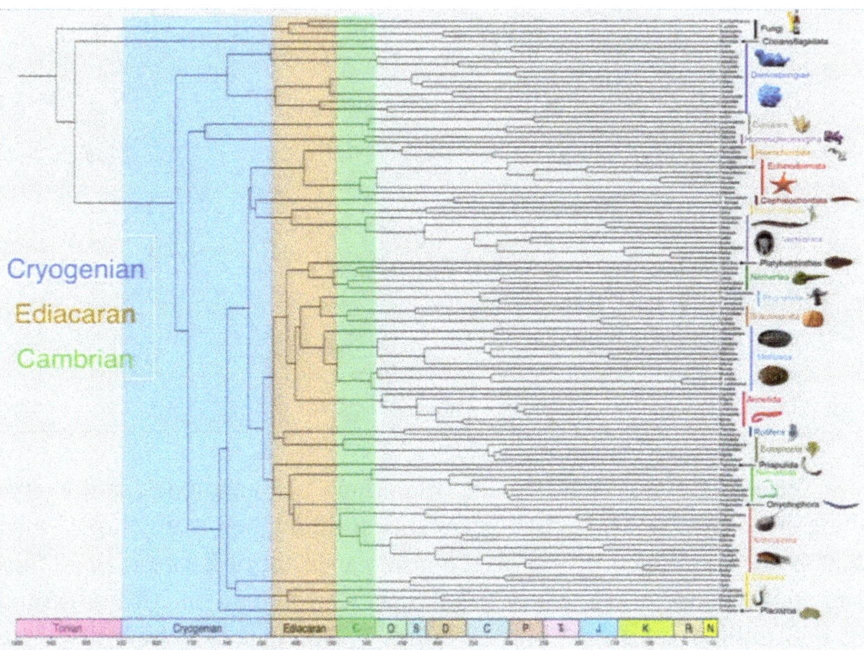

We have taxa represented along here, with fungi represented by beer at the top and vertebrates represented by Linus Pauling, because we didn't have a picture of Murray. This rather rich dataset allows us to infer when these different groups of animals flitted through time. So, within the vertebrates, the shallower branches are relatively recent, and then we go back through time. The Cambrian radiation that I've talked about is between the green and the brown lines here, and then we go back into earlier parts of what we call the Proterozoic. Time is going toward the present, and this is today. If we step through this diagram, the last common ancestor of all living animals looks like it was about 780 to 800 million years ago, long before we see any evidence in the fossil record.

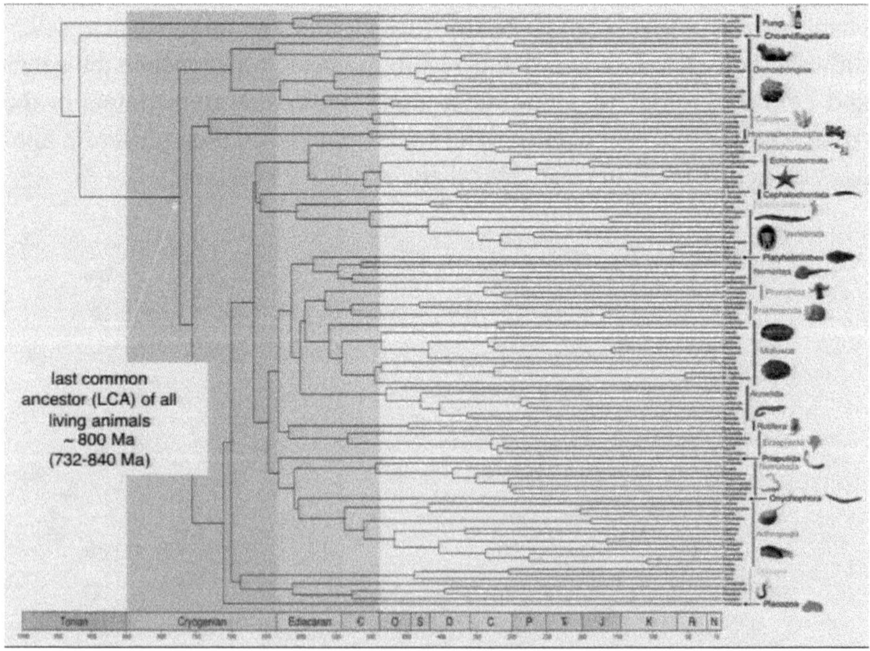

The last common ancestor of Bilaterians, sea anemones, and everything else is about 700 million years ago, still long before anything we've seen in the fossil record. The last common ancestor of Bilaterians — flies and mice, the age of that cartoon that Sean Carroll drew in 2001 — would be 668 million years ago, plus or minus this or that, well before we see any evidence at this level of complexity — 120 million years before we

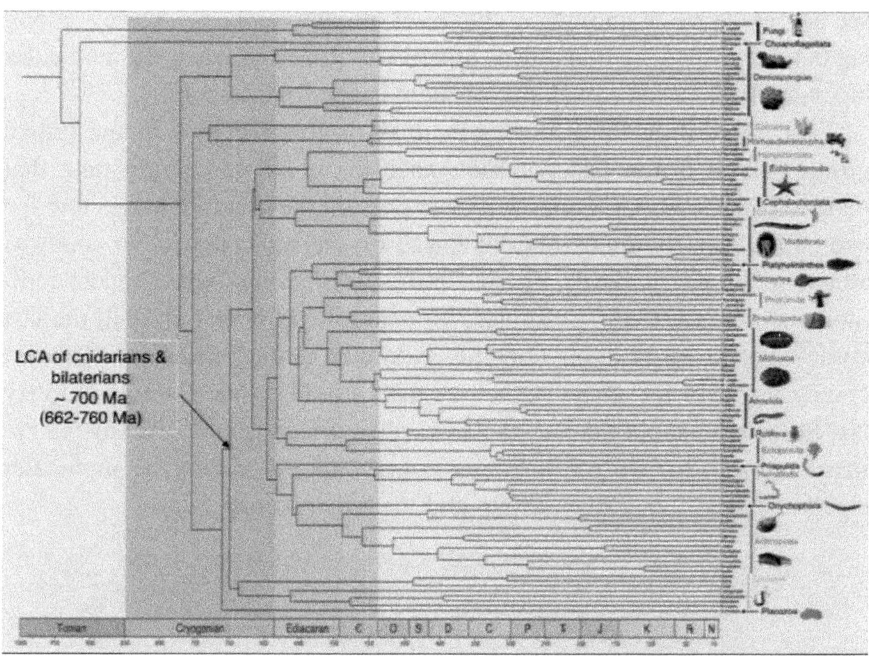

LCA of cnidarians &
bilaterians
~ 700 Ma
(662-760 Ma)
Tonian
Cryogenian
Ediacaran

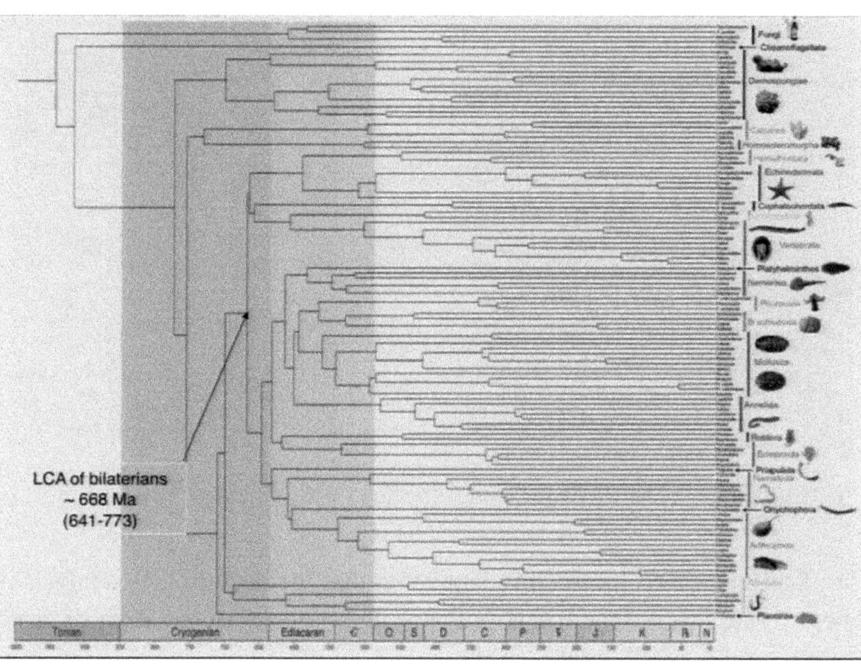

LCA of bilaterians
~ 668 Ma
(641-773)
Tonian
Cryogenian
Ediacaran

see any evidence of animals of this level of complexity. So, there's a long lag between when we think these organisms arose and when, by molecular clock analysis, we see evidence of them in the fossil record.

The interesting thing in this analysis is that we look at the crown group, which is the last common ancestor of all the living taxa that matches the Cambrian quite well. The dots on the diagram here show the age of the last common ancestor based on the molecular clock analysis alone, not based on fossils, except for the calibration points. The last common ancestor for living examples, for example, of nematodes, all the last common ancestor of living annelids, would be in the Ordovician. The last common ancestor of some of the mollusks would be back here in the very last Ediacaran, about the age of Kimberella, which is what we have in the fossil record. So, we see amazing concordance between the molecular clock results for the crown group and the fossil record.

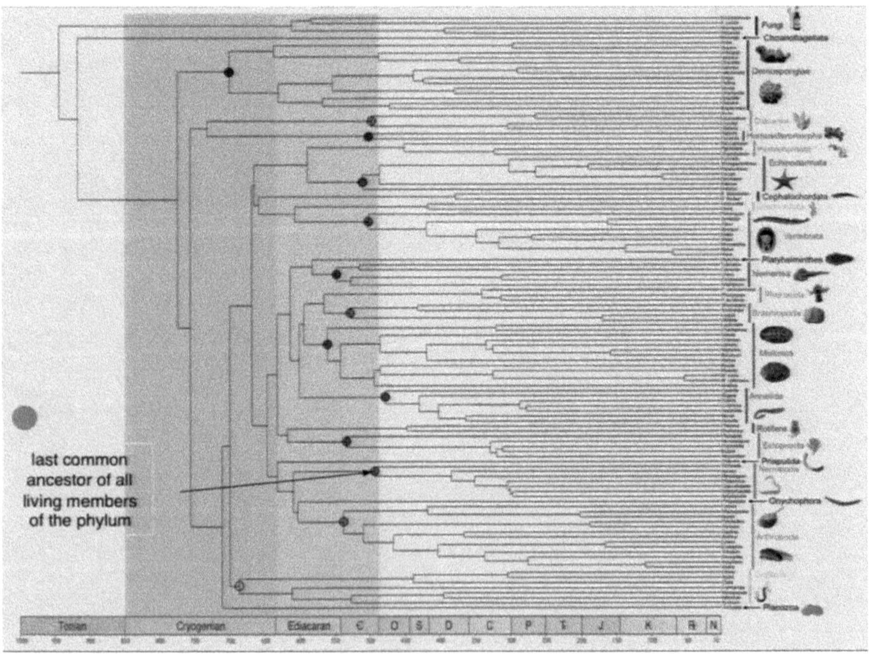

That gives us a lot of confidence that this molecular clock analysis is about right. You see the fossil evidence combined with the analysis from

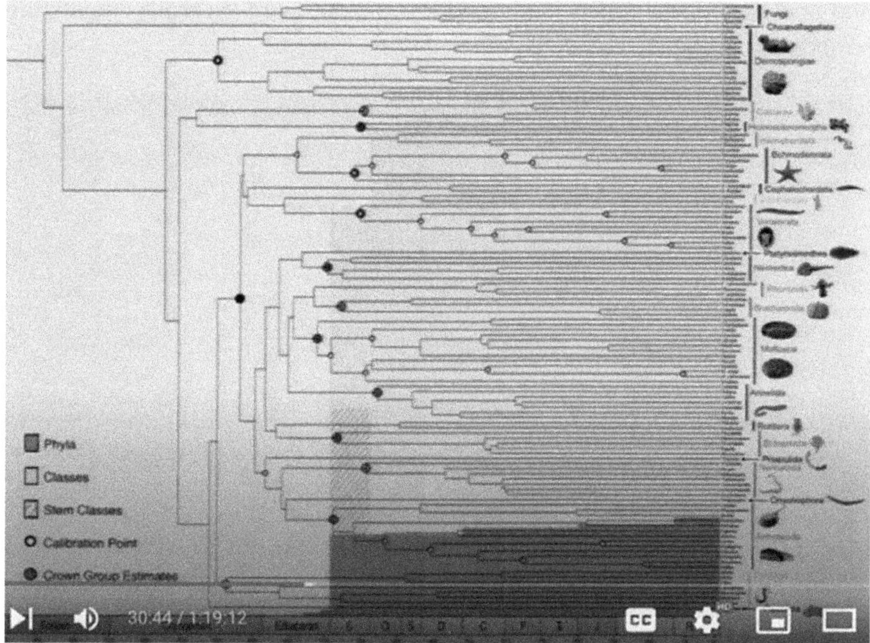

the molecular clock, but that leaves us with a problem — what was happening for 200 million years? Particularly if the animals in that time had a fairly sophisticated developmental toolkit, at least since that node. At that node, there is a last common ancestor of all of these different groups of animals and by all evidence, it had a sophisticated suite of developmental tools.

Let me now move into looking at the structure of actual gene regulatory networks. This is a horrible diagram to show to all of you before lunch. As John Holland said, if you lose half your audience without the equation, you lose 90% of it by showing this diagram. It's a really neat diagram. This is the wiring diagram for how to build the sea urchin. This is from Eric Davidson's lab, and it's all available at biotapestry.org, which is one of his lab's websites. Eric is a developmental biologist who works on fusions and has been progressively refining the structure of regulatory interactions required by the developing sea urchin embryo. This mess shows the interaction between various genes. There are upstream regulatory sequences and how they affect other genes, and it turns out there's a really interesting structure to it.

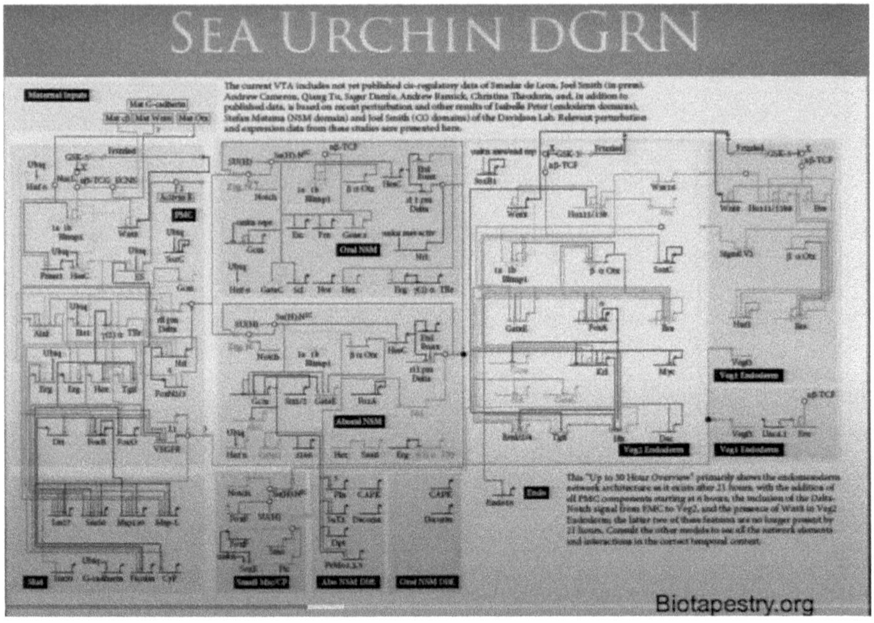

Let me just help you with a schematic here. You have inputs to part of this gene regulatory network of signals and transcription factors like GRN that affect particular genes. The genes are shown in red here. Those then drive other genes. You ultimately wind up driving differentiation genes, which are the sort of things that produce eye spots on butterflies and things like that.

If we just look at one of the earliest components of the network, this is the basis for which this network has been built. You see some maternal inputs up here. Various genes are turned on that activate other genes. A couple of features of these networks are that there are circuits that lock down the pattern of expression of different genes, other circuits that express these differentiation genes, and ultimately, these genes lead to the formation of the skeletogenic system of the embryo. But there's a lot of regulation that has to occur to produce that downstream component.

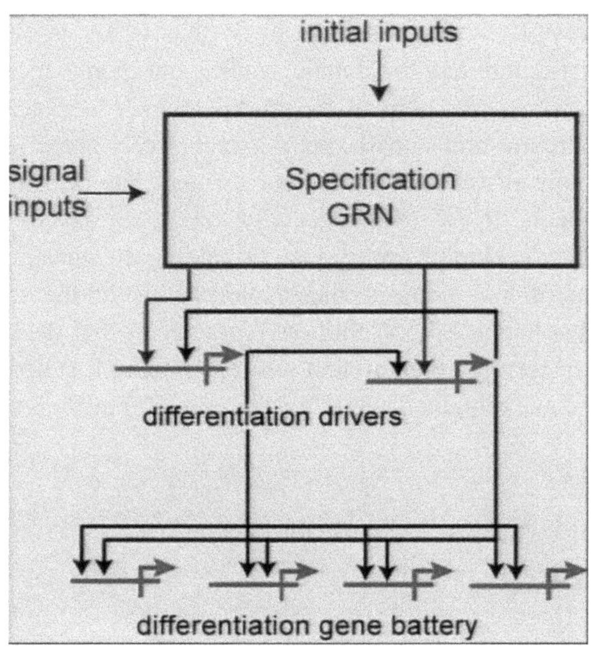

initial inputs
signal inputs
Specification GRN
differentiation drivers
differentiation gene battery

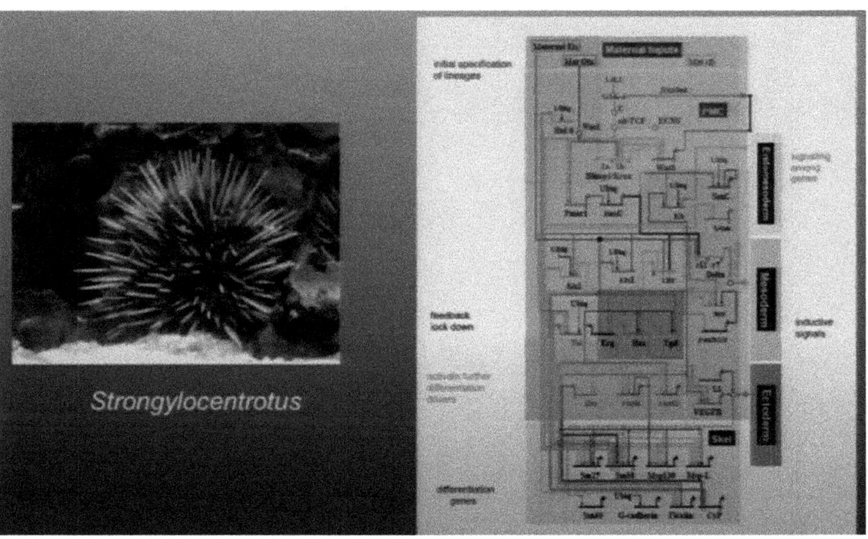

Strongylocentrotus

What's really fascinating, however, is that when you compare and examine the structure and the details, it turns out there's a component of this regulatory network here in the middle which is exactly the same between sea urchins and starfish. So, this recursively wired set of interactions at the core of this network, right in here, has been preserved for 550 million years. If you experimentally knock out any of these genes, you disrupt the developing embryo; so producing the gut, in either a starfish or a sea urchin, requires these interactions in the same pattern. Basically, what happened 550 million years ago is that these genes took over control of forming the gut, and we've, Eric and I, called these a kernel, and they have remained stable for the next 550 million years.

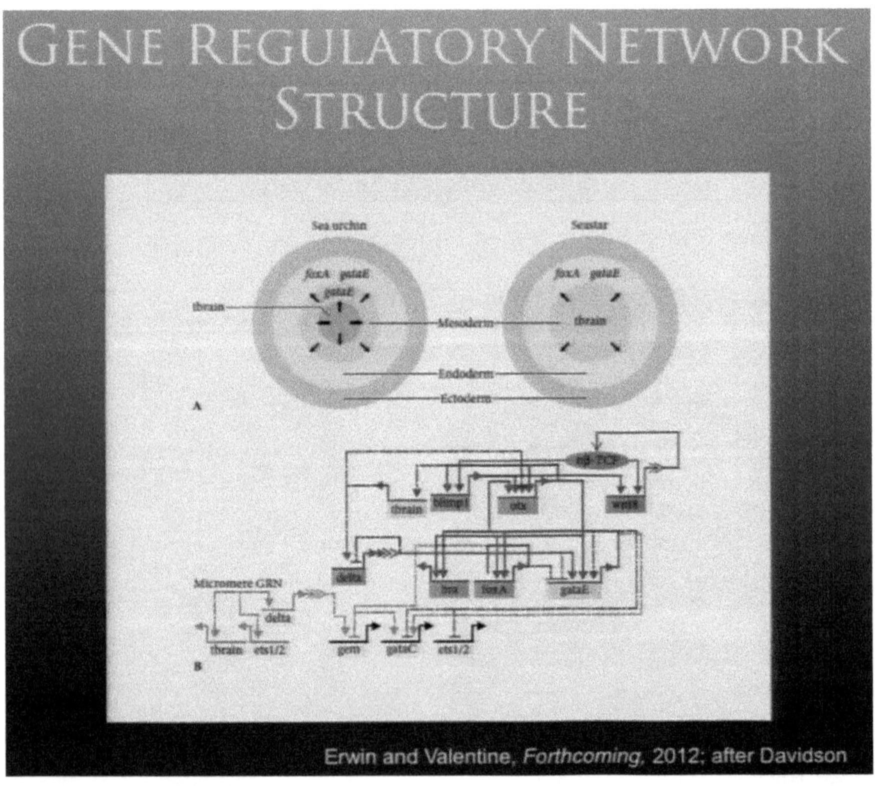

This is exactly the sort of technological lock-in, or biological lock-in, that Brian has described as technological lock-in. Once you get this sort of thing, it's really hard to get rid of it. The only way you can get rid of it is by getting rid of the gut. There are some organisms like Pogonophoran clams, which are worms, that actually do that.

So, that locks things into this particular sort of structure. There are other components in these networks that we have identified. They are highly conserved, in some cases highly labile, which suggests that there's a real hierarchical structure to these developing networks that has allowed the production of all these different morphologies. Some parts of the morphology, like forming appendages or eyes or guts, remain stable for hundreds of millions of years. Other more distal components of the network are more labile and change more readily. But genes aren't enough because most of this toolkit was present 150 million years before what we see it in the fossil record. The other part of this system that we have to look at is the ecological interactions.

Certainly, one of the things that happened in the Cambrian is that all of a sudden, we get a whole lot of things that eat other things. There are no predators within that Ediacaran biota that I described to you at the beginning, but as soon as the Cambrian shows up, there are lots of things that are eating other things. That tells us something about the energy flow within these systems, but it also tells us quite a bit about the complexity of the ecological networks.

We get the same answer incidentally, not just by looking at the fossil record but also by comparing it to this molecular clock analysis that I described, because this also shows us based on molecular data the age of the crown group that includes various predatory groups, some of which are at the base of the Cambrian. Others, for example, vertebrates, are further up in the Cambrian, although we're fairly sure there were predatory vertebrates about 20 million years before that purple dot.

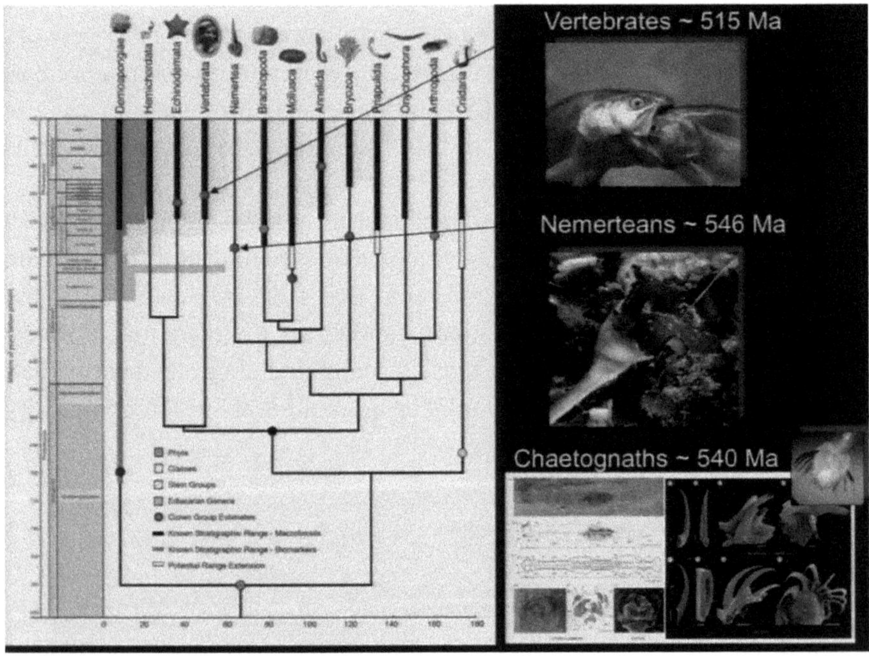

So, there are a number of predatory groups that are clearly a component of these Cambrian faunas.

There's another component to it as well. This is a topic that has indirectly come up in the previous parts of this meeting, and this is how you actually build the ecological networks that these organisms take advantage of or are part of. One of the things that happens is that if you look at the sediment from the late Precambrian into the Cambrian, you go from sediments that are planar laminated — they look just like thin layers of a cake with no disturbance — to very well-mixed sediments, because of burrowing and other kinds of activities. So, even if we had no fossils, sedimentologists

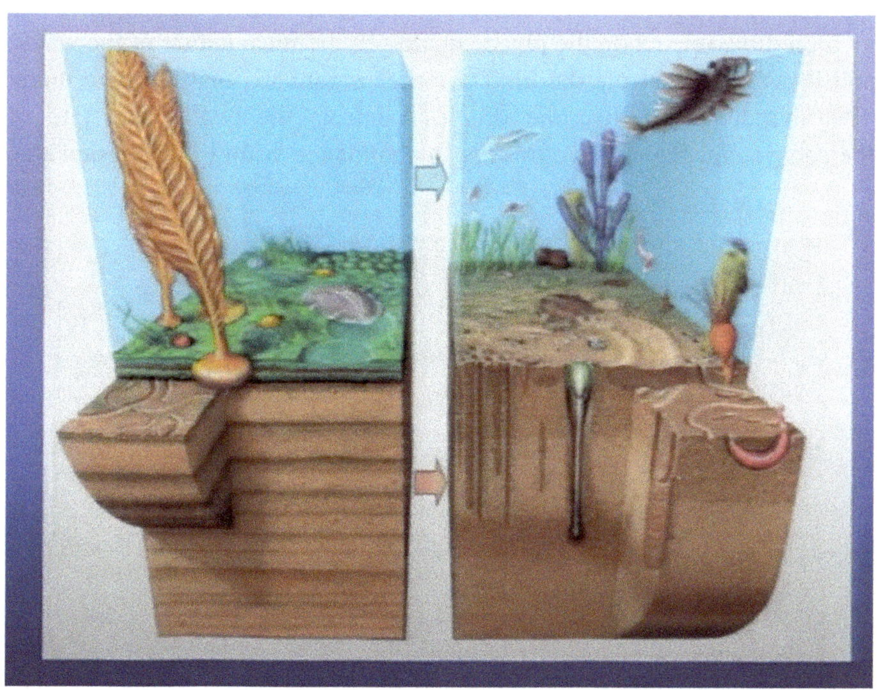

would know that something profound happened, simply because the burrowing activity of organisms changes from insignificant to so important that it completely mixes the sediment, and we no longer get these planar laminated sediments. That has a lot of ecological implications as well.

Let me give you a more recent example of this. This is the American oyster, Crassostrea virginica, and that's Chesapeake Bay on the right in the eastern United States. Washington, DC is up here along the Potomac. This is the mouth of the Chesapeake. It goes up here to Pennsylvania. In 1604, I think, when the first British colonists sailed into the mouth of the Chesapeake to set up their misbegotten colonies down here, in the Chesapeake, basically everything about nine meters' water depth was covered with oysters. Oyster reefs were so common that the British Navy spent a couple of hundred years using them for target practice because they were a threat to navigation. As late as 1870, all the water in the bay went through the filtration of some oysters in two days, so the entire bay was filtered in two days. By 1979, it took over a year to filter all the water in the bay because there were no more oysters — they ate them all. I think I'll skip this, but my point is that the filtering activity of these oysters has an incredible impact on the physical and geochemical nature of the water, and it actually controls the kind of other organisms that can live there. The abundance of this one species controls the ecological state of the ecosystem. If you have oysters in abundance within the Chesapeake,

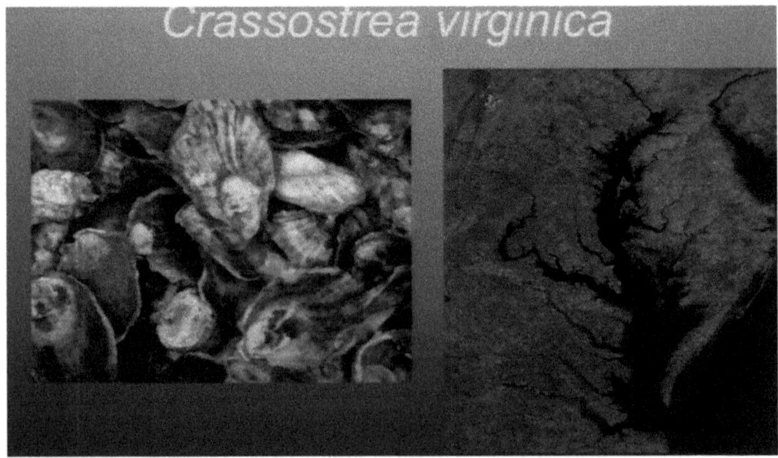

you have fish, mollusks, crabs, and all sorts of other things. If you take the oysters out, you get jellyfish, slime, conidiophores, and stuff that you can't eat. So, these spillovers can have incredible changes on the state of the ecosystem, and one of the things that we've realized in looking at the various kinds of organisms that show up in the Cambrian is that many of them, in fact, are ecosystem engineers now, because we don't have abundant estimates. It's very difficult to figure out exactly what the impact of these organisms was, but it appears that they're also likely, along with predation, an important component of the ecological changes that we see at this time.

The system that I just described is one in which we have a long lag between the origin of the developmental tools that are required to build the complex animals that we see as Bilaterians and the ecological or evolutionary expression of them as fossils. This is not the only macroevolutionary lag that paleontologists have identified. Lags like this are common. This is not what Ernst Mayr told us about evolution. Ernst Mayr thought and wrote during the 1960s that evolution was very opportunistic. When new innovations came along, evolution would take advantage of them. That's not true. If we look, for example, at grasses, grasses are a good idea. Grasslands are widespread across North America, South America, Asia, and many other parts of the world. It's hard to argue that grasses were not a wonderful evolutionary innovation, and it's easy to realize this because of the fossilized records we have.

So, we look again with a molecular tree. This is one from Toby Kellogg from about 2001. More recent ones are about the same.

The origin of grasses is estimated to be about 70 to 55 million years ago, with more recent estimates suggesting it's closer to about 55 million years ago. Interestingly enough, the last common ancestor of corn and rice is estimated to be around 45 million years ago. However, it's notable that there was already diversification into major different groups of grasses by probably 55 or 45 million years ago. Caroline Stromberg, who's now at the University of Washington, has carefully studied the fossil record of grasses, primarily in North America but with similar findings in other parts of the world. In her research, she found a distinct pattern of grass evolution, with grasses evolving at the bottom of the slide and then spreading and diversifying over time.

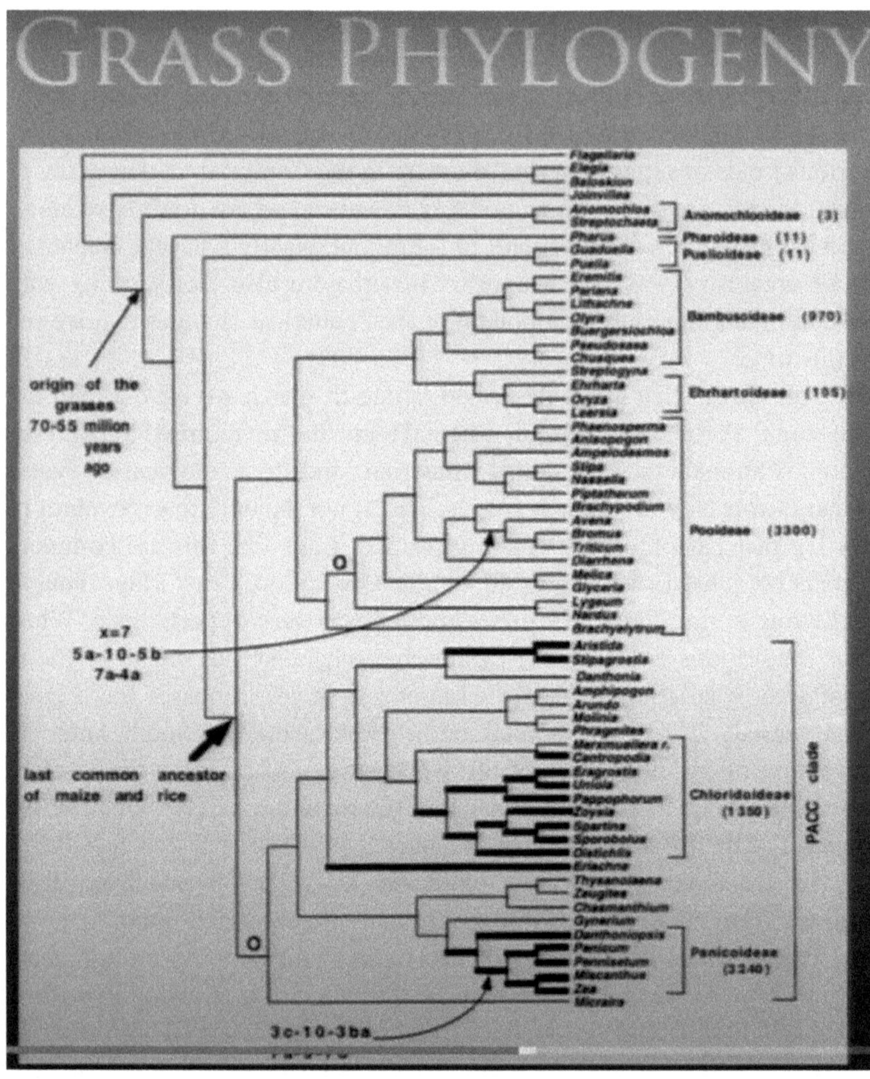

The distinction between invention and innovation, as articulated by Joseph Schumpeter, is indeed relevant not only in economics but also in biology and technology. Invention refers to the creation or origin of something new, while innovation occurs when these inventions become economically or biologically significant. For example, in the case of grasses, the invention could be seen as the diversification of the clade 45 million

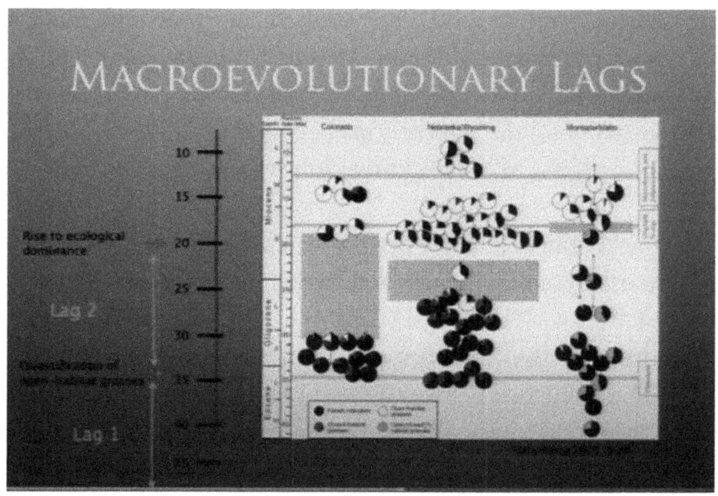

years ago, while the innovation is the appearance of widespread grass-lands 20 million years ago, as evidenced in the fossil record.

This concept of a lag between invention and innovation is not unique to biology or economics. Anthropologists have also observed a similar phenomenon in the origin of early Homo sapiens. According to Chris Stringer's book *Lone Survivors*, the archaeological expression of symbolism among early humans appears to have flickered over tens of thousands

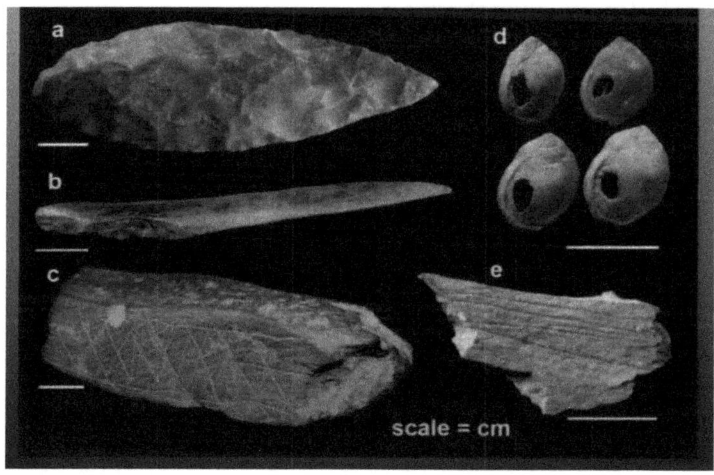

of years in Africa, from around 75 or 80 thousand years ago, until it began to take off more prominently around 45 thousand years ago. This suggests that some innovations may take time to gain traction, and factors such as population growth and the interactions between multiple populations may play a crucial role in facilitating innovation within early human societies.

The artifacts found at the Port Bay locality in South Africa provide evidence of early symbolic behavior among Homo sapiens. These artifacts include red ochre, which may have been used for symbolic purposes, as well as hundreds of thousands of shell beads and finely crafted scrapers and points. The shell beads and engraved lines on the ochre dating back to 75 or 80 thousand years ago are considered some of the earliest evidence of symbolism in Africa.

However, it wasn't until around 32 thousand years ago, as evidenced by sites like the Chauvet Cave in France, that we see a more elaborate expression of symbolic art. The cave paintings found in southern France, such as those in the Chauvet Cave, are remarkable examples of early symbolic expression, depicting intricate and sophisticated imagery created by ancient humans.

Let me, in conclusion, wrap this up into a model, and I'm not entirely wild about the terms. I'm going to take them from a recent paper by Rich Lenski at Michigan State University, who has worked for a long time on *E. coli*. So, how are these spaces created? This is a very simple model that needs a lot more work, but let me just suggest what may actually happen.

The first step is what Rich calls potentiation. You have to create the right environment that can be physical. It can be genetic or developmental. It can be ecological to allow the innovations and the inventions to occur in the first place. Because what really happens in these potentiated environments is that they're then actualized by the actual mutants in a case that Rich was studying, or more generally by genetic and developmental inventions, leading to a new clade, like grasses or some of the things that we saw in the Cambrian. This is a very different model that comes out of the modern synthesis.

This is a very famous figure from George Gaylord Simpson's book *Tempo and Mode in Evolution*, published in 1944. In this model, he has these adaptive zones, shown in white. Time is going in this direction. These are successively more complicated adaptive zones. You can think about this as the origin of flight, for example, from some dinosaur wandering around trying to catch insects, up to fully powered flight up here. But note that these adaptive zones continue off to the left into anatomies.

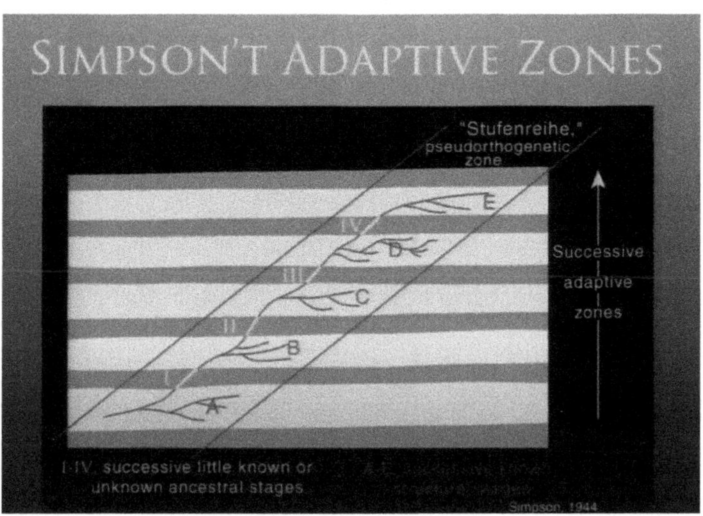

The history of these ideas from the 1940s up to the 1990s showed that people often viewed these adaptive zones as existing independently of organisms that occupied them. In a technological sense, this is akin to thinking that the adaptive zone for the personal computer existed in the time of ancient Rome. That doesn't make a whole lot of sense, so this idea of potentiation that Rich has come up with makes a point that it's the beginning of creating this sort of space. Once the developmental and genetic changes have occurred or been refined by further developmental ecological changes, and we see the same sort of thing in technology, then they're realized as the invention passes to an innovation by ecological expansion, and further evolution becomes successful. So, those are the sorts of steps that come out of looking at some of these events in the fossil record. I'd like to suggest that a similar sort of process and a similar sort of lag may be happening in technological and cultural systems as well.

I'll close with the world's coolest animal. Thank you.

6.2 Discussion

Brian Arthur: Brilliant as usual. Okay. I should say yet again, it was really interesting. You and I have been talking the last few days, and just to bring up a point that came up in our conversations: There are an awful lot of technologies now mentioned too. One is radar and the other is penicillin, where the phenomenon behind the technology has been noticed very early on, just like with some of these, say, grasses where the potential was lurking, you know, millions and millions of years before it was finally actualized. In the case of radar, the phenomenon that high-frequency electromagnetic waves would be reflected by some metal object was known quite early on before 1910. People were trying to send radio waves across the Potomac and Washington, DC. As ships came up the river, they noticed that there was a distortion. That potential for radar was sitting there, but it was at least another 30 or so years before that was developed into full-fledged radar in Britain, in the US, Germany, and a number of other countries. Similarly with penicillin, the famous affected certain spores of penicillium and notatum could wipe out. I am trying to think of the particular bacterium that is Staphylococcus aureus?

Douglas Erwin: Staphylococcus?

Brian Arthur: Yes, yeah, that particular phenomenon had been noticed in 1875, and was noticed again in the 1880s. There's a huge lag. Fleming notices it in his lab in 1929, or thereabouts. But Fleming had been a doctor during the First World War and was very alert to think of such things as possibly being able to cure gangrene, but it still fits when Fleming notices it. Fleming didn't have the tools, the adaptive tools, or the innovation tools, to use Schumpeter's ideas to fully develop penicillin. It was another 13 or 14 years before a team, Florey and Chain and others, was able to fully develop penicillin. So, the bottom line here is that I think that virtually or many of the phenomena or the lags that you've noticed in biology, in the biological record, apply very much to technology. Mechanisms are slightly different, but it's the same sort of thing; lots of technologies, if you trace them back, existed 20, 30, a hundred years earlier, but they're not fully realized until there is a war or some major need and then the same shifts.

Douglas Erwin: Thanks.

Discussant A: I want to tie in what Brian just said with what John was talking about in his talk, which is co-evolution. There has to be some story about how the grasses initially started. There're some soil bacteria that weren't quite right for that, and then it created a home where these particular bacteria could thrive, which then changed the neighboring soil which allowed the grass to thrive even better, etc. — a co-evolutionary story between these different trees and how one's feeding off the other. Can you say more about co-evolution?

Douglas Erwin: Yeah, that's actually the book that James Valentine and I have just published on the Cambrian radiation. Most of it is actually about that ecological dynamic because it looks to us like the primary cause for the Cambrian radiation is actually building out those ecological interactions. It's not development. It's probably not changes in the physical environment, although increased oxygenation was important as well, but it's building those ecological interactions. In the case of grass, it actually may be, at least in part, an environmental change that allowed grasses to spread more broadly…-. It's exactly by trying to understand that those networks of interactions, and particularly the kind of interactions — like as it turns out bioturbation and the spread of oxygen in bottom waters — that affect all members, many members of a community, the kind of things that are normally called spillovers, or other economies have called spillovers, that affect many components of an ecosystem, that I think have a maximum impact on bootstrapping those interactions. That's exactly the thing that I think is actually driving this sort of thing. It's really an issue of how you build and construct niches and interaction networks, rather than just fill them up. Too much of the sort of 1980s view of ecology. 1940s here even was just filling an emptying niche. Most of these niches are not constrained by physical parameters but are concerned about biological parameters. You actually have to figure out how you build that niche.

Discussant B: You didn't mention about the physical environment, but I would argue the physical environment change could also contribute quite a lot. For example, the grassland. If you look at geological history, it is related to the CO_2 concentration in atom's field change. So, if we talk about Cambrian exploration, this was also related at that time. The Earth went back and forth between the ice house and hothouse quite a few times.

Douglas Erwin: Well, yeah. The end of the last glaciation was actually 635. The glaciations… that I didn't talk about… that bracket and these other events, but there is no indication that the glaciations, the Marinoan or the Sturtian, are actually connected to the biological events. There is an increase in oxygen. There are a whole host of other geochemical changes, but there are three problems I'll mention with a purely geochemically driven explanation for the Cambrian. The first is that geochemistry doesn't help you get all those morphologies. It might help you with the timing but not with the nature of the evolution. The second is that with oxygen, it's clear that there's an increase in oxygen at some point in the late Neoproterozoic. The difficulty is that we can't tie down when that is happening, and my view of what we expand in this book is that most of the expansion of oxygen and shallow marine habitats is in fact a consequence of the biological activity. It's biologically mediated, not physically mediated. That's a contentious hypothesis, but I think it's actually within the scope of the data that exist at this point. I also didn't have another half hour to go into all the geochemistry at that time.

Discussant C: It's very interesting, and I understood your model. It was basically the environment changes, but I was wondering to what extent, and maybe the actual structure of the internal changes also, because you said that you know you need to duplicate hox genes. There can't just be one. The potential can be there with one, but you may need duplicates. Could that take some of the time? Going back to John's model, you need to have a number of them, a number of urns maybe, or whatever model you want to think about before you can actually get the explosion so is it...

Douglas Erwin: Well, so the point that [I am] making about the genetic toolkit is that all of that gene duplication, not only in the hox genes but in many of the other transcription factors, has happened by 680 million years ago, so most of the tools, not all of them though. Some refinements happen later, but most of the basic developmental tools are present long before the Cambrian explosion, so it doesn't look like it's the generation of new developmental tools.

Discussant C: No. One other point…, but it could well be in a technological evolution that it is a need to [develop]capabilities as well.

Douglas Erwin: Oh, sure. Yeah, I mean, this doesn't have to apply to everything, but the point is that what you want to do is understand the different components and how they interact.

Discussant D: Hello. I have three questions, and the first two are about the world's coolest animal, so forgive me for a bit of trivia. My first question is, what is that thing's name? Secondly, what exactly is the evolutionary purpose of five eyes? Lastly, is your model about potentiation and actualization correct? So, there is a rather long lag time between potentiation and actualization…

Douglas Erwin: Can be, yeah.

Discussant D: From what I understand of evolution, the potential doesn't directly contribute to fitness. So, is there a danger that the potential will be destroyed during the lag time?

Douglas Erwin: I'll take the last question first. My guess, and that's actually one of the things that Rich Lenski saw in his experiments with *E. coli*, looking at the citrate mutant. They were examining a mutant in which the *E. coli* could feed off citrate, which is not a normal food source for *E. coli*. It turns out that many of the potentiation changes happened in other lineages, but those lineages died out before the actualization changes could occur. So, in many cases, you'll get halfway there and then the whole thing won't pan out. This creature is called *Opabinia regalis*. I have no idea why it has five eyes, but it obviously wasn't a great idea because it's only known from about three meters of rock on one hillside in British Columbia, so it wasn't a great success even if it is really cool. Oh, by the way, since I'm a museum scientist, I have to say this: You all know we have really dumb politicians in Washington, from both parties. We periodically receive these inane questions from the Hill about why we can't get rid of a lot of the fossils. They suggest we could just take pictures of them and save on storage space and costs. We could throw out all the stuff we don't need, which frankly is how many of us feel about our politicians right now. But as it turns out, you can't actually see all five eyes on any single specimen. You actually have to look at a bunch of specimens under different imaging techniques to finally convince yourself that all five eyes [are] present there, which I tried to explain to a couple of congressmen. It doesn't work.

Discussant E: Is there anything we can learn from the so-called failure cases, the evolutions that didn't make it, like the world's coolest animal?

Douglas Erwin: I'm glad you think it's the world's coolest animal. That's good. Yeah, actually, this is a really instructive case.

Let me go back to that slide that I skipped over. This is an evolutionary tree of the lobopodians. These things, like Aysheaia and the living

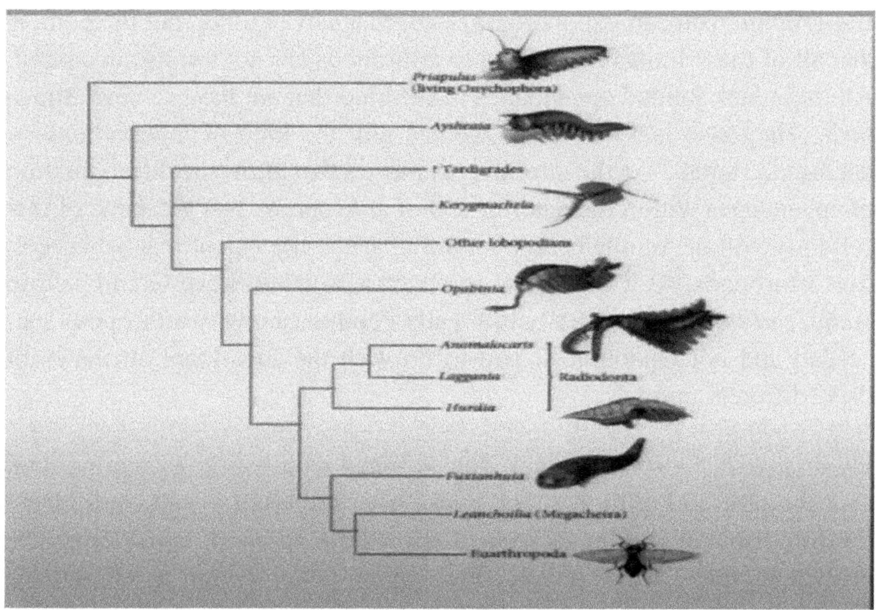

Onychophorans, which are up here, Onychophora, or there's Aysheaia... Hallucigenia would be in here, Opabinia's down here, Anomalocaris, and then eventually you get to the arthropods. That's probably the wrong position for tardigrades, but this is an evolutionary tree showing the phylogenetic relationships between these different groups of animals. The only ones that were successful, incredibly successful, are these things down here — very successful. Tardigrades, restricted, are in these little water rivers that you can get in toy shops. So, they're really good, and they've succeeded in toy shops. And Onychophorans are not noticeably successful. I have to stop telling stories, but this is a good story. The only known photographs of Onychophorans giving birth, they give live birth, were taken in 1995 when the *National Geographic* was doing a story on the Cambrian explosion, and they went down to Australia to take pictures of living Onychophorans. The woman was setting this whole thing up. As he got this thing ready, and he's beginning to take pictures, it started giving birth right, and the woman who actually studies these things had never actually seen them give birth before. So, apparently if you're the *National*

Geographic, you can get photographs of that sort of thing, but the point is that all of these things leading up to arthropods are not totally successful. All these lack jointed appendages of the kind that we have in new arthropods. They also lack, as far as we can tell, the kind of innervations — nervous systems — in the interim appendages that allow the differentiation of appendages within the anterior end of arthropods. You get some of that certainly within Anomalocaris but nothing like the extent that you see in true arthropods, so it looks like in contrast to what Steve Gould said in *Wonderful Life*, there actually is a really good reason why arthropods succeeded and not Opabinia. It had to do with the functional attributes of these different animals.

Discussant F: Good afternoon. You've given us a really fascinating look over the past 500 million years. I was just wondering if you could, from the Smithsonian's point of view, look a little forward, considering the challenges that Earth is facing. For example, we've seen a 30% rise in CO_2, and with a lot of these global threats happening around, how does the Smithsonian see Homo sapiens, the species, progressing in the next…

Douglas Erwin: Well, first, let me clarify that I don't speak for the Smithsonian, especially considering those comments about the congressman a few months ago, since they actually foot the bill sometimes. So, anything I say will be my personal opinion and not the view of the Smithsonian or the government — yes, our government. I've spent a long time studying the end-Permian mass extinction, which is the largest mass extinction in the last six hundred million years. It was not the one that killed the dinosaurs. It was actually a more interesting extinction 250 million years ago that killed about 90% of everything in the ocean. The lesson that I take from that is that you don't want to get into a real mass extinction. There are a lot of people in tech. I think we'll have a talk this afternoon or tomorrow about the Anthropocene. Simon may also be discussing it in his talk. There are many people who have argued that we're already in the midst of a mass extinction. If that's true, I would buy a good case of scotch, because it isn't going to be fun. I very much doubt whether you can actually do a lot if we're actually in a mass extinction.

My caveat, however, is that I don't think we're actually in a mass extinction of a magnitude that we describe as paleontologists. Let me explain that because this is a point that gets missed and there are a lot of the discussions about mass extinction. It's not that we haven't lost tremendous amounts of biodiversity and it's not that we're not going to lose more biodiversity, both with habitat loss with destruction of the environment and with what's going to happen from global warming. What we see in the fossil record — [with] the exception of Opabinia, things like that aside — are widespread, durably skeletonized, and abundant taxa mostly in the ocean. We don't see things on land. We don't see things that have a small population size. We don't see things that are otherwise invisible in the fossil record. Widespread, durably skeletonized, and abundant geographically widespread, also temporally widespread, aren't the things that have gone extinct. By and large, things on islands have suffered tremendously, things in freshwater habitats, a lot of things on land; so if you compare apples to apples, and if you look at the same kind of data, most of the kinds of taxa that we would be able to pick up as paleontologists 10 million years in the future, say, are not the kinds of species that have disappeared yet. That means to me that although we've already done a tremendous amount of damage to the planet, we're not yet at the threshold of a mass extinction of the kind that I study as a paleontologist. It doesn't mean that what we've done is a good idea and it doesn't mean that we won't get there eventually, because we will — maybe not too long a time — but once we've gotten there, I'm not sure that's a whole lot we can do about it. Frankly, my own suspicion is that part of that mass extinction will include humans. I take the fact that we're not yet in a mass extinction as a very cautionary note to make sure that we don't get into one.

Robert Axelrod: In one of your early slides. Go back to where, yeah…

Douglas Erwin: This one?

Robert Axelrod: Yeah. So, can you explain the difference between what had happened about 550 million [ago]versus 520 million [ago]?

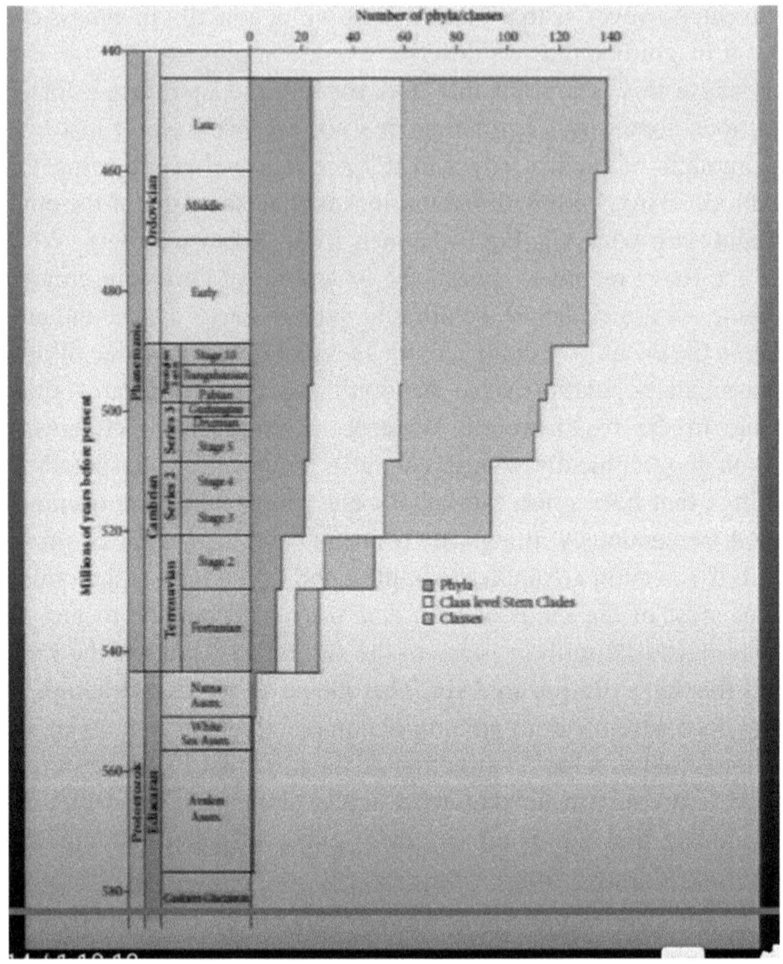

Douglas Erwin: What happens down here is that we have the occurrence of this Ediacaran biota. These are the soft, funny things that have no mouth, no appendages, nothing. They last from the first taxa date to 579 in Newfoundland, and they disappear with 542 in Namibia. We actually just had a paper out on whether or not there's a mass extinction of these things, but with the exception of Kimberella, none of them are complex animals; some sponges are, but not these more complex animals [that appear later]. Then, after 542, we see the rapid appearance of many different kinds of animals. Many of these things that we see up here in stage three probably evolved down here in stage two, and maybe even in stage one, but these body plans, for various reasons, are unlikely to have evolved further down that. It turned out to be three phases of what we see — this first Ediacaran biota here, the small shelly fauna of these plates, tubes, and things in the Fortunian stage two, and then all these other bilaterians in stages three and four.

Robert Axelrod: This is obviously a schematic, but do you think that the diffusion was as quick as this suggests, within a million years, for this to occur?

Douglas Erwin: This is based on the accumulation of data, specifically the first occurrences of fossilized classes. So, it reflects actual fossil evidence. The timeline depicted here closely aligns with what we observe in the trace fossil record, including burrows and tracks. While there may be slight variations in the timing of the origin of these organisms, it's unlikely to deviate significantly from this timeline, typically within a couple of million years. This debate about the rapidity of the Cambrian explosion has persisted since my graduate school days and throughout my career, with the consensus now leaning toward an even shorter time frame.

Robert Axelrod: You think that happens in other forms of innovation?

Douglas Erwin: The best counterexample to this pattern is actually seen in trees and plants, particularly vascular plants. Their evolutionary trajectory differs significantly from that observed in the Cambrian explosion. Instead of a rapid diversification followed by a decline in diversity, trees and plants exhibit a different pattern. In the Devonian period, around 380 million years ago, we witnessed the emergence of various tree architectures, which then persisted over time. While the number of different tree types may decrease, other aspects of plant structure continued to evolve successively. This illustrates that the pattern observed in the Cambrian is not universal across all organisms. Interestingly, similar patterns of rapid diversification followed by stabilization can be observed after major mass extinctions. For instance, a recent study by Anne Yoder in *Nature*, published about two or three weeks ago, demonstrated that placental mammals underwent a similar burst of diversification after the end of the dinosaurs. We also see comparable phenomena after the end of the Permian extinction event, where various taxa such as Ichthyosaurus, Plesiosaurus, turtles, mammals, and dinosaurs experienced a rapid expansion within five to seven million years following the extinction event. As a side note, turtles rank as the second coolest thing after Opabinia.

Discussant G: Good afternoon. Just now I got this impression from your answer that you probably think that this massive extinction will anyway happen no matter what we do. Is this impression correct?

Douglas Erwin: No. I think we still have some hope. There are many people who have claimed that we're already in what's called the sixth-grade mass extinction.

Discussant G: I know. No. My question is this: Even without human beings, do you believe a mass extinction will occur on its own, or might it not simply be a matter of chance, or influenced by different factors?

Douglas Erwin: If humans had not evolved, another mass extinction would still occur at some point. Massive volcanic eruptions would still happen, and another bolide might fall out of the sky. However, it's unlikely that it would be happening right now. What's happening now is because of people.

Discussant G: So, what you're thinking is like this: Mass extinctions are mainly caused by some external force, like something colliding with the Earth, or something similar. The system itself may not really have this kind of mass extinction on its own, just like a collapse without anything colliding with the Earth.

Douglas Erwin: That's a really good question about whether mass extinctions can be endogenous or driven by external factors. Stuart Kauffman has argued for years that some extinctions can be caused by the internal collapse of ecosystems. However, there's actually no evidence for this in the four or five classic mass extinctions that are best studied — all have evidence of external drivers. The two biggest extinctions, the one at the end of the Permian due to the Siberian eruption of massive volcanism in Siberia and the K-T impact, have very significant external drivers. So, while it's certainly possible, we don't actually have any evidence that that's the case.

Discussant H: What was the one that doesn't have an external force? You said four out of the five. The last one didn't have an external factor?

Douglas Erwin: Well, it's the Devonian. We don't really understand whether there's even a mass extinction in the Devonian. There are a series of biotic crises through the late Devonian. Whether they actually amount to a mass extinction is something that a lot of people have disputed. So, there are certainly a series of changes in ocean chemistry during that time that could be driving the patterns we see, particularly in Europe and to some extent North America, but whether that even really constitutes a mass extinction is unclear. That's why I left it out.

6.3 Summary of the Talk

Douglas Erwin's talk explores the enigmatic phenomenon known as the Cambrian explosion, a pivotal event in the history of life on Earth that occurred roughly 550 to 600 million years ago. The Cambrian explosion marked a period of rapid diversification of animal life, giving rise to a plethora of new taxa and anatomical designs. Erwin begins by showcasing various fossils and reconstructions representing some of the taxa that emerged during this evolutionary burst, including sponges, echinoderms, and the iconic Opabinia.

The Cambrian explosion represents the construction of a vast "design space" within a relatively short geological time frame. Erwin poses key questions about how this design space was constructed: What environmental factors played a role? To what extent were genetic and developmental changes involved? How did ecological processes contribute to this diversification? These questions are not only relevant to understanding ancient biological evolution but also have parallels in modern discussions about cultural and technological innovation.

Erwin provides an overview of the timeline of major animal groups appearing during the Cambrian period, highlighting a burst of phyla and classes between 542 and 520 million years ago. By the end of the Cambrian, most major animal groups, including vertebrates, had emerged, with Bryozoans being a notable exception.

The understanding of this ancient interval is greatly aided by exquisite fossil deposits found in various parts of the world, preserving soft tissue anatomy and providing insights into early animal life. Examples include the Ediacaran biota preceding the Cambrian explosion and the Burgess Shale deposits from the Cambrian itself.

One notable aspect of the Cambrian explosion is the rapid increase in morphological diversity, measured as "disparity," which precedes the increase in taxonomic diversity. This pattern suggests a burst of innovation in anatomical forms before a subsequent increase in the number of distinct taxa.

Erwin delves into the role of genomic and developmental changes in shaping this evolutionary event. Molecular clock analyses indicate that the genetic toolkit for building complex animals predates the appearance of fossils from the Cambrian explosion by about 150 million years. The conservation of key developmental genes, such as the hox genes, across diverse animal groups underscores their importance in shaping body plans over evolutionary time.

Furthermore, studies of gene regulatory networks reveal conserved elements that have remained stable for hundreds of millions of years, suggesting a hierarchical structure in developmental processes. These networks provide insights into the long-term stability of certain anatomical features while allowing for variation in others.

Ecological interactions also played a crucial role during the Cambrian explosion, with the emergence of predation and ecosystem engineering shaping evolutionary trajectories. Changes in sedimentary records indicate shifts in ecological dynamics, such as the transition from planar laminated sediments to mixed sediments due to increased burrowing activity by early organisms.

Erwin concludes by proposing a model for understanding evolutionary innovation, highlighting the importance of potentiated environments, genetic and developmental inventions, and ecological expansion in driving major evolutionary transitions. He suggests that similar processes and lags may occur in technological and cultural evolution, drawing parallels between biological and cultural innovation.

In summary, Erwin's talk provides a comprehensive overview of the Cambrian explosion, shedding light on the complex interplay of genetic, developmental, and ecological factors that drove one of the most significant events in the history of life on Earth.

6.4 Relevance of the Talk by Douglas Erwin to the Current Stage of Research

Douglas Erwin's talk on the Cambrian explosion and evolutionary innovation holds significant relevance to the current stage of research in several ways:

Understanding the Cambrian Explosion: Erwin's discussion provides valuable insights into the Cambrian explosion, a pivotal event in the history of life on Earth. By examining the rapid diversification of animal life during this period, researchers gain a better understanding of the processes and mechanisms that drove evolutionary innovation.

Exploring Developmental Biology and Genetics: Erwin delves into the role of genomic and developmental changes in structuring evolutionary innovation. His discussion on the conservation of regulatory networks and genetic toolkits sheds light on how fundamental biological processes influence the emergence of complex traits and body plans.

Integration of Fossil Evidence and Molecular Data: The talk highlights the integration of fossil evidence with molecular clock analyses to elucidate evolutionary timelines. This interdisciplinary approach allows researchers to reconcile discrepancies between the timing of genetic innovations and their fossil record manifestations, contributing to a more comprehensive understanding of evolutionary history.

Ecological Interactions and Ecosystem Engineering: Erwin emphasizes the importance of ecological interactions and ecosystem engineering in driving evolutionary dynamics. By examining the ecological implications of evolutionary innovations, researchers can better comprehend the interconnectedness of biological systems and their influence on ecosystem structure and function.

Macroevolutionary Lags and Innovation: Erwin's discussion of macroevolutionary lags challenges conventional views of evolutionary progress and innovation. By identifying delays between the origin of genetic innovations and their ecological expression, researchers gain insights into the complex interplay between genetic potential and environmental constraints in driving evolutionary change.

Implications for Technological and Cultural Evolution: The parallels drawn between biological and cultural evolution offer intriguing insights into innovation processes across different domains. Erwin's discussion prompts researchers to explore analogous mechanisms underlying technological and cultural innovation, fostering interdisciplinary dialogue and collaboration.

In summary, Douglas Erwin's talk provides a multifaceted perspective on evolutionary innovation, drawing connections between developmental biology, genetics, ecology, and broader evolutionary patterns. By elucidating the mechanisms and dynamics of innovation in biological systems, his insights offer valuable implications for understanding innovation processes in other domains, including technology and culture.

Chapter 7

A Dynamic Constructivist Approach to Culture

Ying-Yi Hong

YouTube: https://youtu.be/wRqxG4x10AE

Speaker: Ying-Yi Hong

Moderator: Cheong Siew Ann

Discussants: Sheila Ronis, Greg Fisher, Discussant A

7.1 Talk by Ying-Yi Hong

I think, today, we all know that we live in a complex world. Here are some pictures I took around the world. For example, here's Hong Kong, and here, guess where? Singapore, at Jurong Point. If you live around this area, you know it's quite complex. When you go there, you see a bunch of people who come from mixed ethnic and racial backgrounds. And in this ever more complex world, there are a lot of things that we need to cope with, so there's a need for us to coordinate, adjust, and change to meet these challenges. In this talk, I want to use the research evidence from a lab pertaining to the complex nature of the world to demonstrate that culture is dynamic, adaptable, and emerging.

I attended the first Complexity Conference last year, *More is Different*, and there's one thing that stuck in my mind for a year. It's an unforgettable video from last year, so I want to start with that. I want to share that with you. Let's see.

Now, I want to use this video to draw an analogy to culture. All these metronomes represent individuals, and the cans serve as a conduit for them to coordinate. So, in culture, the cans can be seen as a metaphor for the shared assumptions and lay beliefs that a group shares, and these provide a conduit for individuals in a group to synchronize and coordinate their behaviors. How persuasive are the shared assumptions or lay beliefs? Why are they important?

Let me give you another demonstration. Look at this surface; it's like the surface of the moon. When you look at it, most of us will see these three bigger dots protruding. You have this perception. Perhaps you see these as concave to the ground. But why do you have that perception? Have you asked yourself? What are the underlying assumptions? Exactly. The light comes from the top, so the upper ends receive the light, while the lower ends are in the shade. These assumptions fit into the ecology we are in, and we use this assumption of the direction of light to infer what we'll see in this three-dimensional world.

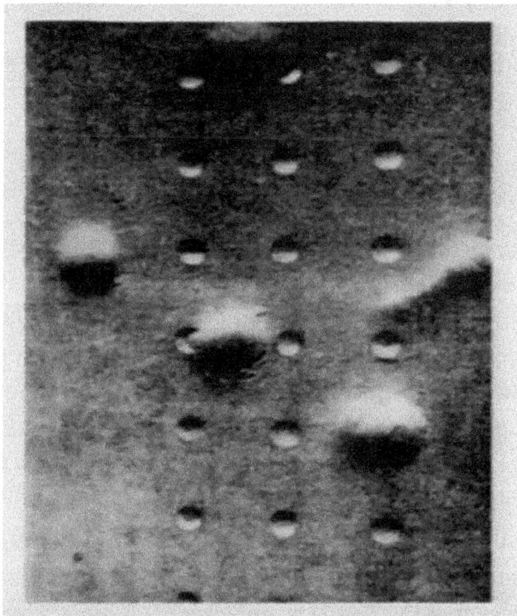

When a group of people lives in the same ecology, they will have the same types of assumptions, and these assumptions also get shared in our daily dialogues, our conversations, etc. However, if you're from another planet or living in another ecological environment where the light source actually comes from below, what would you see? Probably now, especially with these three dots, you'd immediately see them as more protruding rather than concave into the ground.

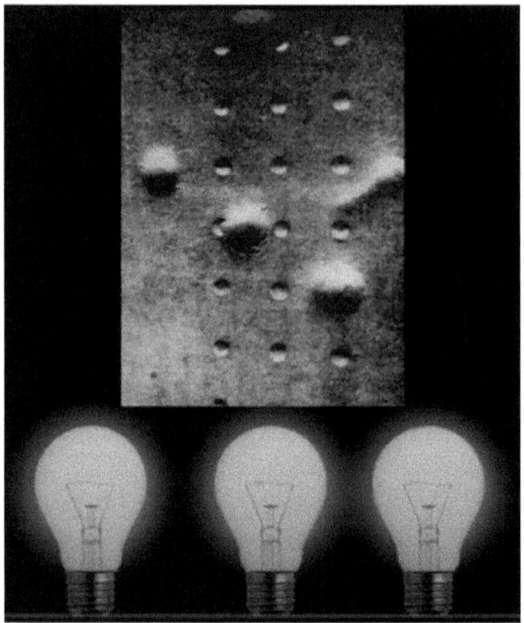

Now, I want to use this as an example to introduce the first two postulates in a theory that we proposed some years ago. This theory is called a dynamic constructivist approach to culture, which we introduced in the year 2000 in a paper published in the *American Psychologist*. Here, we define culture as networks of shared meaning, such as lay beliefs, that are produced, distributed, and reproduced among a collection of interconnected individuals. I know that there are many ways to define culture, and this is one of them. It's also the way that we operationalize culture in our own research, so I want to put that up front.

Additionally, people acquire theories of our world from their cultural groups, as I mentioned in communication. These theories are tools that people use to interpret and understand specific domains of our social world.

In the talk that follows, I want to draw your attention to or make a case that the dynamic constructivist approach can be used to understand and study a wide range of phenomena, from micro-level brain activation to more macro-level social issues, such as intergroup relations. Furthermore, it has implications for social harmony. How so? I will use some cartoons.

At the very beginning of this talk, I'll discuss some brain data and brain research, followed by some interviews. Then, I'll talk about some experiments involving self-reports and also choices people display, along with some skin conductance evidence. Finally, at the end, we will go into the Twilight Zone. I won't tell you just yet what this is, but hopefully it will be a pleasant surprise. We'll see.

Now, let's start talking about the brain. Ten years ago, a professor at the University of Michigan, who also worked with John Holland years ago, conducted extensive research showing that Chinese and Americans have different behaviors in terms of how they look at certain pictures. For example, consider this picture: There is a focal object, an airplane, over a landscape. Extensive evidence, including behavior, memory, and eye tracking data, demonstrates that Americans pay more attention to and focus more on the focal object, whereas Chinese, in comparison, focus more on the context.

My colleagues and I, some years ago, thought, "Oh, you know, can we replicate this? Or can we see this even in neural imaging, using neural imaging techniques?" So, the specific areas that we looked at are the object processing areas, the lateral occipital complex, which are located here and here on the two sides, and they have been widely studied already.

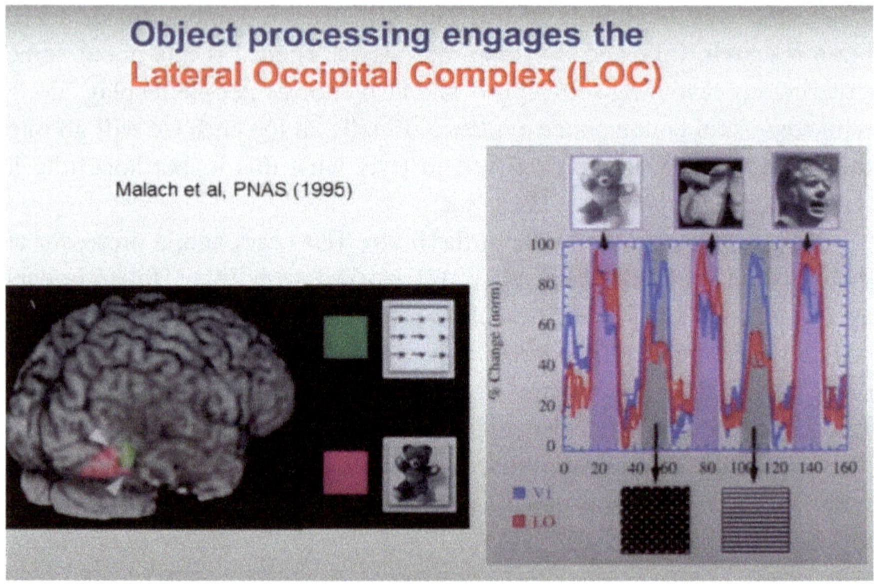

So, we focus on this area. Now, the technique we use is functional magnetic resonance adaptation. It's a mouthful, but actually, the logic is not that complicated. Let me go through this with you.

Let's say, we show you two pictures, the same picture, identical picture, quickly. Then I let you look at the signal at the LOC area, which I just told you is the object processing area. That indicates activation when you see the first picture, shown by the yellow line here. This activation occurs because the brain recognizes the picture as containing an object. Then, when the second identical picture is shown, you see the light blue line, which has a lower magnitude than the yellow one because it's the same picture, so your brain quiets down — it's called adaptation — since it's not interesting anymore, having already processed the information. If you add these signals together for a succession of quickly presented images, you get a resulting curve.

Now, if we show two different pictures quickly, there will be a bigger summation because both instances of activation indicate new information being processed. So, the resulting summation is higher. This is what happens when the images are not repeated. However, if you compare this to the repeated images, you'll see that the non-repeated images have a higher summation than the repeated ones.

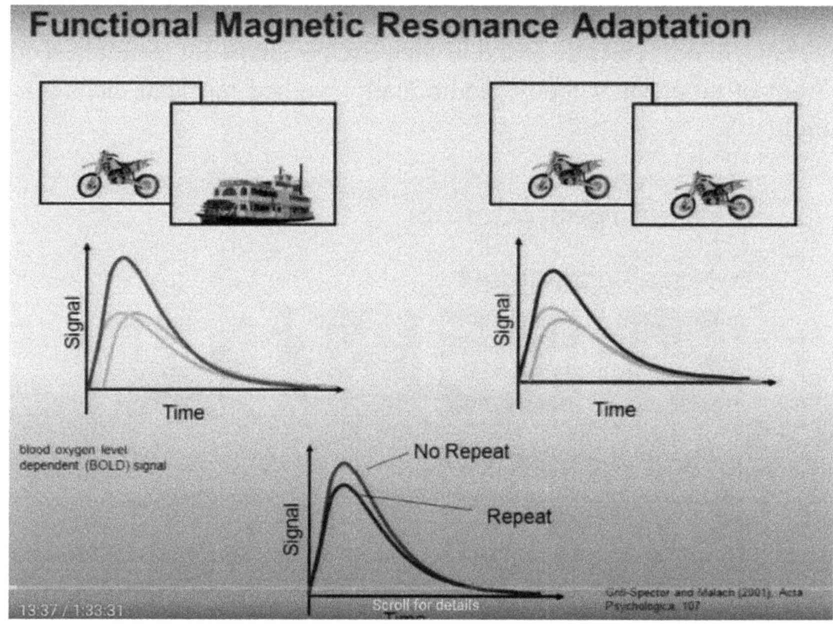

We took advantage of this logic in our research. We showed participants different types of pictures, four in a row, very quickly. One type was
called novel congruent, where the picture is situated in a congruent background — typical backgrounds, like a camel in the desert or a deer in the
jungle. This is non-repeated. We also had a repeated congruent, such as a
pot on a stove.

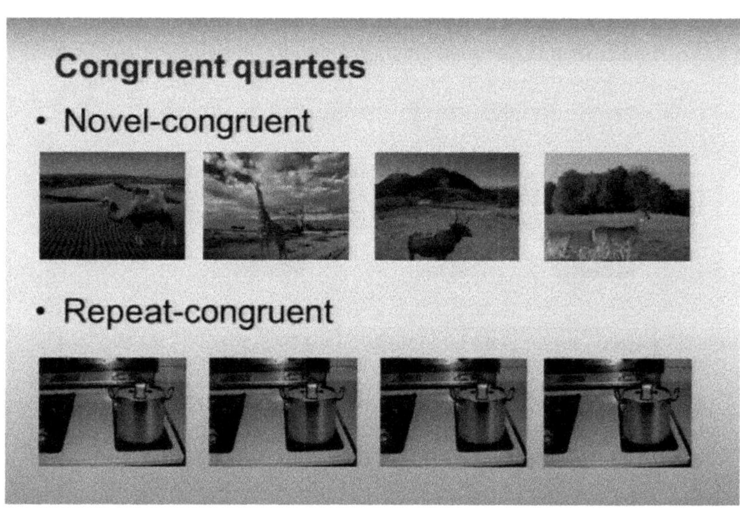

Interestingly, we also created some bizarre stimuli, where the focal objects were not typically found in such backgrounds, for example, a cow in front of an elevator lobby. Additionally, we had repeated incongruent stimuli.

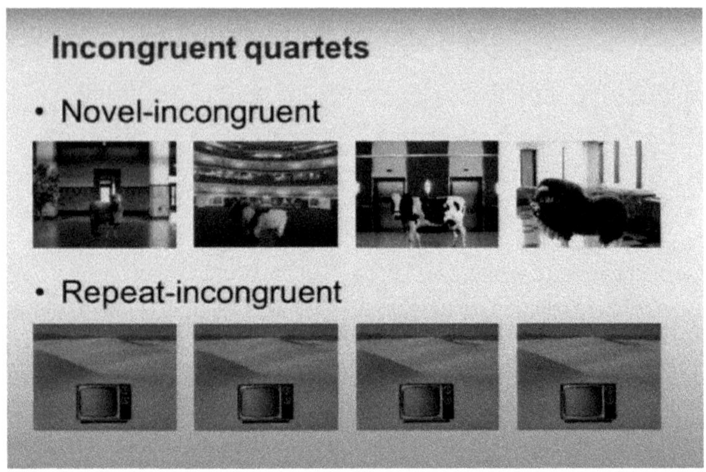

We hypothesized that Chinese participants would show greater adaptation to incongruent stimuli, consistent with their tendency to process contextual relationships more extensively than North Americans. So, in our experiment, we expected these responses to be especially salient for Chinese participants compared to North Americans.

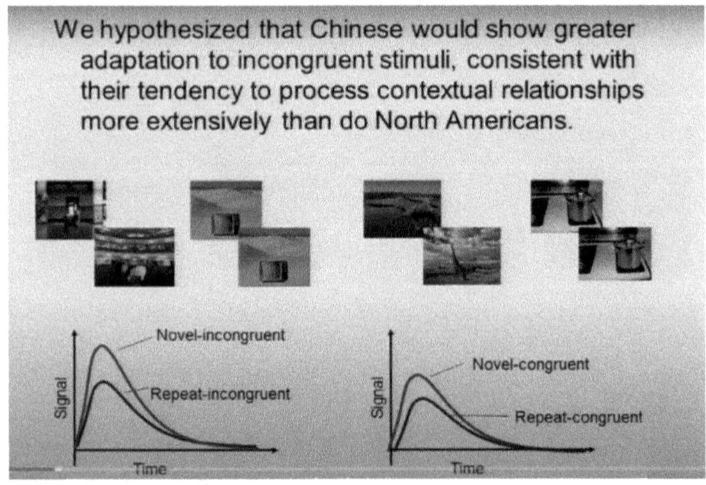

We ran this experiment at UIUC when I was teaching there. We had 16 Chinese participants, mainly grad students who had just arrived, and 16 Euro-Americans who were local. We matched their gender, age, and backgrounds. We scanned them using functional MRI techniques, presenting them with four types of stimuli in a sequence.

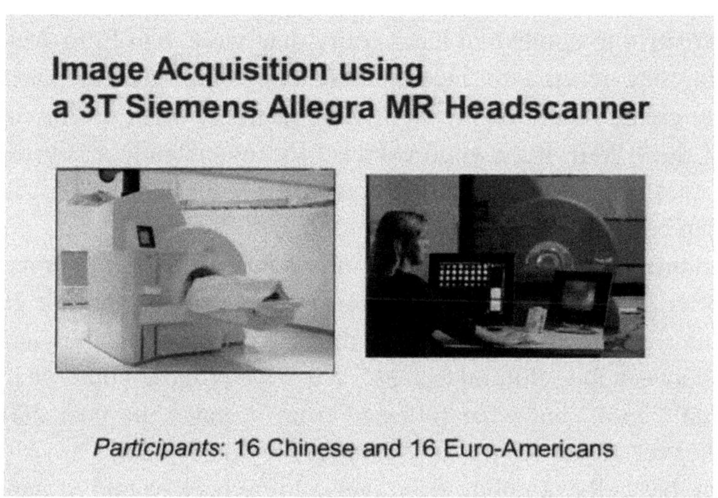

During the scans, participants simply lay down without making any decisions or responses. We just showed them the pictures. In terms of brain structure, both Chinese and Euro-Americans showed no differences, which was good news. We then looked at their brain activation in the LOC area as a function of the different types of context pictures we showed.

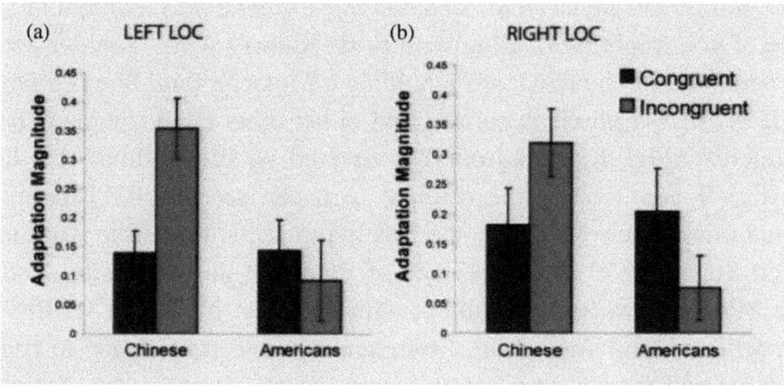

The results were as we predicted. Chinese participants showed greater firing in the LOC area when presented with incongruent stimuli compared to Euro-Americans. This was particularly evident in the left LOC area for Chinese participants, while Euro-Americans didn't exhibit the same response.

In summary, these results indicate that Chinese participants were more sensitive to contextual incongruity than were than Euro-Americans and that they reacted to incompatible object/background pairings by focusing greater attention on the focal object. So, what are the implications of this? Well, it's a good exercise for understanding neuroscience, and it raises interesting questions about whether these neurological differences can be changed.

And this brings me to one of the important postulates in our model. In this model, we contend that after extensive exposure to multiple cultures, individuals can acquire multiple lay theory systems and can dynamically switch between the "cultural frames." When we proposed this, we thought, "Oh, that's cool," but what followed from it, since the year 2000, is a whole host of new ways to study culture. Previously, it was cross-cultural research, basically sampling from one cultural background or nationality and then sampling from another and comparing them. But after this proposal, we can manipulate — we can randomly put people into different cultural frames and look at the consequences. This would afford more causal inference in our own research.

Now, what does that mean experientially? Well, this painting illustrates that very nicely. It's from Frida Kahlo and is entitled *Self Portrait on the Borderline between Mexico and the United States*, painted in 1932. Some of you might be familiar with Frida Kahlo's work. You can see her face here, and in one hand, she's holding a Mexican flag. She is a famous Mexican artist, a Mexican painter, and in her other hand is a cigarette.

Interestingly, the background is covered by illustrations of a lot of industry. If you look at the original, you can see "FORD" there. She painted this in Detroit, where she was living at the time. She lived in the United States for extended periods of time. On the other side, you see more Mexican indigenous culture, with Mexican pyramids, fertility figures, artifacts, and vegetation. From her own correspondence to friends, she explained that the goal of this painting is to express her feeling of falling between cultures and her sense of isolation and loneliness at that

Frida Kahlo: *Self Portrait on the Borderline between Mexico and the United States*, 1932.

time when she was staying in Detroit. We'll come back to this point, but nonetheless, you see this type of consciousness or awareness of the two cultures within an individual.

Now, this is another demonstration of the phenomenology of cultural frame switching, and it's from our own interviews. So, we interviewed Chinese Americans who lived in the United States — for example, in this interview from a student who grew up in Taiwan and came to the United States to study for some years.

> The phenomenology of *cultural frame switching*
>
> "Um… well since I live in both places [Taiwan & the US] and like every year I go back to Taiwan to visit I find myself changing within the two cultures. It's like I go back to Taiwan they sometimes get scared of me because I'm too open and stuff. So I try to be more and more like um like fit into their definition um but when I come back I sometimes feel myself a little bit overwhelmed like oh everybody's so open but then after like a month I get used to it. And then I go back to Taiwan again and then they're not used to me again so it's like a cycle kind of…"
>
> (excerpt from an interview of a Chinese-American, from Chen, Roisman, & Hong, 2005)

So, you see him trying to adjust every time he'd go back and forth. All these are very nice things, but how do we do an experiment? We were fascinated by the phenomenon, fascinated by all the manifestations of that, but how do we do an experiment and how do we study it? So, we developed a way we call cultural priming. Let me illustrate what I mean by cultural priming. Let's say Chinese American biculturals, these individuals have been extensively exposed to both Chinese culture and American culture. According to our theory, because you can learn new cultures, you can apply culture, and all these acquired knowledge systems will be tools for you to see the world. So, here I use a metaphor: This person, a bicultural individual, should have both the Chinese cultural system and American cultural system in the brain. But this is just a metaphor, so I don't mean that American resides here in the prefrontal lobe and the Chinese resides in the occipital lobe. No, it's just like two systems in the same person's mind.

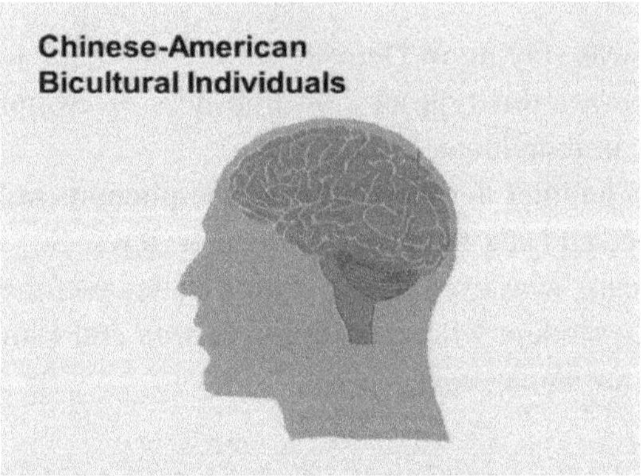

The cultural priming method is through external stimuli. One way is to use cultural icons. For example, the Great Wall here. When a Chinese American sees the Great Wall, this cultural symbol actually reminds them of the Chinese way, the Chinese knowledge system, so it gets activated, or we say the accessibility of the Chinese knowledge comes to the head, like increased accessibility of that knowledge.

The Statue of Liberty, again, similarly, can activate the American knowledge system. So, with that in mind, the general hypothesis we have is that if you expose participants who are bicultural, let's say here Chinese American, to the respective cultural primes, immediately their accessibility of the corresponding cultural system should be higher than the other, and then they should display behavior or thinking or even emotions that are typical of that culture.

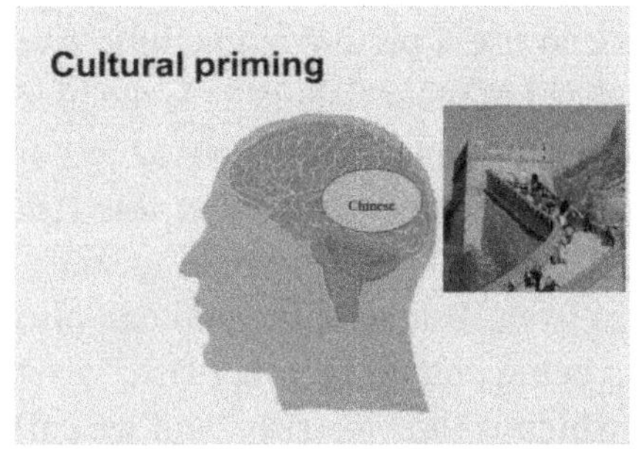

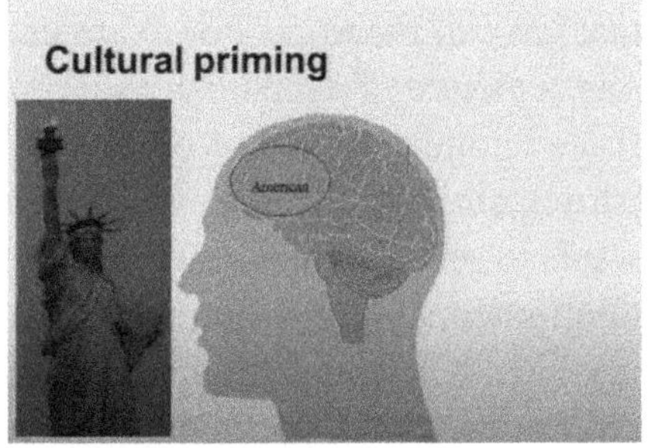

So, in our early study, what we typically do is to find a group of Chinese Americans and then randomly assign them to one of these priming conditions, for example, Chinese cultural priming, American cultural

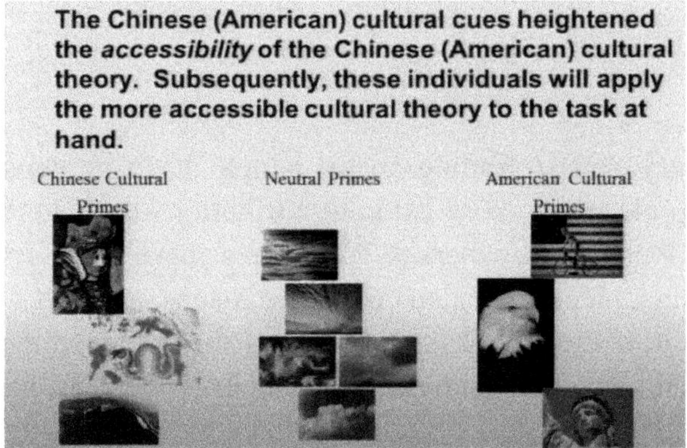

priming, or the control, the neutral pictures, which don't have any meaning in culture. And then they will be shown these icons, and afterward, we measure their responses.

Now, what can we measure? Actually, there's a whole host of things that we can measure. For example, we can use a picture, something like this, a picture of a blue fish swimming ahead of other fish. Why is the blue fish swimming ahead of other fish? That was a question we posed to our participants. Well, there are various reasons. You can say that this fish wants to seek food, this fish is a leader, or this fish might be finding something interesting in front, so it swims in front. But actually, you can also say like, "Oh, this fish, the blue fish, is being chased by other fish," so it can put the causal focus on the target, which is the blue fish. And we call this type of attribution internal causes attribution, internal attributions, where the blue fish is leading other fish, so it's internal to the target, or external causes, for example, the blue fish is being chased by other fish, which is external to the blue fish. Do you see the locus of causality, different here?

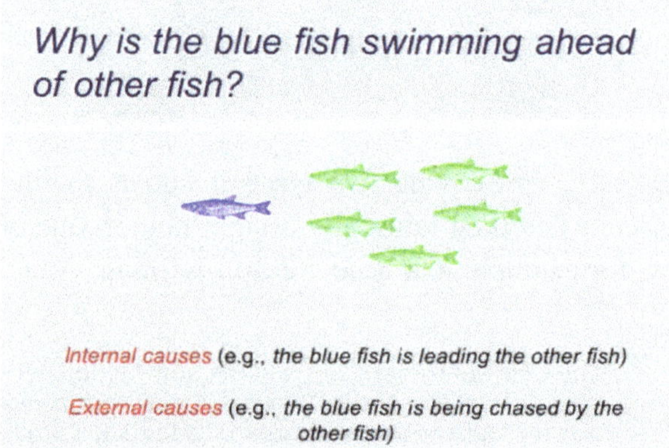

One very robust finding many people have demonstrated is that Chinese people make more external attributions than do Americans and also other Westerners, whereas Westerners make more internal attributions than do Chinese and also other East Asians, and that's quite robustly found. So, now the question is, if that is the evidence already, can our priming method actually activate that and show that? We have done many versions of this, and indeed, it's shown in the different ways that we measured this — using fish, using a picture of a pig, horse, all right, stories, other sorts of scenarios — we still can find this, and so this is an illustration of the outcome. So, three groups of Chinese Americans we randomly assigned to these positions: American picture, neutral, and Chinese. And then the participants in the Chinese picture conditions made more external attributions than did the American participants in the American conditions. Okay, and so far, actually, that evidence is quite huge already. So, we have demonstrated cognition in terms of external versus internal attribution, behaviors. We can put people in prisoner's dilemma games and observe their competition versus cooperation, but still we find cultural priming; in fact, cooperation versus competition or, relational versus egocentric, elicit an emotional projection. So, we find this evidence in that.

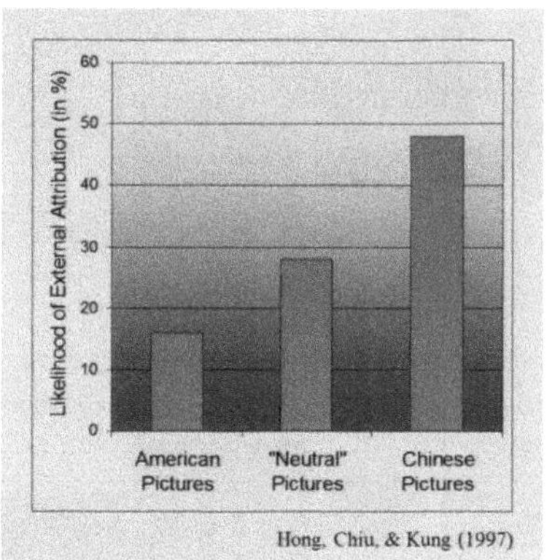

Hong, Chiu, & Kung (1997)

Now, the question is this: Does it work on Singaporeans? Singapore is a place where East meets West, right? So, Singaporeans are arguably bicultural as well, just like Chinese Americans or Hong Kong Chinese. Actually, we have also done it on Hong Kong Chinese and it works. Does it work in Singapore? Yes, it does. Colleagues in the Business School have demonstrated it, and colleagues in the Psychology Department here at NTU have also demonstrated it. So, this is one example: My colleague Sharon Ng from the Marketing Division at Nanyang Business School... So, she has icons like these Singaporean icons, and you see Singapore

Airlines here, the Singaporean flag, yeah, also featured here. And then, US corresponding icons. The dependent measure was consumer behaviors, and she also found it very consistent. In fact, we have dependent measures on managerial practice or Chew Ling's research on internet behavior. You can look at people's behavior on *Facebook* or *Ren Ren*, which is the counterpart of *Facebook* in Mainland China.

Now, how does this relate to brain research? If we tie it back to the brain, does the brain change or can the brain adapt to priming? Okay, let me show you something. Now, Asian and Western cultures, there are many theories, and on their website is Markus and Shinobu Kitayama's theory about different self-construal. So, in their theory, they thought, typically in Western cultures, individuals are more independent, so the self and close others might be disjointed or somewhat disconnected, whereas in East Asian cultures, individuals are more interdependent, so the self may overlap with close others. Under these different conditions, when you evaluate the self, in the Western concept, the independent self-construal concept, you avail yourself in more general terms. For example, "I'm talkative." Now, in Eastern, because yourself also includes particular others, you are giving yourself more contextually, and so, you would think, "Okay, I'm talkative with my close friends," which is specific to a certain relationship. So, the two types of self-construal might give rise to different relevancy of the self.

This study was conducted by Joan Chiao, a professor at Northwestern University. She based her research on a theory and conducted functional MRI studies. She recruited participants from Japan and Caucasians from Illinois, USA. The participants performed different tasks while lying down and looking at a screen. They had to make decisions based on the font of words, indicating whether they were italic or not, and press a button for yes or no responses.

This served as the comparison control. This is like the translation when talking to my mother, the Japanese and then the English. The participants were asked the same question, but needed to respond in a more contextual manner, specific to a relationship. Now, the general: In general, how would you describe yourself? This is more relevant for an interdependent or collectivistic self, while that is more relevant for an independent or individualistic self.

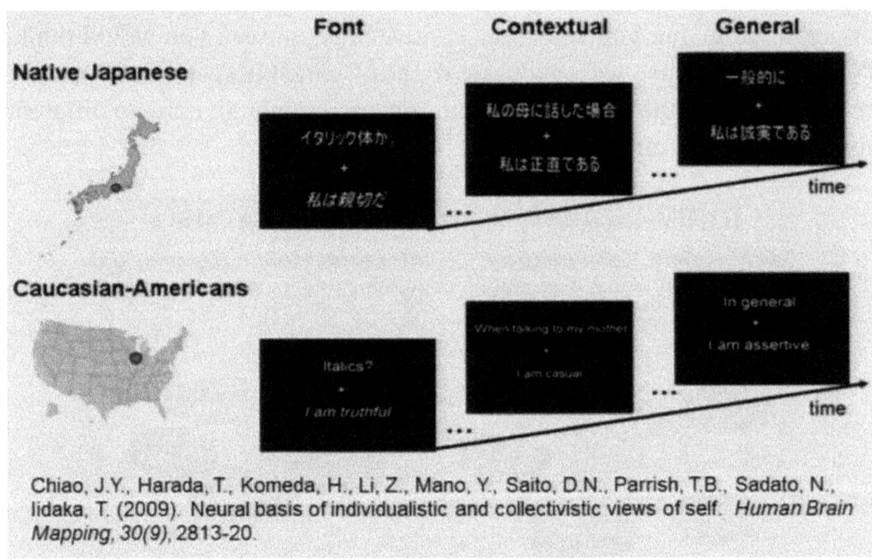

Chiao, J.Y., Harada, T., Komeda, H., Li, Z., Mano, Y., Saito, D.N., Parrish, T.B., Sadato, N., Iidaka, T. (2009). Neural basis of individualistic and collectivistic views of self. *Human Brain Mapping, 30(9)*, 2813-20.

Chiao looked at the brain area, the medial prefrontal lobe, which processes self-related information. Interestingly, she found that individuals' value orientations, whether they were more collectivistic or individualistic, predicted their medial prefrontal lobe recruitment. From the chart, we can see that individuals with more collectivistic values recruited this area more in the contextual task, while those with more individualistic values did so in the general task. This suggests that geographical location didn't constrain people; rather, individuals' own propensities, whether they were more collectivistic or individualistic, predicted their responses.

Now, that is not all because this is just between-group or cross-individual associative research; it's correlational only. Now, she used the priming method, and in another paper, she primed people first, and she

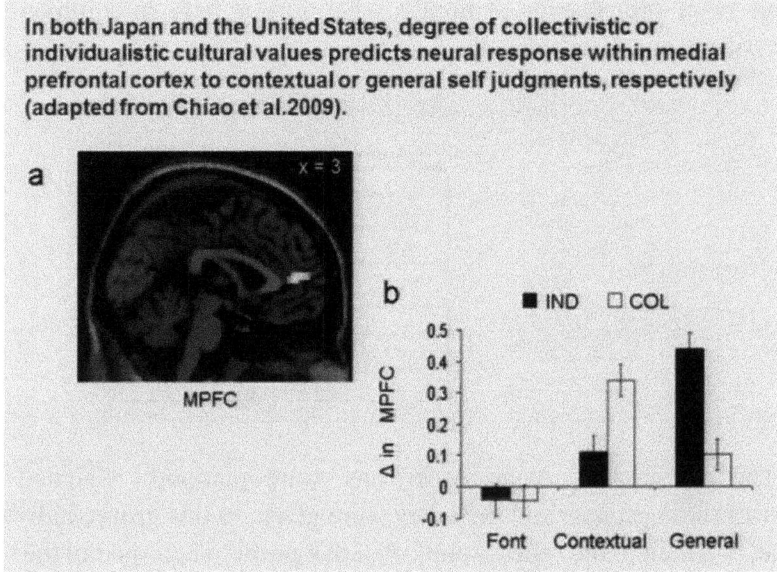

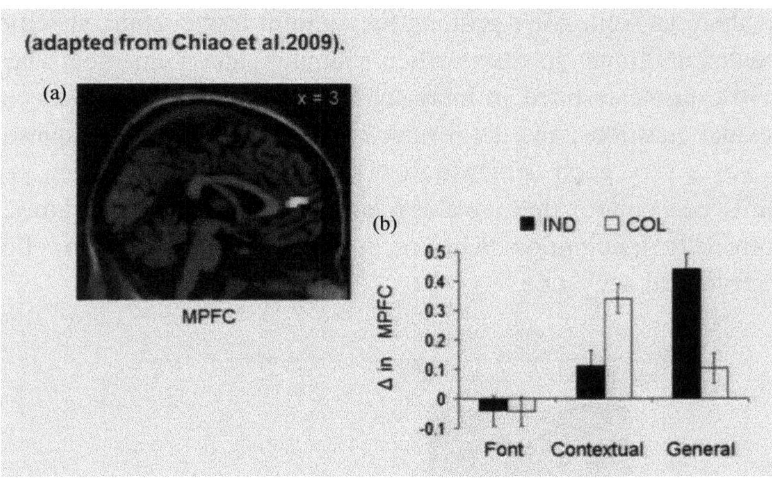

used a story or a question. These are different methods to prime. For example, in the individual prime, the participants read a story that tells about a certain target person. So, the story, for example, here it's a warrior who is chosen based on his or her merit versus if he was chosen based on his family relations, and then the right as is what makes you unique from your friends or family, what do you have in common with your friends and family.

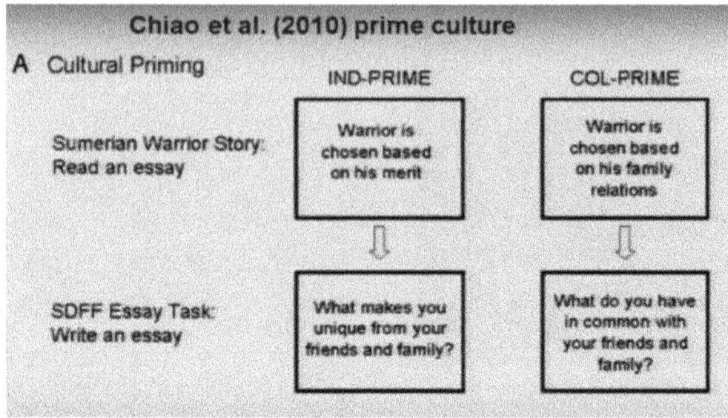

The participants, Asian Americans, were randomly assigned into either of these groups, and then they were given, in this group, individual prime, and then in the other group collective prime. Then she ran the same procedures as she did in the last study, and indeed she found the switching effect there as well. After priming the cultural frame again, specifically the contextual frame you saw earlier, a similar pattern emerged. The collectivistic prime resulted in more medial frontal lobe activation on the contextual questions, and the reverse was true for the general questions. This was a very good demonstration of cultural priming. When people acquire a new culture, they are able to switch between cultural frames, and this affects their cognition, behavior, and emotions. This is even reflected in neurological activation.

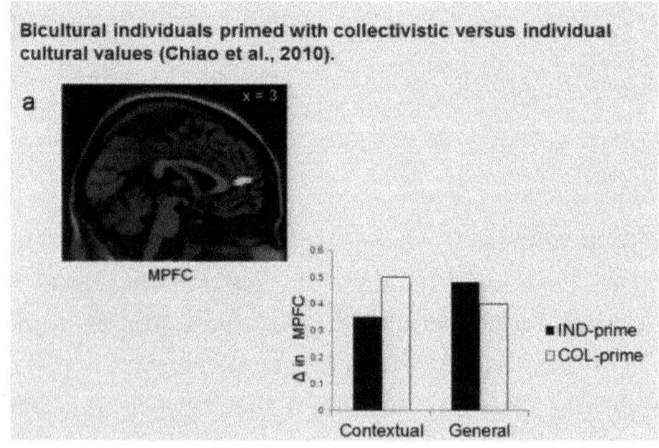

This summarizes that the multicultural mind we acquire can represent multiple cultural knowledge. Pretty cool, right? It's cool in the sense that it's a nice finding.

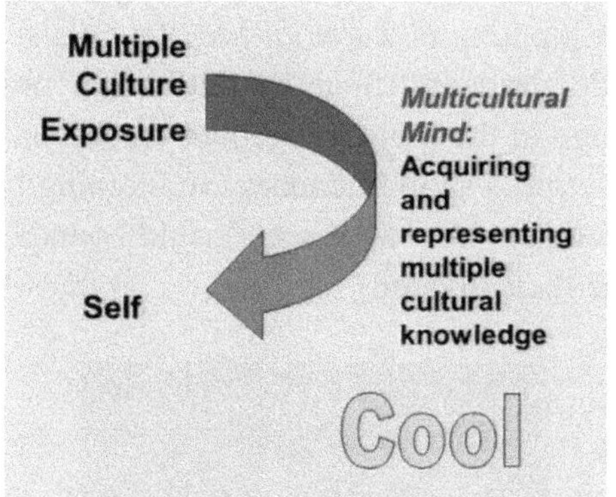

However, this is just half of the story; it's just half. Why? Because we are just passive receivers of these multiple cultural imprints. The other half is what I call the multicultural self, which is how we react toward the implications of this new culture to the self.

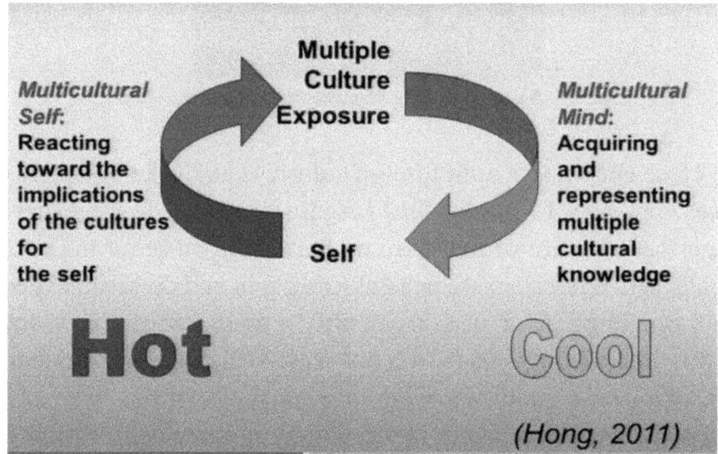

We ask, "Who are we? Are we really — for example, many Chinese Americans were asked — are we really American? Would I become a true American?" So, I think some of these things, or if you see all these things, like in a culture which now has new culture coming in, and people might see it as a contamination of their indigenous culture. So, the receiver of the new culture might not necessarily be cool.

Now, this is another illustration. So, this is a Filipino painter artist, and the title of her painting is *A Racial Identity Crisis*. So, she portrayed herself in Spanish garments, Spanish outfit, but on the inside, very divided, is her image in the more indigenous traditional outfit, and she… but it's all divided with a rigid boundary. So, coming back to John's talk earlier on, don't you see she indeed sees a rigid boundary to the two parts of the self, and she feels divided?

Pacita Abad: *A Racial Identity Crisis*, 1992.

So, close encounters with foreign cultures can evoke hot exclusionary responses, but how to predict that? Again, we need to do experiments. Although there is a lot of literature on that, and you see it in the arts and in people's self-reflection, we need to bring it in and do some experiments on it. So, one thing is we need to identify a predictor of who responds in a cool way and who responds in a hot way. Years ago, my students and I thought, okay, how about we bring in a concept that has been studied in anthropology and discussed in anthropology quite widely, which is racial essentialism. Racial essentialism is defined as race reflecting inherent essence or disposition, indicative of a person's ability, and unalterable. These are the features of these types of beliefs. In our own lab, we need ways to measure or manipulate it. There are two ways you can measure participants' chronic beliefs: their endorsement of racial essentialism using questionnaires. We use self-report, for example, items such as, "To a large extent, a person's race biologically determines his or her ability and traits." People read this statement and tell us honestly or spontaneously whether they believe or not believe. Races are just arbitrary categories and can be changed if necessary. Racial categories are constructed totally for economic, political, and social reasons. If the socio-political situation changes, the racial categories will change as well.

> ### Racial Essentialism Scale
> #### (No & Hong, 2005)
>
> *Sample items*:
> To a large extent, a person's race biologically determines his or her abilities and traits.
> How a person is like (e.g., his or her abilities, traits) is deeply ingrained in his or her race. It cannot be changed much.
>
> Races are just arbitrary categories and can be changed if necessary.
> Racial categories are constructed totally for economic, political, and social reasons. If the socio-political situation changes, the racial categories will change as well.
>
> *Ratings: from 1, strongly disagree, to 6, strongly agree.*

So, on this scale, the higher score means more endorsement of racial essentialism. What's the distribution like? Actually, I discussed this with many colleagues in the United States and they couldn't believe it. They thought that people wouldn't agree with this, but no, actually it's quite beautiful. It's normally distributed. We've extensively surveyed undergraduate students at the University of Illinois at Urbana-Champaign, and yeah, it's widely distributed in a normal distribution, meaning there are a good portion of people who agree and a portion of people who disagree.

How about in Singapore? So far, we're still collecting data, but again, we replicate, and this is Singapore data, and these are real data points from 101 Singaporean students we have collected so far, and you can see again, it's a normal distribution. And people can reflect and tell you whether they agree or disagree with these types of beliefs.

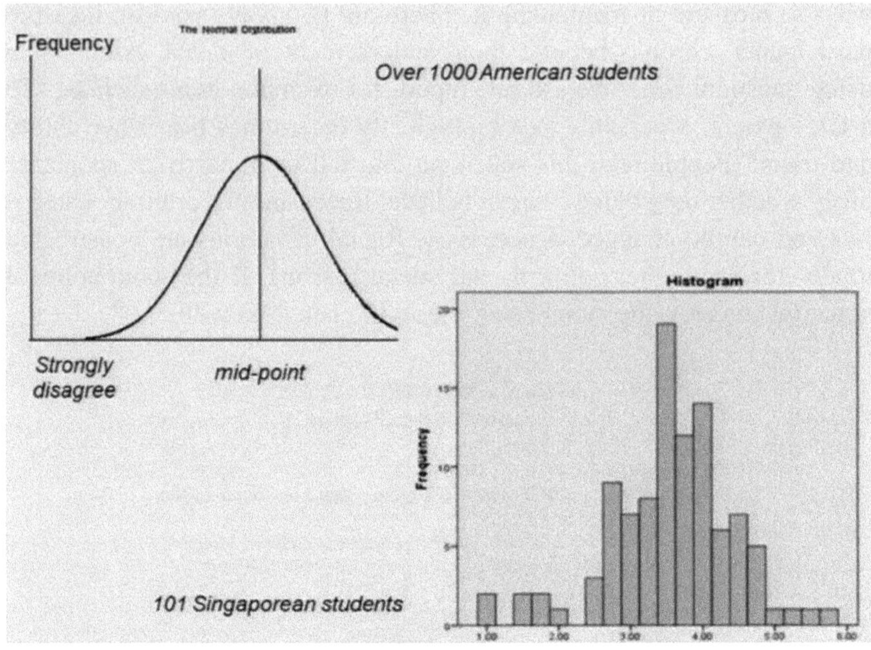

So, this is one way we can measure, but also we can manipulate, we can activate the temporal accessibility, that is, the temporal awareness of these theories, and also convince people that this is true, this is the reality; how so? Usually, as psychologists, we use some deception, we lie a little bit, for the sake of science, but at the end of the day, at the end of the session, we will debrief and always tell the truth to our participants. So, we lie a little bit. When participants come in, we will say, "Okay, now, in this study, we have multiple tasks for you to do, so we begin with a reading comprehension task. We are interested in how people process information, reading essays. So, in the following, I will give you an essay that we got from *Time Magazine*. Please read it carefully, and we will ask you questions afterward." So, participants, half of the participants were given this article, "The Mystery of Race: There are accents underlying racial groups. The concept of race has divided humankind into meaningful social groups based on differences in their innate qualities." So, the details were very convincing about experiments, etc., all supporting this view. Half of them read this.

The Mystery of Race

"There is essence underlying racial groups. The concept of race has divided humankind into meaningful social groups based on differences in their innate qualities."

By WILLIAM HILTON

IN OUR DAILY LIVES, WE ALWAYS come across the concept of race, such as Black, White, and Asian. We often categorize people based on their language use, their physical characteristics, or their cultural tradition (e.g., Latinos, African Americans). We can even use geographical regions to divide people into different groups such as Europeans, Asians, Africans, and many others.

Does *race* really exist? Are the qualities of different races exclusive, unique, and unchanging?

After conducting extensive field study and archival research, the Ethnology Research Center in the University of Geneva concluded that humankind is indeed made up of different races. Each racial group possesses unique and inalterable attributes, known as racial essence. A well known cross-cultural researcher, Professor George Levinger, reviewed a large number of research findings from biology, archeology, and cultural anthropology on race and social development. The results suggested that, in fact, the members within each racial group share similar traits; they all shared common languages, resided in areas with similar social [...]

[...] the skin color of individuals who are White remains fair, even if they are born and grow up in tropical region. Similarly, individuals with dark skin color remained dark-skinned even if they are born and grow up in colder regions. This suggested that different races have unique genetic makeup. The amount of melanin is genetically determined and is expressed in variations of skin color (a phenotype). In line with this evidence, racial differences are reflected in visible characteristics (phenotypes) and these differences cannot be changed easily.

Another example is blood-type. There are four basic blood-types: A, B, O and AB types. Research has found that these 4 types of blood are distributed differently among various racial groups. For instance, biologists discovered that the blood-type of *all* members of the Yleta, a tribe that lived along the Amazon River, is O. Experts suggested that this biological characteristic has been maintained within that tribe because its people have lived in a secluded area and there were almost no cross-tribal marriages.

Archeological research has found that the ancestors of mankind first appeared in the Mid- [...]

[...] our parent's genetic makeup. Individual differences in genetic makeup lead to their unique characteristics. Different races are unique entities with their own genetic combinations, resulting in innate differences between racial groups. In one of their study, genetic scientists at MIT randomly selected 350 White participants in New York City and 400 Black participants in Botswana, Africa. Their analyses revealed that the genetic makeup of 96.8% New York City participants was vastly diverge (over 80% dissimilarity) from the genetic makeup of Botswana participants. The extent of disparity in the genetic makeup between the New York City and Botswana participants resulted in the distinct physical characteristics between these two groups. In addition, genetic scientists found that the genetic makeup of individuals within the same racial group is highly similar. In sum, there are innate and inalterable differences between racial groups.

The human society is divided into racial [...]

The other half read this: "The Mystery of Race: The meaning of race is socially constructed. It is used to categorize different racial groups," with equally convincing evidence, and actually, we wrote this so they were made up by us, and we could test it extensively. You know, undergraduate students really believe that these are taken from *Time Magazine*, and so we know that students will be convinced, and then these different mindsets will be active in their mind, at least immediately after reading the articles. So, what's the effect of reading these articles?

In one study, we asked Asian Americans to read these articles, and then immediately we looked at their Asian identification and American identification, and we see a drastic difference here. So, if you read the social construction article, the Asian and American identification was really high, but if you read a racial essentialism article, the Asian identification remained high; however, the American identification with American culture dampened. Why? It's logical, because if you think that race really rigidly characterizes a person and you cannot change, what's the use of identifying with American culture? As American culture is predominantly

a White culture, and certainly not Asian culture. So, for these individuals, there's not much use to identify with American culture.

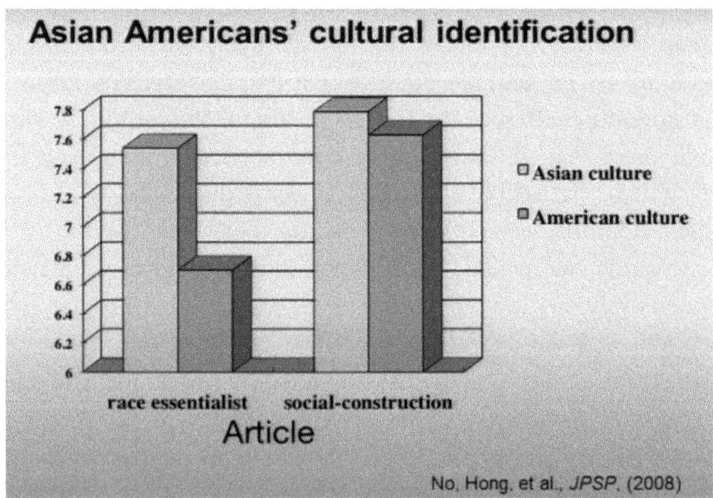

Now, another task: You might say, "Okay, those are self-reports, how about indirect tasks?" Again, we asked participants to read this article, and immediately after, we gave participants the following instruction: "In the following lists, among the three things listed together, please indicate which two of the three are most closely related by circling each of those two words."

They were given three things, which could be names or objects, and then asked to circle two that are closely related. For example, consider the first set: Jimmy Carter, Martin Luther King Jr., and Barack Obama. Which two would you circle? You could circle Jimmy Carter and Barack Obama because both are presidents, or Martin Luther King Jr. and Barack Obama because both are African Americans. Got the idea? Similarly, consider this set: Michael Jordan, Denzel Washington, and Bruce Willis. You could circle Michael Jordan and Denzel Washington because both are African Americans, or Denzel Washington and Bruce Willis because both are famous actors.

The study examined whether people became more aware of race and used race as a basis for categorization after reading an article on racial essentialism. Indeed, the findings indicated that reading the article resulted in a higher likelihood of using race as a basis for categorization.

Now, I claim that this process is "hot," but I haven't yet demonstrated its "hotness." So, how do we demonstrate this? By showing that you would actually sweat more. How so?

So, there was a study in which we interviewed Chinese Americans. We asked them to come into the lab, then we conducted an interview with the attachment protocol, adult attachment protocol. Let me describe what that is for you. So, they first came into the lab one by one for individual interviews. They were asked some warm-up questions that were not related to race or culture at all. Then, in Phase Two, which was our target phase, we asked them to talk about their experiences with Chinese and American cultures. According to the protocol, we asked them to generate five words and then use each of these words to relate to their personal experience with Chinese and American cultures. Our hypothesis was that this would really elicit the hot process because the hot exclusionary process makes it hard for you to switch between the two cultures, especially for people who hold racial essentialist beliefs. Finally, there was a rest period at the end of the interview. While we interviewed them, we hooked them up with skin conductance measurement, so they were wired in terms of how much they sweat in comparison to their baseline when they first entered the lab. And guess what? Indeed, did the racial essentialism predict an increase in skin conductance, which is sweating, in the warm-up phase? No, because it's not related to their cultural experience.

However, quite dramatically in the second phase, when they were entered into talking about their cultural experience, switching between then the racial essentialism, some belief predict and predict in a way that if you the more you believe in racial essentialism some you have high increase in skin conductance, meaning you sweat more, more effort for stressful for you, and then final phase rest again, this disappears. So, it's very specific about their own experiences and how they explored the physiological response to it, and this pattern holds even when we control for their language proficiency, length of stay in the United States, and also the type of experience they recall, whether it's positive or negative. So, the whole pattern stays.

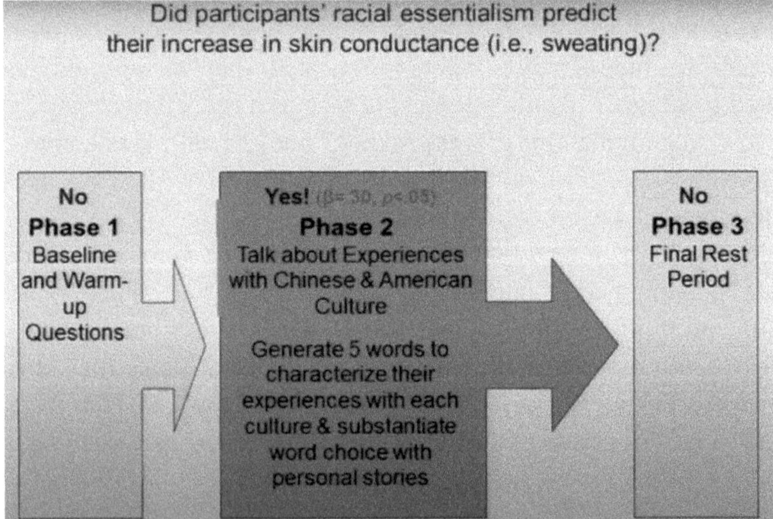

An essentialist race theory sets up a framework within which the Chinese and American cultures are seen as discrete and non-overlapping. Therefore, it is more stressful for them to pass between these cultural frames. So far, in the past few years, my students and I have conducted a series of studies, and now the evidence is quite strong that racial essentialism, just by believing in it, or activating it, leads to higher race-based categorization and cognition. You're more sensitive and ready to categorize based on race, identity, and identification, and contrast from the mainstream culture. This is about Chinese Americans or minorities in the country compared to the mainstream.

Additionally, cultural frame switching results in more emotional and cognitive difficulty in switching between cultural frames. I've shown you the emotional part, and there's also a cognitive aspect; we have a reaction time paradigm that shows it takes them longer in terms of reaction time to switch.

The latest paper we just published in *Psychological Science* also demonstrates that if you hold this belief, you are actually less creative. We have evidence for that, mainly because of the rigidity in categorization. Creativity relies on the ability to take concepts from one category and combine them with concepts from another category in order to synthesize new ideas. If you perceive rigid boundaries and have rigid categorization, it becomes harder for you to combine different concepts and become creative. We have demonstrated this entire process and more.

So, to summarize, there are more parts to this, including an acquisition part, but in this acquisition part, people are not passive; they also react because we are reflexive. We constantly think about what the meaning of that is for us and for our culture. Certain beliefs that you hold will affect the path that you go down. So far, that's what I think from the evidence that we gathered, and we are quite confident that's the case.

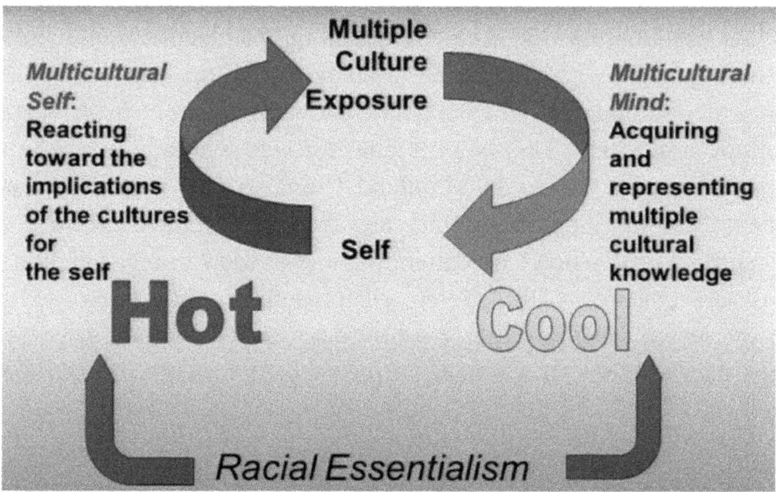

But now, I want to introduce or lead you into a Twilight Zone. We're not yet firm on the findings, but we are very, very excited about this new development. What are they? There are two things: One is psychological research in Mainland China that we are actively doing and the other is culture gene and molecular biology that our group is also doing.

Okay, now psychological research in Mainland China. If you know the history of Mainland China, the education was disrupted for ten years during the Cultural Revolution, and psychology, especially social psychology, was restored very slowly after the end of the Cultural Revolution. So, now, there is a whole layer that is disconnected, and many young people are very inspired and talented, but they don't have mentors who are well trained to mentor them. So, my husband and I wanted to teach in different parts of China five years ago, and then our goal was to help the junior faculty members in Mainland China and train them on social psychology methods and research. At the same time, we learned a lot from these students because there are so many interesting social-cultural topics that can be studied. Also, just to remind you, the Chinese population is one-fifth of the world population, so 20% of these individuals is the group that you can study under the Chinese culture.

Now, let me give you one example. This was from a class exercise that we did during one of the teachings when we were there. Each time we went, we went to different areas of Mainland China and usually stayed for about a month, teaching intensively. One time, one of the groups in my class was interested in ethnic relations in Mainland China. China has 56 ethnic groups, with Han being the majority group, and we wanted to explore the relationship between ethnic and national identification among the ethnic minorities. This is a big question and would give rise to a lot of social harmony issues in Mainland China. At the time, we were in Yunnan, teaching. Yunnan province has 26 ethnic groups, and these were the pictures I took when I was attending one of their village celebrations. It was very colorful, with different customs and traditions. So, we were very interested in how these people identified with their own ethnic groups and whether, at the same time, they identified with the nation, China.

We used a very simple method because many of them might not be able to read characters, the words, so we used pictures. For example, here's ethnic identification *"minzu rentong," "tong zu."* This is yourself *"zi ji"* and this is the ethnic group *"tong zu."*

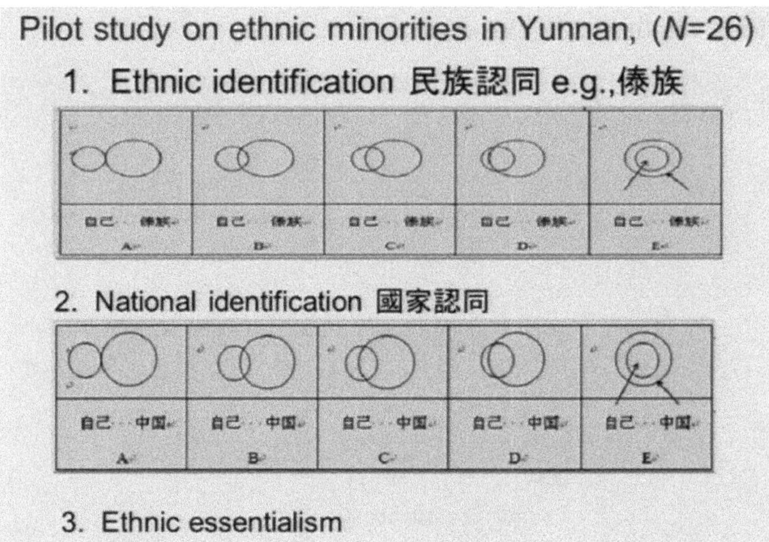

What we did was, we asked participants to tell us which one represented them. If you see yourself and your ethnic group as separate, you would pick this one. But if you think that you are slightly close, very close, or embedded in the ethnic group, then you would pick that. So, each person picked only one picture. Similarly, for *"guojia renting"* national identification, in the same way, they needed to pick one. We also modified our ratio in our racial essentialism — those items, whether they agreed with it or not.

Our prediction was that if you believe in ethnic essentialism, that is, you believe that race is fixed, unalterable, and since Han is the majority group, then China is mainly Han, then you wouldn't see yourself as compatible with that. So, you would see these two types of identification as antagonistic to each other. So, either you are highly with this or highly with that, and you should see a negative correlation between the two. But if you don't believe in this, then you can be both high, both low, you know, and you don't have this antagonistic relationship. And that's indeed what we found. So, red indicates high essentialism, and this is national identification on the vertical axis, with ethnic identification on the horizontal axis. So, for weak ethnic identification, it would be related to high national identification. For the high, and then the reverse for strong ethnic identification for high essentialism, but weak national identity, so you see the negative. And this is the slope.

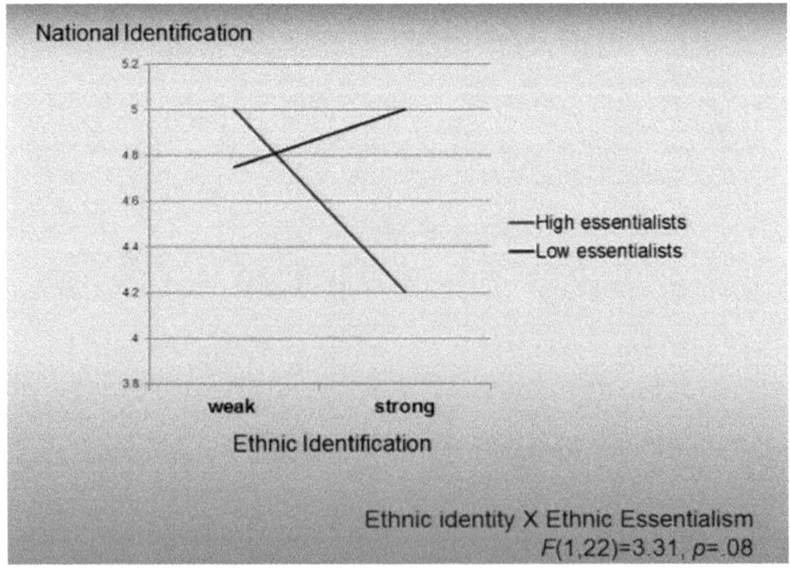

But that's a class exercise, they did it in one day. They just went and grabbed people as many as possible, so it's a small sample. So, this is the Twilight Zone. That's why I said this is a small pilot, but it's promising, and actually now we are collaborating with the National Academy of Social Sciences to do more systematic research in this regard.

Now, culture, genes, and molecular biology — how do I combine them? The general models that we work on look something like this: We think that there are personal tendencies or personal dispositions that might affect the outcome, including genetic imprints or the beliefs one holds, which create factors within the individual. Additionally, there are contextual factors, such as situational pressures, the past history of one's group, one's own past experiences, and cultural norms, all of which may mediate through a processing system characterized by sensitivity and reactivity to situations. All of these factors could have implications or could lead to different intergroup relations.

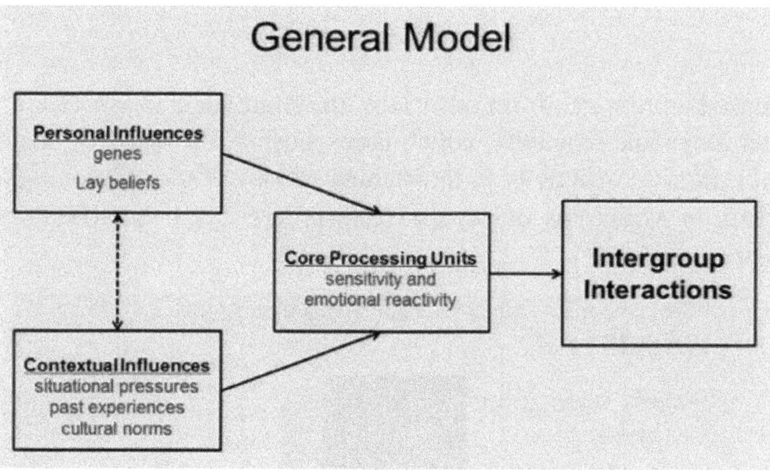

More specifically, what we look at are genetic markers, one of which has been widely studied: serotonin transporter gene polymorphisms. In this polymorphism, which is related to serotonin genes and synapses in our brain, serotonin is a neurotransmitter that passes from one neuron to another. After it's passed, there's a mechanism for it to be reabsorbed; otherwise, it would be flat. In each individual, there are two alleles, each with variants, either short or long, resulting in three different types: short-short, short-long, and long-long. The short allele has been associated with less efficiency in serotonin reuptake, meaning there's more serotonin flooding or concentration in the synapse for individuals with the short allele than for those with the long allele.

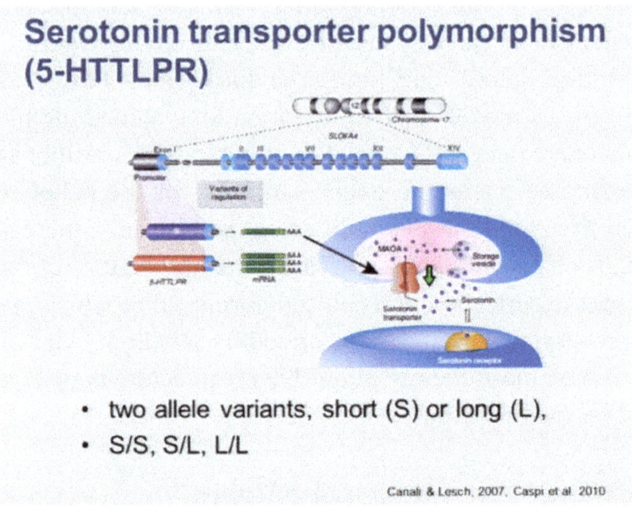

Based on research from other labs, the short allele is associated with greater amygdala reactivity, which is responsible for many of our emotional responses, reactivity to threatening stimuli, fear conditioning, neuroticism, or symptoms of depressive disorders. So, it has been widely studied.

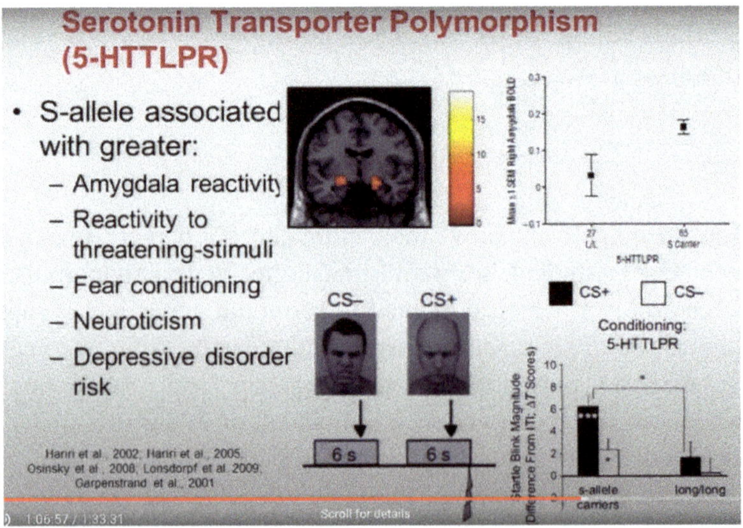

Now, our model looks something like this:

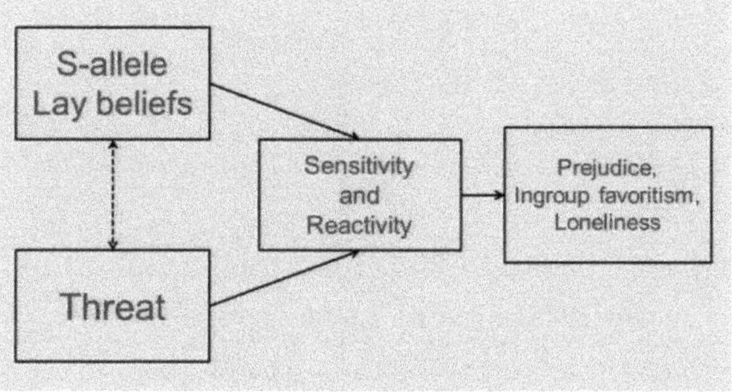

We think that the short allele might respond to some internal conditions, such as the lay beliefs that I have been showing you, and also external threats, whether physical or social. This could give rise to greater sensitivity and reactivity among short allele carriers than among long allele carriers, leading to a host of outcomes, such as prejudice, in-group favoritism, or loneliness in a new culture. Do I have evidence? We are getting more and more evidence along these lines. For example, in a study conducted by Toby, my postdoc, and his former advisor, Joan Chiao, they collected data from Caucasian Americans in the United States. They measured belief in a dangerous world, which is already a widely used scale, distinguishing between low and high believers. They also genotyped these people, categorizing them as long-long or short allele carriers. What they found was that for groups usually seen as threatening, such as Arabs or schizophrenic patients, those who also held a high belief in a dangerous world and had the short allele actually responded more extremely — they were more sensitive. This means that if you held this belief, you showed the highest amount of bias toward both groups.

Now, how about in experiments? We also ran another study in which we created groups randomly and had them play behavior games with each other. There was a group that consistently cheated participants, not giving them a fair share. How did participants respond to this group? At the end, participants were asked to allocate resources to a member of their own in-group or a member of the group that cheated or was unfair. The results showed that those with the short allele, again, showed more in-group favoritism — they allocated more resources to their in-group at the expense of the out-group.

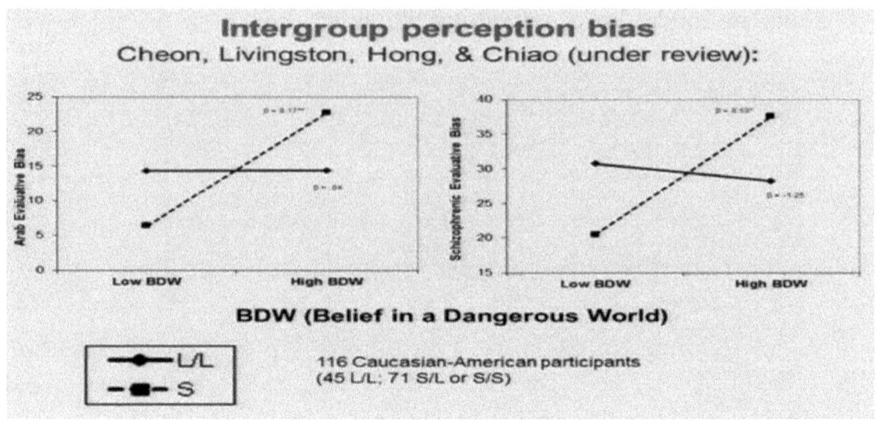

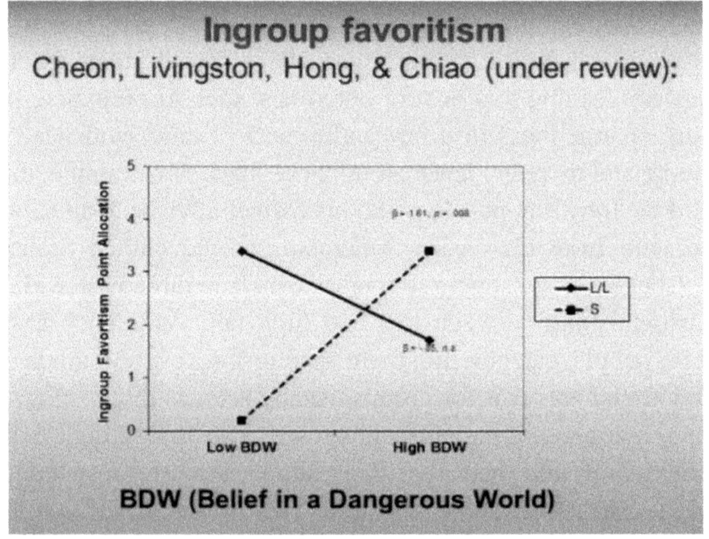

To clarify, the model focused on sensitivity and reactivity. For instance, if individuals were in a condition where they didn't believe the world was dangerous, they wouldn't be as sensitive to this outlook. What the allele did was exaggerate the impact of both internal and external conditions. So, if you have a strong allele and hold a not-so-favorable belief, and are also in a threatening situation, your responses would be the most negative. But if you don't hold this belief, you may not show such extreme responses, or even reverse them if conditions are favorable. The point is that it will be sensitive. Individuals would be more extreme in their responses. It's important to clarify that our proposal doesn't suggest we've identified certain deterministic genes. No, we don't believe in that. More importantly, we've identified factors that could put someone in a more vulnerable situation when combined with other vulnerable factors or circumstances, such as lay beliefs or being under threat from anyone.

Now, this is just fresh from the oven. We also collected data here in Singapore with our undergraduate students. In this project, we tracked undergraduates going on an exchange program. We were very interested in them, so we obtained cheek cell samples before they left, and now, actually, in these past few weeks, we have been obtaining cheek cell samples again after they returned. In the middle, we also measured their stress levels while overseas through the Internet, assessing whether they were feeling stressed or not. This is from Desiree, who has been working in the lab, extracting DNA, and attempting to genotype the serotonin transporter gene. Here, you can easily see the extracted alleles traveling from here to here, with the longer one heavier, so they travel slower and for a shorter distance. These are different individuals, and these are the alleles only. So, we looked at whether they carried long-long, short-long, or short-short alleles. For example, this person is long-long because both travel slowly, and this person is short-long because you can see both. This is short-short. This is an easy way to look at it, but we are now running more sophisticated procedures, although this is good for demonstration.

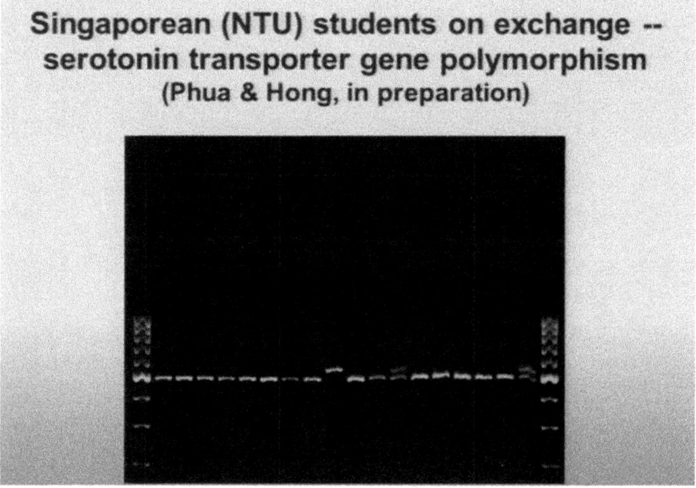

We also have measures of racial essentialism, as you may remember, and we asked them whether they have a need to belong. Do they have a high need to belong or a low need to belong to a group? Do they care about being accepted in a group? Again, you can see a differential, and this is the pattern for short-short. We didn't find any systematic pattern for long-long, but the short-short allele carriers show a very interesting pattern. If you have a high need to belong, you really care about whether you belong to a group, and those who hold racial essentialism are the people who show the highest levels of loneliness when they're overseas. We measured loneliness when they were overseas through the Internet, asking them how they felt. Similarly, if you don't have a high need to belong, you associate with the lowest levels of loneliness. As you can see, we don't have the whole puzzle yet. There are still gaps, but we see some promising results. This is based on a very small sample, and we still need to recruit more participants. If you have ways or sources through which we can obtain participants' cheek cell samples, please come and talk to me. We are really in need of help. This is why I say it's the Twilight Zone because we are going into areas that few people have gone to before. It's really the Twilight Zone, integrating all these. We see some promising results. We don't know at the end of the day whether we can put the puzzle together, but we are very excited about it.

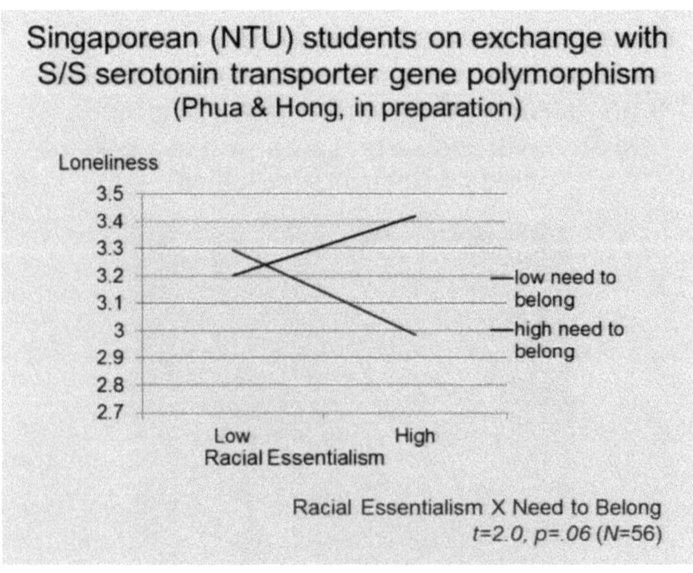

Now, it is near the end. I want to wrap up by emphasizing that culture is dynamic. It's not a static entity; it undergoes change and adaptation to new environments. Throughout this talk, most of the data has been aimed at illustrating the adaptability of people. Individuals within each culture can acquire new cultural meaning systems, such as lay beliefs from another culture, and can switch between cultural frames. Moreover, emergent cultural phenomena can arise as a result of cultural mixing, and this process can be nonlinear.

Important social events can sway people's responses and produce different behavioral trajectories for groups. Although I don't have time to talk about this now, I've also worked on the psychological effects of events like Hurricane Katrina in 2005 and the earthquake and tsunami in Japan in 2011, where we observed shifts in people's responses following natural disasters.

To end my talk, I'd like to leave you with a metaphorical image: the map of science. It's a truly beautiful map. When I first saw it, I googled it and read an article about it. As you look closely, you can see each branch of science represented, from business to social and personality psychology. My journey has been about navigating from one area to another, from social and personality psychology to brain research and genetics, and then back again. It's been a process of taking small steps and learning along the way, all in an effort to integrate these different fields. Thank you, for your attention.

7.2 Discussion[1]

Sheila Ronis: My real question is, having traveled all over the world and experienced many different cultures, I think that the United States is changing so dramatically in its cultural makeup. I'd be curious if you are considering intergenerational studies because nobody other than American Indians is indigenous, so everybody is an outsider. The American culture, although many people think of it as emerging from the British, is really gone. So, I would be very curious as to how you define an American Caucasian, who could be from 50 different nations on the planet. What I think begins to change is with the 2nd, 3rd, 4th, 5th, and 6th generations, when it is very difficult to distinguish who's who.

Ying-Yi Hong: A very good question indeed. How do you define American? Well, that's already a big subject matter by itself, and there are different theories about that. In our own research, actually, we don't define an American for our participants. It's their own subjective experience, how they define it, and how that guides their own responses. In social psychological research, there were studies that looked at this very concept of American using the Implicit Association Test (IAT), which examines how people associate in an unconscious way, more like the collective unconscious of what an image of an American is like. From that research, findings show that many participants still associate a strong image of White

---

[1] One discussant's question was inaudible and the speaker's answer was removed.

American men as the prototypical American. But your point about change, I think, is an adaptive system. With demographic changes, that will change as well.

It's a great point about the demographic changes, but you know, tying back to the subjective experience and subjective beliefs, we have data that show if you grow up or live in an environment that is more diverse and where people engage more in discussions of interracial issues, then you're less likely to hold a racial essentialism theory. But if you are in an environment that is more homogeneous in terms of race and there's little discussion on that, then you're more likely to hold such a belief. So, even the belief [held by]a layperson is affected by the environment they are in and the social discourse that surrounds them.

Greg Fisher: In an earlier slide, you had a title of "evidence," and you mentioned the prisoner's dilemma game. I would be really interested to know a little bit more about that.

Ying-Yi Hong: I have many extra slides. So, this is the prisoner's dilemma game. It was published in *Psychological Science* some years ago. We ran a typical prisoner's dilemma game paradigm, where we told the participants that if both of them cooperate, they get three points each. However, if one cooperates and the other defects, the defector gets four points while the cooperator gets zero. If both defect, they both get one point. That's a typical prisoner's dilemma game, but we put a twist on it. We ran this in Hong Kong with Hong Kong Chinese college students, arguing that they are biculturals due to their background. Before participants entered the game, we showed one-third of them a Chinese prime, one-third a Western prime, and kept one-third neutral. Then, we varied who they played the game with — half with friends and half with strangers they had never met before, though they were in the same section, as they just saw these people but didn't receive individual feedback. It's a two-by-three-by-two design in terms of timing conditions and whether the partner was a friend or a stranger.

The vertical axis here shows the likelihood of choosing the cooperative strategy in the prisoner's dilemma game. We observed that when playing with friends, participants primed with Chinese culture cooperated

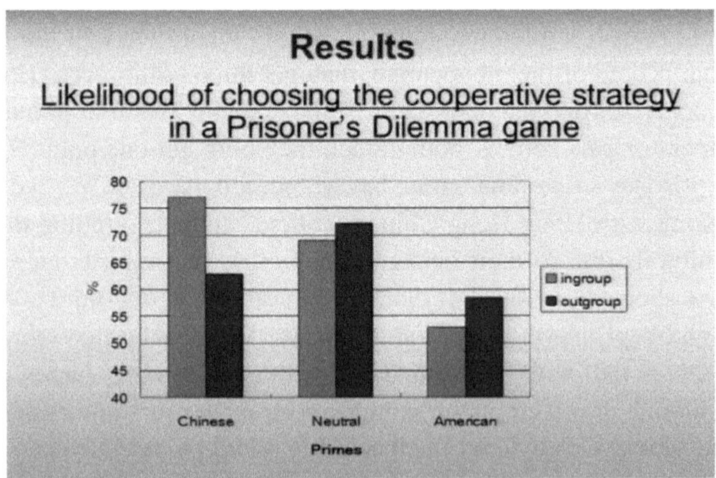

		You	
		Cooperate	Defect
Partner	Cooperate	3 (You) 3 (Partner)	4 (You) 0 (Partner)
	Defect	0 (You) 4 (Partner)	1 (You) 1 (Partner)

more than those primed with American culture. However, this effect was not observed when playing with strangers. It's very specific in terms of how accessible lay beliefs are applied, demonstrating the applicability of these beliefs. This finding aligns with the well-known difference between Chinese and American responses toward friends versus strangers.

Discussant A: Hello, I found all of this really fascinating, especially the way you bring together genes, memes, artifacts, and context to understand responses. I'm very interested in inclusionary and exclusionary responses, and I have two questions related to specific contexts. In South Africa, through scenario exercises, people have identified inclusion and exclusion as major uncertainties for the country's future. Today, we know that power is in the hands of one ethnic group, known for having the strongest internal locus of control. My first question is this: What would you suggest measuring in that society to understand the propensity for inclusion and exclusion? My second question is this: What do you think could be the effect of cultural priming through publishing stories and articles in newspapers or other media to influence the system in a particular direction, and what would be the timescales involved?

Ying-Yi Hong: Fascinating, fascinating. I don't know that specific context, but one thing is clear: When you put a group of people together, they often find ways to divide themselves into "us" versus "them." This is because people have a basic need, according to Marilyn Brewer, called the need for optimal distinctiveness. On one hand, we want to be connected to others, but at the same time, we want to differentiate ourselves and be unique. These opposing forces are in constant action within each individual. So, returning to your question, the first thing I would ask is how they are actually divided up internally. I know they can be divided up in different ways, but in order to get a sense of how I can prime them, I need to know their background, their icons, and then I would suggest we need ways to maximize or symbolize these different cultural systems, and then work on people who already know both. However, sometimes it's hard if there are emotional tensions in the group already. Just like what I showed, if there are things that trigger emotional responses, that might block the way they recognize the symbols of another culture, and they intentionally contrast away from it. And we have evidence, like with some Korean Americans who have this reactionary response: If you show them American cultural icons, they behave even more in a Korean way. So, that could go back to the opposite direction. So, I think it's very important to understand the nuances in the culture and then try to use symbols, but also use them in a way that is safe. This is interesting and tied to research that we are conducting, which we call cultural attachment. We borrowed John Bowlby's theory of attachment. The idea is that when we grow up, we attach to our caretakers. Culture, in a way, could be some kind of symbolic safe place, a safe haven for us, and then we develop emotional attachment to it. So, oftentimes, especially for people who travel abroad, when they feel stressed, they want to eat the cuisine of their native culture, or watch movies, or listen to their own cultural heritage music. So, that's a way for them to get back to the emotional attachment. I think, in that case, you can also look at the emotional attachment basis, because that will add more on the emotional side, rather than just the cool cognitive knowledge, so both might be interacting.

Discussant A: Thank you, very much.

7.3 Summary of the Talk

Ying Yi Hong's talk explores the relationship between culture, cognition, and behavior, focusing on how individuals from different cultural backgrounds perceive and react to their environment. The following is a summary of the main points discussed in the talk:

Cultural Differences in Perception and Attention: Hong presents research showing that individuals from different cultures, specifically Chinese and Americans, exhibit distinct patterns of visual attention. While Americans tend to focus more on focal objects, Chinese participants pay more attention to the context surrounding the objects.

Functional Magnetic Resonance Adaptation (fMRI): Hong explains the use of fMRI to study brain activation patterns in response to different types of visual stimuli. By measuring brain activity in the lateral occipital complex (LOC) region, researchers can investigate how individuals process congruent and incongruent visual stimuli.

Cultural Priming: Hong discusses the concept of cultural priming, where exposure to cultural symbols or icons can temporarily activate cultural knowledge systems in individuals. This method allows researchers to investigate how cultural priming influences cognition, behavior, and emotional responses.

Racial Essentialism: The talk explores the concept of racial essentialism, which is the belief that race reflects inherent and unalterable qualities. Research shows that individuals who endorse racial essentialism tend to have more rigid categorizations of themselves and others based on race, impacting their identification with different cultures.

Cultural Adaptation and Creativity: Hong discusses how cultural beliefs and attitudes, such as racial essentialism, can influence creativity. Individuals who hold rigid beliefs about race may have difficulty in creative thinking due to the limitations imposed by fixed categorizations.

Psychological Research in Mainland China: Hong describes efforts to promote psychological research in Mainland China, where there has been a historical lack of infrastructure and mentorship in social psychology. Collaborative efforts aim to explore cultural dynamics and social behaviors in Chinese populations.

Culture, Genes, and Molecular Biology: The talk concludes by discussing ongoing research into the interaction between cultural factors, genetic predispositions, and neural processing. Understanding how these factors intersect can provide insights into individual differences in intergroup relations and cultural adaptation.

Overall, Ying Yi Hong's talk highlights the complex interplay between culture, cognition, and biology, emphasizing the dynamic nature of cultural adaptation and its impact on individual behavior and perception.

7.4 Relevance of the Talk to the Current Stage of Research

Ying Yi Hong's talk touches on several themes that are highly relevant to the current stage of research in psychology and related fields:

Cultural Neuroscience: The intersection of culture and neuroscience is a rapidly growing field of study. Understanding how cultural factors shape neural processing and cognitive functions has important implications for our understanding of human behavior and intergroup relations.

Cultural Differences in Perception and Attention: With increasing globalization and cultural diversity, there is a growing interest in understanding how individuals from different cultural backgrounds perceive and attend to their environment. Hong's research sheds light on how cultural differences influence visual attention and cognitive processing.

Cultural Priming and Behavioral Effects: The concept of cultural priming, where exposure to cultural symbols or cues activates corresponding cultural knowledge systems, has implications for consumer behavior, intergroup relations, and decision-making processes. Hong's work demonstrates how cultural priming can influence behavior and cognition.

Racial Essentialism and Intergroup Relations: The discussion on racial essentialism and its impact on identity formation and intergroup relations is particularly relevant in the context of increasing diversity and discussions around racial identity. Understanding how beliefs about race shape attitudes and behaviors can inform interventions aimed at reducing prejudice and promoting inclusivity.

Cultural Adaptation and Creativity: In a globalized world, individuals often navigate multiple cultural identities, which can impact creativity and problem-solving abilities. Hong's research highlights how cultural beliefs and attitudes can influence cognitive flexibility and creative thinking, with implications for education and workplace diversity initiatives.

Psychological Research in Non-Western Contexts: There is a growing recognition of the need for more diverse and culturally sensitive psychological research, particularly in non-Western contexts. Hong's efforts to promote psychological research in Mainland China and other regions contribute to the broader goal of globalizing psychological science and understanding cultural variations in human behavior.

Overall, Ying Yi Hong's talk addresses key issues at the forefront of contemporary research in psychology, neuroscience, and cultural studies, highlighting the importance of considering cultural factors in understanding human cognition, behavior, and social dynamics.

Chapter 8

A Complex World from a Virus's Point of View

Peter M. A. Sloot

YouTube: https://youtu.be/50b6GCgBFSk?si=YTWihRfUKkY0sxj3

Speaker: Peter M. A. Sloot

Moderator: Michael H. Lees

Discussants: Simon Levin, Greg Fisher, Kevin Xiao, Discussants A, B, C, D, E, and F

8.1 Talk by Peter M. A. Sloot

Thank you. I came up with this title. We've been doing a lot of work on modeling; I think Murray Gell-Mann was talking about world modelers, and I said, "Yeah, I think I feel like being a world modeler." So, I came up with this title because we did quite some work on looking at the way information is being transferred from a virus's point of view. Now, there are two ways I can tell this story. You can really sit inside viruses and look from the virus's point of view, or you can take it from, let's say, an epidemiological point of view. What I'm trying to do is I will try to do both, just to give you a feeling that when you look at things slightly differently, you might eventually come up with good answers.

So, this is the world clock, the one that I like to look at every now and then. It gives you a kind of feeling of where we stand. This has been counting since the beginning of this year, and so we have a world population that is just past 7 billion, as you know. So, what you see here are a couple of interesting numbers. I'd like to point you to the diseases here, from tuberculosis to communicable diseases, hepatitis, malaria; you see they're really huge numbers. They account for a serious amount of death that we have in the world. There are other interesting numbers here, and it's still counting, of course. Being a Dutchman, I'm also very much interested in the amount of bicycles produced. Actually, you see the number comes very close to the population growth, so for every new kid, we have a bike, so that's good. But the message of this slide is that we are with a lot of people, there's a lot of stuff happening, and a lot of the stuff that is happening is because we interact. And that will, of course, result in problems.

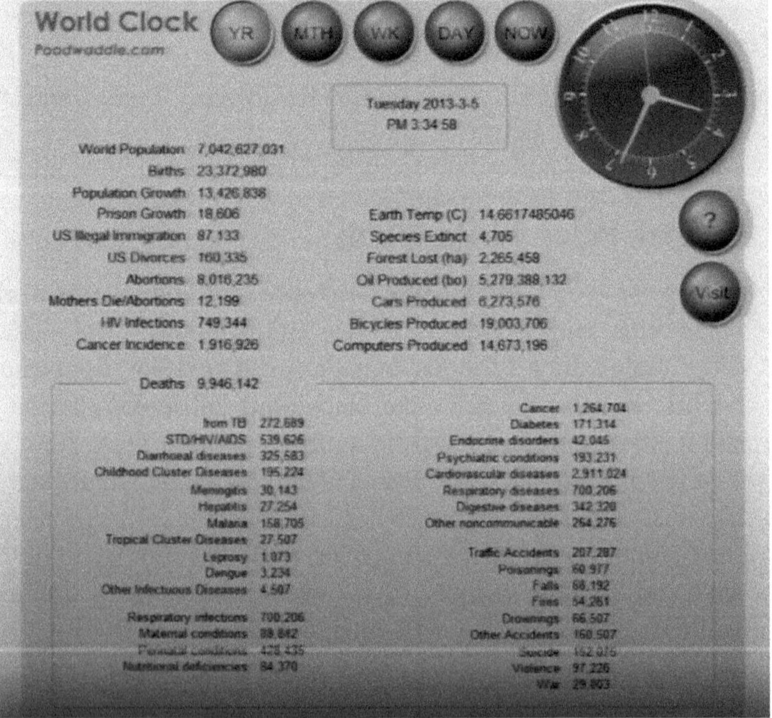

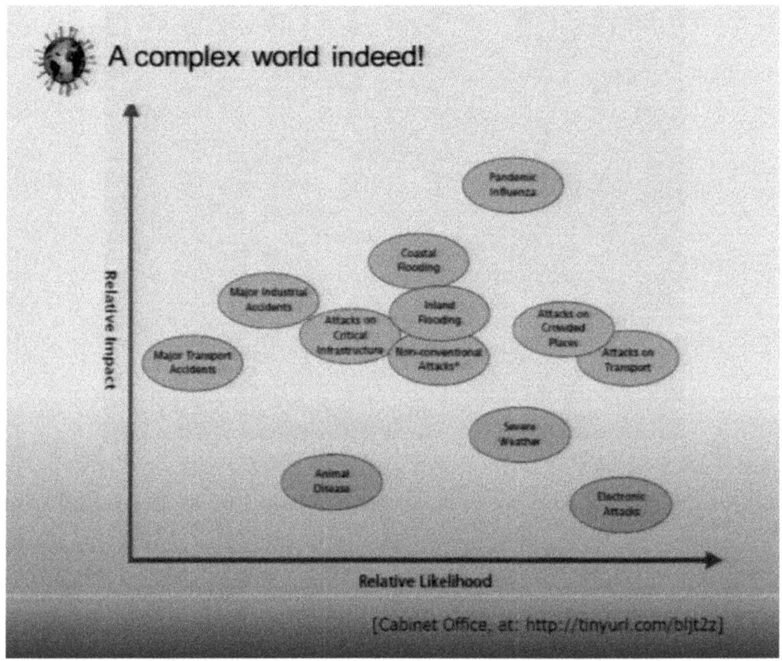

These were laid out in this graph, where we look at the relative likelihood of some kind of disaster happening against the impact that we expect it will have on the world. So, there are things like severe weather; well, we have seen some, but the only thing that really stands out is the pandemic influenza. There's a very high likelihood that we're going to be hit by yet another pandemic, not only influenza but epidemic diseases in general, and the impact will be very strong. We have had that, you know, through the years. We have had, of course, the plagues in Europe, we have had this huge influenza in the nineteen hundreds in Europe; we had a tremendous number of comparable disasters, I would say, that really hit us hard.

And there is a pandemic that's going on, which actually urged the European Union to ask this question. They put it out in the wild to say, "If we were to have 1 billion euros that we could spend on stopping the pandemic, HIV, AIDS, what should we spend it on? Should we spend it on getting better medicine, or should we spend it on changing behavior?" And these are the kinds of questions I like, you know, because when you start to think about that question, within a few seconds, you realize that when we talk about better medicine, we talk about molecules, right? We

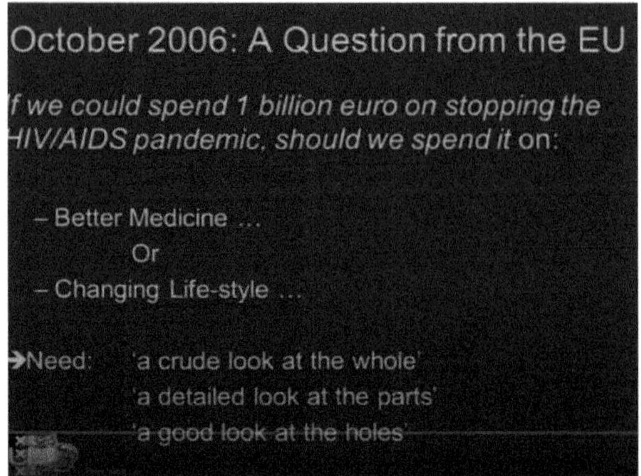

talk about chemistry and we talk about the immune system, and toxicity and whatever you have. When we talk about changing lifestyle, changing behavior, we will talk about sociology, policy, money, and psychology. So, the question is, I mean, if you look at this, you think, "How can we ever find an answer to a question like this?" It's really comparing apples and pears and cows and whatever you have. So, in order to get an answer to that, you really need to take a crude look at the whole, and then you have to take a detailed look at the parts, and then you have to take a good look at all the holes that you have missed, right? So, I'm going to take a crude look, a detailed look, and a good look at the holes.

Okay, so this is the situation, this is the HIV pandemic as it stands, well, as it's recorded up to 2010. The dark areas are, of course, places where the highest incidence is, and we see that the number of people infected with HIV is about fourteen million people in the world. Now, this is a dramatic underestimate; this is what we really know that is true, but it's probably much higher. The reason why it's not higher is, you already guessed that, of course, because many people do not know whether they are infected, right? And there are many areas where we don't have correct samples, so this is a very strong underestimate.

Now, look at this. In reflecting on the first slide that I showed you, that we are, you know, seven billion people, we're very much connected. We're flying from one place in the world to another place in the world, so everything is connected to everything, and everyone is connected to

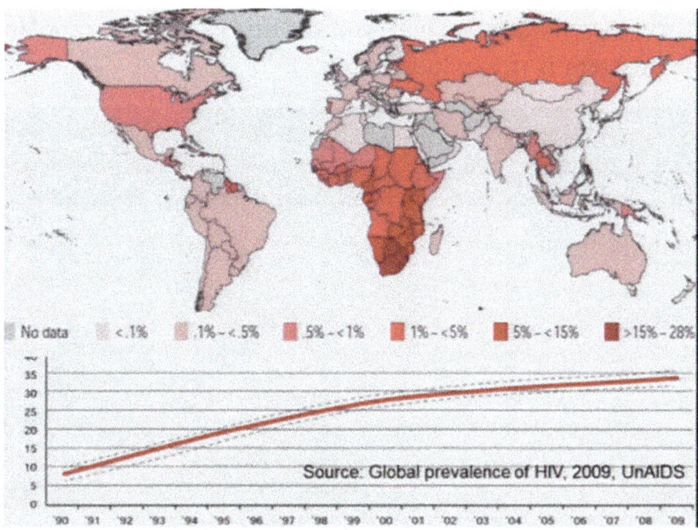

everything. So, these kinds of pandemics are very, very hard to stop and very, very hard to really assess and to understand.

So, why does it take... places of the world, and you might recognize this:

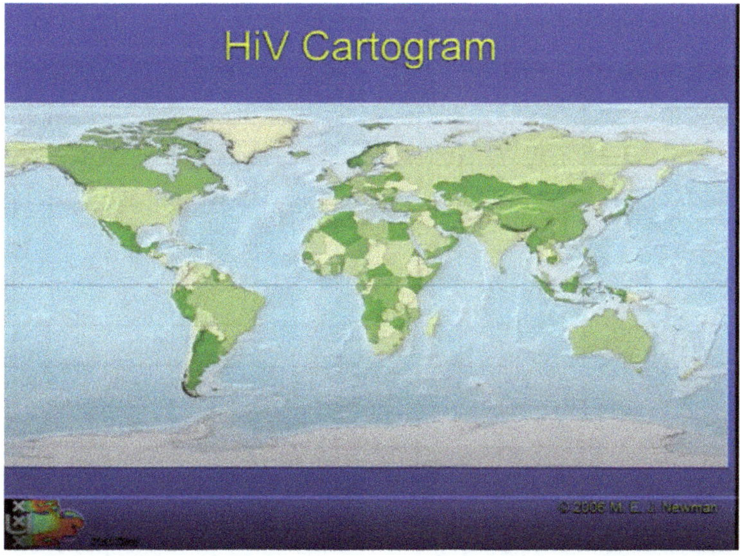

And this is happening if you scale each country with the amount of HIV that's in the country. So, here you see this is what the problem is and where the problem is if you look into it.

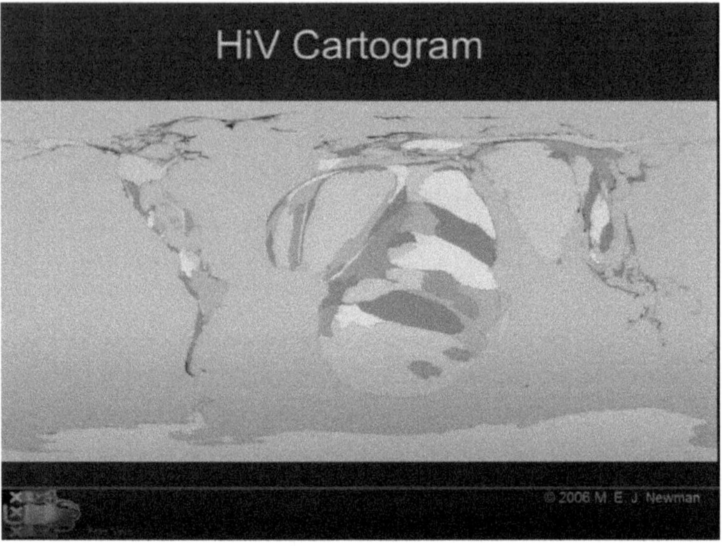

If you now scale every country with the amount of money that's going into HIV research, the world would look like this. So, the money is going where the problem is not.

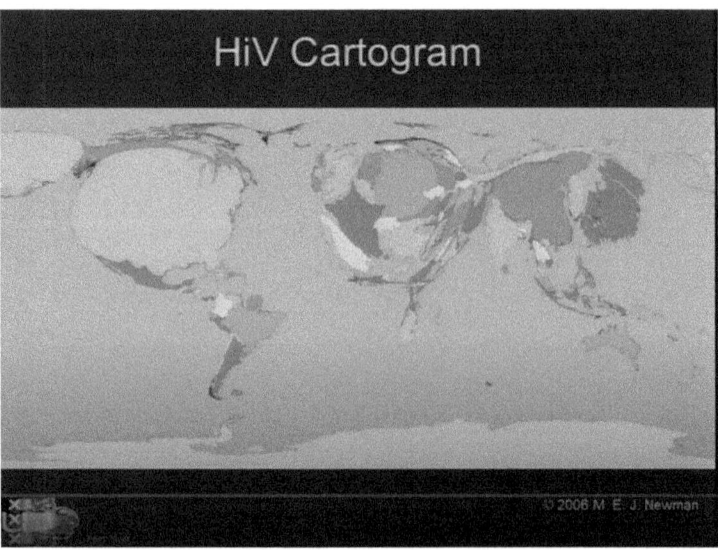

So, I think it's our task to really make sense of the amount of money that we're spending to see if we can get good answers to those questions and to the problems that are there. Now, HIV is tremendously complex, and I'm going to take you through it in a couple of minutes, and I'll be very concise about it, but I need to bring a couple of things across. Like the previous speaker, who at the beginning of her talk said, "You know, at the end of my talk will be the Twilight Zone," that really keeps you on the edge of your chair, which is a very good trick. I'm going to do the same trick. At the end of my talk, we'll talk about sex, so…

Okay, but before we get there, this is basically the problem. There are many aspects here. Let me just take you through it very briefly. This is an artist's impression of HIV. We have all kinds of drugs, about 24 drugs. I'll say a bit more about it later. There's drug resistance, which means the virus won't work anymore at some point in time. There is an immune response, which we hardly understand. There is an infection process in which we probably understand only 40%. There is no understanding at all at this point in time about toxicity, degradation, and inhibition effects. So, this is the problem, and this is what we have to really get an understanding of.

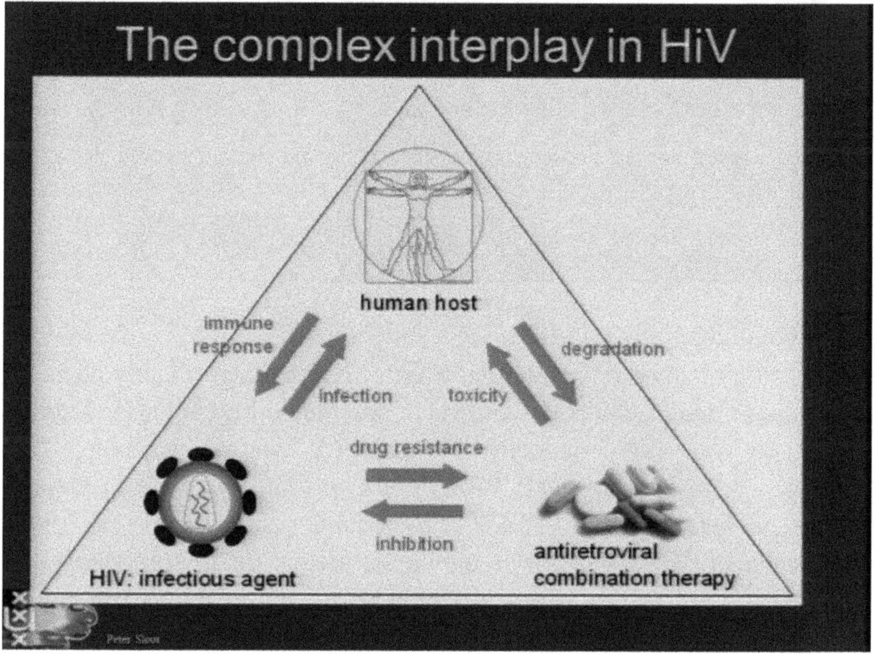

Apart from that, there is the problem of the spreading of the virus through the population. So, yeah, it's a truly complex problem, and the only way to address that is, like Neil Ferguson said — he's a very famous epidemiologist — he said, the only way to actually look at these kinds of problems is to take a complete holistic view. You know, you cannot view these things in isolation; you have to look, well, of course, you have to look at them in isolation, but at the same time try to make the connections to all the different parts that constitute that system that you want to understand, which is, I think, more or less the definition of a complex system. So, that's what we're going to do and size matters in that case.

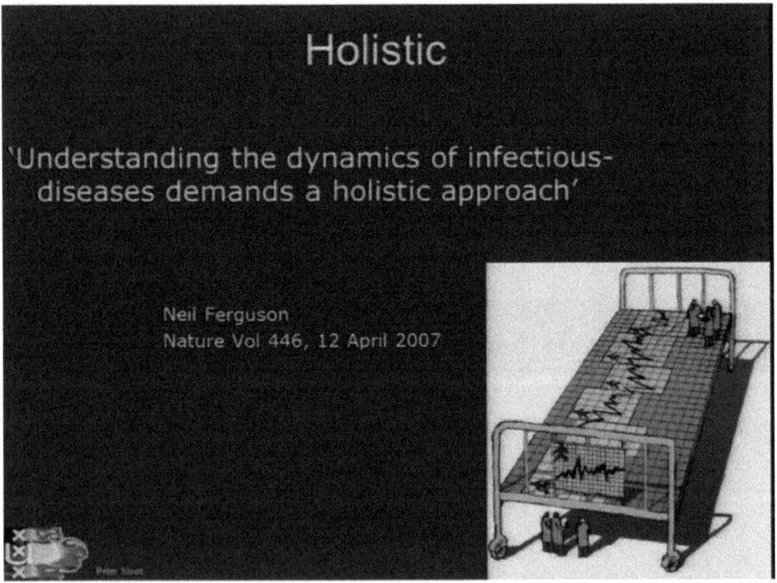

If you look at the scale of things that you want to address, if you want to have a bit of an understanding of what's happening, you really go along many, many scales. Murray Gell-Mann was quoted by John Holland as saying that you get a completely new science... What was it? Every 10 to the power... Maybe you said it, I forgot it, it was like every three orders of magnitude. So, look at this. We're talking about processes that take place on a scale of 10 to the power minus 10 meters up to something like 2 meters if you like, and timescales of 10 to the power minus 14 seconds on the DNA replication level, all the way up to a lifetime.

So there, you know, there's not one science that can cope with this. We cannot just design an equation here, right, and then track that thing all the way through. There's no way. So, we have to think of a more integrative way to deal with this. Now, I do believe in systems studies and system dynamics, if you like, but I don't think that system dynamics makes sense here, because there will be no governing equations or something like that. You really have to break it down into bits and pieces, try to understand these bits and pieces, and then see if you can find mechanisms to glue them together in a sensible way that you can actually validate. And that part of validation is crucial, which is quite often skipped.

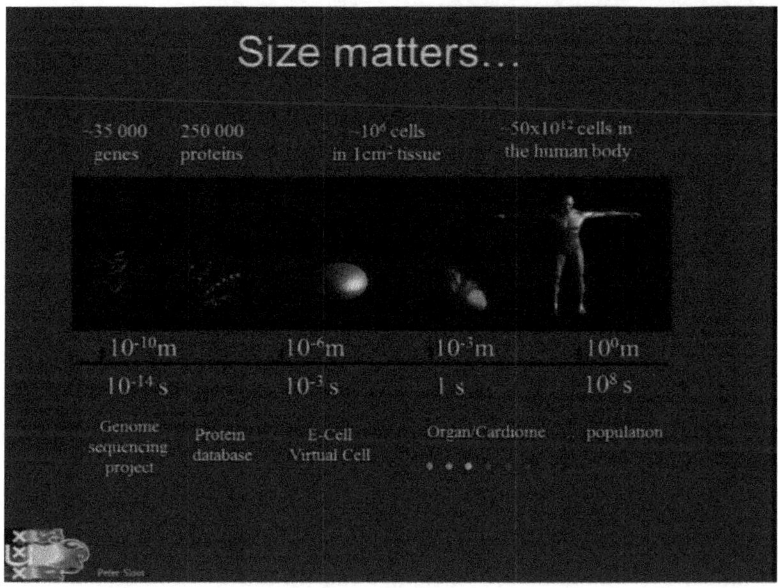

So, what I'll do is I'll take you through the essential parts so that you can hopefully appreciate the complexity. The virus, again, is an artist's impression, and once an individual gets infected, the individual produces about 10 to the power of 9 new viral particles per day. That's quite a lot, so per day, 10 to the power of 9 new viral particles.

Lots of those viral particles are not fit in the sense that they will actually be eradicated by the immune system, but lots of them are. So then what happens is something very special that I really want to draw your attention to. HIV is an RNA virus; it's a retrovirus. It doesn't contain

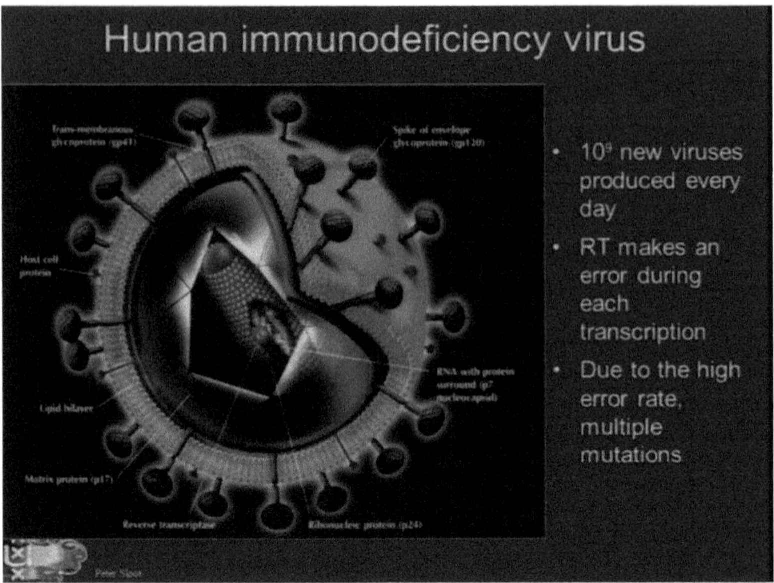

DNA, but it has RNA in it, and a virus is a dead thing. It only becomes alive, if you like, because of us, right? So, what it does is it hijacks the reproduction mechanism that we have in ourselves. It hijacks the fabric, if you like, the machinery in our cells, to be replicated. Now, our cells work on DNA; they don't work on RNA, so what it needs to do is transcribe the RNA, which is in the virus, to DNA. Now, that transcription is done by a molecule called reverse transcriptase, and that's a kind of a stupid molecule, in the sense that it makes mistakes. So, while it's reading RNA to produce DNA, which then will be inserted into the cell, it's making mistakes, and these mistakes are basically the mutations that cause the virus to change. We'll come back to that later.

But that's the biggest problem, that's really the big flaw. So, once you design a medication, because of this replication error, the medication might not work anymore. I'll show you. So, what we want to do is to basically track this whole problem of HIV, from molecule to man, as I like to say, and we're going to do that in four steps. I won't take you through all the steps in detail, but just give you a flavor of how we do this kind of research, what we have obtained so far, and some hot new results that we have. So, we're doing something unlike molecular simulation, something called agent-based simulation, something like cellular automata, and something like complex networks, and then somehow we have to make sense of all these different things in a good way.

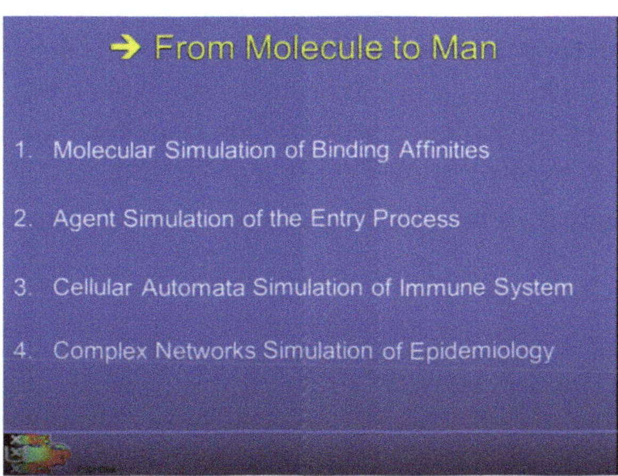

So, here's a kind of animation of the viral process. The green thing is the virus, these are your white blood cells, your T cells, T lymphocytes, and you see different steps and different drugs and inhibitors. A drug or medication actually blocks the first attachment of the virus to the cell, but these medications are not used very often.

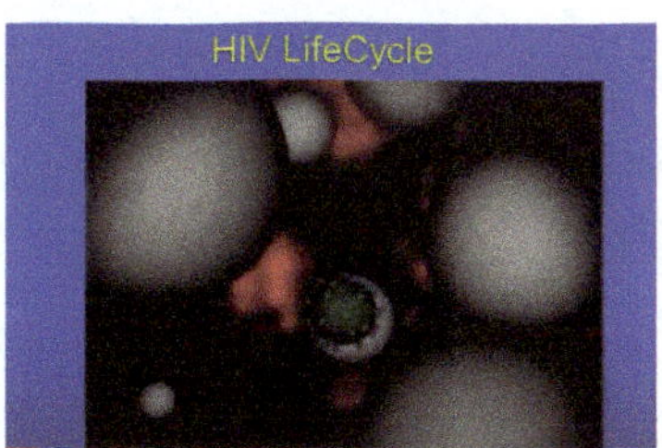

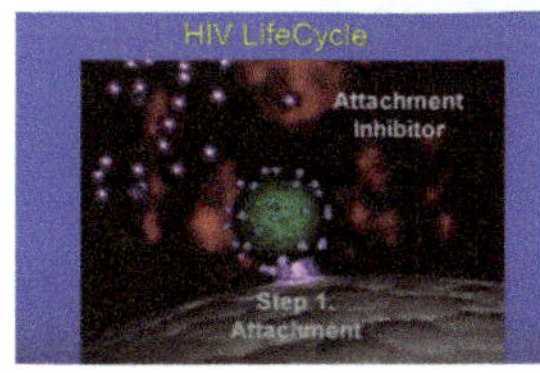

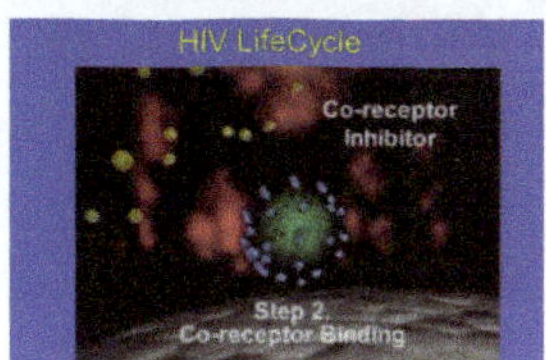

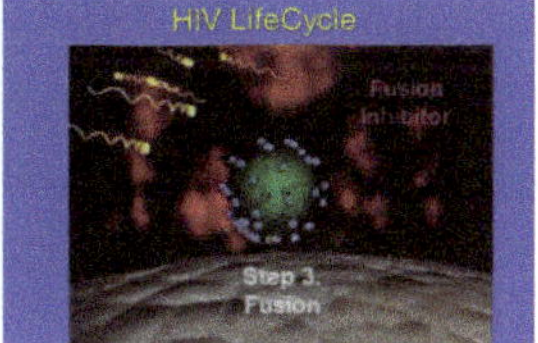

If that fails, then there is a process going on called co-receptor binding, which actually makes the virus more strongly connected to the cell. If that process is not blocked, then we get fusion, which is actually kind of like melting of the virus membrane with the membrane of the white blood cell.

Then what happens is that the virus essentially acts like a broker, now just giving its code to your cell to be replicated. Here is this RNA, along with all kinds of supporting molecules, like for instance the reverse transcriptase, the molecule that remains throughout the transcription.

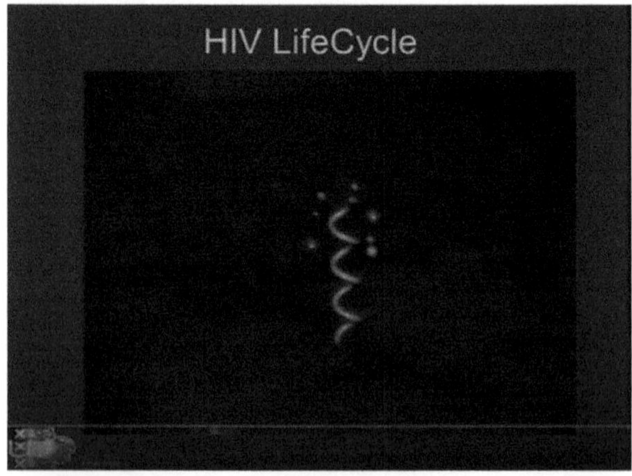

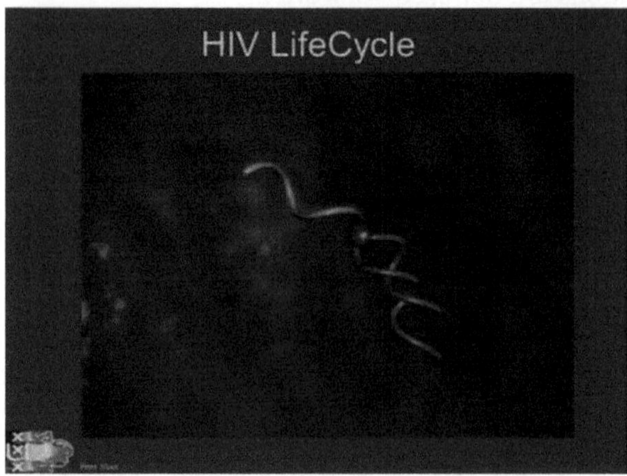

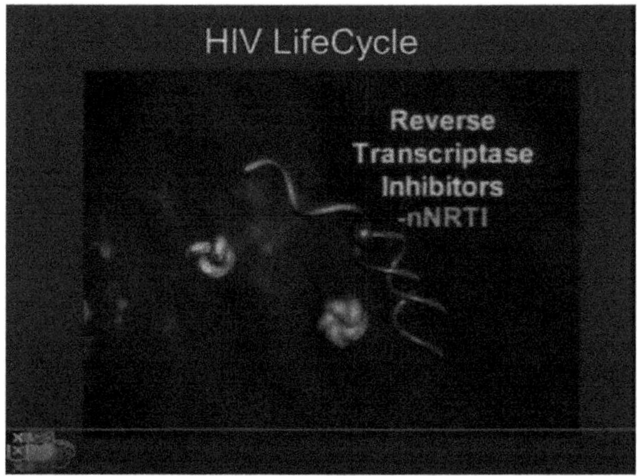

And again, we have blockers there, and this is the important blocker that's being used mostly nowadays, this reverse transcriptase that tries to block the process of transcribing from RNA into DNA, where all these mistakes are being made.

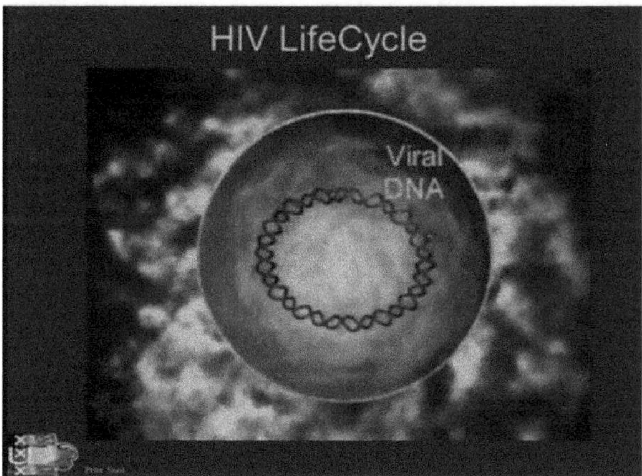

Now, if that blocker doesn't work, then the virus is actually entering its final DNA into the DNA of your white blood cell, which then happily starts to copy because, you know, that's the thing that cells do.

They want to copy, and by copying themselves, they also copy this information, this program. They run this program of the virus and actually

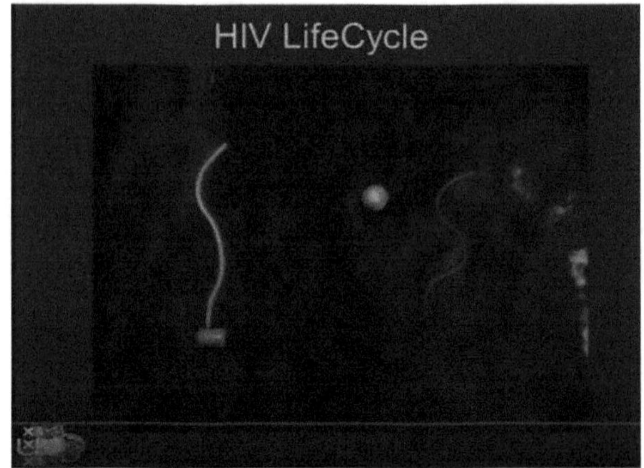
HIV LifeCycle

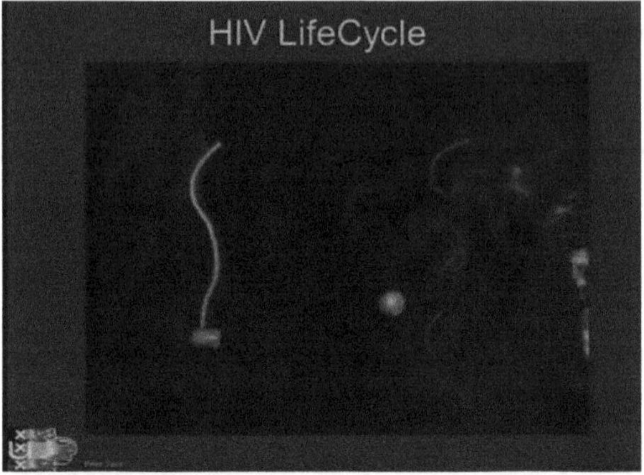
HIV LifeCycle

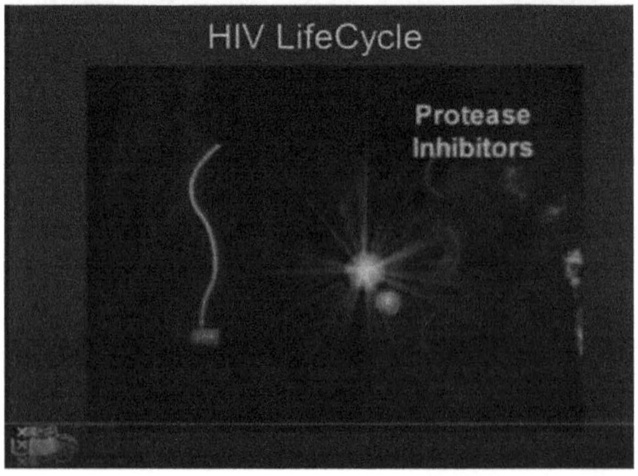
HIV LifeCycle
Protease
Inhibitors

produce the virus particles. And that's happening, there are new virus particles.

But I need to compartmentalize and think of the story of John Holland, about boundaries. It needs to have a boundary to actually do something. The animations are still working.

So, I need to have a boundary, and that's what's actually happening. It's going to aggregate, and once it's aggregated, it creates this polypeptide boundary, and then it actually releases again, with a production rate of 10 to the power of 9 viral particles per day per patient of the primary infection.

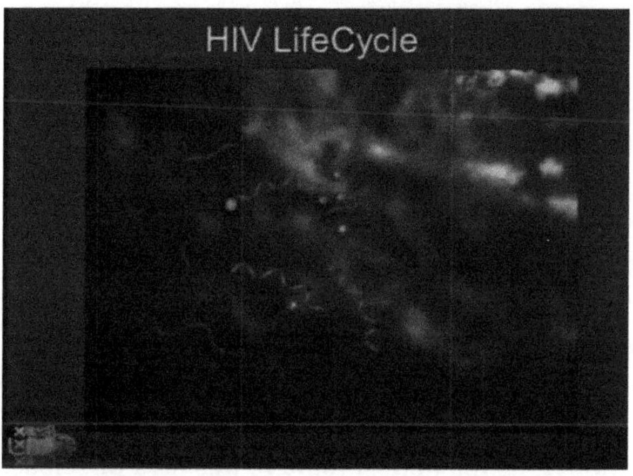

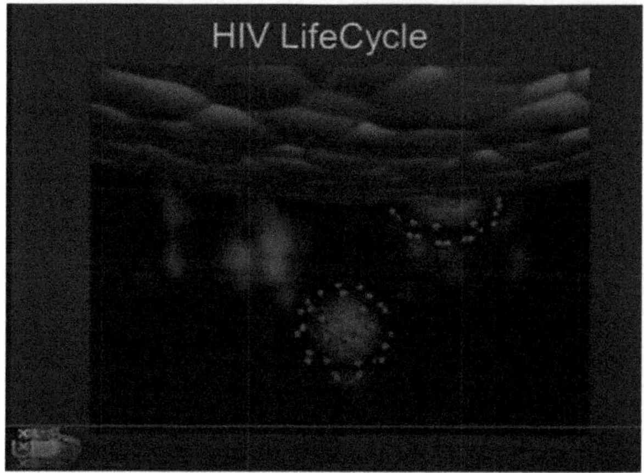

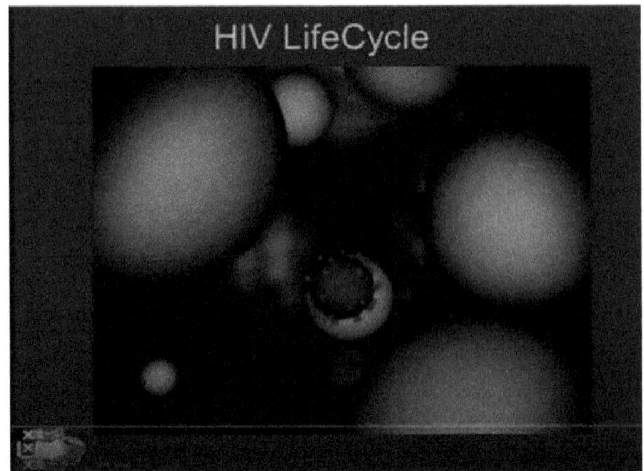

Okay, so you've got the idea. Now, the problem is these mutations. What we know is that if we have a drug that is working for a certain virus that's in the body, and then these mutations occur, then after two mutations, let's say two fit mutations, we already have fourfold resistance to the drugs. So, the drug is working only for 1/4 of its power, so to say, and with three mutations, there's already a tenfold resistance. And like I said, there will be many mutations that lead to resistance. This is the biggest problem of HIV, and understanding the transmission of HIV — understanding this resistance, and understanding the transmission of resistance in the population — is the hottest topic in this area of science.

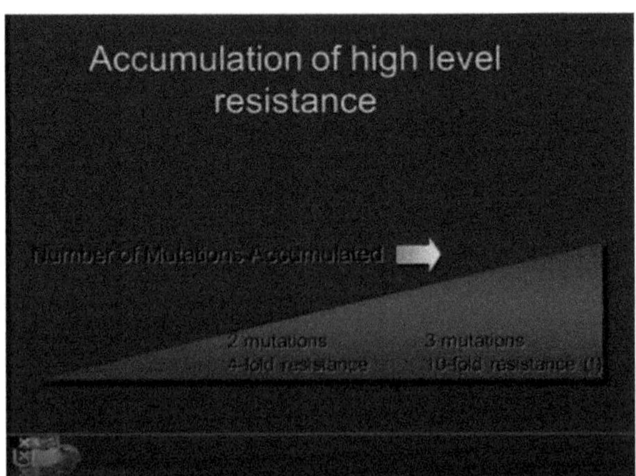

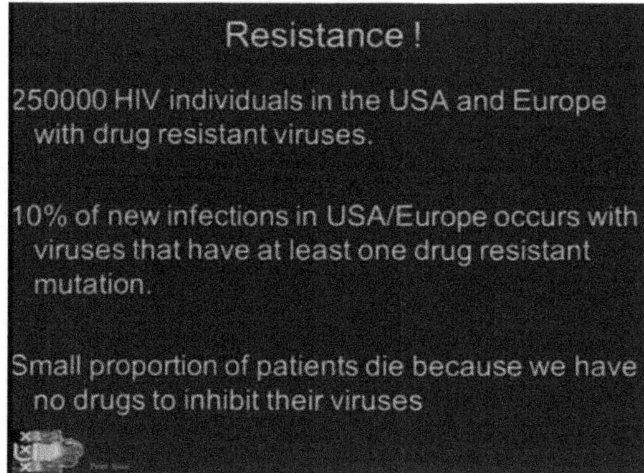

In this situation, we have about a quarter of a million people in the US and Europe living with drug-resistant viruses, and 10% of these new infections are with drug-resistant viruses. A small portion of these patients die because we just don't have any cocktails that we can think of that work because they are resistant to all of them.

This is a really interesting topic to get a good understanding of that.

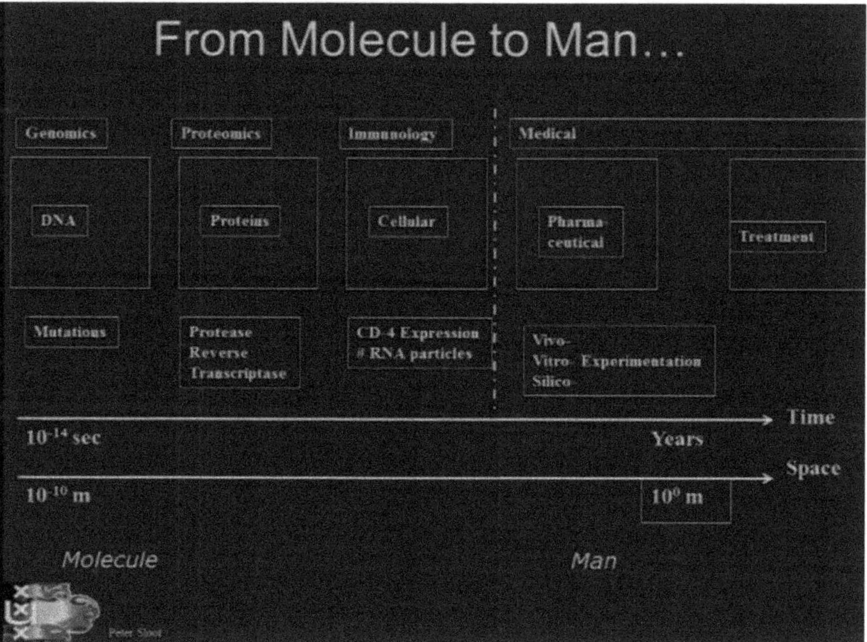

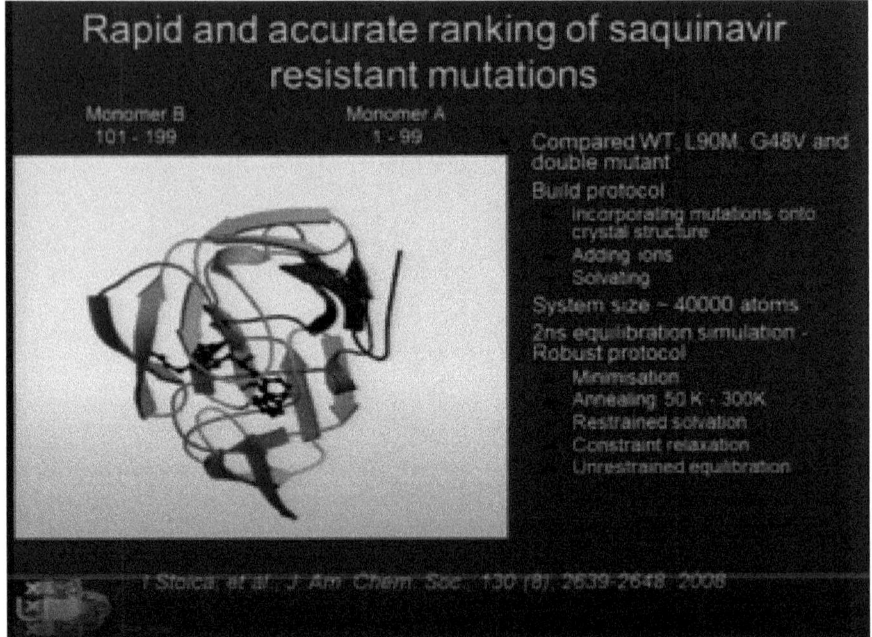

Now, in order to come to this question of the European Union, we need somehow to get from this molecular description all the way up to this description here.

So, I'm going to do that in a couple of steps, and the first is on this medical stoppage step, if you like, this drug step.

So, here, what you have is a reverse transcriptase again, that famous molecule that transcribes an RNA to DNA, and here is the drug, this medication. The way these things work is, you probably remember that from high school, the reverse transcriptase tries to read this part of the RNA and then produces its DNA, right? But it reads with a small part of the whole molecule, so a small part of this whole thing, this whole spaghetti string, is actually the active part, and that's the part where the green, by the yellow thing, resides. Now, if you would kind of block that part, if you would put a cork in it, which is the yellow thing, if you would just block it, then it cannot read anymore, right? So, that's the whole way these medications work. You just block the active side of the molecule that is the culprit here. So, what you do is you design a molecule that fits perfectly in that beaker,

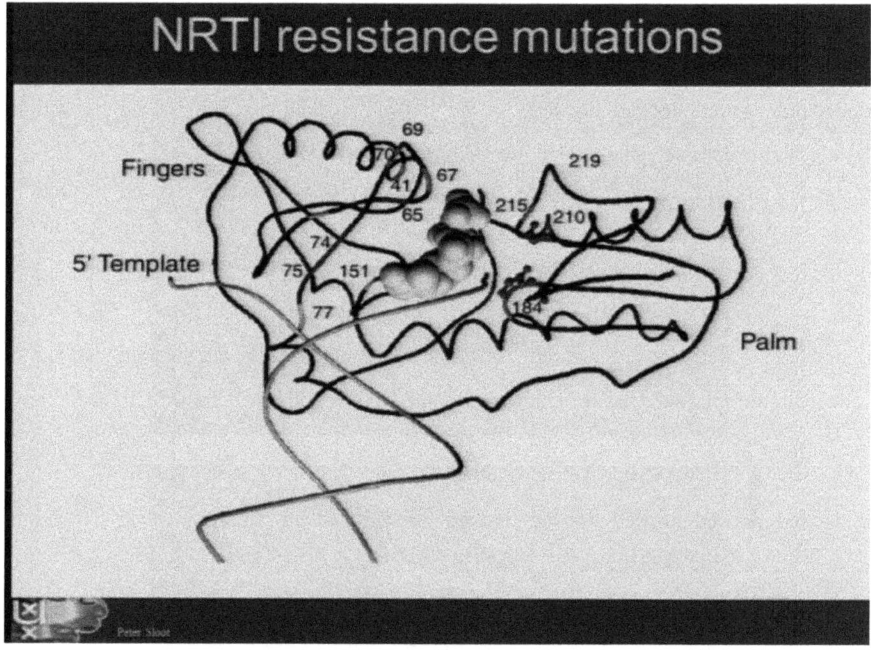

that reverse transcriptase molecule. If you do that, you know, you're in business, if you can manage to do that.

And so, we've been studying that in quite some detail, together with the group, a company in UCL, and here's an example of many simulations that we've done. So, here is reverse transcriptase again, and here is the molecule that we insert into the active site. In the red spot is the active site, and the thing that you see here is the thing that you do not want, right? Because what you want is to put a cork in place, and it should stay there.

But what's happening — I'll run the thing again if I manage to do that — so what's happening is that we put it, you know, in the computer, we put it in place, but then it walks away. And the reason why it walks away is because of mutations.

So, there are mutations here, mutations there, you don't see that because it's too small, but because of the mutations, you change to the spherical interactions and the spherical hindrance processes in that big molecule, and then basically the medication walks away. Now, the only thing that you really need to remember up to this point is one thing: that

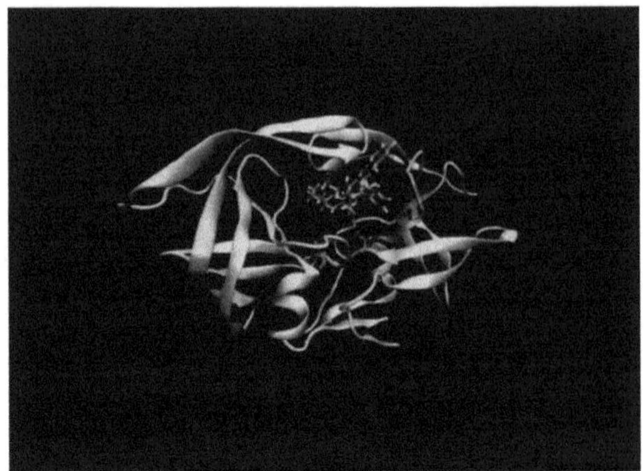

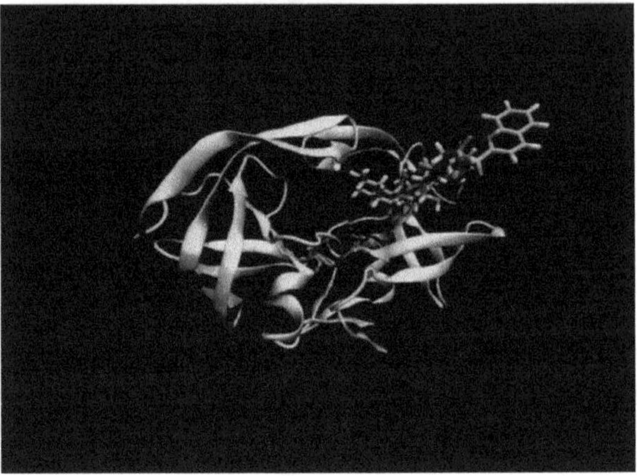

we can actually calculate the binding affinity of this medication, of this reverse transcriptase. So, we calculate how likely it is that the medication will work, okay? That I call the binding affinity. If the binding affinity is 100%, that means the medication works 100%, but it's not 1 or 0, it's something in between, right? So, it can be 30%, 20%, whatever. Now, this binding affinity is important because if the binding affinity is very high, that means that the viral load is very low, that means that the probability of transmission will be very small. So, there is a direct link. Here, you already see how simple actually this work is because you already see the link between sexual interaction, the transmission of the virus, and that molecule. So, if the binding affinity comes down, the viral load comes up and the transmission probability goes up. So, we will make use of that later. This is the thing that helps me connect models together. Here, I have basically solved Newton's equations, and here, I would be something else. You got this idea? Good.

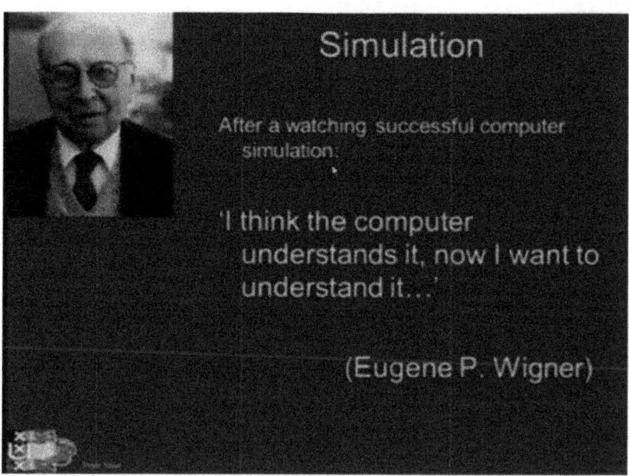

But that was a simulation, right? So, a point about simulation, I'd like to give you this quote; who recognizes this guy? Murray, Murray, he would probably know. This is Wigner, a very famous physicist and mathematical physicist, I would say, who said, after watching a successful computer simulation, "I think the computer understands it, now I want to understand it." Right, and this is, of course, the trick that we have to be very much aware of. By the time we're doing all these beautiful simulations, you know, we still have to be sure that we understand what we're doing. But then, on the other side, most of the complexity science studies are with computers. You know, we are building models because this is the only conceptual framework that we actually have, so we need to do that. Then, we need to always go back to this point, you know — do we really understand what's happening here?

Well, I won't go into the details of what we did, for instance, for the specific case of calculating this binding affinity. We actually did laboratory experiments in which we calculated the amount of free energy, and we compared that with the simulations, and it was pretty, well, you can see the numbers are very close.

Complex	$\Delta Gsim$	ΔG_{expt1}	ΔG_{expt2}
WT	-13.76(1.46)	-14.30	-13.76
90M	-11.55(1.31)	-12.51	-13.09
G48V	-11.31(1.06)	-11.28	-12.16
90M/G48V	-10.64(1.25)	-10.21	-10.04

And this is for different mutations in the reverse transcriptase, so we can actually do it.

So, having solved that, we have a binding affinity. Now, we come to the real hard part, well one of the real hard parts, and that's a part where we have to take into account the immune system. John Holland here said, "The most complex system I know of is the immune system," and I think I agree with that. But even then, he said, even though it was the most complex problem, with one vaccine, we all can already act on it, right? Even if it's complex, very complex, there might be some simple principles underneath it that we might want, whatever we can make use of, which is what we did. So, what we actually did, I won't go into the details, I could just give you the result, but we actually made a cellular automata model, a three-dimensional cellular model. This is a two-dimensional introduction to that three-dimensional cellular automaton model, in which we just said, "Okay, if the cell is here, if a virus is here, what is the probability that it can actually be depleted by the cell, what is the interaction between the cells," etc. So, there are very generic principles that you can do.

Then you can just run all kinds of simulations by looking at how virus particles will be depleted by the cellular system. We worked through that, we worked through that. And if you do that, after a couple of years — because it took us really four years to do this — you are actually able to reproduce the dynamics of the virus.

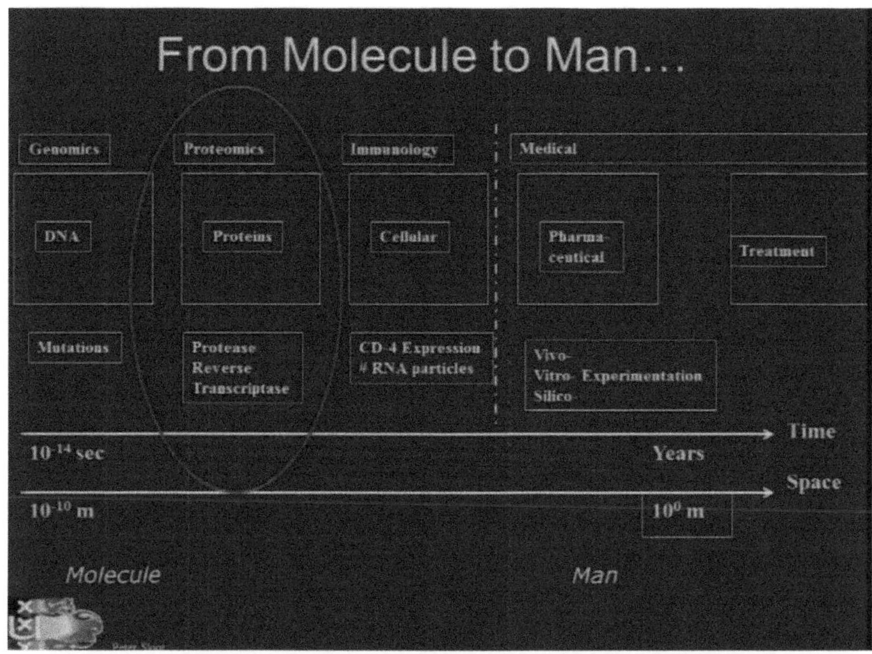
From Molecule to Man…
Genomics
Proteomics
Immunology
Medical
DNA
Proteins
Cellular
Pharma-ceutical
Treatment
Mutations
Protease Reverse Transcriptase
CD-4 Expression # RNA particles
Vivo- Vitro- Silico- Experimentation
Time
10^-14 sec
Years
Space
10^-10 m
10^0 m
Molecule
Man

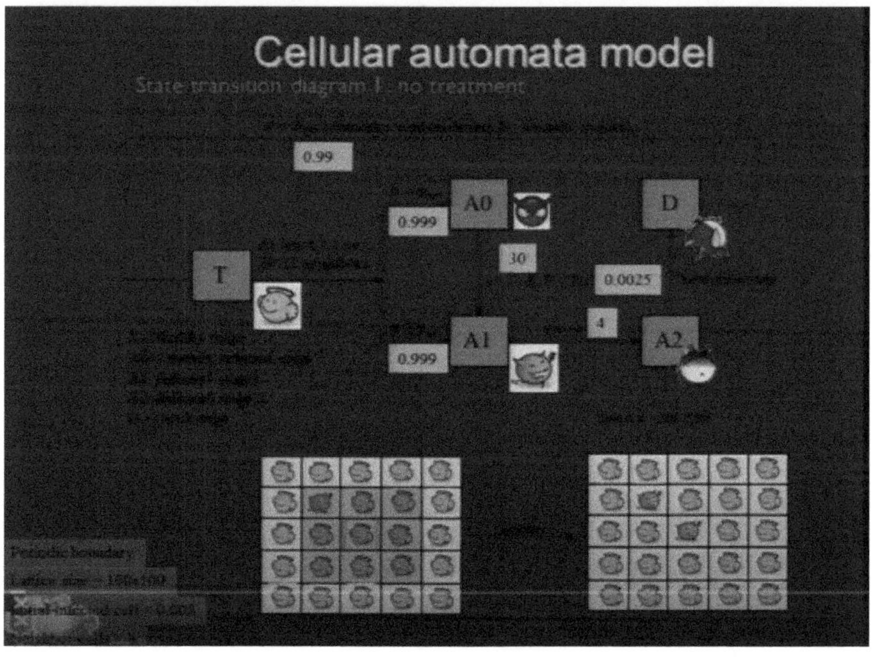
Cellular automata model
State transition diagram I - no treatment
0.99
A0
D
0.999
30
T
0.0025
4
A1
A2
0.999
Periodic boundary

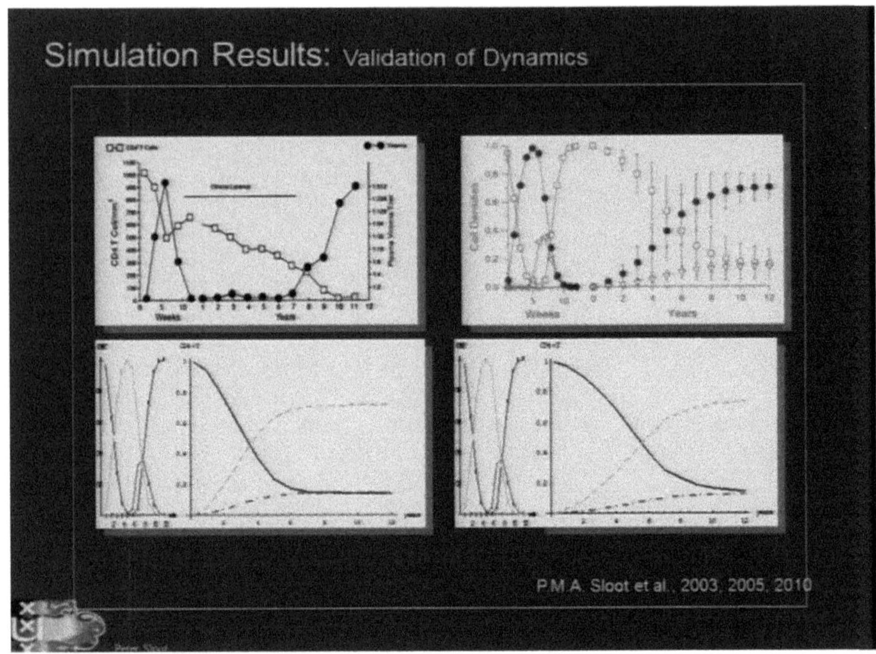

This is a slide, the second point that you want to remember from this talk. So, what you have here is a patient, and what you have here is our simulation. What we're looking at is two things: Those black dots represent the amount of virus particles in the body and the white squares represent the CD4, that's the molecule that sits on the T cell, and let's say that represents the health of your immune system. So, the higher that white square is, the better it is. Now, what happens — and these are weeks, and here we go to years — what happens is the following: After the primary infection in an individual affected by HIV, there's this explosion, a real explosion of viruses. Remember, 10 to the power of 9 per day. And then the immune system starts to kick in, and it actually starts to basically reduce the amount of viruses, but at the cost of something. So, the immune system starts to collapse. You don't feel that well at that point in time. Then the immune system recovers and the virus seems to go down. If I were to stop this picture here, at this point, you would have basically influenza. This is the way our immune system works, right? It just depletes the viruses and it recovers. But the point with HIV is it doesn't recover, and the point is it doesn't deplete the virus completely. I'm not going to tell you why, but that's what's happening. And then what happens is that the virus actually starts to hide in the lymph nodes, and it really hides. One of the ways it hides is actually by taking away some molecules on its surface, so the immune system cannot see it anymore. The immune system becomes completely blind to the virus. It's an amazing trick. And then the virus slowly starts to increase again, and now we're talking about years suddenly. In the meanwhile — this is called the clinical latency phase —the immune system starts to collapse. Right? And then we come into this — this is the primary infection phase, this is the clinical phase, and then the AIDS phase. It's the moment where the amount of virus is so large and the immune system is so compromised that you get all kinds of collateral diseases coming up. If you then have a cold, that might actually kill you because there's no immune system to recover. But also, you know, our immune system protects us from cancer, and so a lot of cancers actually come up here, in the moment that the immune system is collapsing.

Okay, so the second thing you remember from my talk so far is this model of a real patient. This is another model, it's a real patient, but you see this very good comparison. And the second thing you remember is this peak going down, going higher up, going down again, and this collapse of the immune system.

That's a good point. So, even though the virus becomes mostly invisible to the immune system, it doesn't become completely invisible. You get all kinds of what are called breakdown products that will actually trigger the immune system, and it cannot actually handle that. A very good question.

Okay, so with these two things, we actually now have something that we can start to think about. Okay, that's all happening in the body, but now we have to take the thing outside of the body. So, we need to talk about sex like I promised. And because our study is done in different cities, I'll talk about sex in the city.

So, I will be talking about how we take information from social and behavior data, medical information, doctor records, etc., and how we use that to actually combine with this model. And we do that for different cities: Amsterdam, Rome, and San Francisco because that's where the data

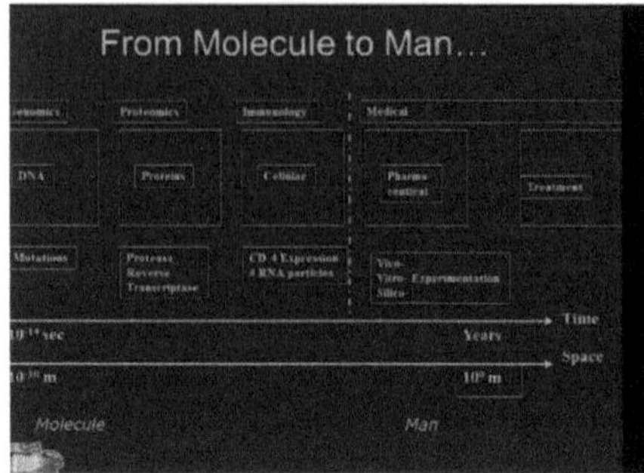

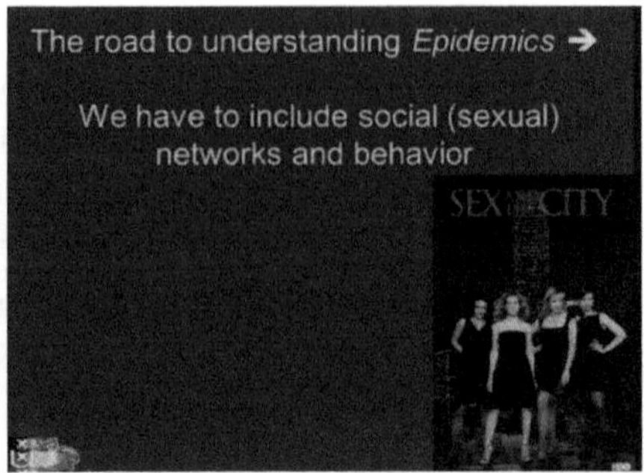

are. Okay, so we have to include, now, we have to start thinking about what's happening there. So, what I'm going to do is digress now. I'm going to digress and tell you about why we chose this specific model for the transmission of viruses. Because all complex systems research is about, and where I spend all my time, thinking, "Okay, I got this complex system, this is the data, what is my model, where does my model sit?" If I have a model which is too highly granular, with too many details, and there're no data to support it, then that doesn't make sense. So, I have to kind of fit my model with the data that I have. And then again, I need to make a model such that I can link it up to other models that I have. Because if I have a beautiful model of transmission — and there are beautiful models well known in epidemiology — and I cannot combine that with, for instance, treatment models, I cannot combine it with behavior, or I cannot combine it with the binding affinity, I can't just throw it away. It doesn't make sense to me. So, I need to look for models that actually fit this sequence of models, if you like.

So, a way to study that, as I'm trying to convince you now, is making use of something called complex networks. Now, this is a complex network, and it has nothing to do with sexual transmission. This is a dating diagram, or maybe it has got something to do with it. This is a dating diagram from interviews done at a high school in the United States. The blue ones represent boys and the red ones represent girls. The interview was about whom you are going to date in three weeks, right? So, from that interview, you can actually find connections. And that's the message to get from here. The message you get from here is that this boy was dating two girls within a certain period of time. Now, you can ask yourself, there are some very happy boys here and some pretty active girls also. I forgot where they are, but there are some pretty active ones. You can ask yourself where you yourself were. I think I was around here. But you got the message, right?

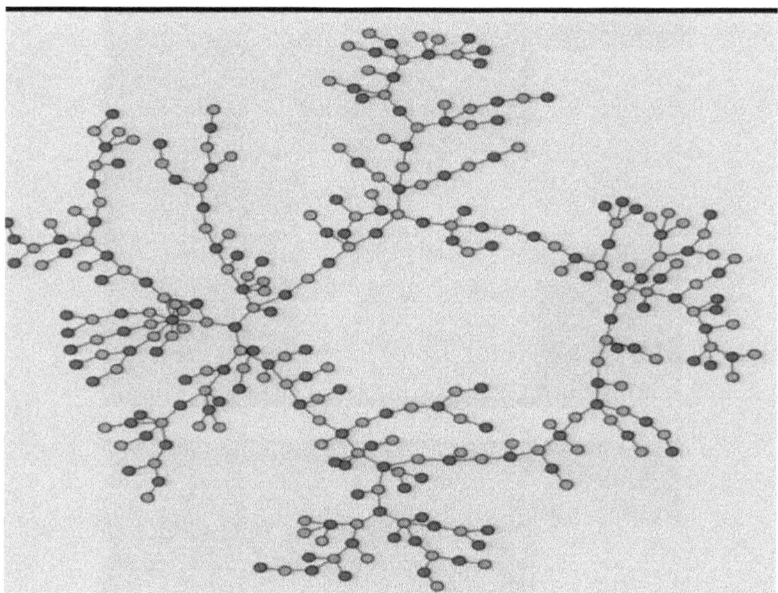

The message is that we can actually make these kinds of graphs to express two things: individuals and the interactions between individuals. And that's exactly what I want because up to this moment, I was talking about what's happening in an individual, and what I need to start talking about is what's happening between the individuals. So, this is actually a nice way to go from the virological description to the epidemiological description and connect them, which again brings a crucial question: How do we connect all these models? Because I don't believe in abundance; I don't believe in one model that does it all. I have to connect models. These things can get very heavy — let's skip that one.

I have to say a few more things about these networks for you to appreciate what we can do with them and how we can make them work for us. You've all heard about small-world networks. The basic story is that we are all connected, right? So, if I want to send a message to anybody anywhere in the world, I might know you, and you might know him, and him, and then within six steps, we can cross the whole world. That's the message.

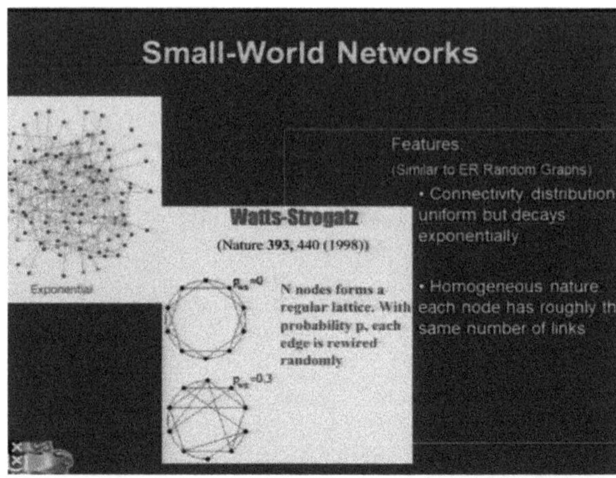

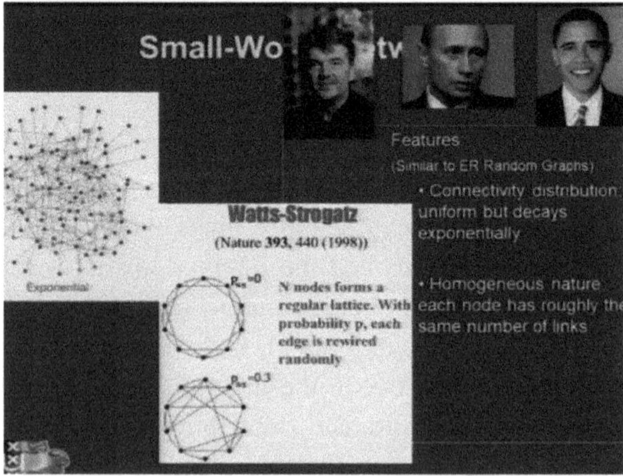

That's like a small-world vision of the world. So, even if we're talking about seven billion people, we're only six handshakes away, right? That's more or less proven. I'm careful here, but let's just say that it's proven. So, that means, in my case, I had an opportunity at some point in time to shake hands with this guy, and this guy shook hands with that guy. So, that means that I'm two handshakes away from Barack Obama. So, after this lecture, feel free to shake my hand. And that said, that gives you the idea.

The point is that we always thought that's how these networks worked. So, we get this idea: The K stands for the amount of connections an individual has, right? So, we always thought it would be like this or it will be, as everything you were saying, that also most things that we look into the world are somehow normally distributed. So, this is what we thought it would be. The probability of having K connections would be something like this: There would be the majority, who would have let's say ten connections, and a few would have a hundred, and a few would have one.

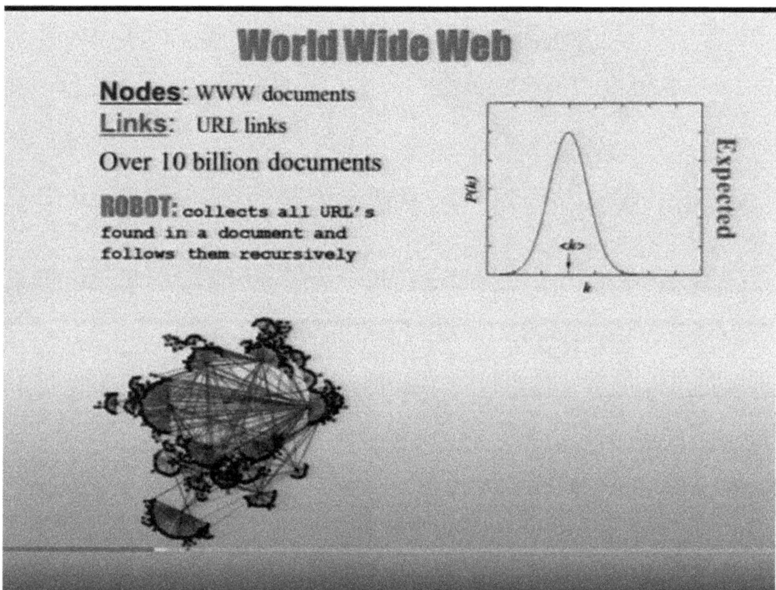

Unfortunately, or maybe fortunately, if we look into real-world connections, they don't behave like that. They behave like this, and this is scale-free, which shows that the system is scale-free. We call this scale-free complex network. What it means is that the ones with a few connections, there are most of them, and the ones with a lot of connections, are few of them.

This behavior you will see throughout all kinds of networks, all kinds of real networks, whether it's networks that come from plane connections, the flight connections in the US, or from science co-authorships, something completely different. Again, they have this scale-free structure.

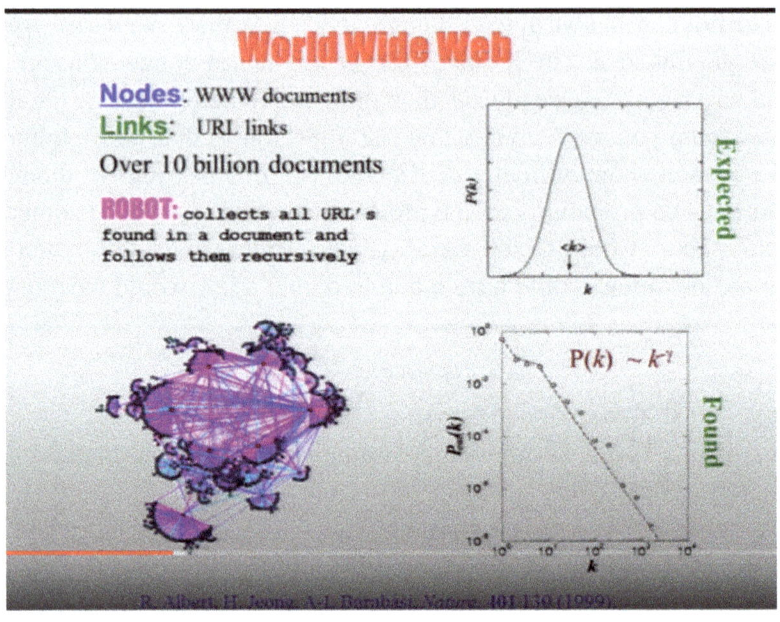

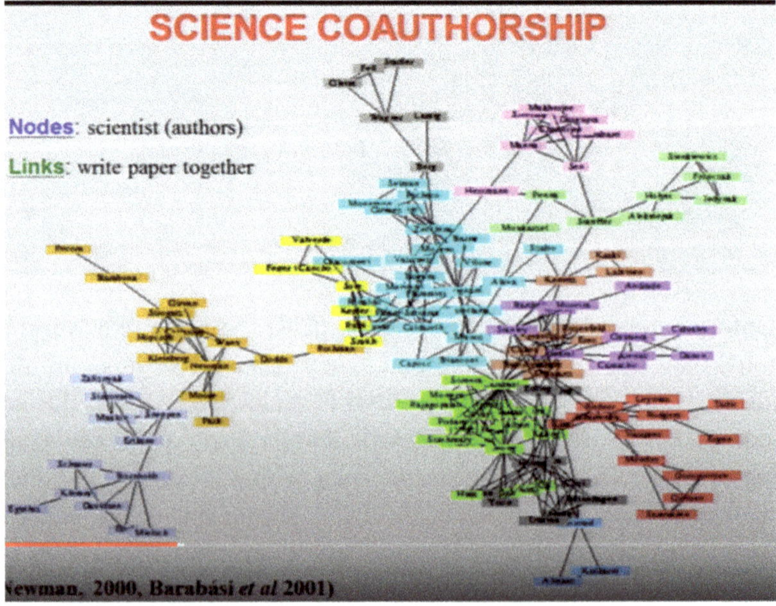

If you look at another one, the structure of organizations, which is an emerging property, right, it seems that the emergent property of these kinds of interactions is always scale-free.

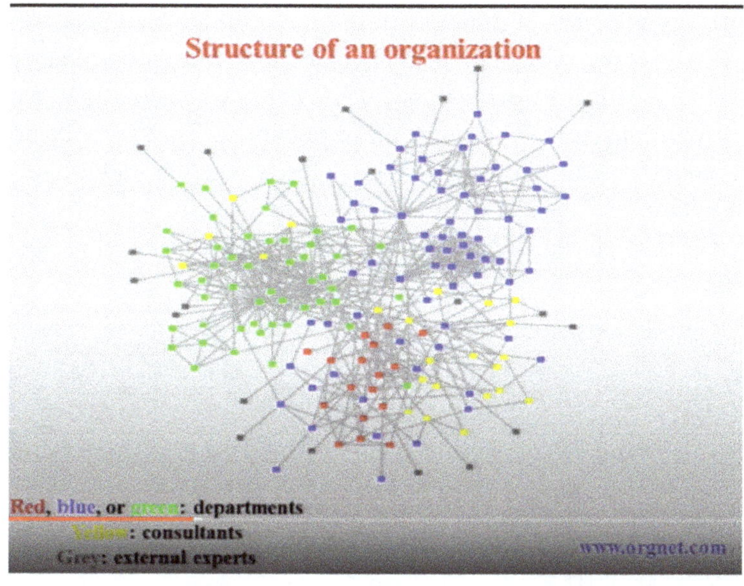

Structure of an organization
Red, blue, or green: departments
Yellow: consultants
Grey: external experts
www.orgnet.com

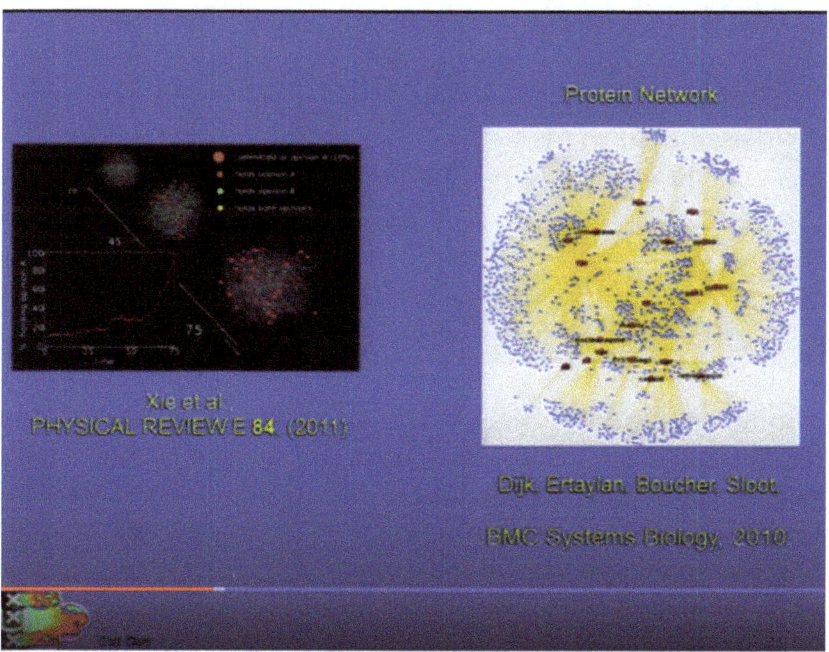

Protein Network
Xie et al
PHYSICAL REVIEW E 84 (2011)
Dijk, Ertaylan, Boucher, Sloot.
BMC Systems Biology, 2010

But also, you can do things with it. Once you have an understanding of those networks, you can actually see how recent opinion spreads through them. This is research done by Xie *et al.*, who actually looked into the way opinions spread through those networks. That's good; now we have a model we can play games with. This is what we did ourselves on protein networks. You can really map out the protein networks and then again see that these things are scale-free, exhibiting this lock-step behavior.

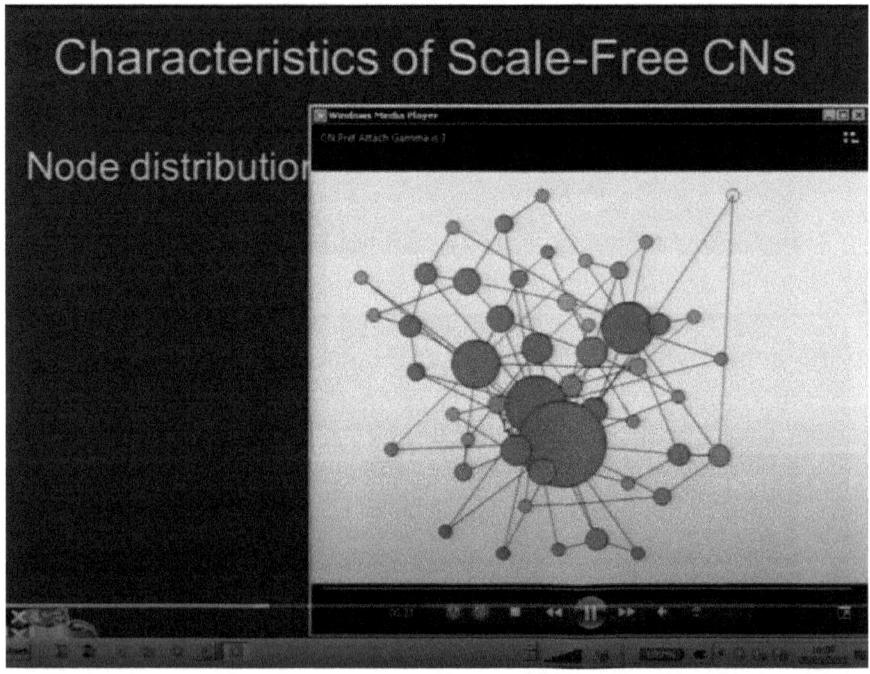

So, let me say a few more words about this, again, to get a better understanding. These models, these networks, can be modeled like this. This is the reason why they're called scale-free, because if I change K for a K, I basically keep the same function. So this has the probability of having K connections as K to some kind of power. Right, so this means, and here I need your attention, this means that with one parameter, gamma, I can describe the network. So, if the network is characterized by this, with one parameter, one parameter basically describes the interaction between us. I think that's nice, especially from a modeling point of view. But as John also mentioned, you know, these networks are not static. They are dynamic, they're growing, they're rewiring, etc. And so just a way for you to start thinking about how these things look, what they look like, this is of course a toy model, but it just gives you the feeling of how these things actually emerge. So, the size of these balls tells you something about how many connections they have, right? And the more connections they have, the higher the likelihood that they get more connections. So, the rich get richer and the poor stay poor, that's the idea.

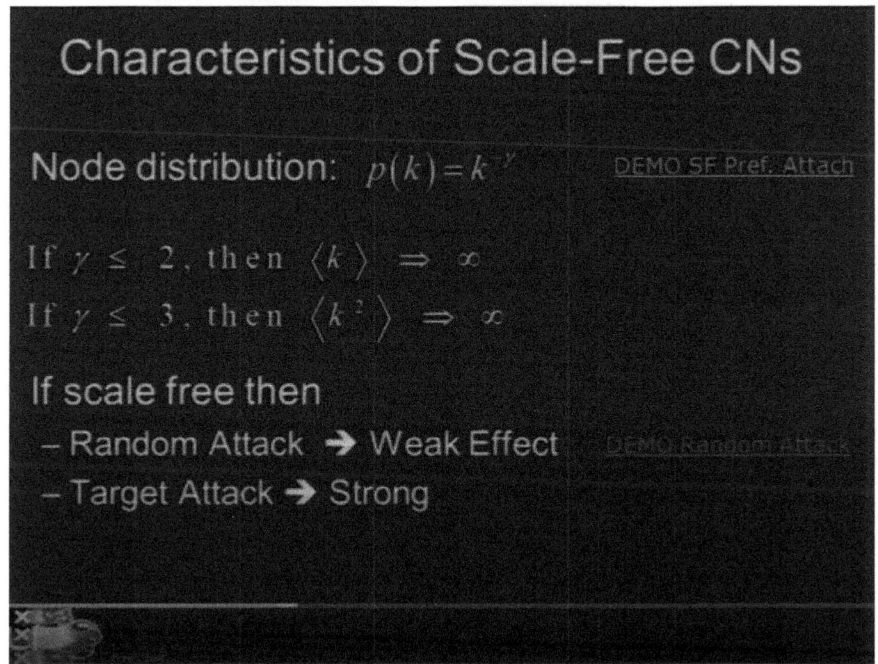

So, we call these the hubs and the other ones are, let's say, the peripheral nodes. And this is typically the way these networks emerge. So, it's like I said, the rich get richer model. It's also called preferential attachment. Now, if you remember high school mathematics, you will see that if you integrate this thing to calculate the average, nothing will diverge for a value about gamma equals two. And if you look at the variance, so the spread in this function, you will see that its first four values are three. So, that means there're lots of strange dynamics happening there. It also means, I'm not sure if I can say it, it also means that because the average actually doesn't exist at these values, all our learning algorithms will not work on this, because learning algorithms expect an average somewhere; how else can you calculate the cosine function? So, this makes these things incredibly difficult to study. Okay, so these things are scale-free.

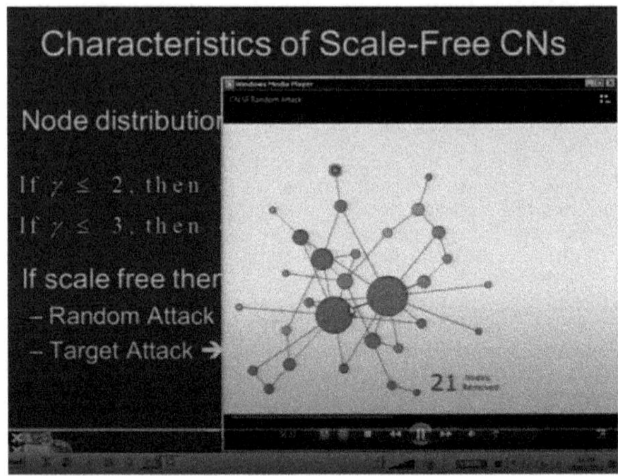

Now, the point is, if I have a thing like that, and now I would like to disrupt that network, and that could be a criminal network, but it can also be a sexual network, or it could be any other network…

If I want to disrupt it, what I can do is basically start shooting at it. So, that's what I'm doing here, I'm starting to shoot, and this is a random shot job, just randomly shooting at these balls; what you see is that even after seven nodes are removed, the whole thing is more or less intact. I can still get information from here all the way up to there. It's still somehow connected; the network's integrity is still preserved, if you like, even after 21 removals. That's if I do a random attack on the network.

On the other hand, if I do a targeted attack on the network, meaning I start to hit the richest ones, so I'm going to kill those that have the highest connections, after only 7 nodes are removed, you see the whole thing basically falling apart. Here you see the importance of understanding the structure of these networks in terms of controlling them, changing their behavior, or changing the way information flows through them. So, that's the message you want to take home here: The structure of these networks is pretty easy to manipulate, and if it's like that, we might be able to first of all model them because we can easily generate networks like that from a computer. That's a simple thing, and we can actually interact with them.

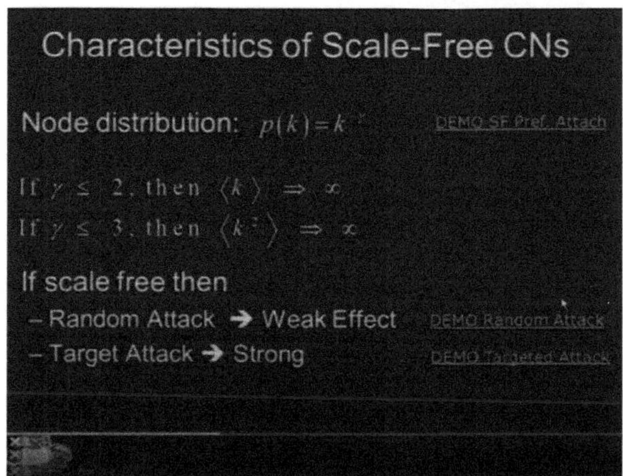

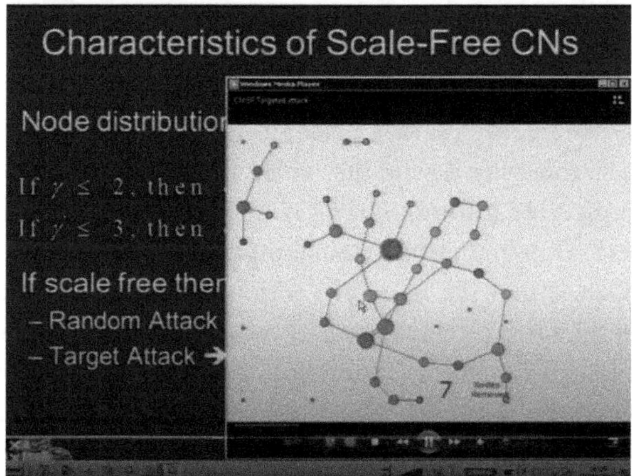

What we discovered ourselves, and published two weeks ago or so, is that in these networks, if you apply noise to it, they seem to propagate information better. Now, listen to what I'm saying here: I'm talking about information. Well, in the beginning, I talked about viruses, right? So, this is the way I look at things. You know, a virus is just a bit of information that you want to propagate, or an opinion is a bit of information; so information is the fundamental unit here.

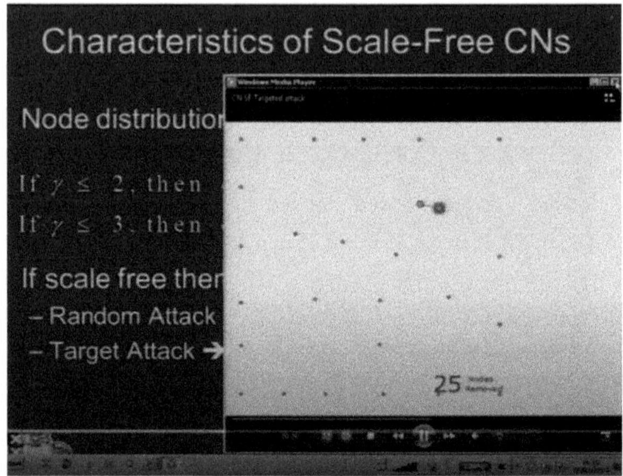

Now, what you're looking at here is a kind of artist's impression of something that is more deep, and it tries to give an answer to the question, "What happens if I have a network, like I just showed you, and I put information here in that network, and I measure how much of that information appears at the end of the network?" I measure that. And then what I do is I superimpose errors, so instead of taking the shortest path, you know, that's just rewiring going on that I have to go like that, okay. So, by introducing errors in that way, I send that virus — or that opinion, or whatever you have, the piece of information — from A to B. By introducing errors, which is noise, doing them on the temporal situation, actually what's going to happen is that the network, which is somehow in some kind of optimum, can then just move to another more optimum. But the point here is that I have to drive that noise somehow in coherence with the dynamics that's already in the network. That's why this thing is called stochastic resonance. It's a resonance effect because of the stochasticity in the system.

This, of course, is a two-dimensional thing, but the network that we were studying is highly dimensional, and the thing is that you have to drive it in such a way that this switch that you see actually synchronizes with the amount of uncertainty that you have in the system.

So, we did that for many systems, and one of the things that we did that for was an email network. Here's what we did: This is a very famous email network that's publicly available from a company, Enron. Enron

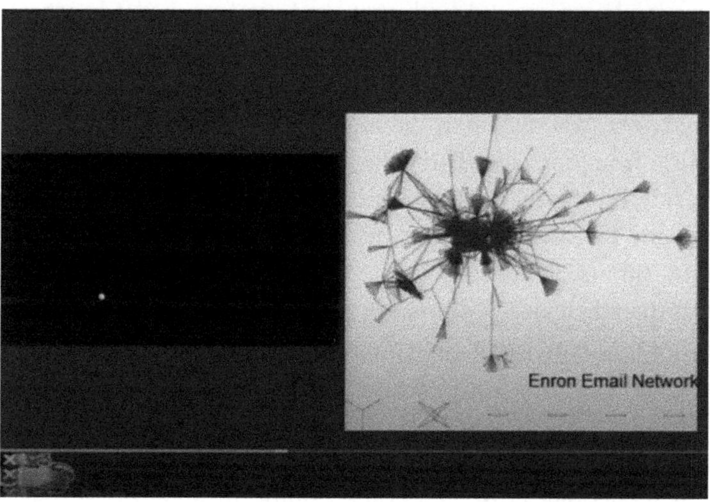

was a company, and maybe for that reason, they shared their email network. So, the question you can ask yourself is this: What would happen if I sent the information from this part to that part, and then disrupted that network slightly, and see if I can still get that information from one individual in that network, from one individual in that company to another individual in the company?

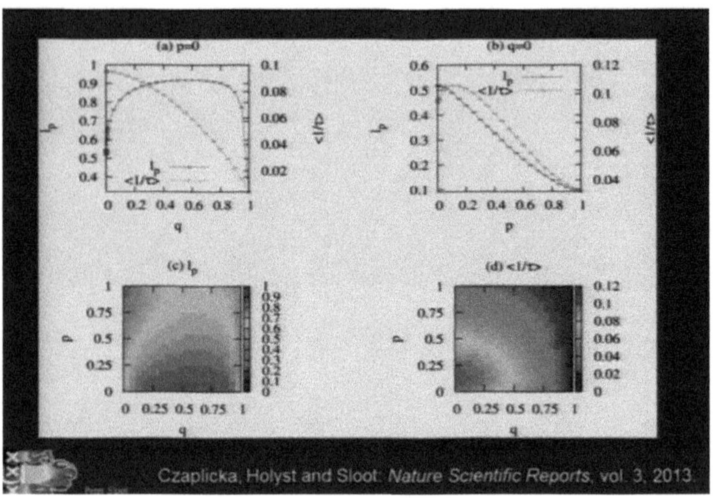

And we played around with that, and we found that it actually goes through an optimum. So, what you see here is it goes through an optimum. The red spots are the optimum. So, what I would expect when we started to do this work was that introducing noise would make it harder to get information from one side. But what happens is, by introducing a little bit of noise, it will actually make it easier. So, noise helps you to get information from one side of the network to another side of the network.

I think this is a fundamental aspect of complex systems, but let me finish this part because it's crucial. This is just a conjecture, right, this remark. It might well be that if you put too much noise in it, then actually the whole thing doesn't transmit anything anymore. So, with no noise, you have a certain efficiency; with too much noise, you have low efficiency; and in between, you get better efficiency. That's basically what's happening. So, it seems that noise is a crucial thing. It might be — and this is a kind of conjecture, we didn't study that yet, but what I'm doing now is it might be — that the reason why these networks are the way they are is because they have this aspect. It might be that they are in some kind of phase space, that they constantly need to be able to adapt to noise, to changes, to errors. So, we're doing our work on calculating this for protein networks.

So, I was talking about the characteristics of those complex networks, right? This is another new characteristic that I would like to share with you because it's a high result, I think. Now, this whole research on complex networks is really booming, and it's one of these paradigms that we want to use to study complex systems. Chaos was another one and catastrophe theory was another one; and what else do we have? Neural networks, fractal descriptions, self-organized criticality, and other paradoxical stuff. These were all paradigms that suddenly popped up and were being taken over by all kinds of researchers in the field of complex systems. Whether we're talking about ecosystems, the brain, or social-cultural systems, these conceptual frameworks were very rich and very much used.

What you see here is the kind of publications that these different theories and methodologies produced, and if I were to superimpose onto that the publications that were available up to 2010 for network research, you see that it's completely exploding. The reason why it's exploding is because it's such a tremendously rich concept. You can model opinion spreading, you can model obesity spreading, you can model, well, whatever. So, I think that's really one of the reasons. I'm not saying that it's a panacea, but it's something very useful, and we're going to use it.

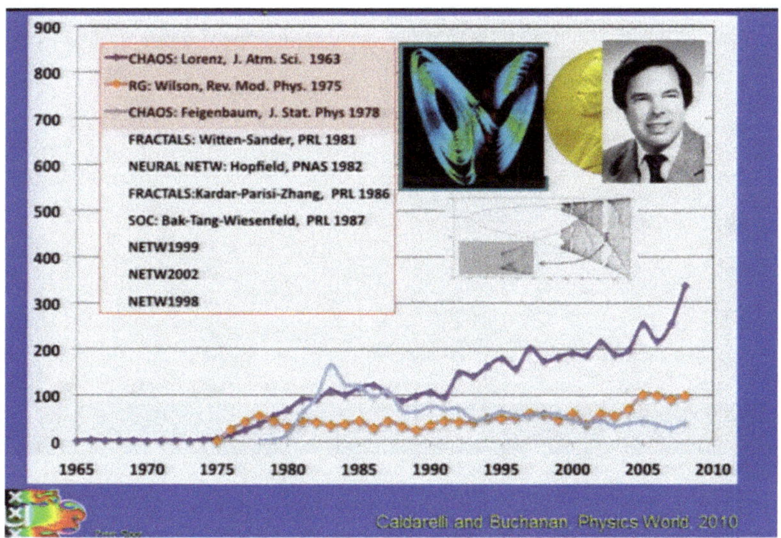

So, the steps ahead, right? What I need to do now is link the whole story I was telling before about what's happening in the body with this network story somehow, to make sense of it. So, what I need to do is infer

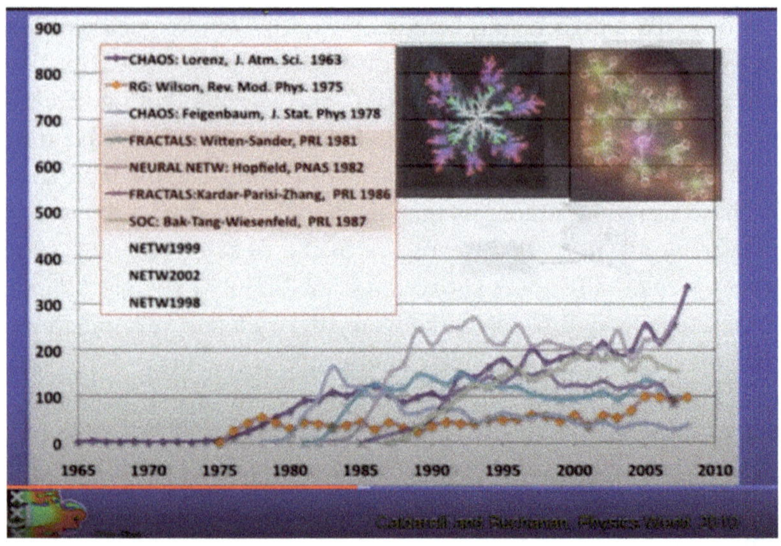

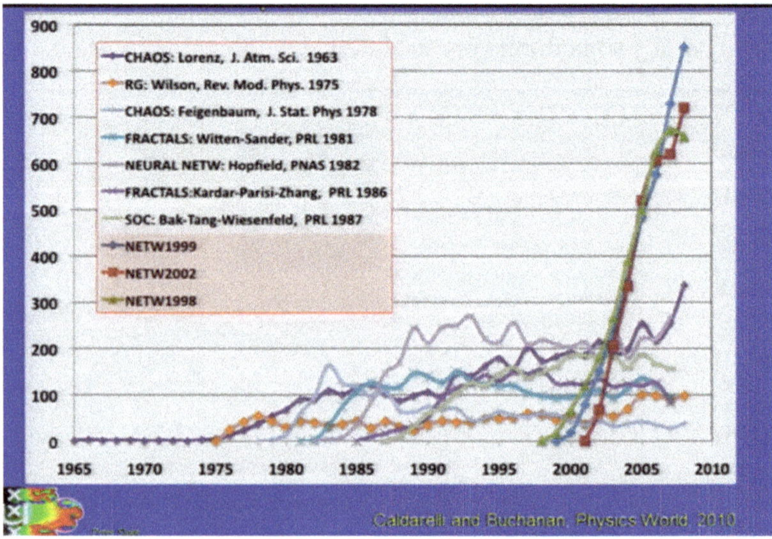

the sexual networks from the data that I have. I need to describe the nodes in the networks and the links in the network, and somehow I need to take into account all the data I can get my hands on. There are no shortcuts here. I really need to gather as much as I can to make sense of it. So, any genetic information, medical behavior information, whatever I can access, and then hopefully have a mechanism to glue that together. And then, of course, it's crucial to somehow validae it because making models doesn't make sense if you cannot validate them. So, it's crucial to somehow validate it. And then if you can validate it, you can only hope that it makes sense. If you can validate your models and explain the historical data, then maybe it's also predictive in the sense of telling you something about the future. That's maybe there, of course. And then the third question is, once you know all that — the things that decision-makers would like to know, that medical doctors would like to know, that the government would like to know — "Can we actually control it then?"

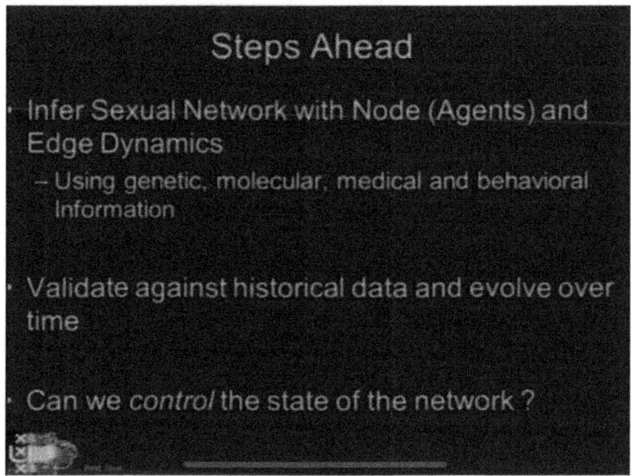

So, if we know how this spreading is taking place, can we control it? My first reaction when we start to do these things is, of course, you can control it. Once we have it, we just play around with the parameters and we can control it. Well, that's not true. That might not be true, let me put it like that. And it actually reflects on a remark I made yesterday that there is a certain arrogance we have. By the time we think we understand how to describe a complex system, we also think that we can control it. But that might not be.

Okay, so these are the kinds of questions we want to ask, right? How do we infer these networks, how do we validate them, and if what we're doing makes sense, can we control them? So, here we have things that are also called ridiculograms, but basically networks. This is actually from when we went to the Center for Disease Control in Atlanta and we dug into their data. We came up with the heterosexual and the homosexual — this is men having sex with men, that's the formal terminology; people are using MSM, so from now on, I will use MSM because it's a better denominator. So, for the MSM and the heterosexual population, you can actually dig into the data there. I'm not going to tell you how to do that, but the bottom line is that we found that for the heterosexual population of the United States, the gamma is about 2.7. So, again, remember the equation that describes this function says the probability of having K connections is K to the power of gamma. Gamma is 2.7 for heterosexuals and it's 1.6 for homosexuals, and the K max is about 70. So, that means this is about the maximum amount of sexual contacts per year, and that can go up to 250 different sexual contacts. I have to be careful here, so with different individuals, the amount of different sexual contacts per year goes up to a maximum of 250 typically in MSM in the States and about 70 for heterosexuals.

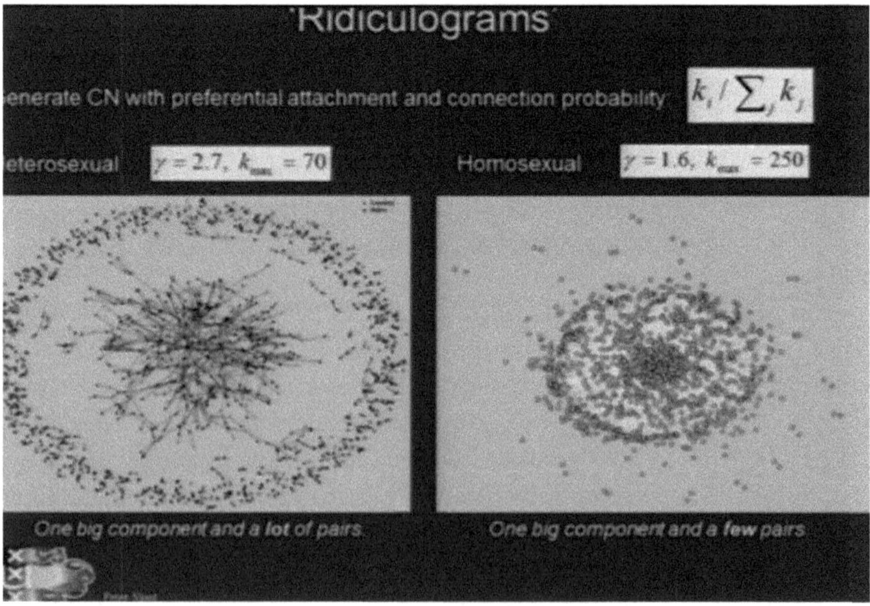

The thing is, with two parameters, I characterize those dendrograms, right? I characterize those graphs to some extent, of course. This is a simplification, we all know, and it's a snapshot, we all know. I think they're changing all the time, we all know, but it's the first good approximation. Now, the thing that you can do is actually add to that information that you have from social interactions, on something or things that, on average, we know that in heterosexual populations, the connection between two individuals is actually less when the age difference gets larger. To say it differently, people prefer people for whom the age difference is not bigger than, let's say, forty years. So, you can actually filter out those that, if you look at the total population, do not comply with these kinds of social rules, or with some probability.

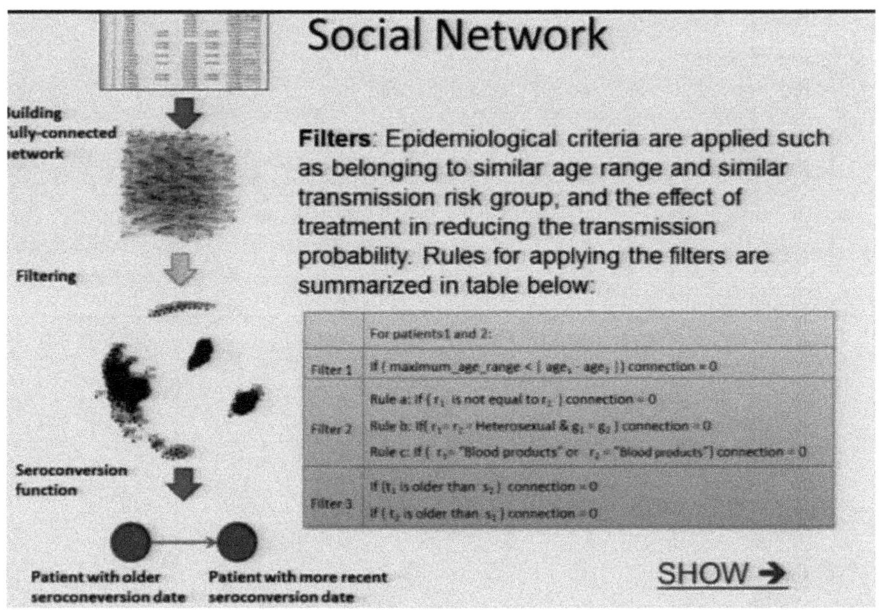

	For patients 1 and 2:
Filter 1	If (maximum_age_range < | age$_1$ - age$_2$ |) connection = 0
Filter 2	Rule a: If (r$_1$ is not equal to r$_2$) connection = 0 Rule b: If(r$_1$ = r$_2$ = Heterosexual & g$_1$ = g$_2$) connection = 0 Rule c: If (r$_1$ = "Blood products" or r$_2$ = "Blood products") connection = 0
Filter 3	If (t$_1$ is older than s$_2$) connection = 0 If (t$_2$ is older than s$_1$) connection = 0

So, you can do that, and if you do, you can actually get a good idea of how these networks fall. Here, we start out with everybody connected to everybody, and then we just apply those kinds of filters: social filters, behavioral filters, and of course, there's also a sexual preference filter. Then you sort it out, in this case, into three populations. This is the population of heterosexuals, this is the MSM population, and this is the drug users' population. These are population [groups] all infected with HIV, so it's not the total population [infected with] HIV, because we have patient data there, so we have all these additional demographic information.

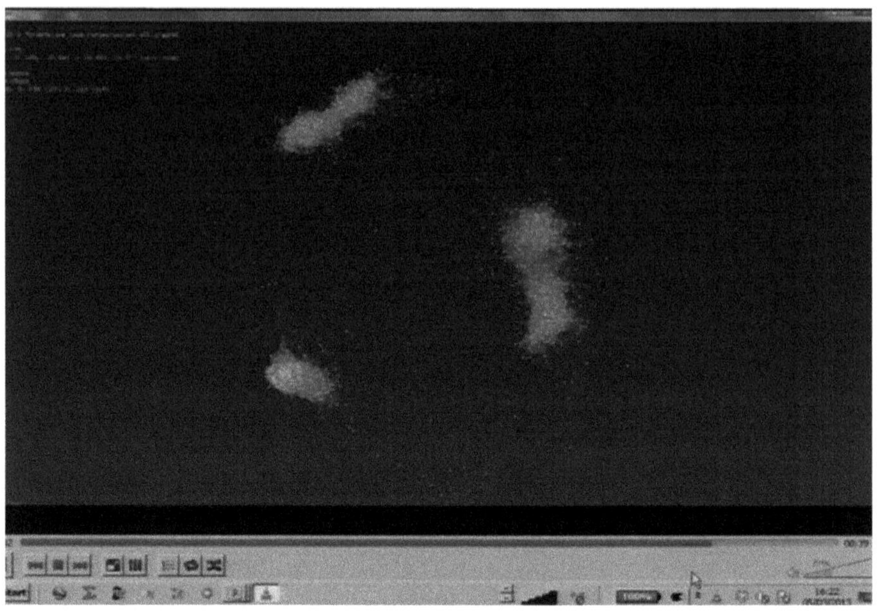

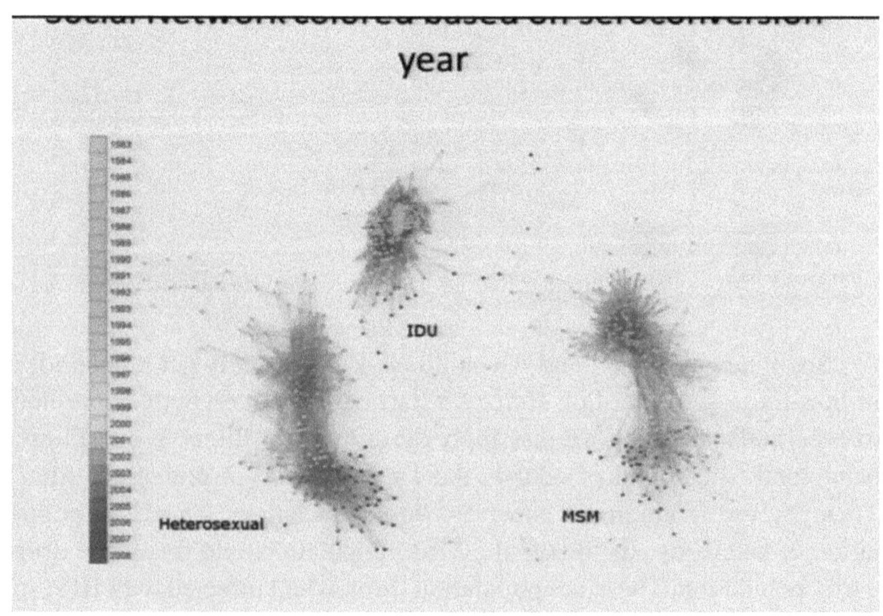
Social Network colored based on seroconversion year
IDU
Heterosexual
MSM

Okay, that's what you can do, and it's pretty useful because then you can start doing simulations and analysis of that. Remember, I was trying to extract all kinds of medical and behavioral parameters out of the system so that they can be plugged back into my models. Here, you see the time of seroconversion, meaning the time that people got infected, and that's for the drug users, for the homosexual population, and for the heterosexual population. Drug users were already infected with HIV much earlier in time, so that was really a peak, that's before the peak of the homosexuals and heterosexuals. So, these are the seroconversion times.

These are all individuals, and these are real data, real patients, real connections. So, that gives us some confidence in what we can do.

What we did last year — or rather it took us a couple of more years, but last year we published it — is that we combined this social network, which mostly consists of social information like sexual preferences and ages, with genetic information from phylogeny. So, for every individual, we have the genetic information of the virus. This combined information allows us to get a much clearer estimation of this famous scale-free function. It fine-tunes the whole dendrogram, or ridiculogram, with genetic and social information. This is for the homosexual population that we studied.

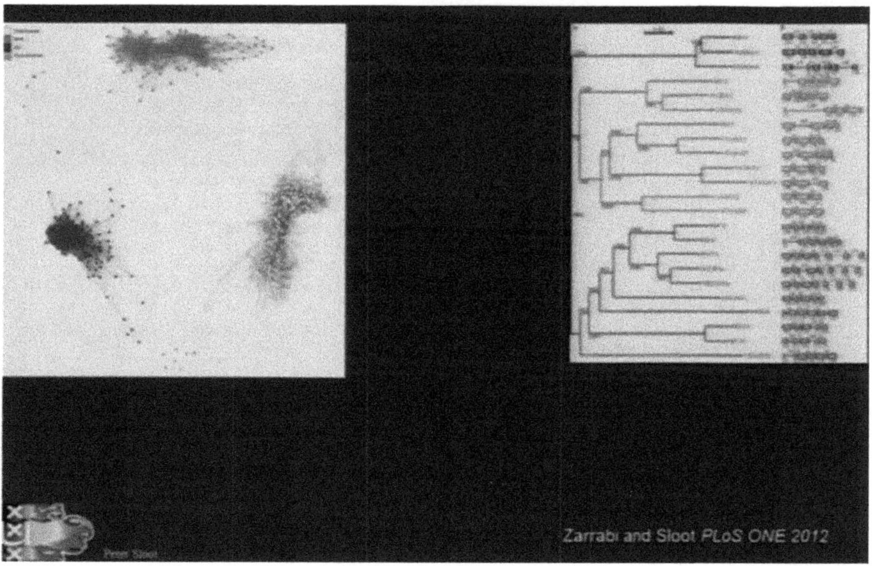

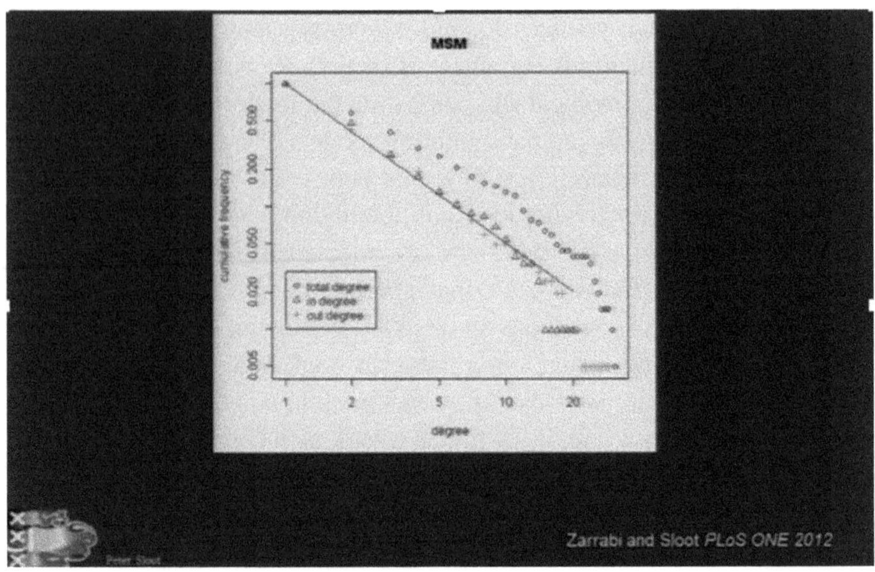

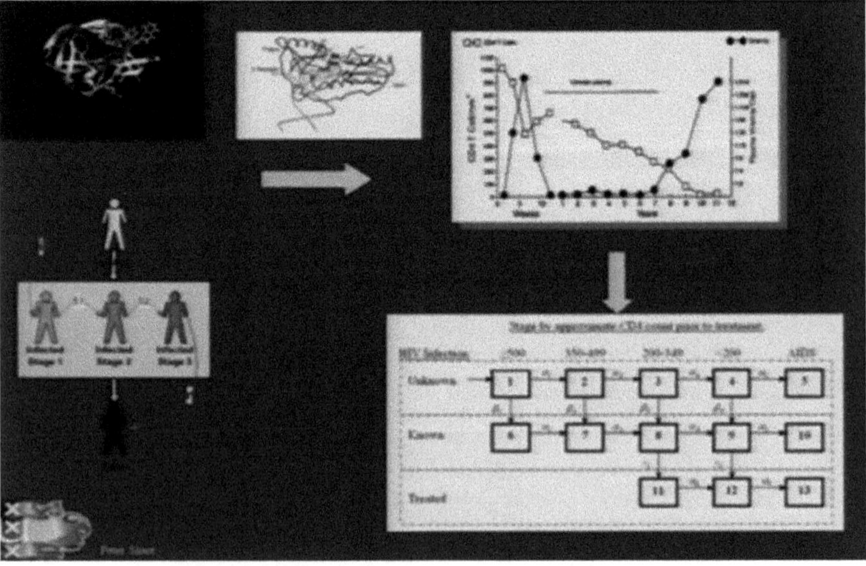

So, again, that gives us a number, which we can then use in our simulations. Now, we can bring things together. We can ask ourselves, remember binding affinity? That's one thing which tells me something about the success of applying a certain therapy. Immune system response tells me something about how much of the virus will be depleted, recovered, etc. Also, where the patient is because if the patient is here, and if this is an important thing, if the patient is here let's say at point five, then if there is a sexual contact and unprotected sexual contact, the probability will be very high, right, because you have this high number of virus here, whereas here it's much, much lower. Then here it will go higher again.

So, that's, if you like, the second thing, the third thing if you like. What we can do is devise strategies for infection, if you like. We can say, if the viral load, if the white blood cells, the CD4, come down here, you are in the AIDS phase, right? That comes down to different regimes, if you like, of the healthiness of the individual, which is associated with the viral load. We can actually say, okay, there are different groups. You just say okay, there's a group of 500 which is the CD4, especially 500, etc., all the way up to AIDS. Then we could say, okay, people might not know that they were infected or people might know that they're infected, and people might be treated, right? Not all that are treated know that they are infected, that's an important thing, but of course, all those that are treated know that they are infected. And so here you already see a starting point of some behavior emerging, I hope you see that is, because if I look at people that do not know that they are infected, they might behave differently than people that do know that they are infected, right? So, what we can do if you can overlook all these thirteen different populations is that we can actually start to study how this whole process is happening.

But before we can do that, we have to think about how the transmission of that virus occurs within that sexual population, especially in the case of unprotected sex. So, what you do is say, "Okay, if I have a virus somewhere in the network, and I take a snapshot of that network, what's the likelihood that the virus, given the network that I know, will be transmitted to another individual and just propagate through that network?" Well, you can write down equations for that, and if you do that, you'll find that it doesn't work.

The reason why it doesn't work is because you're taking too systematic an approach, too much like you're back in the basic SRI models. I don't know if you're familiar with the basic partial differential equations, but what you need to do is take into account behavior. You need to

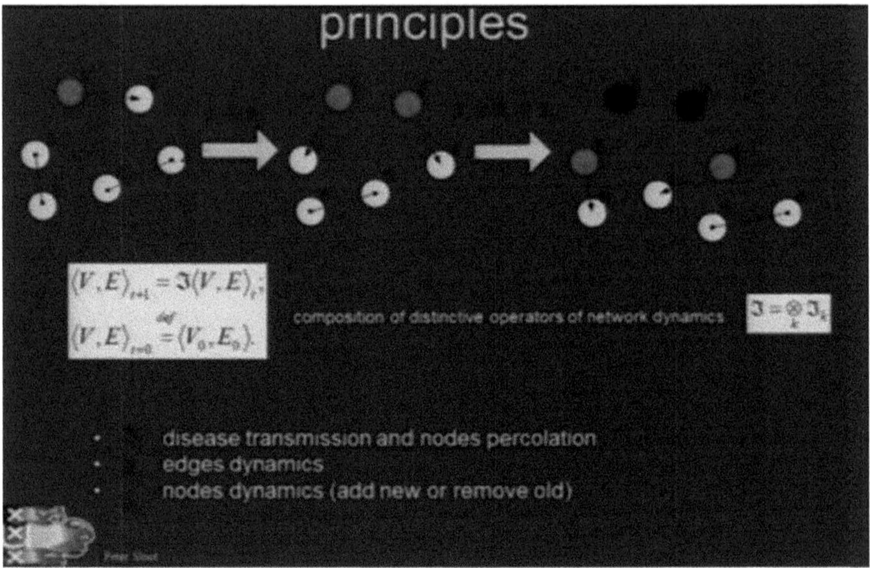

consider the choices that people have, like whether they have unprotected sex or not, and the choices they have in terms of their network structure. So, that's what we're trying to do. One of the things you can explore is how people vary their relationships. But then you get into a very difficult area because you have to start interviewing people and asking them.

We were very lucky because, by the time we reached this point, a new study had come out, called the MCM cohort study, where MSM were followed for 20 years. During these 20 years, they were interviewed almost every couple of weeks, especially when they were getting new treatment, and they were asked about all kinds of behaviors, like their friendships, and so on. Some of them were very open, while some were lying, of course, as you know, that's what you have in interviews. Nevertheless, you get data. This is 20 years of study on a significant population.

From these data, you can ask yourself questions like how people change their relationships. For example, if someone is in a permanent relationship and claims they have a partner, what's the likelihood that they will still engage in unprotected sexual contact with another partner? These are the kinds of things you can measure, deduce, or infer from the data, and that's what we did.

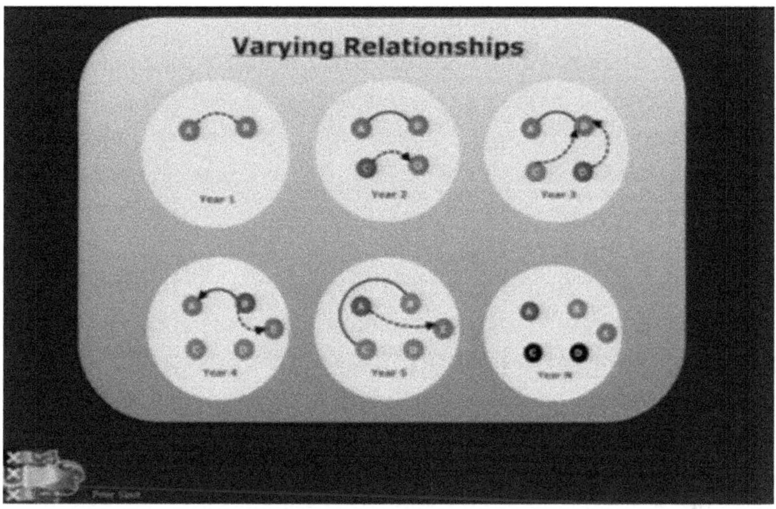

I won't go into the details here because some of them are very gory, and others you simply don't want to know. However, the bottom line is that by delving into these interviews, you can discover interesting numbers regarding the probability of transmission in the state. Here's something important: the concept of steady partnerships for homosexuals. I will revisit this number later, but the key point is that even if an individual in a homosexual MSM population claims to have a steady partner,

Parameters and their values from ACS

from Amsterdam Cohort Study	
Duration of steady partnerships	1.5 years
Duration of asymptomatic stage	13 years for failed treatment, and a mean value of 22.5 years for successful treatment.
HAART uptake after 1996	81%
Proportion of individual who are diagnosed and successfully treated	70%
Frequency of URAI and UAI between steady partners	15 per year
Steady partnership factor F_{sp}	0.84
from literature	
Proportion of diagnoses among patients	0.6-0.4 from 1996-2004
Probabilities of transmission per URAI/UAI act at PI stage	0.22/0.044
Probabilities of transmission per URAI/UAI act at asymptomatic stage	0.011/0.0022
Therapy efficacy (assumed)	Moderate: 50-90%, Pessimistic: 10-50%, optimistic: 90-100%

Table 2: The yearly values of risky behavior factor for HIV negative and positive MSM with the risky behavior for HIV negative MSM in 1997-1999 as a baseline*

	1985-1986	1987	1988-1991	1992-1995	1996	1997-1999	2000-
HIV−	3.50	2.50	1.50	0.80	0.90	1.00*	1.30
HIV+	2.80	1.61	0.42	0.88	0.78	0.70	1.30

there is still an 84% probability of having sexual contact with another person without protection within one year. To reiterate, if someone in an MSM relationship mentions having a permanent partner, there is an 84% chance that they will engage in sexual contact with someone else within the year.

These are the numbers that we can utilize to create new models, which is the essence of our work. Our goal is to grasp these concepts even if they may not all be immediately clear.

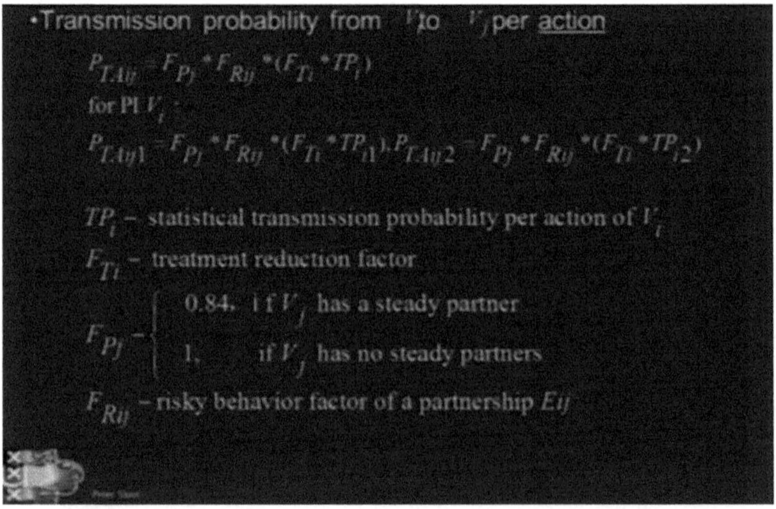

Here is a very simple equation that allows us to do just that: It describes the probability of transmission after engaging in unprotected sexual contact between individuals I and J. This probability is a function of various parameters, one of which includes treatments. When individual I is treated, we observe a decrease in viral load. By understanding the treatment, we can calculate the binding affinity and determine the viral load. Additionally, we consider statistical probabilities derived from the literature. Just because there is contact with an individual carrying the virus does not guarantee infection; there is a probability associated with it. The concept of steady partnerships serves as an example, along with factors such as the amount of risky behavior, which essentially involves the lack of condom usage. By incorporating all of these elements into a complex network, we are able to model the propagation of the virus within that network and assess whether we can retrodict the information known from existing literature.

Now, I will show the result because I'm almost at the end of this talk. I will show you that result, but before I do that, I want to come back to this point about controllability. And there is a nice paper written by Albert-László Barabási and others, published in *Nature* under the title *Taming Complexity*.

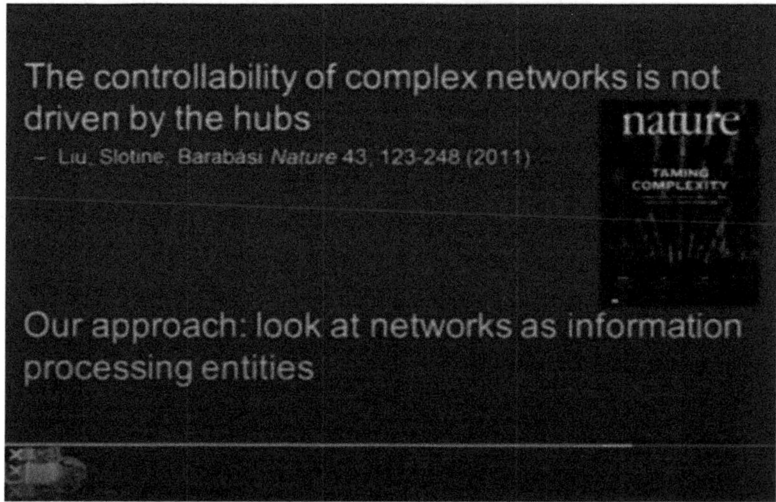

So, the idea was there; the question was, "Can we control complex networks?" And now, what we all expected — I hope that you expected, because I was priming, and we just learned that word — I was priming your brain that you expect that if I kick out the hubs, the whole network will collapse, right? So, I did that on purpose, of course. And everybody was thinking at that point in time, in 2011, that by taking out the hubs, you can actually manipulate the whole network.

But it seems it's not like that, and it seems that it's not the hubs that drive the dynamics of the network, which is strange. I always thought about this as very counterintuitive. They did that in a kind of analytical fashion, but I don't have an explanation, and I'm going to offer you an explanation for that. For what it's worth, and the way we want to do that is, again, and this might be the red thread of my story, coming back to a basic entity, rather than free energy, rather than entropy, well maybe even entropy, which is information processing. So, what we're going to do is look at the networks as an information processing unit and then ask ourselves the question, "Can we direct that information?" Can we predict that information, the way it goes through the network, depending on what we do with that network, which is controlling it?

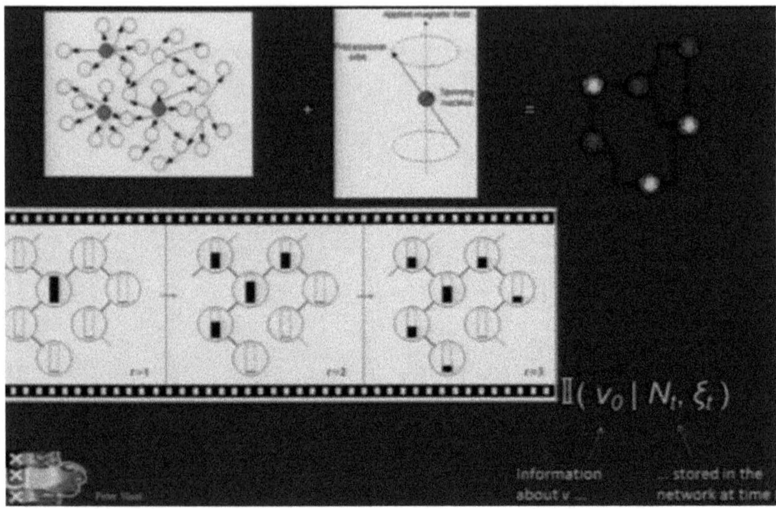

So, the way we look at it is this: We have a network, a slice of a network, containing an amount of information. This information could be viruses, opinions, or anything else, but I quantify it. Our basic unit here is information. Then, I observe how that information propagates through the network. This is a symbolic representation of that process.

Now, you have to figure out how to do that. You can't just put a package there and hope it arrives somewhere; you can do it a little more structured. That's what we did. Essentially, for the physicists in the room, we took an Ising spin model and put that across the network. For those who are not physicists, what we did is that we created such a network and put on the network an agent, like a spin, that can collapse. The spin is like a magnet; it can be up or down, north or south, or south-north, or whatever you want. It influences other spins. So, if I have a spin here that's up, and this one doesn't know what to do, we can actually align them — that's basically the idea.

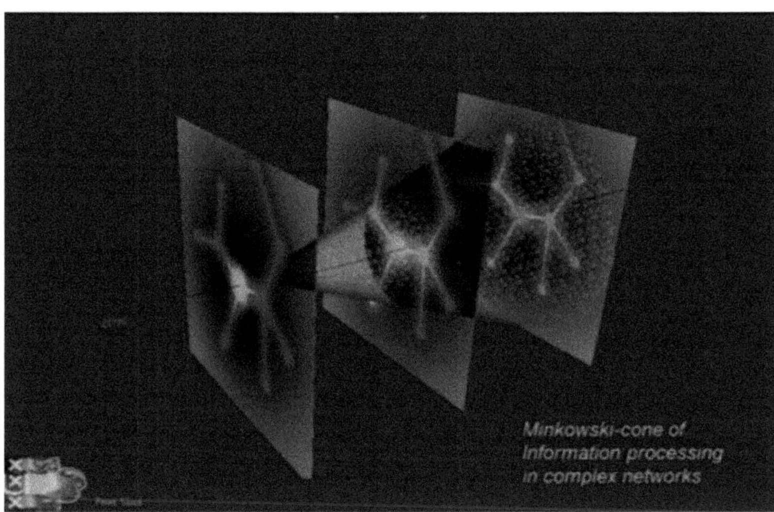

So, what you can then do is that by using this concept of Ising spins (and I'm going to go into the details there), it actually calculates something, like how much information about a certain part in the network is known and how that propagates through the network — these are

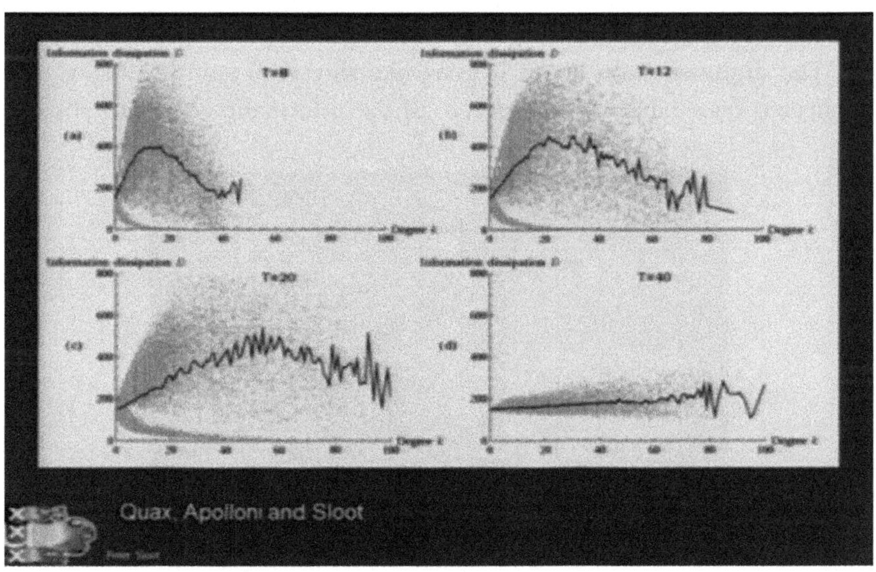

propagation functions — and how it stores in the network. Here, you see the picture: This is the information in a node, depicted by this Ising spin that's flipping up and down. Depending on the connections in the network, that information can spread through the network. So, we look at the dynamics of, if you like, correlation in the network, where we input the pulse of information and see how it spreads.

So, that's what we're doing. And then, if you do that, this is basically what you have to consider. Let's look at this one. So, if you do that, you can vary it for different temperatures. That means the speed with which you fluctuate these spins. But if you do that, then you'll see that it actually reaches an optimum. What you see here again is the degree of connection, so the amount of connections that one of the nodes has, in terms of sectioning that network, the amount of contextual connection that the individual has. And this is then the thing that we call information dissipation, telling you something about how much that information is spreading through the network. So, what you see is that what I thought — you know, that the ones with the highest connections will be the ones that are spreading most of the information — is not happening. So, the ones that really spread most of the information relatively are the ones that are kind of intermediately connected. And this might be a kind of explanation for this paradox of Barabasi, that these interconnected ones are more able to actually transmit information.

The argumentation about it goes like this: You say that the highly connected ones have already stored all the information in that network.

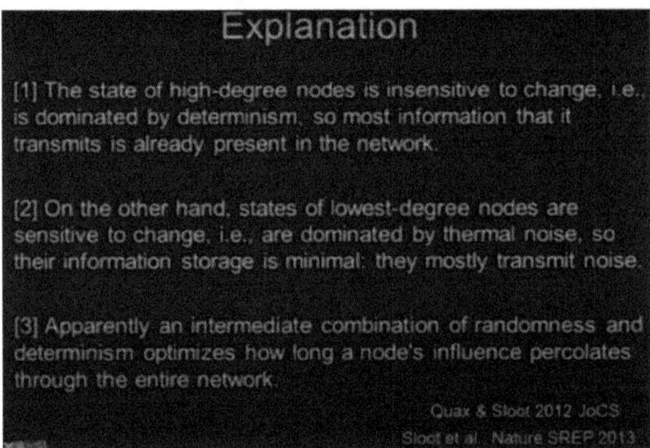

If I take them out, their knowledge is still there, so to speak; their information is contained in the network. On the other hand, the ones that are on the periphery, those hardly connected ones, are noisy. You know, it doesn't matter what happens to them; they are so noisy that they just disappear anyway. So, it's almost like, when I was in Santa Fe in 1995, there was this whole thing about life at the edge of chaos. Complexity happens at the edge of chaos. It's a bit like this because if it's too noisy, the whole thing will not do anything; if there's no noise, the whole thing wouldn't do anything. It's actually something in between.

Okay, so then we put all the pieces together, I'm almost done. What we did is we incorporated all the different simulations. We included the molecular dynamics simulation, remember the calculation of the binding affinity. We integrated the immune system simulation, the transmission probability, and the sexual networks. All these parameters from these different models were filled with real patient data. Then, we found this relationship here. The reported data are from eight cases spanning from the 1990s up to somewhere in 2009, and the red line represents the

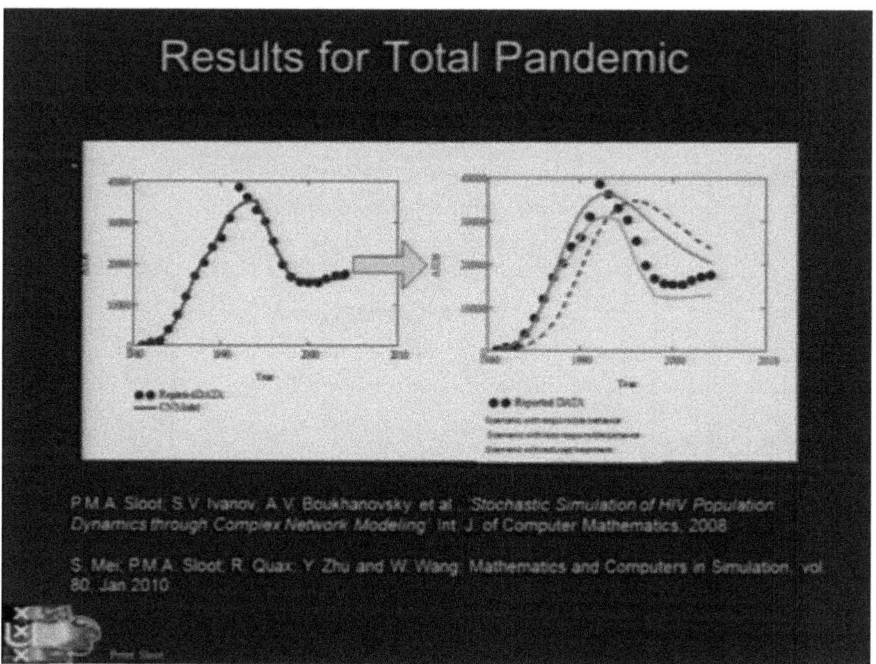

simulations, the outcome of all these combined models. The blue lines represent the actual data.

I was pretty happy with this because it seems that we are able to retrodict the AIDS outbreak in the early 1990s. This gives you confidence in this tremendously complex phenomenon, where behavior and molecules basically interact with each other, and yet you can still make some kind of sense of it. And then, what you can do — and that's the whole reason why we want to do these things, the whole reason why we want to do complex systems research in the first place — is to address "what if" questions, to run scenarios. Peter Ho, who is here, that's what he wants, right? He wants to run scenarios. You say, "Okay, you give me advice. What would happen if I do this? What would happen if I do that?" And if you have some confidence that you're making sense, then you can start to play games, to say, "Okay, what would happen if I change the scenario to involve more responsible behavior, or if I get better medicine?" And hey, that's the exact question we started out with, right?

So, we did that for a couple of big cities: Rome, Amsterdam, and San Francisco. This is the Amsterdam case. Of course, the data get noisier there because you have fewer individuals, around 2,000 individuals.

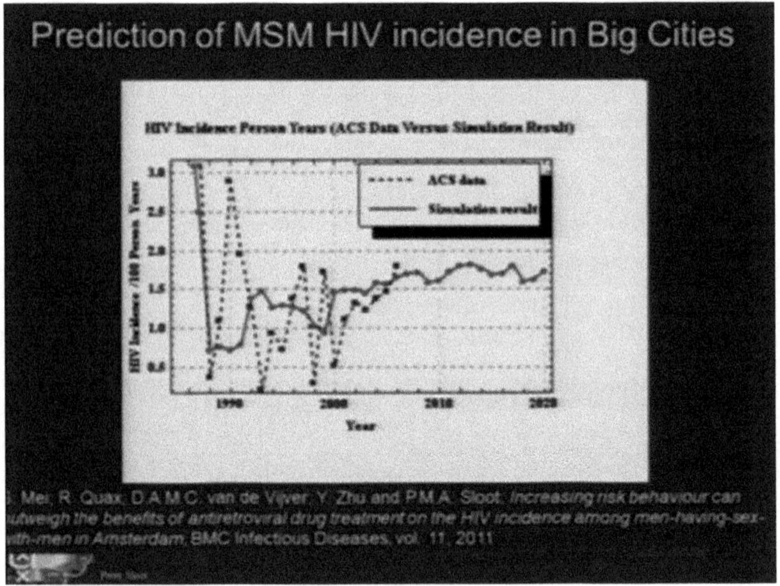

Here, the red line again represents the simulation, and the blue line represents the data from the cohort. We're actually predicting this upstream, and that's happening. As new data come out, it seems that our predictions are working out well.

So, we're pretty happy with this, and it shows that you can actually do these kinds of things. Now, you can play around with different scenarios, as I said, because that gives you some confidence that you can make predictions in the first place. We published this — I won't go into that.

So, the answer that we can give to the European Union is, when we played around with this, what do you think? Why?

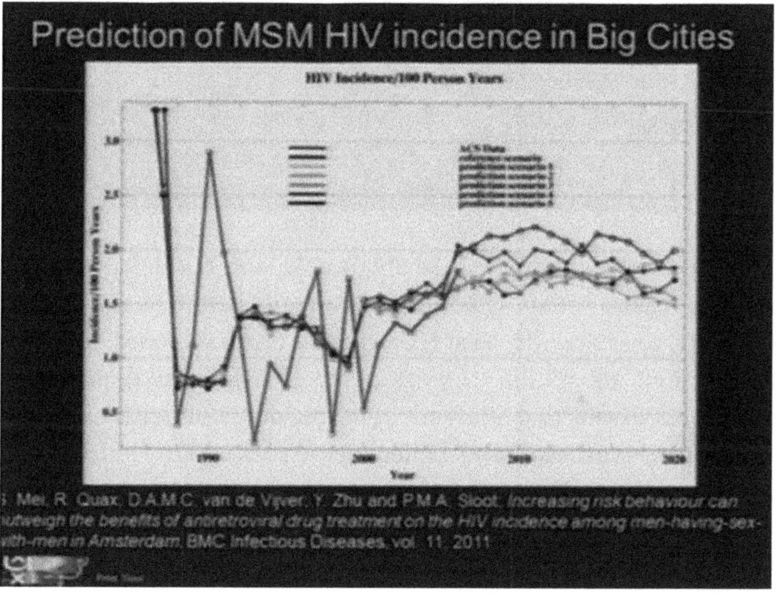

Discussant A: If behavior doesn't change, the drugs become progressively less effective.

Peter Sloot: I never thought of that one. That's a good one. So, you made a link between resistance and behavior. That's interesting. I'll put it aside for now. I don't think it's the answer, but it's a very nice one that I want to think about, and give even more time for that.

Peter Sloot: So, behavior is one, who votes for behavior, and who votes for behavior? Okay, so that's about half. Let's say, I'm not sure it's half, because there are always shy people who never put up their hands, so I cannot do the statistics here.

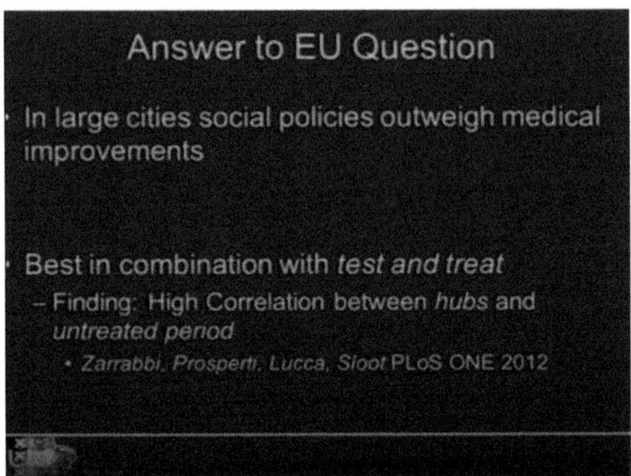

Okay, then the story is indeed about behavior, but the reason is quite subtle. The reason is that if the medication gets better — and it's getting really, really good now because HIV is not a death sentence anymore and the medications are getting very effective... We have all these nice combinations of cocktails that allow for very low doses of molecules, yet they still work very well. So, with better medication, your viral load comes down faster, and you feel better. But when you feel better, you might forget that you're ill. So, if you forget that you're ill, your risky behavior increases. There's a real trade-off here. Feeling better may lead to forgetting about the illness, or pushing it away. It makes a difference whether you wake up every morning thinking, "Oh my god, how am I going to live through this day?" or just needing to take a pill, and that's it — it's a big difference. The whole attitude toward the disease changes, and because of that, risky behavior might increase, leading to an increase in contacts. However, the viral load is lower. So, if you do all these things, and if you take into account all these factors, the conclusion is that eventually, if we were to decrease risky behavior by 30%, we would stop the pandemic. That's a really spectacular result — a 30% reduction in behavior could stop the pandemic. This applies especially in large cities, with a caveat: In large cities, social policies far outweigh medical improvements. But it's the combination that's crucial. Better medication combined with behavioral changes — it's really like in physics, we call it a first-order system, where you can make changes and things happen accordingly. That's what's happening here.

We conducted more studies, which I don't have time to delve into, where we combined models of tests and treats. If we were to test everybody in the whole world and immediately start them on treatment, then within a couple of years, the pandemic would be gone. However, we don't have the resources for that.

Anyway, this is my overview slide, if you like, to summarize what we've done. There's one part that I didn't talk about yet. Here's what we do: We get patients in, and we measure their CD4 count, which is a telltale sign of immune system activity. We also measure their viral load, and we look at mutations. These three pieces of information are crucial.

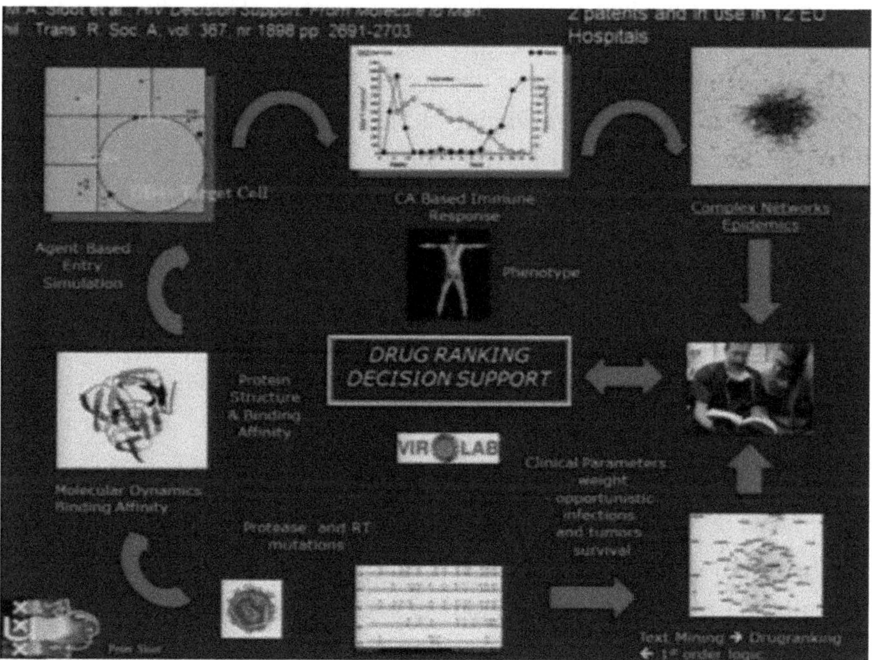

Then, we move to the left here. What we do is calculate the protein's structural binding affinity. This tells us how well the drug will work for that specific patient with that specific set of mutations. It's really patient-specific. By the way, this involves a tremendous amount of calculations. We've developed special techniques to utilize supercomputers in the States and Europe, connecting them to perform these calculations. But I won't delve into that today. Then, we conduct an immune system simulation for the patient based on their CD4 count. We make predictions about the immune system, and then we can determine which treatment would be best for them and how it would affect the population, incorporating population details. Meanwhile, an important aspect that I didn't talk about is text mining. We delve into all the literature — 32 million papers — related to HIV diseases and documents. So, while I've been talking, there's a program running in Amsterdam that's browsing the web for all kinds of information on the relationship between viral load and given drugs, determining the probability that the viral load will decrease. We put all of this together and provide the whole package to medical doctors, who can then make informed decisions. That's how all these components work together. At this point in time, we're running tests in 12 hospitals in Europe.

We aim to release it worldwide at the AIDS conference in December this year. I hope we will make it; it depends on many factors, including a couple of other patents to safeguard our intellectual property. Of course, there are many different sources from which we get our funding, and I definitely want to acknowledge them, including you, because you really play a part in it. I think the complexity program is really helpful there. So, my last slide then: If you want to know more, just type my name into Google, click "I'm feeling lucky," and you might find me. Thank you for your attention.

8.2 Discussion

Simon Levin: So, thanks for a really stimulating talk, I wanted to explain why I voted for behavior, and see if you can see what your reaction is to it. If you look at the simplest epidemic models, most of the terms in those models are linear; the only one that's not is the transmission term, the nonlinear term. So it seems to me that affecting the transmission is much more likely to have knock-on effects, nonlinear effects, and reduce the spread. So, yeah that was basically my intuition.

Peter Sloot: Yeah well, that's very good intuition, but there is this connection between the nonlinear part and the linear part, right, so that makes the intuition… that it needs a bit more support for the intuition, so to say, because you're right, the transmission part is more or less linear, more or less, it's depending a bit on the different type — kind of sexual contacts, you have a slightly different behavior.

Simon Levin: Well, it's bilinear. That's the quadratic term.

Peter Sloot: You know, there are more aspects here. The drug-resistant virus has a different transmission probability than the one that's not drug-resistant. So, there are all kinds of lateral effects, if you will, that play a role. And then there's the behavior that comes into play, which actually feeds back into that dynamic. So, yeah, you're absolutely right. That's a good point. But you have to take the whole thing into account nevertheless.

Discussant B: Professor Sloot, thank you very much for a really fascinating overview of your networking and modeling studies, You've made a very fundamental observation in the beginning, maybe not an assumption, but a statement that the RNA transcriptase is a very error-prone enzyme, and it mutates — there's an error in transcription every time, right? And therefore, this underlies almost all our efforts in AIDS drugs, because effectively, you're saying that whatever molecule you devise, you will actually get mutations which will then dislodge it, as you've shown in the modeling with your colleague at UCL. So, your data clearly show, also with your correlations with the population data and real patients, that behavior does dominate. Does it mean, therefore, that we may have over-emphasized our resources in drug research? Because if your initial statement about mutations is correct, then really, it's mutating every time — it's just like how we are attacking malaria and as all, you know, we are always fighting a moving goalpost. So, maybe we shouldn't just forget it, you know, let's…

Peter Sloot: Be careful, be careful — you bring up a good point. We really have to start thinking here as complex systems scientists, because everything we do will affect something else. If we don't have optimal drugs today, if we don't work on getting better drugs, then viral load will be higher. And if viral load is higher, and we have the same behavior, then the transmission of the virus through the population will increase dramatically. That might change our behavior, because then we'll see, like in the situation with the plague in the Middle Ages, we actually changed our behavior, but we didn't know why we were doing it — we changed our behavior because of what we were seeing. So, you see, there's this feedback loop that's happening there.

Now, regarding reverse transcriptase, yes, it introduces mutations, but it's not the only interaction we have. We also have drugs — I showed you all these inhibitors. And one of the drugs that's coming up now is the co-receptor blockage. It's almost like a detective story; it was discovered in sex workers in Brazil that some of them are actually resistant to the HIV virus, and they showed a certain switch in the co-receptors. So, there are a lot of drugs being developed to stimulate that co-receptor switch and block it. So, there are many more ways to attack it. But the bottom line is, if we don't do anything — if we just say, "Well, we have a solution to this whole problem, we'll just isolate infected individuals" — that's another way of looking at it, and of course, it's nonsense. What we need to do is find a balance between what is happening in the population and what's happening for the individual. If you say we don't treat because then the disease will die out, for instance, because people get so ill that they can't interact with each other anymore, that's not a solution if you're a patient. So, there's also this trade-off between what is good for the population and what's good for the individual. We still have to develop medication that makes people feel good, but then change the behavior that makes it difficult.

Discussant B: Yes, can I conclude then that attacking reverse transcriptase is something we should veer away from, you know, and focus on other sites like co-receptors and other targets? Because that's the conclusion that I draw from your statements.

Peter Sloot: I personally think that looking for inhibitors that prevent the virus from entering the cell is the approach we want to take. Therefore, I would very much put my bet on this CXCR4 co-receptor switch. But technically, it's easier to attack the reverse transcriptase because we know much about the molecular structure and the spherical hindrance there.

Discussant B: But it's changing all the time, so it's really not very effective.

Greg Fisher: If I understand this correctly, the EU asked this question: Should we be putting money into changing behavior or drugs, and you concluded behavior. This goes back to the question about control and influence. Did you look at the extent to which, in general, governments can influence behavior? Because I work in a think tank, and I take the view that governments as a whole historically have thought they can influence behavior more than they actually can. In fact, in some countries, the government is so illegitimate that they can actually do exactly the opposite of what government wants them to do. So, did you consider that when you say they are good?

Peter Sloot: Excellent remark. So, two remarks there. The conclusion that we draw is that you should devise policies to change behavior [that] should trickle down to the workers in the field, to the people, to the doctors who provide prescriptions. You have to realize that for this very complex disease, it's really complex. You have to look at the genetic pattern and the history of treatment because it's always changing. The patient you see on day one is completely different from the one you see ten days later. You have to reassess the whole thing. And you know how much time they have for patients? Six minutes. So, what you could do as a government is somehow devise something so there's more time to actually have a faster assessment. And these tools that I was talking about give you that faster assessment. It's not going to work from the government level if the government just says on a webpage, "You have to change your behavior." That's not going to work. Although in some countries like Singapore, things might happen. There's a funny story. I don't know how much of it is true, but when dengue was coming up in different clusters in Jurong Point, just next to here, there was a story that they wanted every individual to measure their temperature because it would be a telltale sign of what's going on with that epidemic. There was a very high response. More than 70% of people actually did that. I imagine if I tried that in the Netherlands, I'd get shot. There's no way. There is no way. But indeed, it would impose the opposite reaction. I'm not saying that policy change should come from the government, but the government should drive to facilitate ways to change behavior.

Discussant C: Just a very brief point: I think it makes the point that your model shows where the leverage is, but it doesn't necessarily say where the most effectiveness lies. Yeah, you know, this is about the effectiveness of policies; the model shows where the leverage is.

Peter Sloot: Yeah, but you need these kinds of models in the first place to convince the government that they can do something.

Kevin Xiao: I noticed that you mentioned Barabasi's paper on the contribution of the network. Shortly after his paper was published, there was another paper published in *PLoS ONE*. The author, I think also communicated with a pharmacist group; so their argument is that Barabasi's analysis is totally correct. But the argument is basically that they make assumptions on a network model which may not be so realistic. It's regarding the network model where on each node, there is no self-loop, meaning that the next moment's status is basically decided by the neighborhood's status and their input. Yeah, the algorithm in real life for most networks creates nodes like that — it's like what we are going to be tomorrow is quite strongly related to what we are today. So, if you put in a self-loop, you still use the analysis, and you will find that perhaps those high-degree nodes come back. You'll have control with those several nodes; you'll have control over the whole network. However, I've found that this conclusion doesn't, in my understanding, really affect your study on transmission. You know, if looking at those low-degree nodes, I'm thinking that maybe controlling the network and [controlling] the transmission of information are probably still two different things.

Peter Sloot: I wouldn't know because I didn't look at that experiment that these guys did. But what I do know is that we applied this whole concept to something completely different, which I had no time to talk about unfortunately, because it would have been fun. It was about criminal networks. We have lots of data coming from the Netherlands on drug use and criminal networks. One of the things that we did is analyze the criminal networks and then ask ourselves this question: What would intervening in such a network amount to? If you want to stop the production of cannabis, for instance, how do you do it? So, then you come to the same kind of question because you want to control it — do we take out one that is connected to all the other ones? It could be the one that produces the plants, the one that does the harvests, the one that takes care of the electricity, etc. So, do you take out the hub, or do you take out the in-betweeners — the ones that have the highest betweenness — or the ones with the specific function? It's almost more than ten times more complicated, and we think it's not just about the topology; the policy is one aspect of it. It's also about the functionality of the individuals in the network, and that policy is changing over time. So, I think the answer is you have to really look at each network separately, and then I think this information propagation can help you to assess where the hotspot or the sensitive spot in the network is, but you have to do it for each realistic network, time and again.

Discussant D: Thank you, Professor, for your talk. It was very interesting again, as usual. I was very curious about the diagrams you built, which represent the different population rates: heterosexual, homosexual, and drug users. I wanted to ask a little bit about drug users. Is it both homosexual and heterosexual together? And I was also very interested in why it's so separated. Is it just because of metrics, or is it seriously so separated from both groups?

Peter Sloot: Okay, for the drug users, whether they were homosexuals or heterosexuals, it didn't matter. But once they were drug users, we disconnected them from the other clusters, not just to study the structure by itself. So, we took them out from both the homosexual and heterosexual populations.

Discussant D: I see because it seems very strange that they have no connection.

Peter Sloot: Of course, there are connections, sure. But because we wanted to understand the structure of those separate networks, it's a subnet. These networks drive the dynamics actually, so we wanted to isolate that circuit. And the connection is very poor because the number of individuals who are both drug users and infected with HIV, and homosexuals, is very small. So, the overlap is pretty small.

Discussant D: Yes, the diagram was built for a particular city. However, there wasn't a crucial difference, for example, from the US to Europe. The differences observed are typically not as stark as that. They may vary depending on factors such as cultural norms, demographics, and healthcare infrastructure, but they are not typically continent-specific.

Peter Sloot: Ah, I see. So, what you were looking at was actually a combination of the three cities: San Francisco, Rome, and Amsterdam. In terms of the city differences, they were not that significant. Also, the gamma of the three populations was almost equivalent, but within the error bounds.

Discussant D: And through the time also the evolution of…

Peter Sloot: Ah, that's another good question. So, we have time points, because we have, you know, like, over 20 years, as I said, so we can look at different points in time. But the difficulty is that when you go further in time, you get more data, it's richer, and the errors are smaller. So, how I correctly compare these different longitudinal moments is very hard. But what you see is that the gamma starts to kind of asymptotically go to a certain value over time, so it seems it's a stable factor.

Discussant D: Thank you very much, yeah.

Discussant E: I come from computer science, as you do yourself, so I was wondering, you just said that the cities had similar networks — Amsterdam, Rome, and San Francisco. But wouldn't it be very different in countries where AIDS is most prevalent, in Africa and Asia, where the culture is completely different, right? These networks would have a very different structure, I'm assuming; and would the spreading effects and all be the same?

Peter Sloot: Yeah, so that's a very important point, and thank you for bringing it up. It's very relevant. How much of what we have done can actually be translated to countries like African countries? We have collaboration projects now going on, we just started it like three years ago, with Zimbabwe and Kenya. What we're actually trying to do is the same thing. The point is, the data are very poor there, so we're now setting up posts, done together with Erasmus Medical Center, in these countries to actually get those kinds of data, which are very, very poor. Also, the cultural differences are very much different, of course, than what we have in Europe. That's one thing we're doing. Another thing we're doing is in the slums of Bangalore — 36 slums. We're getting data from that now. That's all for the past three data points, where we get all this kind of information. I'm very much looking forward to mining those data to see what's going to happen there, so it's data, and it's completely different. You might have completely different effects there, so we're looking into that.

Discussant F: I'm sorry, I'm not sure whether this question has been asked before. Well, you say that noise has to transmit information within your network…

Peter Sloot: Could you say it again, I said that?

Discussant F: Noise, noise has to transmit information in your network.

Peter Sloot: It helps, yeah, yeah.

Discussant F: Yeah, it could help, but how much noise should be added into your network so that you really maximize the efficiency of the transmission process?

Peter Sloot: I don't know. That depends on the network. So, we did that for five different artificial networks; it's in the paper, you can actually look it up. These five different artificial networks, which you can generate, just have different gammas. Then we found that, you know, up to something like 20% of noise will actually still enhance the signal-to-noise ratio, if you like, of transmission. So, that's just pretty good. And then we did it for this email network that I was talking about, as a realistic network. We tested the gammas that are close to the homosexual and heterosexual populations, and we took this hierarchical with a slightly different type of network of the email network. For the email network, it's pretty small. I don't remember, like 5% of cells, so you can add 5% noise and the transmission of the information increases. You go beyond 5% and it decreases.

Discussant F: But should it depend on the topology or the functionality of different noise, when you're adding noise, I mean?

Peter Sloot: This was homogeneous in that sense. So, the network topology, we changed because of this gamma, but the activity of the individuals in the nodes or the agents was all the same. So, they all had the same rules there. Yeah, if you changed it, of course, if you change that, it might also have an influence.

8.3 Summary of the Talk

Peter Sloot delivered a comprehensive talk addressing the complexities of HIV transmission and resistance, framed within the context of global health challenges and the interconnectedness of our world. He emphasized the need for a multidisciplinary approach, combining molecular insights with epidemiological perspectives to gain a comprehensive understanding of the HIV pandemic.

Sloot began by highlighting the global impact of diseases and disasters, underscoring the need for informed resource allocation. He discussed the HIV pandemic, its distribution across regions, and the disparity between affected areas and research funding. To address this, he advocated for a holistic approach that considers various scales and disciplines, echoing epidemiologist Neil Ferguson's call for comprehensive analysis.

Delving into the molecular mechanisms of HIV replication and drug resistance, Sloot showcased the complexity of the virus and the challenges it poses for medication efficacy. He presented simulations and models that elucidate the interplay between viral mutations, medication effectiveness, and immune response dynamics.

Sloot emphasized the importance of understanding simulations and integrating experimental validation to ensure accuracy and reliability. He discussed the development of cellular automata models to simulate virus depletion by the immune system, highlighting the progression from primary infection to AIDS and the collapse of the immune system.

In conclusion, Sloot underscored the necessity of bridging molecular insights with broader epidemiological perspectives to effectively combat the HIV pandemic. He stressed the importance of understanding the complex interplay between viral dynamics, medication efficacy, and immune response dynamics to inform targeted interventions and resource allocation strategies.

8.4 Relevance of the Talk to the Current Stage of Research

Peter Sloot's talk remains highly relevant to the current stage of research in several ways:

Multidisciplinary Approach: The talk emphasizes the importance of a multidisciplinary approach to understanding and addressing complex issues like HIV transmission and drug resistance. In the current research landscape, there is a growing recognition of the need to integrate insights from various fields such as molecular biology, epidemiology, computational modeling, and public health policy to develop effective strategies for disease prevention and treatment.

Technological Advances: Sloot discusses the use of advanced computational models and simulations to study HIV dynamics, drug efficacy, and immune response. As technology continues to advance, researchers now have access to more sophisticated computational tools and data analysis techniques, enabling them to generate more accurate models and predictions. This aligns with current research efforts focused on leveraging big data, artificial intelligence, and machine learning to improve our understanding of disease mechanisms and develop targeted interventions.

Global Health Challenges: The talk highlights the global nature of health challenges, emphasizing the interconnectedness of populations and the need for coordinated efforts to address diseases like HIV/AIDS. In today's interconnected world, where emerging infectious diseases and pandemics pose significant threats to public health, there is a renewed focus on global health security and cooperation among nations to mitigate the spread of diseases and strengthen healthcare systems.

Drug Resistance: Sloot discusses the challenge of drug resistance in HIV treatment, a problem that continues to be relevant today. With the ongoing evolution of the virus and the emergence of new strains, drug resistance remains a significant concern in the field of HIV/AIDS research. Current efforts are directed toward developing novel antiretroviral therapies, understanding the mechanisms of drug resistance, and implementing strategies to prevent and manage resistance in clinical settings.

Overall, Peter Sloot's talk provides valuable insights into the complexities of HIV transmission and drug resistance, offering a roadmap for future research directions and interventions aimed at combating this global health challenge.

Chapter 9

Collective Phenomena, Collective Motion, and Collective Action in Ecological Systems

Simon Levin

YouTube: https://www.youtube.com/watch?v=KSShtZ7zlY8

Speaker: Simon Levin

Moderator: Chew Lock Yue

Discussants: Guaning Su, Chris Monterola, Gregg Fisher, Marcus Karner, Discussants A, B, C, and D

9.1 Talk by Simon Levin

Thank you very much. Okay, we may think of ecosystems as enduring parts of nature, but in fact ecosystems and the biosphere are dynamic, with lots of turnover in species composition in any particular place, especially on local scales.

But despite that, when we think about particular ecosystems and what we want to sustain and protect about them, we're thinking about characteristic regularities and macroscopic patterns that characterize them, whether we're talking about forests, aquatic systems, or marine systems.

These crude regularities, indeed, provide a basic framework for understanding the whole. They characterize what are called biomes. Starting with Holdrege and going through Whittaker and others, there

357

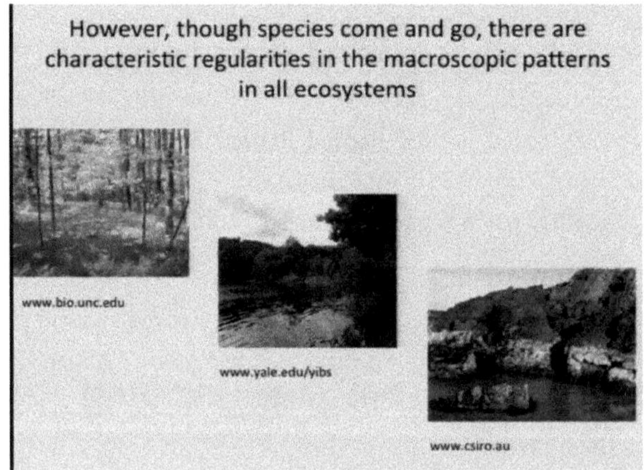

have been various attempts to characterize where you would find particular kinds of vegetation in relation to temperature and precipitation. For example, you know where you should find deserts, tropical rainforests, savannas, and other grasslands. However, as you'll see, this is only an approximation. Firstly, it doesn't specify which species to expect to find, only the vegetation type. Secondly, there's still some uncertainty within these regions.

The characteristic patterns that are referred to are things like species abundance relationships and stoichiometry patterns of nutrient cycling. These are emergent properties that make the ecosystem operate, but they are the collective properties of many different species cooperating together. Yet, the ecosystem hasn't been selected like an organism has for

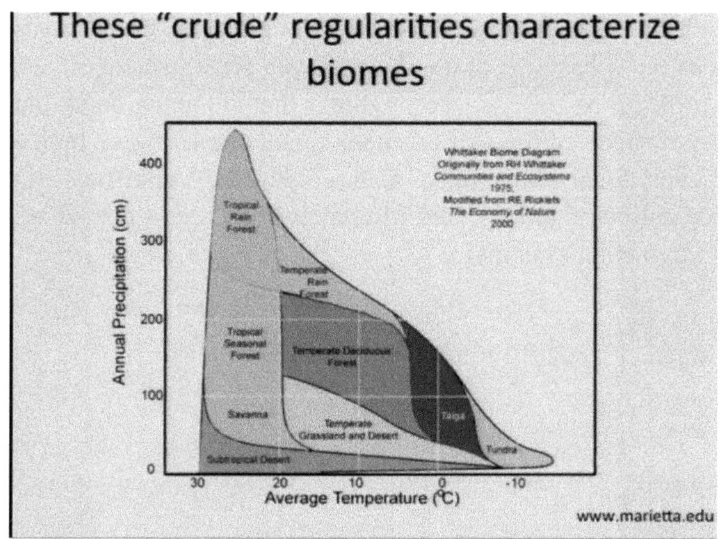

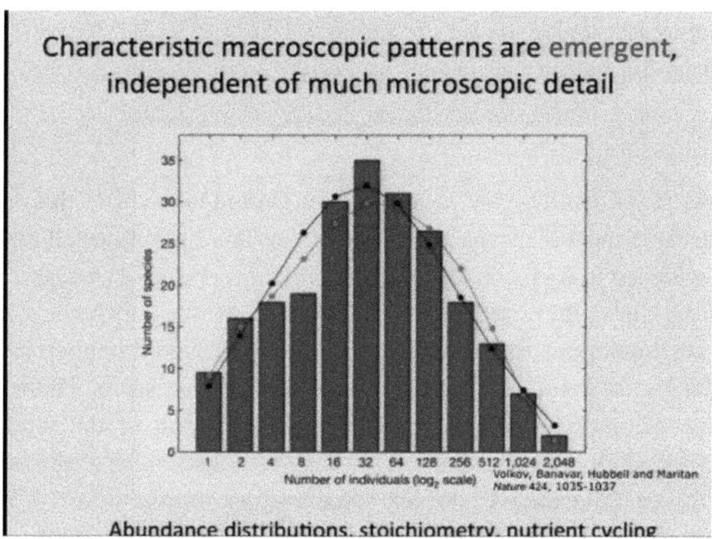

its particular properties; this is rather the transformational evolution of the system into something that works.

So, this implies, at least to me, a need to relate phenomena across scales: from cells to organisms, some of which we've already seen at this meeting; from organisms to groups of organisms; from groups of organisms to ecosystems; ultimately, to the biosphere. We need to ask questions like this: How robust are the properties of ecosystems that we care about? How does the robustness at the macroscopic scale, the robustness of the macroscopic properties, depend on things that are going on at much finer scales — of space, of time, of organizational complexity, on both ecological and evolutionary timescales? Can we develop a statistical mechanics of ecological communities, and ultimately, you see, a couple of human and environmental systems?

This implies a need to relate phenomena across scales, from

- cells to organisms to collectives to ecosystems and the biosphere

and to ask

- How robust are the properties of ecosystems?
- How does robustness of macroscopic properties relate to ecological and evolutionary dynamics on finer scales?

- Can we develop a statistical mechanics of ecological communities, and of coupled human-ecological systems?

Now, sustainability, we've heard a good deal about it at this meeting. It's something most of us are concerned about. If we're interested in sustainability, we need to focus on the macroscopic properties, but we also have to understand how they relate to phenomena at the microscopic scale. Something analogous to this is a cartoon of what would happen if I had a particular gas or a liquid to which I was applying pressure. There will be phase transitions that will occur; these are the result of the interactions among many individual molecules, but they're not sensitive to the details of what any one molecule will do. So, somehow, we have to develop, as physics has, the statistical mechanics and the thermodynamics that take us from the individual interactions to the collective properties, the emergent properties, that really can't be inferred from just the details of local interactions.

So, I think a perspective from complexity theory, mathematics, and physics can help with many features of this. First of all, in developing the statistical mechanics of ecological communities and understanding when they are likely to undergo or are about to undergo critical transitions — the collective phenomena that give rise to patterns like the ones I showed you at

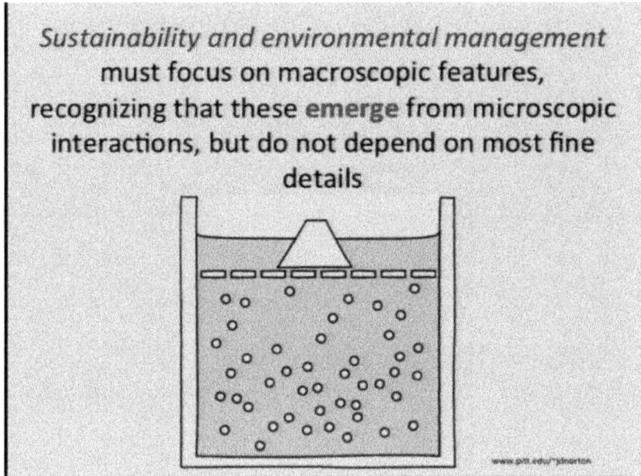

the beginning, of bird flocks and fish schools, collective motion, which also exhibit emergence and pattern formation — we require a statistical mechanical approach, but it's much more difficult than when we're dealing with molecules because individuals are moving actively and there are nonlinear interactions. And finally, dealing with what I think, and as we heard from some of the speakers in the last two days, dealing with global problems — the problems of conflict and how we take collective action to address problems that concern us all — this was something that Murray raised in his opening remarks.

So, let me turn to these topics one at a time.

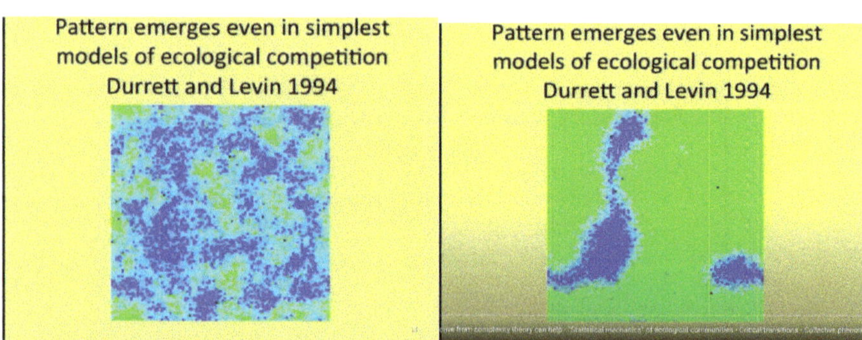

Statistical Mechanics of Ecological Communities: Even with the simplest models of competition, it's very easy to observe the emergence of patterns. For example, in a stochastic model like an interacting particle model or stochastic cellular automata, which Rick Durrett and I looked at years ago, of competition between two species, all that's going on are local interactions. But as a result of that, patterns form. The system leads to the aggregation of the two types of competitors into groups and, in this case, the ultimate elimination of one type by the other.

More realistic models have been studied for a long time, especially in dealing with forest communities. My colleague Steve Pacala, along with Dan Botkin, Hank Shugart, and others, has been building individual-based models of forest communities.

This is a simulation that we published in *Science* some years ago, with my student Doug Deutschman and me, in which we grew temperate forests. The way these models run is by taking detailed measurements that have been made on individual plants — how they grow, how they shade each other out — and from that, running simulations. Patterns emerge that show great similarity to what one actually observes in real communities. Our objective in looking at that was to ask this question: How can we reduce the complexity of these interactions of many species to capture the emergent properties?

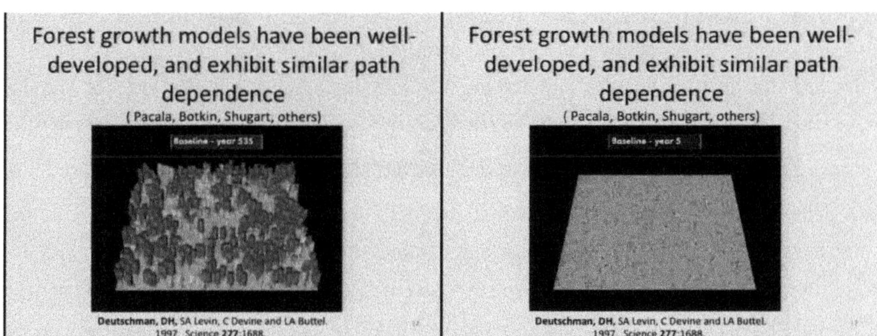

Models of this sort have been extended to much broader scales, from regional to national to global scales, producing maps that tell you where you would expect to find particular kinds of vegetation. At this scale, you wouldn't expect to be able to specify which species to expect there, but you should be able to group species into aggregations and make robust predictions.

Similar sorts of models have been developed, particularly by Mick Follows's group at MIT, working with Penny Chisholm and others. What Follows and his group have done, both through simulations and analytical models, is to write down equations for the dynamics of interactions between species. They embedded this in a fluid dynamic framework that reflects what's known about the oceans. This standard model is broken into three categories: nutrients, phytoplankton, and zooplankton (no fish in this model). Ultimately, fish needs to be included, but each of these types is broken down into individual species. What Follows and his group did was to take essentially a menu of different kinds of species, hundreds

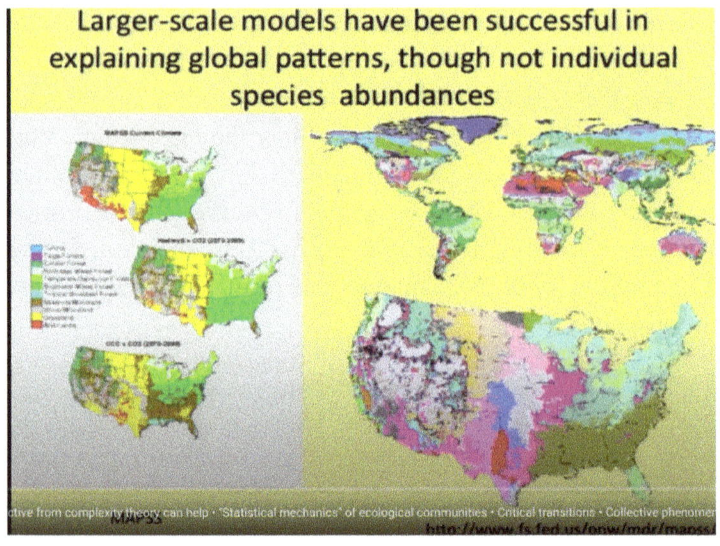

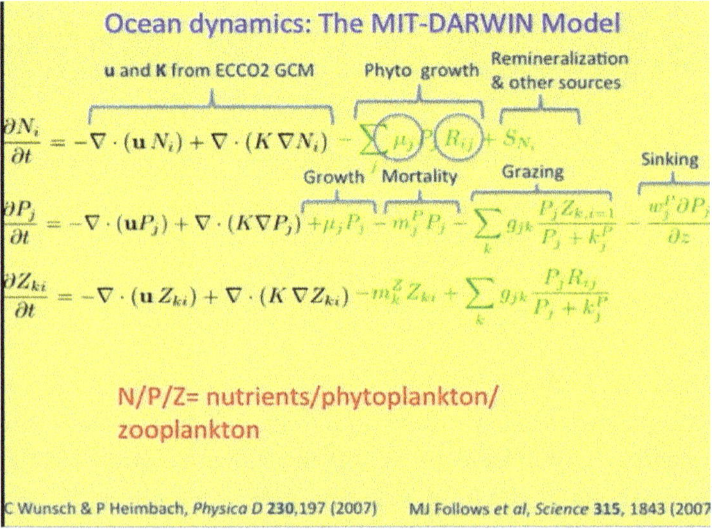

in total, that could be there and put them into competition with each other. They allowed them to compete and asked what patterns emerged.

And indeed, repeatedly, although they would get different details regarding which species (though they weren't real species anyway, but rather particular characteristics of species), they were able to predict not where individual species would be found, but where individual ecotypes would be found. They predicted where the diatoms would be found, where the large eukaryotes would be found, and where the group Prochlorococcus, an important bacterial group in the oceans, would be found.

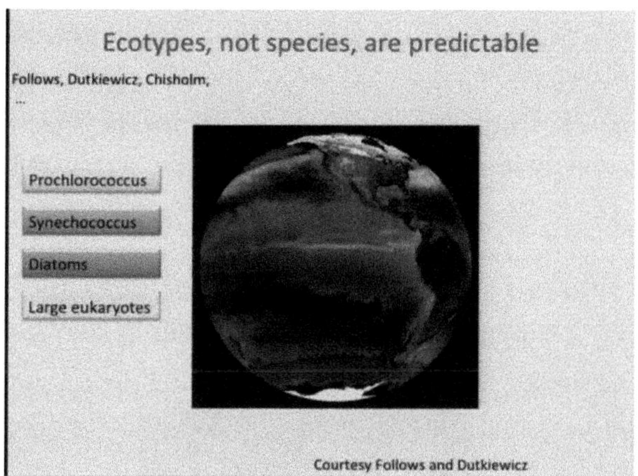

And we've been working with Mick and others on these models to try to simplify them. Whether we're dealing with forests, oceans, or similar models in other systems, the challenge is to simplify these descriptions using aggregation and other coarse-grain methods. These methods will help us take the detailed models, develop statistical mechanics, and try to predict what sorts of emergent patterns we will see.

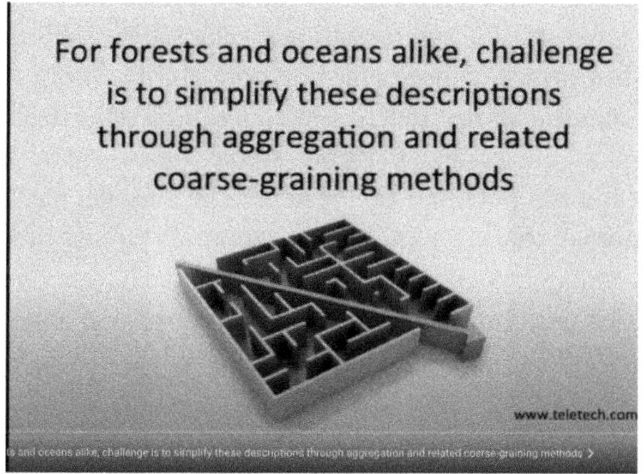

All of this is because, as we've heard since the beginning of this meeting, ecosystems and the biosphere are what are called complex adaptive systems. These are systems made up of individual agents that interact with each other on local scales. As a result of these interactions, there are changes in behaviors or even changes over evolutionary time. Patterns emerge from these interactions and those patterns then feed back to affect individual behavior. So, there are dynamics that occur over multiple scales. It's not just ecosystems and the biosphere; the socio-economic systems with which they are linked are also complex adaptive systems.

And that means that the dynamics of these systems play out on multiple spatial, temporal, and organizational scales. They self-organize, patterns result, and there's consequent unpredictability. Brian Arthur taught us years ago about path dependence in economic systems. These systems have multiple stable states, multiple stable basins of attraction, path dependence, and the potential for hysteresis if they get shifted from one system to another. As we saw in the economic crisis, there's also the potential for contagious spread and systemic risk. In other words, there's the potential for destabilization of the regimes one's in and regime shifts either as a result of exogenous changes or as a result of slow, time-variable evolution endogenously.

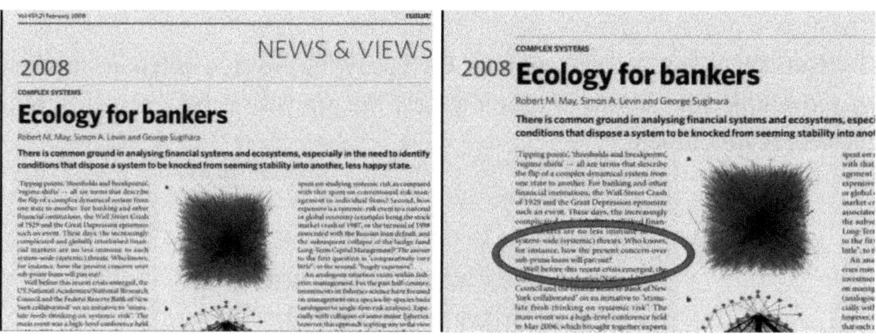

In 2008, about six months before the financial crisis, Bob May, George Sugihara, and I wrote a paper in which we said, "We don't know much about financial systems, but we've been working in food webs, and the more interconnected those systems get, the more unstable they become." There's something about the nature of interconnectedness in financial networks that doesn't look good to us. We said, "Who knows, for example, how the present concern over subprime loans will pan out?"

Now, I wish I had read this paper, and not just written it, because I didn't do much about it. But in fact, we know what happened six months later, and this paper called "Ecology for Bankers," which was published in *Nature*, made all of us popular with consulting firms. It seemed to imply that we knew something that we didn't.

Critical Transitions: So, the problem of critical transitions is evident in this example and has generated a lot of recent interest. Certainly, whenever I go to the Santa Fe Institute, there's a discussion of the question, "Are there indicators that tell you when a system is about to lose its robustness?" This is a paper I was involved with, published last year, I think in *Nature*. I can't remember if it's *Nature or Science*, but I guess it's *Science*, called "Anticipating Critical Transitions."

But this is not a new problem. Certainly, in physics, interest in when systems will flip among various states, the transition between different phases, is fundamental.

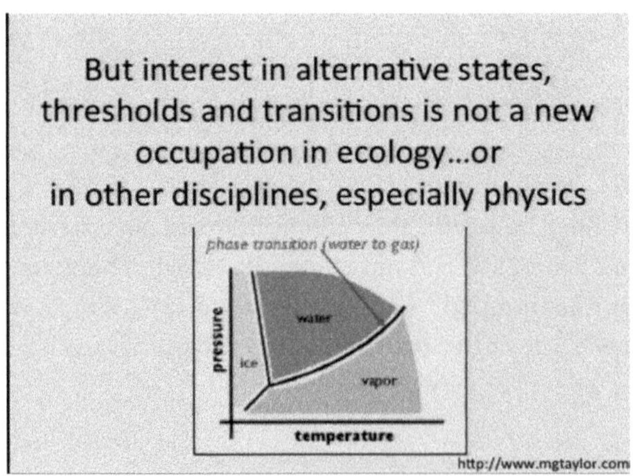

In the 1960s, when I was first starting out in these fields, René Thom, the famous mathematician, developed an approach called catastrophe theory. He wrote down a system of equations that could easily imitate many of the bifurcations that were observed. Indeed, this was a gradient system, where the right-hand sides were gradients of polynomials of a low degree. It was the lower-degree polynomials that seemed to generate the most interesting phenomena. But, as with many of these subjects, catastrophe theory got oversold. If people saw that A could produce B, and they saw B, they thought therefore A. Well, it doesn't work that way. A similar thing happens with the attention to power laws; when people see them, they expect some simple explanations. And so, catastrophe theory fell into disfavor.

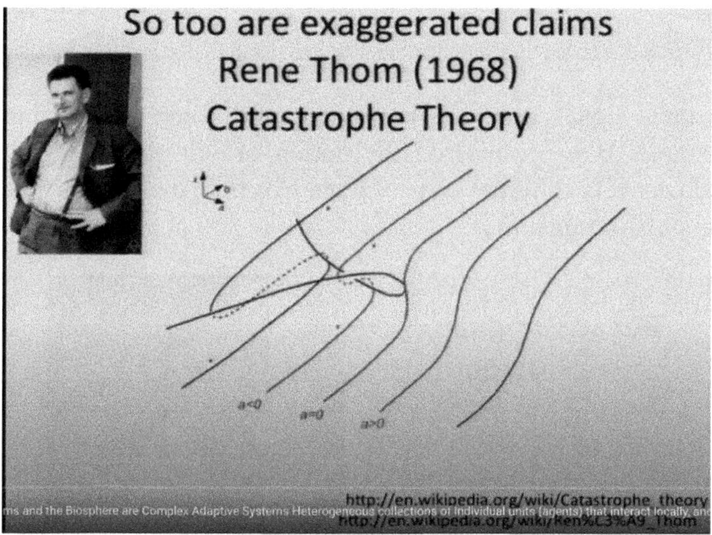

But despite the problems of these approaches, it's true that many of the sorts of transitions of interest have characteristic signals. Critical slowing down, for instance, is where the rate at which systems return to an equilibrium seems to get slower as the depth of the basin of attraction gets less. There are increasing variants, as we probably are seeing in climate systems now, increasing autocorrelation, and perhaps even flickering between different basins of attraction.

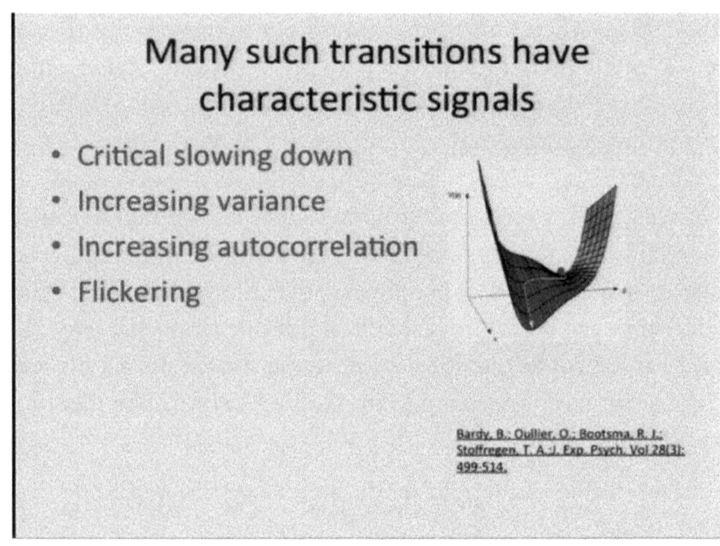

So, these signals are there, but signals of this sort are not ubiquitous. For example, Bak popularized the notion of self-organized criticality, which shows very different sorts of dynamics than the dynamics that lead to these sorts of transitions.

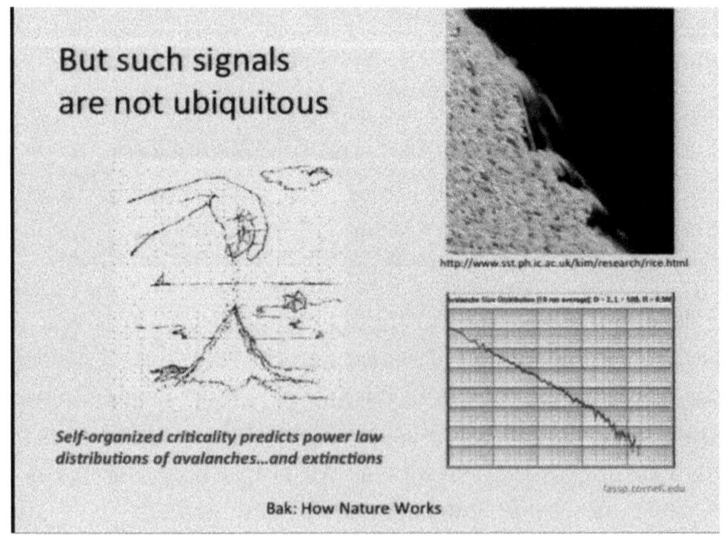

I woke up this morning to find out that the Dow had closed at its highest level ever, so I was inspired to pull out a snapshot that I took from a paper off the web a few years ago when I decided to check what was happening with the Dow. I looked at it and saw it had lost a hundred percent of its value, dropping down to zero. Now, this was a mistake, fortunately, but it was a little shocking to look at. Surprises and big drops can occur, and we know there are days when we've had drops of 10% or so on the Dow; they're not always predictable.

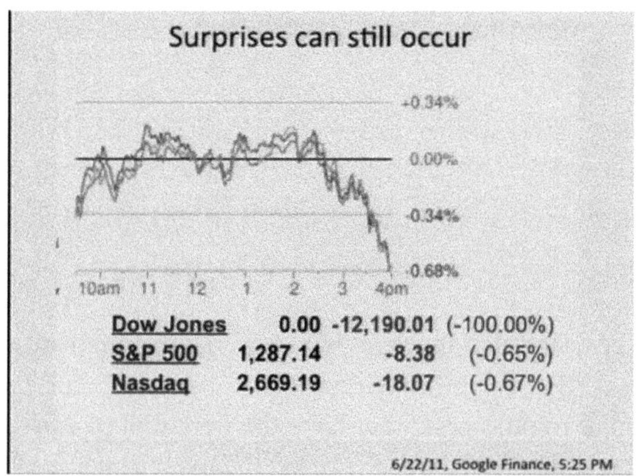

Nonetheless, it's well documented that many systems have alternative stable states. Certainly, in the systems I've been interested in, ecological systems, this is well documented. There's a paper Bob May wrote in the 1970s on thresholds and breakpoints, documenting some of the cases. The Resilience Alliance has on its website a listing of critical transitions — indeed, system flips in a variety of ecological systems.

We've been interested in savannas and forests, primarily my student Carla Staver, who recently completed her degree and has been working in South Africa. Together with Sally Archibald, a graduate student in South Africa, Carla began looking at remote sensing datasets, the so-called MODIS datasets. Going back to the Whittaker and Holdrege sort of diagram, they tried to determine where one ought to expect to find savannas and where one ought to expect to find forests. There were regions that were deterministically savanna, and indeed the data showed that they were

savanna. There were also regions that were deterministically forests, and the data showed that they were forests. However, there are broad regions, as you see in the middle here, that have the potential to exist as either forest or savanna. If you look at the vegetation cover, you see a bimodality to

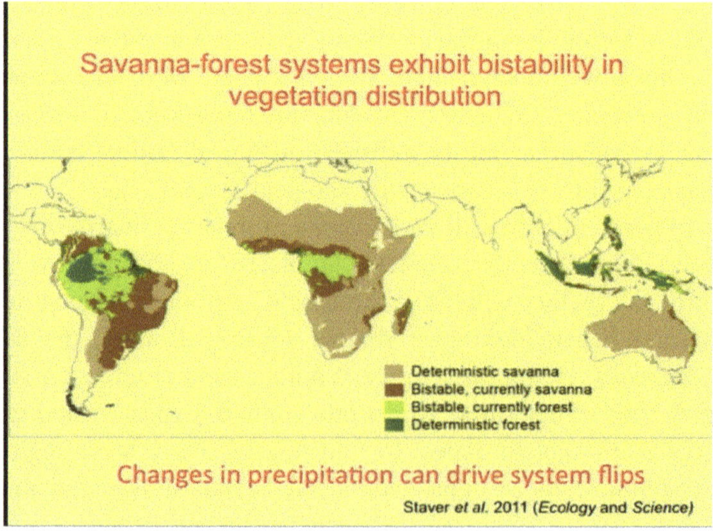

the distribution; it's not a smooth transition. These are two very different states. Of course, there's a geographical factor involved here, namely, if you're next to a savannah, that's going to bias you toward savannahs. But there are these boundary regions that could flip between states. What we think is going on, by the way, is that changes in precipitation have the potential to drive these flips.

This is what those areas look like. This is a savannah with some isolated trees within it. This is different from a forested system. There are trees in the savannah, but fire plays an important role. Working with Emmanuel Schertzer, a former postdoc of mine, we are building percolation models to try to understand the spread of fire, so we can understand what the potential is for the system to flip from one state to another.

Now, why is fire important? Well, grass burns and the young saplings that have the trees that you saw in the middle of the grassland burn. But once the trees get above a certain stage, they're no longer subject to burning, and therefore fire becomes a self-sustaining state. Indeed, it's an adaptation for grass to burn, so that the grass state becomes self-sustaining. If the trees get above a certain amount, the system flips into a tree-covered system. The frequency with which fire occurs is going to be a function of the amount of grass that's there. So, we can document this sort of function, which will tell us the relationship between fire and the amount of grass, and write down a system of simple differential equations. At the start, the whole landscape is broken up into grass, saplings, and trees, just to make it simple. These are savanna trees. We document within the equations what the transition rates are, with the critical term being this nonlinear term, given by that function up there, which says that the frequency of fire depends on the amount of grass.

Grass
$$\frac{dG}{dt} = \mu S + \nu T - \beta GT$$

Saplings
$$\frac{dS}{dt} = \beta GT - \omega(G)S - \mu S$$

Trees
$$\frac{dT}{dt} = \omega(G)S - \nu T$$

$$G + S + T = 1$$

Staver et al. 2011 (*Ecology*) and Staver & Levin (Amer.Natur.)

When you look at the dynamics of this system by examining the two isoclines, you'll find that you can either have a state that's very low in grass cover, covered with trees, or you can have a state that's very high in grass cover and self-sustaining due to fire. But at intermediate levels of precipitation, there are two stable crossing points: the big red circles. Where these lines cross are equilibria, but where the red circles are, these are stable equilibria, and that's easy to show.

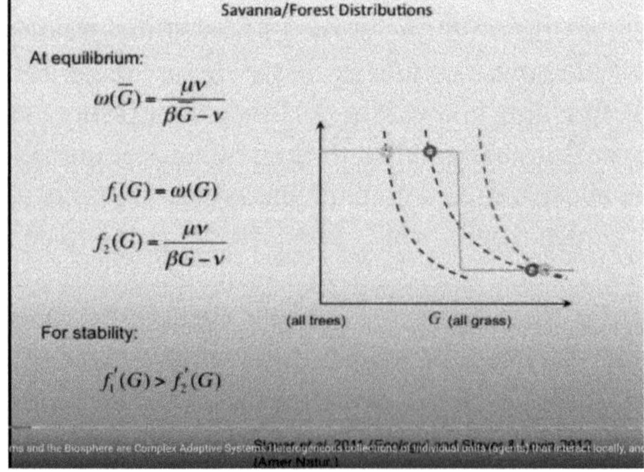

That means the system has two stable states for the same parameter values. If we draw it in this way, borrowing and modifying a diagram of Martin Scheffer's, then the system can exist either at low vegetation cover or at high vegetation cover. At high levels of precipitation, the grass level is low. At low levels of precipitation, the grass level is high because there are lots of fires. The system has the potential, if precipitation were to decrease, to lose stability and flip here. You'd see critical slowing down here. And if precipitation increases, it flips from the high grass state to the high vegetation stage.

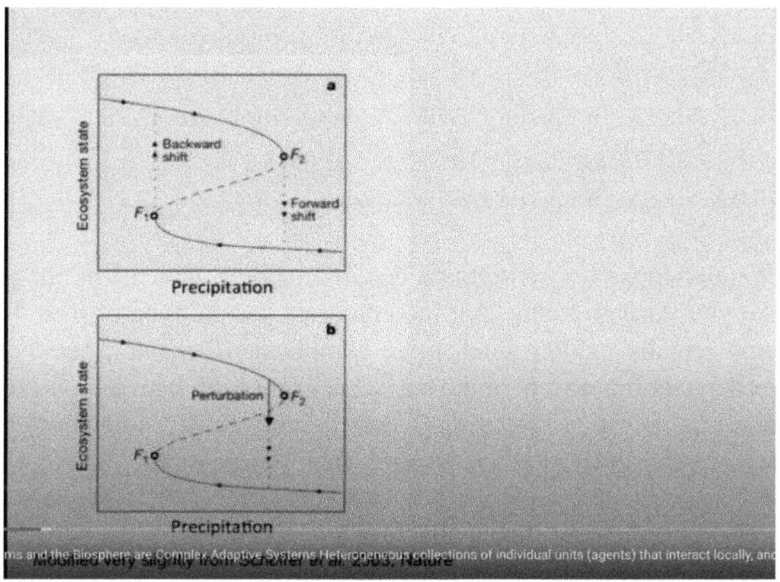

So, responses to changes in rainfall status are going to be rapid, they're going to be threshold transitions, and they're going to be difficult to reverse. We see the same sorts of phenomena in other systems. Steve Carpenter, Martin Scheffer, and others have demonstrated that lakes can flip between eutrophic and oligotrophic, or much healthier systems. We certainly see it in pathogen systems, forest pests, for example. In Buzz Holling's initial work on the spruce budworm, it has the potential for this sudden transition.

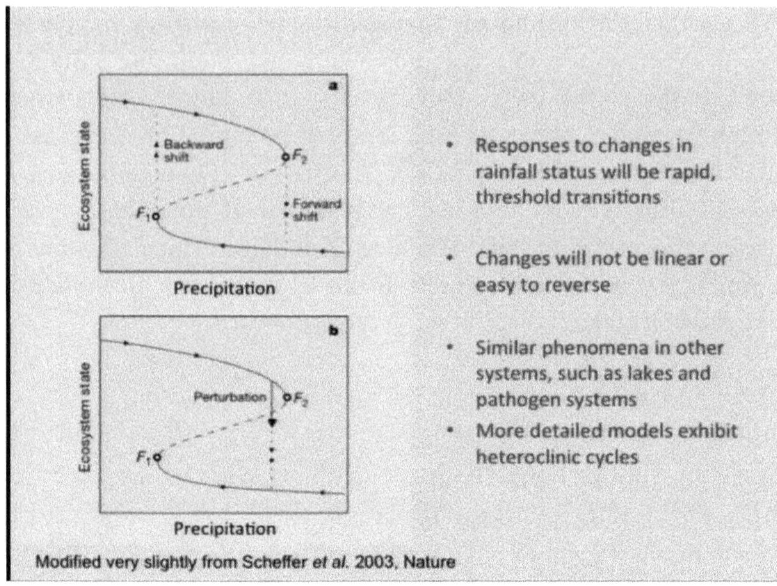

We also looked at some more detailed models, which I'm not going to take you through today, that incorporate a second type of tree. These systems actually exhibit much more complicated behavior, heteroclinic cycles that can flip on a periodic basis, back and forth between two states. We don't know yet; we're trying to determine from data whether this is a mathematical artifact or whether this is really meaningful.

Collective Phenomena and Collective Motion: Now let me turn to the third topic, which is collective motion and collective phenomena. Many speakers in this meeting have already mentioned Phil Anderson's paper that was the subject of last year's meeting, *More is Different*.

The idea is that putting lots of things together will lead to emergent properties that are difficult to predict; two postdocs, Kelly Caylor and Todd Scanlon, together with Ignacio Rodriguez-Iturbe and me, have been interested in vegetation patterns close to those regions I already showed you in Africa. This is in the Kalahari, though. If you look at satellite data and examine a cluster diagram showing the number of patches of vegetation of different sizes, they have a power law sort of relationship. If you build simple models that just put disturbances and allow growth locally, then you don't reproduce these patterns. But if you allow those disturbances to

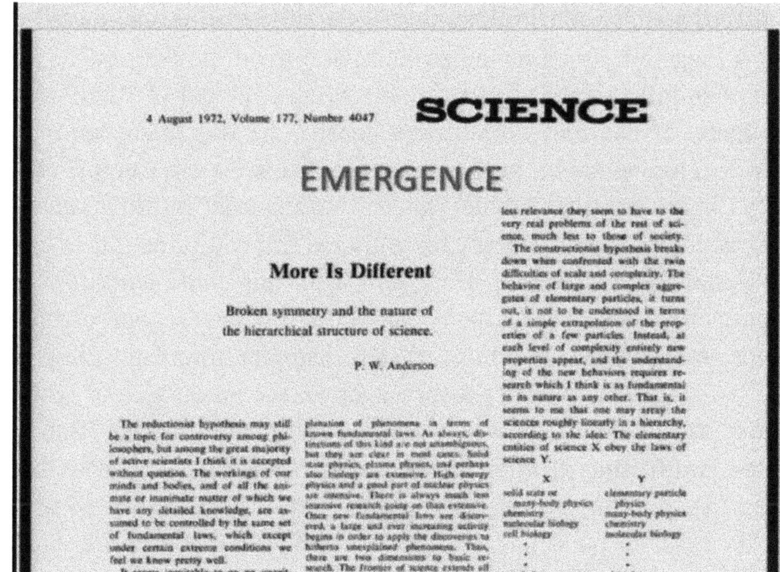

spread locally, then you can reproduce the patterns quite well. So, very simple local models can lead to these emergent properties, in this case, the power law phenomenon.

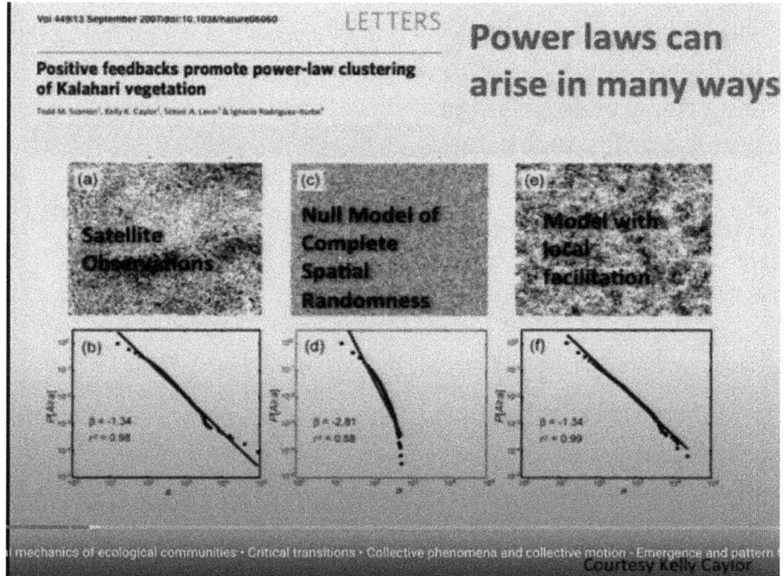

Indeed, there are other patterns in vegetation that have attracted interest for a long time, such as the striped patterns in the tiger bush. People have successfully modeled this by borrowing a model of Alan Turing's, which many of you may know about because he was interested in problems in developmental biology. Turing asked how pattern can form endogenously, where there's no blueprint and no external information. In the case of developing an embryo, the only external information is gravity, which is not enough to do it. He said, "Well, this could work if you had two interacting morphogens, two interacting chemicals, one of them an activator and the other an inhibitor." The activator stimulates the production of both, and the inhibitor dampens down the production of both. In a well-mixed medium, they reach some sort of equilibrium. But if the inhibitor is diffusing at a higher rate, then what can happen is that the activator, by random perturbations, gets a little bit above at some local point. It begins to build up, it stimulates the production of the inhibitor, which begins to build up. But then the inhibitor diffuses away, not doing its job to dampen down, and the activator keeps going. Meanwhile, the inhibitor is in places it shouldn't be, and that can break symmetry. You can show that it leads to non-uniform stationary patterns. Basically, what's going on there is short-range activation and long-range inhibition. You can see these phenomena in other models, and people have applied this to those vegetation systems with a good deal of success, water being the element that's moving there — probably more success than has been shown in developmental biology systems, where the search for these morphogens has not been terribly successful.

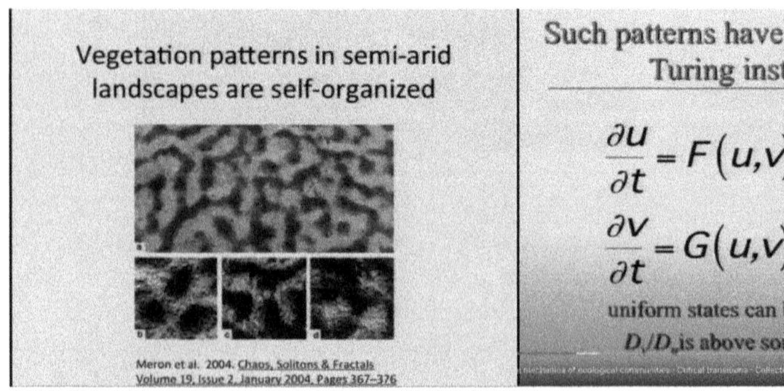

$$\frac{\partial u}{\partial t} = F(u,v) + D_u \nabla^2 u$$

$$\frac{\partial v}{\partial t} = G(u,v) + D_v \nabla^2 v$$

So, Lee Siegel and I, and separately Akira Kubo, got interested in applying these ideas of Turing's to plankton systems, where one sees patchiness on almost every scale. We said, "Well, maybe the phytoplankton, the small plants, are the activators stimulating the production of zooplankton, the inhibitors, which will dampen us down." So, we published a paper in *Nature* which showed that this could happen. Unfortunately, when you look at the patterns themselves, you see that this doesn't work. If you look at the patchiness in the system, you'll find that the zooplankton are more patchily distributed than phytoplankton, which is not what should happen with the inhibitor because the inhibitor is diffusing at a higher rate.

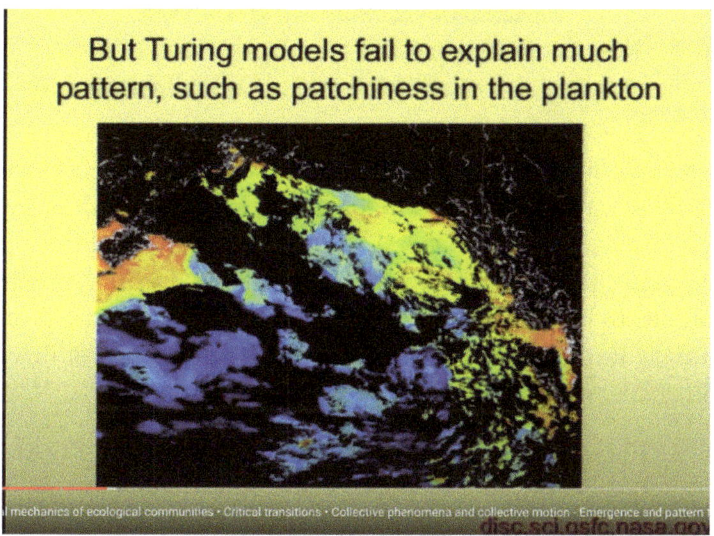

So, something was wrong, and my student Danny Grunbaum worked on this for his thesis. The problem is that plankton don't passively diffuse, so the diffusion model doesn't work for them. They aggregate, and similar problems of this sort have been looked at for a variety of these systems in which collective motion is important to pattern.

Many people, including us, have worked on this for bacterial pattern formation, bacterial fronts that can grow. The study of slime molds is one of the classic examples in developmental biology, and my colleague John Bonner at Princeton did the seminal work, but Lee Siegel and

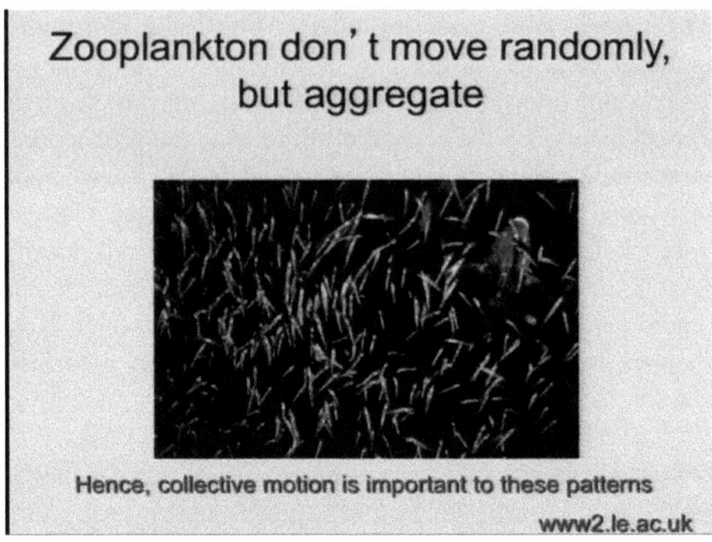

Evelyn Kellard also did a good deal of work on this. Insects swarming, krill, which are small organisms in the ocean, birds, as I'll show you, fish, and ungulates like these wildebeest. You see these remarkable frontal patterns that arise, that seem best modeled as a hydrodynamic instability of the front, due to local attraction and repulsion. So, this is a front of wildebeest moving through the Serengeti, but you'll see patterns of this sort for many other species.

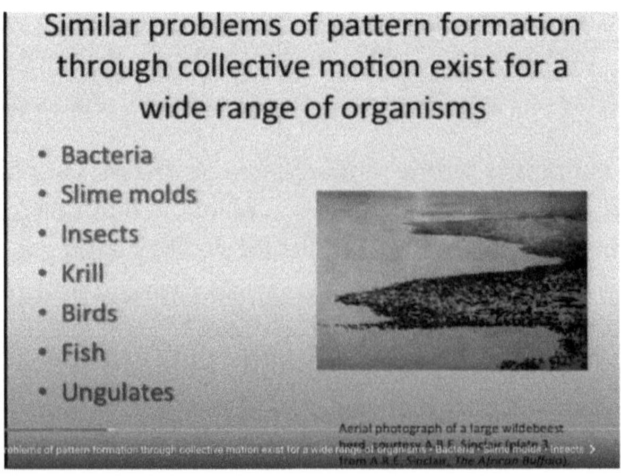

In these cases, the problem again is this scaling problem: How do we take this crude look at the whole and understand it in terms of the behaviors of individuals? How do we relate the macroscopic patterns to the microscopic rules?

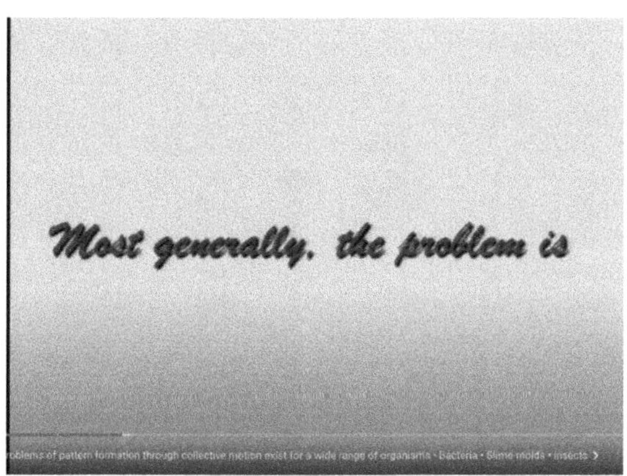

Richard Feynman was interested in this: How is it that ants that are moving from point A to point B end up doing so in a straight line? He had a very simple set of local rules for it, which is not far from true. He said, "Well, suppose they're moving along on a path of this sort, and each one is only trying to follow the one in front of it. Then this one here eventually will turn and straighten out this line, just moving toward the individual in front of it." So, local rules can give rise to emergent patterns that strike us as remarkable.

mishilo.image.pbase.com

Onlipix.com, original source unknown

The null hypothesis for movement is a random walk, but the random walk obviously doesn't work very well.

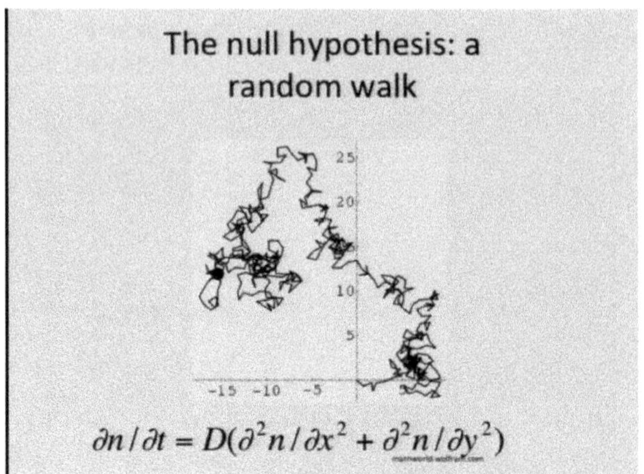

$$\partial n / \partial t = D(\partial^2 n / \partial x^2 + \partial^2 n / \partial y^2)$$

Here, inter-individual interactions are essential. These are reindeer, and you can see that some are going in one direction, presumably the right direction, while others are going in different directions. Some are going the wrong way because they're following individuals, but eventually, they'll end up in the right place. What else do they have to do? So, these sorts of patterns are emergent from local interactions.

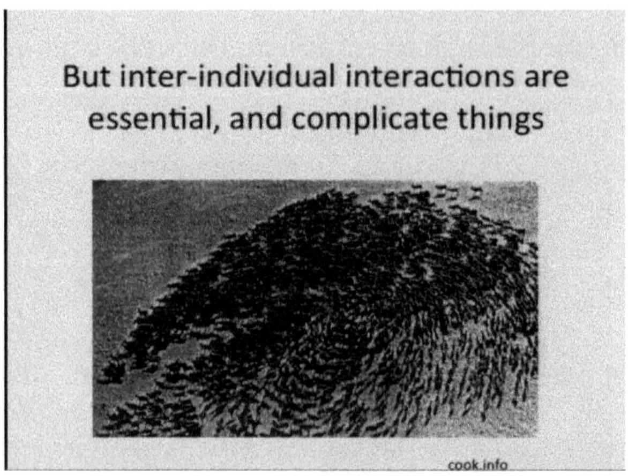

Now, Danny's thesis and the later work we did with Glenn Flierl and Don Olson tried to get at this problem by developing proper statistical mechanics. I'll just give you a hint of it here. The way this works is you write down a density function for the velocities and spatial positions of each individual, of all the individuals in the system. This density function changes in a time step Δ-t. It's the integral of what all the positions were, and densities and velocities at the previous time step, multiplied by the transition functions that tell you the likelihood of going from this position to that position. That either comes from data or from your hypothesis about how things are playing out in this system.

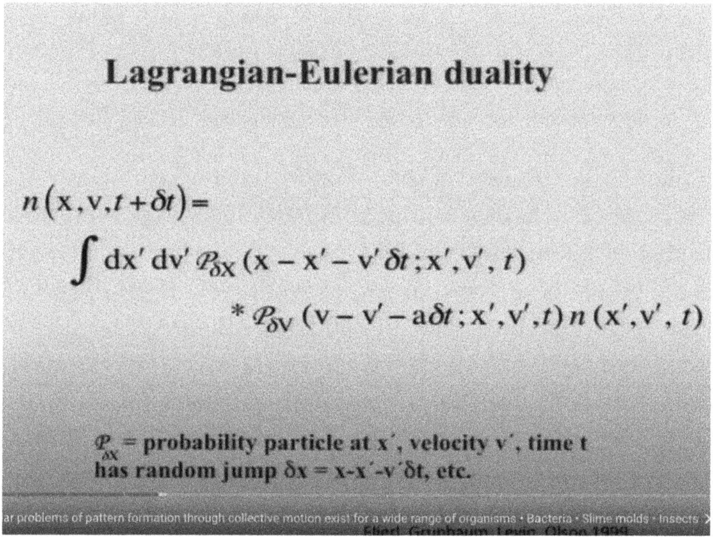

$$n\left(x,v,t+\delta t\right)=$$

$$\int dx'\, dv'\, \mathcal{P}_{\delta x}\, (x - x' - v'\delta t\,;x',v',t)$$

$$*\, \mathcal{P}_{\delta v}\, (v - v' - a\delta t\,;x',v',t)\, n\, (x',v',t)$$

Now, if you've got local rules of interaction, and most information you get, for example, the data on those birds I showed you before and on fish, says that the birds respond, say, to their six or seven nearest neighbors. I'm not going to talk about that, but there's a whole literature of what rules people have used. You've got to somehow translate that into a density. So, you take this, you think you start from an individual-based rule, about a Lagrangian rule about how individuals move and how they respond to their neighbors. You develop a statistical description of that, and then you've still got to close up the system in some way by translating a density into the probability that I'm going to see seven or six or five or no individuals in my neighborhood. So, I have to use some closure scheme like assuming that the occurrences are local or locally plus all. So, these work pretty well in many cases in giving you a continuum description of the system.

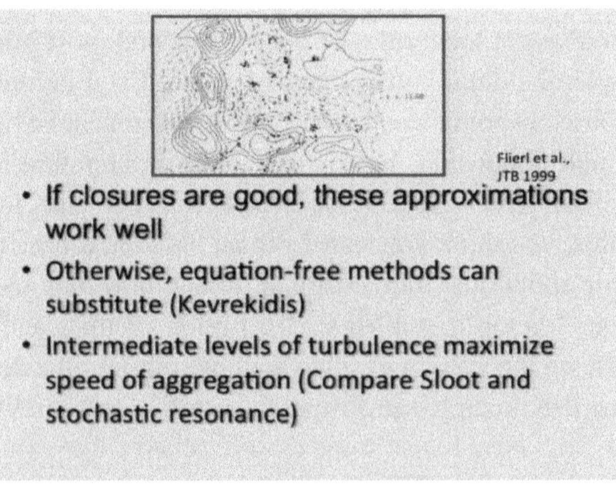

$$\frac{\partial}{\partial t}n(x,v,t) = -\frac{\partial}{\partial x_i}[v_i n(x,v,t)$$

$$-\frac{\partial}{\partial v_i}[a_i n(x,v,t)]$$

$$+\frac{1}{2}\frac{\partial^2}{\partial v_i \partial v_j}[\gamma_{ij}\, n(x,v,t)$$

In a picture like you see at the top here, what we did in our paper was to put aggregation schemes within some Navier–Stokes type flow. There are a number of interesting conclusions that occur here. First of all, if the closures are good, these approximations work well. If the closures are not good, then there are some very imaginative schemes designed by Yanis Kevrekidis and others called the equation freeway methods. What Kevrekidis does is he uses his microscopic simulator to generate the tangent planes, etc., that you would need to do bifurcation analysis. So, he operates with the microscopic model as if he had the closure to generate the information the closure would give you.

One of the things that we observed, and I was really struck by the resonance with what Peter Sloot talked about yesterday with stochastic resonance, is if we put low levels of turbulence into the system, into the fluid dynamics, that made it easier for individuals to find each other. Aggregations formed more quickly because individuals were moving around and much more likely to come into contact with each other. But if you tuned up the turbulence to much higher levels, that began to break up the assemblages. So, just as he reported yesterday, the most efficient aggregation in this case came about at intermediate levels of turbulence. Very similar phenomenon. But these models work well, but real aggregations are heterogeneous assemblages of individuals, like these birds that I showed you at the beginning. These are starlings, this was given to me by Claudio Carere. These pictures were taken in Rome, but you can see them in other places. I just came back from Australia, where I saw the flying foxes, they're huge bats that are going through patterns of this sort. Now, if you're a fluid dynamicist, it's hard to look at this and not think there's statistical mechanics, there's a fluid dynamical description. Oh, by the way, they're not all starlings, this guy right here is a hawk, one that's driving a lot of these activities, and it may be that this motion has collective benefits that actually confound the predator. But in any case, there's no obvious leader; individuals are moving in response to each other.

So, with my postdoc Ian Couzin who's now my collaborator — remember the faculty at Princeton — I set out, and that really means, he said if I say I, I mean we, okay? If I say we, I probably mean he, okay? We set out to look at them. In his case, it's mainly fish that he's interested in, how these patterns can form, and what the role of leadership is. So, we built a simple model.

This is a stochastic model, although it won't be evident from what I'm saying. It's a stochastic model, and the idea is that every individual in this has a direction in which it's moving, and every individual updates its direction at different time steps in relation to two kinds of information: its intrinsic information about where it thinks, say, the food is or where it thinks it ought to go. It doesn't have to be the correct information in this model. That's this G and D, which is information about its neighbors. Information about individuals in some neighborhood, their positions, their velocities. The positional information averages over the positions. I'm sorry, the social information averages over the positions and velocities of neighbors.

And we update this at each time step according to a rule of this sort. The updated velocity vector is some weighted average of the social vector and the informational vector, normalized. So, if Omega is equal to zero, you do nothing but follow. If Omega is infinite, you do nothing but lead. And we did look at some versions in which Omega was adaptive. Namely, if you have a high Omega but nobody's following you, you lower your Omega — not something most of our political leaders do.

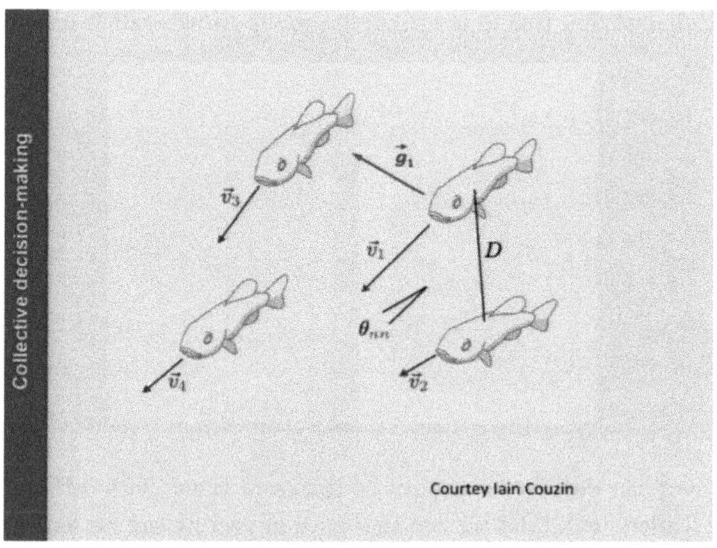

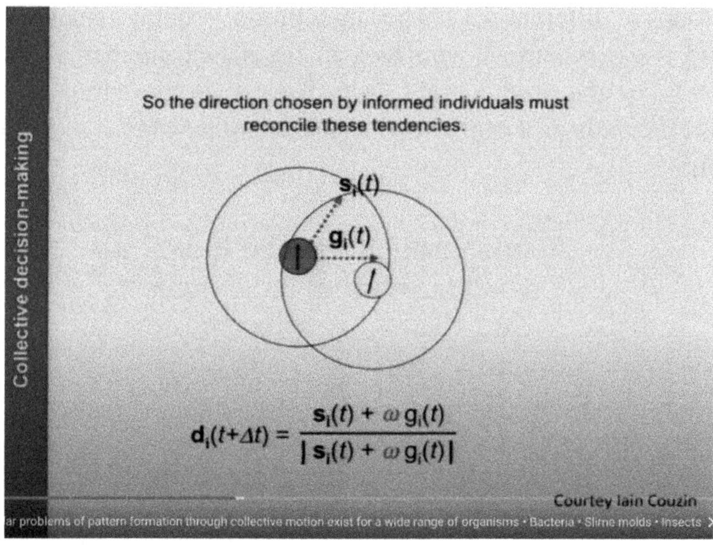

So, in this first video I'll show you, there are a hundred individuals, 99 of whom are pure followers, and one who wants to go up to this point here. So, that one is the white dot, and there are 99 green dots. The group doesn't move very effectively. But if I change that one to five, then the five find their way to the front, and the group begins to move up there.

But if I change that five to ten, then the group moves rather efficiently to the front.

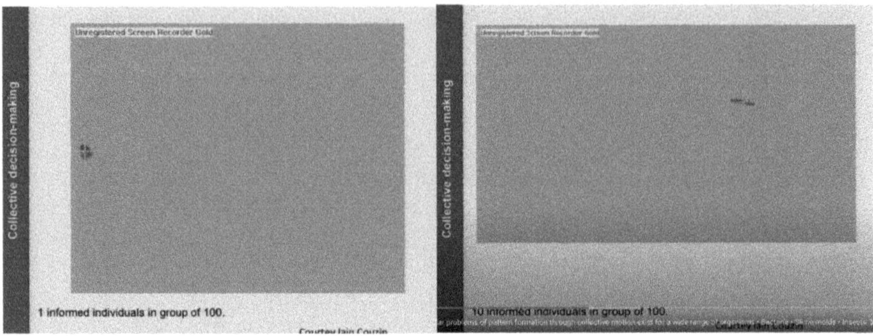

So, we can do this for groups of different sizes, with different numbers of leaders, etc., and we see that animal groups can be led by a very small number of leaders. This is the proportion of informed individuals in these groups of different sizes. The larger groups require a smaller proportion, and that's because if you look at the actual number of informed individuals, by the time you get up to five or ten, the group is moving about as efficiently as it can. This is a measure of how fast it moves toward that front.

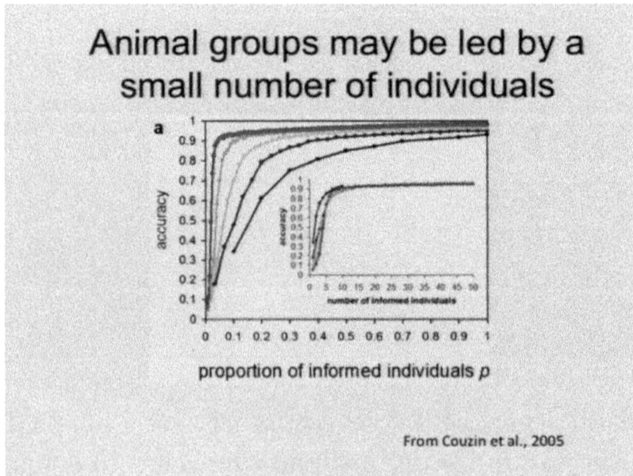

Now, that's if all the leaders want to go in the same direction. What if they don't? If you have competing preferences, but they're not too different, then the group will split the difference initially. It'll go down the middle.

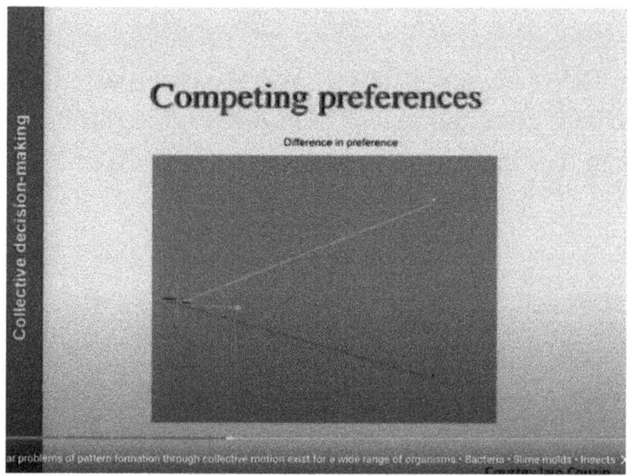

But at some point, it's going to have to make a decision, or else it's going to have to split, and a bifurcation occurs. That's what you see here: An equal number of individuals, five in this case, wanted to go in each direction, and at some point, half the time, the group was going one way and half the other. But you can also get this sort of behavior in which five individuals want to go north and five south.

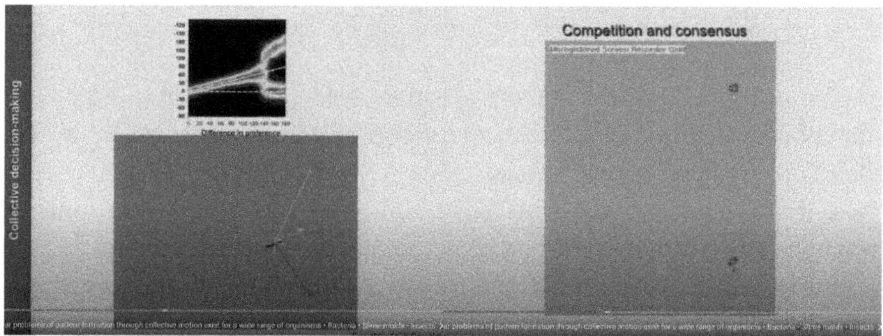

I never understand why everybody finds this funny, but it must tell us something about human nature and indecision. So, the group split, as you saw, but you can also get this behavior. Remember, this is a stochastic model, in which the group goes up there for a while and then it heads down here. The reason for this is the model's stochastic nature. When you get up close to the two one target, all of the vectors that are pointing to that target are no longer coherent, whereas the ones that go in the other direction are all pointing down there. So, this can actually lead to periodic behavior. It's something like what one expects in a two-party political system, where once the group is in power, infighting becomes more important than getting to the target because you're so close to the target.

What happens is that if there are more leaders that want to go in one direction than in the other, then the group will almost always go in that direction in these simple models. So, we set out to try to understand these models, and worked with Naomi Leonard, who's known to some of you and is a control engineer interested in putting robots out in the field to gather information. We modeled this most simply by ignoring explicit space and using a model due to Kuramoto's model of coupled oscillators.

Ignoring the summation terms for the moment, what this says is that these individuals want to go in Direction $\bar{\theta}_1$. So, I write those equations as θ_j dot equals sine of $\bar{\theta}_1 - \theta_j$ for all those in N1, the individuals in that group. If that's all that went on, those individuals would all end up in that direction. Then, we have a second group that wants to go in a different direction, and then we have a bunch of followers. In this model, there's also all coupling, although we relaxed that later, and everybody is influenced by the average of everybody else in the group.

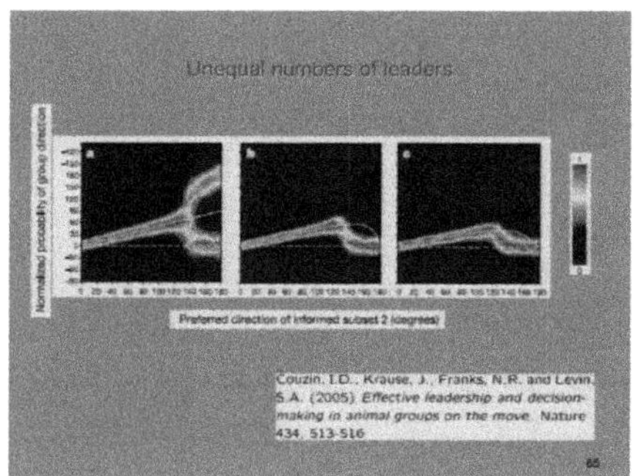

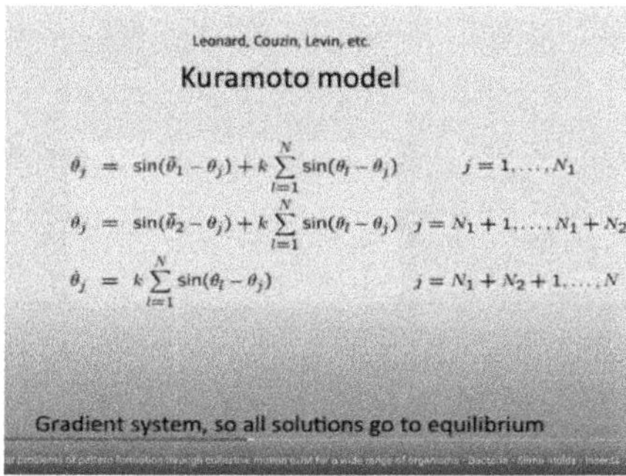

$$\dot{\theta}_j = \sin(\bar{\theta}_1 - \theta_j) + k \sum_{l=1}^{N} \sin(\theta_l - \theta_j) \qquad j = 1, \ldots, N_1$$

$$\dot{\theta}_j = \sin(\bar{\theta}_2 - \theta_j) + k \sum_{l=1}^{N} \sin(\theta_l - \theta_j) \qquad j = N_1 + 1, \ldots, N_1 + N_2$$

$$\dot{\theta}_j = k \sum_{l=1}^{N} \sin(\theta_l - \theta_j) \qquad j = N_1 + N_2 + 1, \ldots, N$$

So, this is a very simplistic cartoon; ignore space and you're coupled to everybody. Turns out this is a gradient system, so the system goes to equilibrium. Just to give you a cartoon movie of what's happening, there are five individuals who want to go here (that's the heading, not a position) and five that want to go there. On the fast timescales, individuals organize themselves into groups, and then on the slow timescales, you can look at the dynamics of the group.

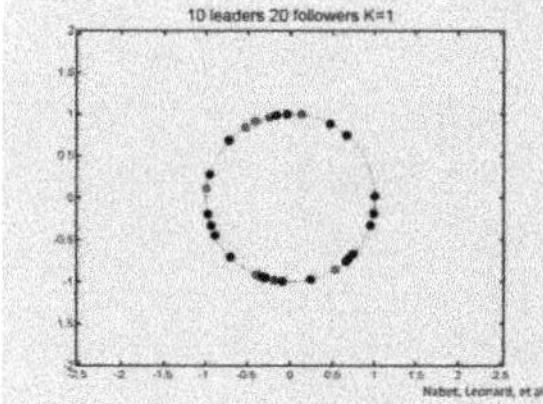

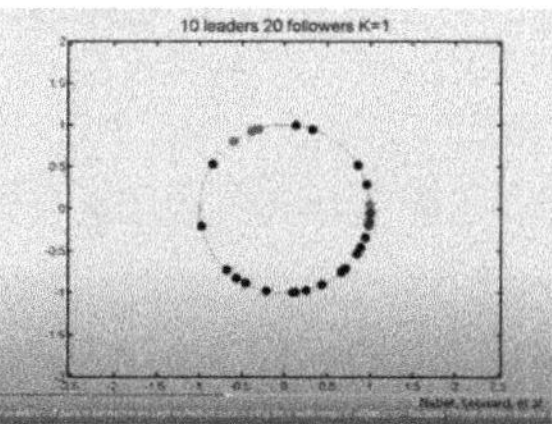

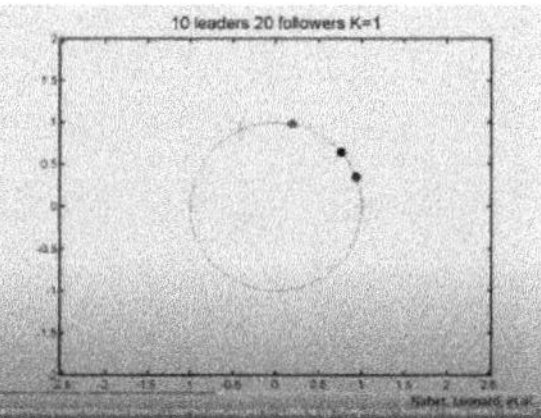

And so, we write this down. For any particular group j, $e^{i\theta}$, C_j is the summation of all of the individuals in Group j, so all the ones who wanted to go in Direction one would be $e^{i\theta}$. Then, we differentiate, and then you can show that on a fast timescale, these form themselves into groups. Then we can write down the slow timescale equations, which I won't do for this system; I'll come back to that in just a moment.

We write the dynamics for ψ_1, ψ_2, ψ_3 the average heading of respectively η_1, η_2 and η_3.

$$r_j e^{i\psi_j} = \frac{1}{N_j} \sum_{l \in \eta_j} e^{i\theta_l} \qquad j = 1, 2, 3$$

$$\dot{r}_j e^{i\psi_j} + i\dot{\psi}_j = \frac{1}{N_j} \sum_{l \in \eta_j} i\dot{\theta}_l e^{i\theta_l} \qquad j = 1, 2, 3.$$

During the second time scale

$$\theta_l = \psi_j$$

Everyone in cluster has same heading

$$r_j = 1$$
$$\dot{r}_j = 0$$

Courtesy, Ben Nabet

And, there are a couple of conclusions: We found that naive individuals, meaning uninformed individuals, those that don't have any preferred direction, are crucial to consensus, but the non-spatial models explain some of the behavior, but not everything. We needed multi-scale analyses, and this reminded me of John's talk yesterday; we ought to look at models in which there were changing signals and boundaries. Individuals tended to change who they paid attention to, and the network changed.

Preliminary conclusions

- Naïve individuals are crucial to consensus
- Non-spatial models miss key detail
- Multi-scale analyses also essential
- Changing signals and boundaries are important (Holland)

And so, this is a paper we published a year and a half ago, in which we did this. Naomi Leonard is the first author in this, in which we set out to look at uninformed individuals, but the model was a little complicated.

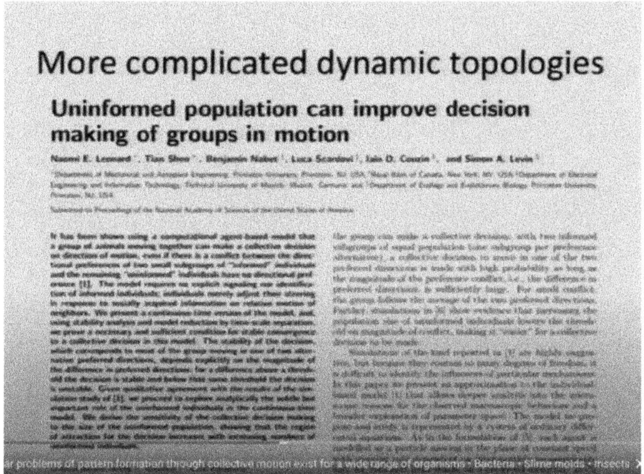

This looks exactly like what I wrote down before, except the theta J's can change as a function of time now. Well, they could in the other model as well, but these coupling parameters *AJ*, the degree to which you pay attention to other individuals, now change as a function of time. How do they change as a function of time? Well, I won't take you through the mathematics here, but I'll tell you what it says. Basically, it says that you tend to strengthen the connections to individuals in your neighborhood and weaken the connections to other individuals. What you find is that the group self-organizes now into a number of different clusters. These are the dynamics on the slow timescales; these are the equations for the directionality of that, or actually for that summation function for that group, and the *A*'s, which are the coupling terms, tell you how much you depend on the other groups, tend either to be zeros or one.

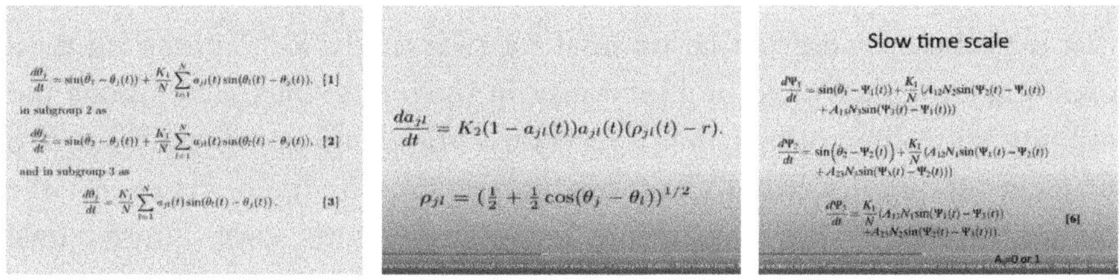

$$\frac{d\theta_1}{dt} = \sin(\theta_1 - \theta_j(t)) + \frac{K_1}{N} \sum_{l=1}^{N} a_{jl}(t)\sin(\theta_1(t) - \theta_j(t)). \quad [1]$$

in subgroup 2 as

$$\frac{d\theta_2}{dt} = \sin(\theta_2 - \theta_j(t)) + \frac{K_1}{N} \sum_{l=1}^{N} a_{jl}(t)\sin(\theta_2(t) - \theta_j(t)). \quad [2]$$

and in subgroup 3 as

$$\frac{d\theta_3}{dt} = \frac{K_1}{N} \sum_{l=1}^{N} a_{jl}(t)\sin(\theta_3(t) - \theta_j(t)). \quad [3]$$

$$\frac{da_{jl}}{dt} = K_2(1 - a_{jl}(t))a_{jl}(t)(\rho_{jl}(t) - r).$$

$$\rho_{jl} = \left(\tfrac{1}{2} + \tfrac{1}{2}\cos(\theta_j - \theta_l)\right)^{1/2}$$

$$\frac{d\Psi_1}{dt} = \sin(\theta_1 - \Psi_1(t)) + \frac{K_1}{N}(A_{12}N_2\sin(\Psi_2(t) - \Psi_1(t)) + A_{13}N_3\sin(\Psi_3(t) - \Psi_1(t)))$$

$$\frac{d\Psi_2}{dt} = \sin\left(\theta_2 - \Psi_2(t)\right) + \frac{K_1}{N}(A_{12}N_1\sin(\Psi_1(t) - \Psi_2(t)) + A_{23}N_3\sin(\Psi_3(t) - \Psi_2(t)))$$

$$\frac{d\Psi_3}{dt} = \frac{K_1}{N}(A_{13}N_1\sin(\Psi_1(t) - \Psi_3(t)) + A_{23}N_2\sin(\Psi_2(t) - \Psi_3(t))). \quad [6]$$

$A_{ij} = 0$ or 1

And therefore, we can study the dynamics of the system, and it turns out it's very complicated. There are eight possible stable invariant manifolds, and as many as five of them can be stable at any one time. So, the system is very sensitive to initial conditions.

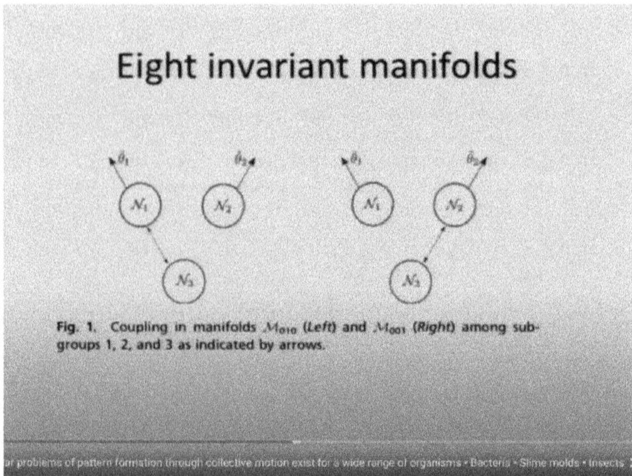

Fig. 1. Coupling in manifolds $\mathcal{M}_{010}$ (*Left*) and $\mathcal{M}_{001}$ (*Right*) among subgroups 1, 2, and 3 as indicated by arrows.

Conflict and Collective Action: Let me just touch on the last point, as Helga addressed it in her comments, and it was implicit in what Murray had to say as well. Systems of this sort are subject to conflict when they ought to be engaging in collective action, so they serve as models for our thinking about how we can address environmental problems and other issues we face. Imagine this cartoon, which Danny Grunbaum developed when he was writing his thesis, although he didn't include it in the thesis, and I had to give him a copy last year. These are the krill that he studied. The idea here is that suppose you're searching for resources on some landscape, and this landscape has multiple peaks. The way you would probably do it is by hill climbing, just moving up the gradient. That's good, it might get you up to this peak or that peak. But if you shared information, maybe everybody could get up to this peak. So, that tells you that it probably wouldn't be optimal for everybody to be going on their own. On the other hand, if everybody in those previous simulations did nothing but follow others, then everybody would end up as you'll see in a nice cluster, but it wouldn't have anything to do with the resource. So, the optimal approach lies somewhere in between, with some intermediate level of leadership and following, some intermediate level of foraging and scavenging, and either with some subsets who are leaders and some who are followers, or everybody doing a little of each.

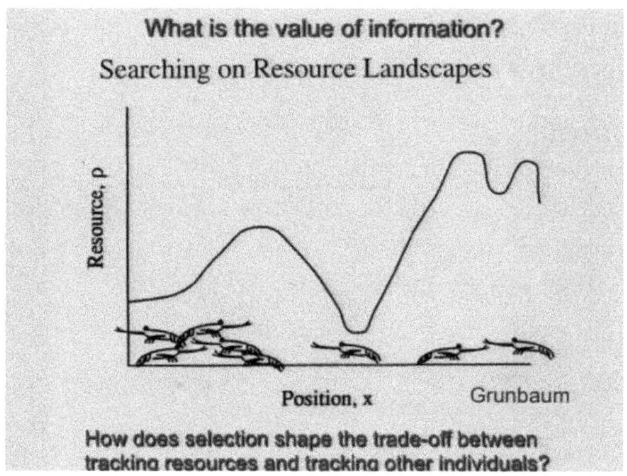

But it's not the right way to look at this problem, because this is a game theoretic problem. I'm going to freeride on you; I don't necessarily care about what's good for the group. So, we asked how selection shaped the trade-off, but in this game theoretic situation, between tracking resources and tracking individuals, how many leaders, how many followers, what is optimal from the viewpoint of the group, and I will justify that in the next slide. Naomi Leonard has a student who has looked at what rules individuals ought to use to maximize the robustness of the group, and she found — the student found — that individuals should pay attention to their six or seven nearest neighbors, which fits very well with observations. But the trouble is that, and we're looking at this now, there's no reason to believe that individuals will do what's in the best interest of the group. So, we look for game theoretic solutions, and there are lessons here for cooperation in public goods situations. Other work I'm doing focuses on public goods resolution, in which these insights are helpful.

Now, as I told you, Naomi puts robots out in the ocean, and those robots exchange information with each other. She's trying to optimize information in her system, so it's perfectly reasonable to ask what's optimal. But she got interested in how animal groups do this, so we began working on models of fish populations, where the optimization is at the local level. In reality, in groups, there's probably some level of what economists call pro-sociality, caring about other individuals. So, the optimum is somewhere in between. We're trying to look at what the consequences are of having selection at those different levels.

We, referring to the local group, have published a series of papers that try to address this question. I won't talk about my paper with Colin Torney, but this is the paper that Vishu Guttal and Ian Couzin, and maybe some others, have published. First of all, they just said, well, there are two parameters now in this model: One is the investment you make in finding out where the resources are and the other is the investment you make in finding out what others are doing. First of all, we can vary these parameters, and if all you do is invest in following others, then you have a swarm, as I mentioned, but it's Brownian, and it's going to move around with no relation to resources. On the other hand, if you don't make an investment in either, then all the individuals will be randomly walking. If everybody puts their energy and resources into detecting the gradient, then you'll get solitary migrations. It's only when you have investment in both that you get this collective motion in relation to the gradient. Then, they put this in an evolutionary model, where they rewarded individuals based on their payoffs and allowed them to evolve.

According to some replicator schemes, they ended up with a bifurcation of the population into leaders and followers, so the system self-organized. Another way to look at it was to examine it in this landscape.

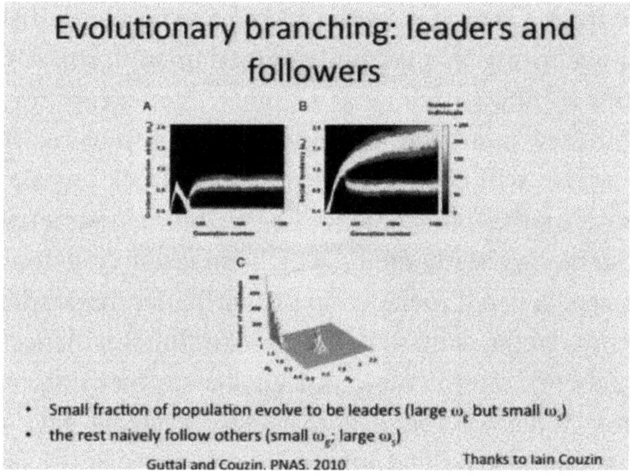

Collective action is indeed the fundamental problem in dealing with sustainability. William Forster Lloyd discussed the Commons nearly 200 years ago, and Garrett Hardin popularized this concept in the 1950s by referring to the tragedy of the Commons. He proposed a solution that was later refined by Lynn Ostrom and others, known as mutual coercion, mutually agreed upon. This concept was implied in the talks given by Murray and Helga.

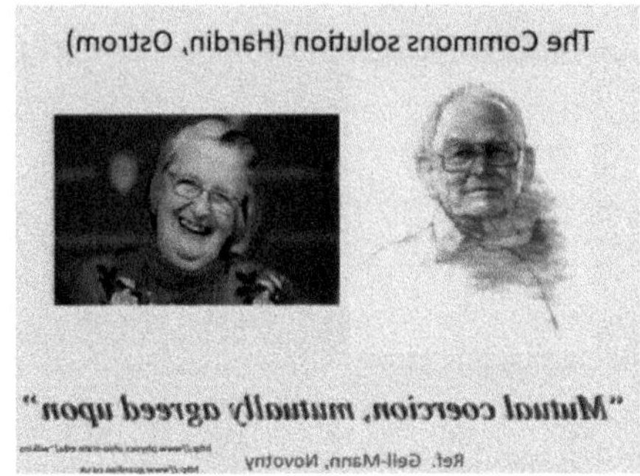

How do you achieve collective action for the common good? In his introductory remarks, Lock mentioned a paper in which I am indeed an author, but Alessandro Tavoni and Maja Schlueter are the lead authors. In this paper, we explore how norms arise in a particular system, drawing from a vast literature on the topic. We also investigate how these norms can foster cooperation in the Commons. The basic idea is that individuals can withdraw from a shared resource at different levels: selfishly, at a very high level, or according to a mutually agreed upon norm. It's evident that they fare better if they withdraw at a higher rate, except for the fact that those who withdraw at a lower rate band together to ostracize and punish them. However, we still need to incorporate, as Lock has already pointed out, some cost to the punishment. For instance, ostracizing someone means not interacting with them, which incurs a cost for me as well, although this aspect is not included in the particular model discussed here.

By the way, Maja, who's at the Stockholm Resilience Centre, has conducted fieldwork on this topic in Uzbekistan, focusing on water utilization systems. What we found is that these systems have the potential for multiple stable states. The graph illustrates the level of selfishness,

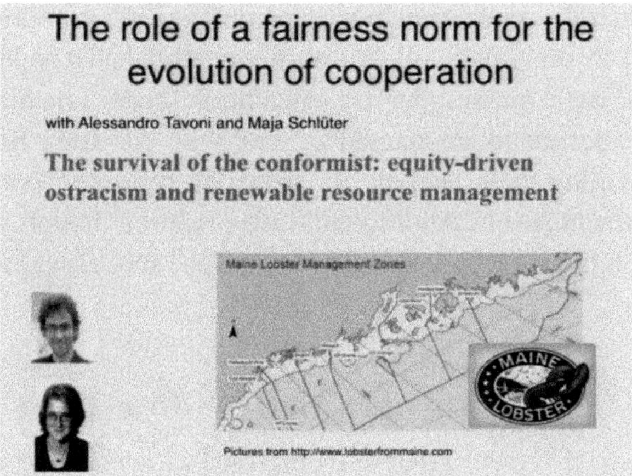

indicating how much more selfish individuals withdraw, and the frequency of cooperators. It becomes evident that the norm can only be sustained if it surpasses a certain threshold, leading to stability in the system. This is why many international agreements are structured to come into effect only if the number of signatories exceeds a specified threshold.

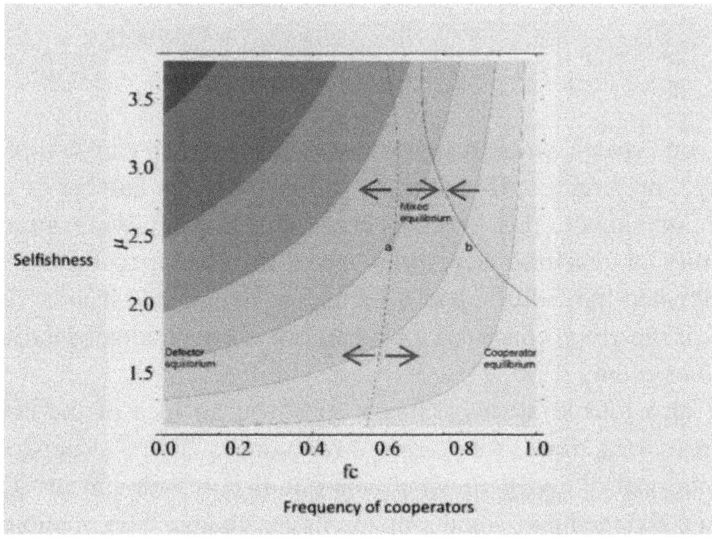

More generally, consensus may emerge in systems. With Ian, Naomi Leonard, and several other collaborators, we published a paper last year in *Science*. In it, we examined the fish orientation model I mentioned earlier, along with experiments conducted by Ian, who can train fish to targets, effectively making them leaders. By adjusting rewards, he can modulate their commitment levels. Additionally, we explored models of consensus formation in human groups. Remarkably, all the models exhibited similar behavior.

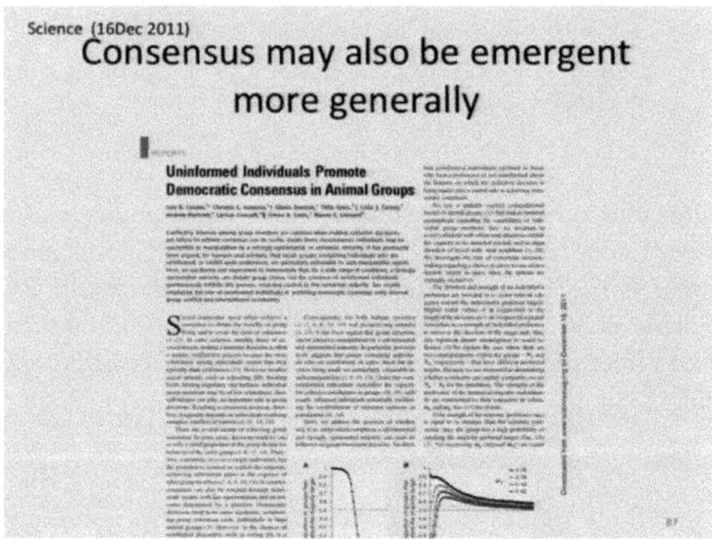

If you have a strongly opinionated minority, it can dominate the majority's preferences. However, as you increase the number of unopinionated individuals, this becomes more challenging. We examined the probability of reaching a majority target in relation to the number of unopinionated individuals, and it increases. Eventually, it must decrease because if the group consists almost entirely of unopinionated individuals, it becomes evenly split.

We also looked at this. I'll just give you an idea of the collective decision-making model we examined for humans. There's a network with a large number of nodes, in which you initialize it with a mean degree of ten — ten connections — and individuals can change their opinions. They change their opinions in relation to what their neighbors are doing, and

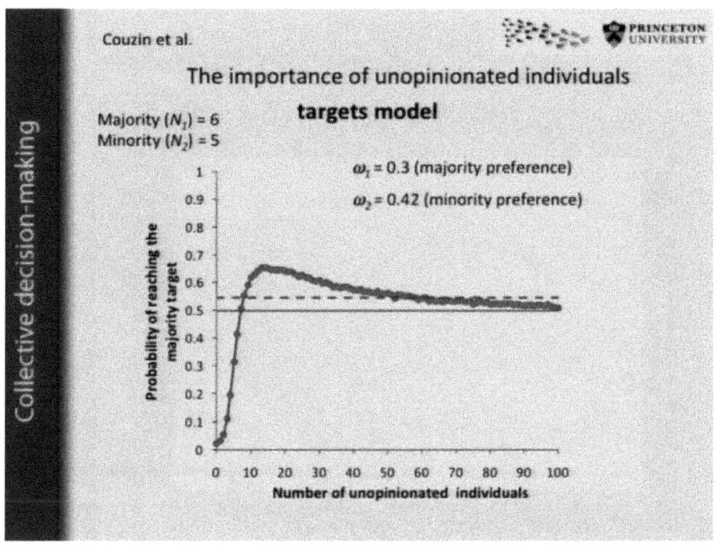

there's a nonlinear relationship. The more your neighbors have the opposite opinion, the more likely you are to change. However, you have an intrinsic opinion — at least some individuals do — and you're much more likely to spontaneously flip back to that. Finally, just as in the fish model, you can make or break links depending on your similarity to your neighbors. So, this model gives rise to the same sort of behavior. Indeed, I don't have time to go into it, but it turns out there's bi-stability in this system as well.

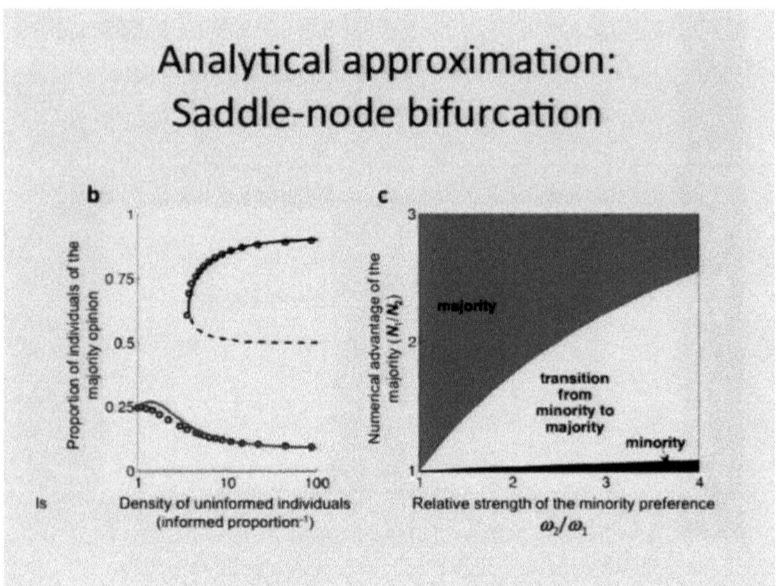

Let me finish up and draw some conclusions. Collective phenomena and emergence characterize a wide variety of systems, from microbial committees to entire ecosystems. A fundamental challenge lies in scaling from the microscopic to the macroscopic, from individuals to gaining a crude understanding of the whole. Conflict and cooperation present challenges in all systems, and there's extensive literature on how these dynamics are resolved, not only in human systems but also in collective motion across various domains. Multicellularity, for instance, represents a form of collective action and is considered one of the major transitions in evolutionary theory. I found this information on the web yesterday to illustrate that conflict and cooperation occur even within our bodies, where selfish DNA and competing elements exist. Finally, I hope I've convinced you that approaches from complexity theory, particularly techniques from mathematics and physics, can not only inform our resolution of these problems but also stimulate new avenues of examination, as has historically been the case, inspiring mathematicians to tackle new types of problems and provide fresh perspectives.

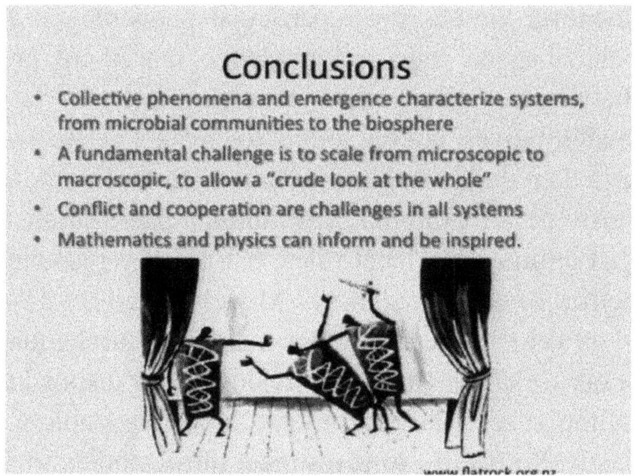

Thanks very much.

9.2 Discussion

Discussant A: The mathematics is indeed quite beautiful when we can generate local rules and then simulate to observe the emergent global behavior. However, particularly in robotics and engineering disciplines, there are instances where we desire the emergence of global behavior and seek to derive the local rules accordingly. How effective is mathematics in solving such problems? Consider a scenario where a group of individuals possesses specific capabilities — they can only see a certain distance, move at a particular speed, and so on. Now, the goal is to derive the rules of their local interactions to achieve the desired global outcome.

Simon Levin: So, there are really two possible interpretations of your question, both interesting. One is reverse engineering, where we observe a phenomenon and inquire about its underlying mechanisms in the system. The other possible interpretation is how we can engineer a system to achieve a desired phenomenon. Both of these are intriguing. In the first case, we must recognize, as I've alluded to, that there are multiple pathways to achieving the same behavior. This is why catastrophe theory collapsed; whenever we observe a particular phenomenon giving rise to a specific behavior, we must acknowledge that there are alternative routes to achieving it. I don't know of any simple way to exhaustively explore these alternatives, but the aim is to create a catalog of the various approaches and then design experiments to distinguish among them.

Let me provide an example with animal aggregation patterns. We understand that animals use local rules, but do they respond to the positions or velocities of their neighbors? Models based on positions versus models based on velocities can produce many similar phenomena but also some distinct ones. For instance, if individuals only respond to the velocities of others, the groups they form tend to be less stable and can break apart more easily. So, we try to investigate these models to provide ways to differentiate among them. Another question to consider is whether individuals respond to all other individuals within a geometric neighborhood, as traditionally assumed, or within a topological neighborhood, such as the seven nearest neighbors. Therefore, we design experiments and measurements to distinguish among these possibilities. Such models serve as a starting point for thinking about these problems and for designing the necessary experiments or measurements.

I can't provide a general answer to how effective mathematics is at addressing these issues; it depends on the context. The second question is also intriguing, but in that case, you're not necessarily concerned about having the right mechanism. In fact, the first models of this kind I encountered were at Santa Fe made by a person named Craig Reynolds, who developed a program called "Boids." He was primarily interested in building movement models for, I believe, Lucas Studios. He would carefully examine the model, incorporate certain rules, and if the resulting behaviors were unsatisfactory, he would adjust the local rules until he achieved the desired outcomes. Naomi Leonard faces a similar situation with the robots she deploys in the field. It's up to her to determine the rules to implement. Her hope is that studying how evolution has addressed these problems over millions of years will provide insights into which rules to utilize, and she's quite content with that approach.

Guaning Su: You mentioned the power law during the discussion, and although I didn't fully grasp it, I'd like to follow up with a question related to Freddie's earlier inquiry. Geoffrey West has extensively discussed power laws in relation to cities, GDP, and various other economic factors. Do you have any insights into what fundamental behaviors might underlie this observed power law relationship? I recall he also discussed living organisms in this context.

Simon Levin: Do you mean why I get the power laws in our model or why Geoffrey West gets the power laws…

Guaning Su: Why Geoffrey West gets the power law, starting with how you got the power law.

Simon Levin: Let me emphasize again that there are numerous ways to obtain power laws and the allometric scaling relationships addressed by Brown, West, and Enquist have been of fundamental interest in ecology for a long time, with various theories proposed to explain them. They developed a compelling model of branching networks, such as bronchial trees, which gave rise to scaling networks under certain constraints regarding the smallest branch size. I find their explanations persuasive in this context, as bronchial trees are subject to natural selection and optimization processes.

However, when extending these models to other branching networks exhibited by real trees, such as above-ground branches or roots, additional factors come into play. For instance, the purpose of developing branching networks in trees is not solely to gather resources but also to shade out neighboring trees. Therefore, I believe these scenarios require a game theoretic approach to consider how to interfere with neighbors effectively.

West has also explored extending these ideas to parallel relationships observed in the dynamics of cities and banking systems. This introduces another level of abstraction, as optimization processes may occur at much more local scales, making the optimization arguments less persuasive in these cases. As for why power law relationships emerge in these systems, there are various potential explanations, such as self-organized criticality or certain phase transitions, as demonstrated by Bak's work. Mark Newman at Michigan has provided a comprehensive analysis of the different ways power law relationships can arise.

In any specific situation, distinguishing among competing hypotheses requires careful consideration, as highlighted by Freddie's question. Unfortunately, I don't have a general answer for how to do this.

Guaning Su: Thank you.

Discussant B: Hi, this might be a small detail, but I think it's an important question: Do your models incorporate risk preferences, or do you assume that all agents are risk-neutral? Additionally, do you allow for variability in risk preferences?

Simon Levin: In which models, you mean?

Discussant B: All animal models or models involving agents have a value function.

Simon Levin: So, the answer to your question is that in some of our models, and I'll specify which ones, we have incorporated risk preferences. However, in the aggregation models, there are no risk preferences because there's no optimization involved; individuals simply follow local rules dictating their behavior. It's natural to inquire about optimal behavior, and the extent to which we've explored this, as I mentioned earlier, is through simulations where individuals compete for rewards and produce multiple copies based on their performance. Any emergence of risk aversion in these simulations is not explicitly programmed but rather emergent.

There's one exception to this. We conducted a study, which I didn't discuss earlier, where animals could invest in information to aid their movement in a particular direction. Stochastic noise was introduced into this system, and the behaviors that emerged didn't include risk preferences but reflected reaction times to uncertainty. A similar scenario applies to the Tavoni–Schlueter model.

I've also been collaborating with Avinash Dixit, a prominent game theorist and economist, on a project examining how individuals deal with uncertainty and develop insurance arrangements. We focused on herdsmen in East Africa who share grazing grounds. We computed the social optimum for moving cattle and examined its stability in a game theoretic sense. We also explored potential second-best solutions, such as limited exchanges of cattle, to address the issue of risk. Risk aversion is incorporated into the utility functions in these calculations, reflecting individuals' preferences, and the acceptance of deals depends on their discount rates, reflecting their attitudes toward future outcomes relative to present ones. Therefore, risk aversion manifests in two distinct ways in these models. However, it's important to note that these models are not primarily optimization models, so this issue doesn't always arise.

Discussant C: Thank you for this very enlightening speech. I have one question: Have you ever observed nepotism evolving from any local behaviors? Nepotism is a classic phenomenon in human societies, where it's considered part of social equilibrium. I'm studying nepotism in various contexts, including the financial industry, mutual funds, and similar areas. Let me define what I mean by nepotism: Essentially, a weaker, dominated species, despite its inherent weaknesses, becomes the leader merely because of its close association with a more powerful species. It's based on this relationship rather than any other criteria related to achieving an objective function. Have you observed such nepotistic behavior in any other species?

Simon Levin: So you said this is nepotism?

Discussant C: Yes.

Simon Levin: Certainly not in our models. However, I'm aware that there are species in which leadership is determined, to some extent, by a hierarchy. This hierarchy could be influenced by previous degrees of contact among individuals in the group and relatedness, so it's not a uniquely human phenomenon. While I can't recall any models specifically examining this behavior, it would be an interesting area to explore. Perhaps researchers studying social insects or primate groups, where the leadership structure is known, could provide more insights. Unfortunately, I don't have much information on that, but I'm certain that such behavior occurs in primate groups.

Discussant D: That was a wonderful talk, and I feel a bit like a goose fattened up for pâté de foie gras — there's so much to absorb from it. I work in the area of public policy, and I'm always interested in translating works such as yours into an analogous form, perhaps like Agricultural Extension bulletins, which could be distributed to congressional staff members or used for congressional testimony. For example, Elizabeth Paddock Cornell's work at Stanford was applied to analyze risk considerations in the Concorde crash and the oil well disaster in New Orleans. I'm curious if there is an extension service affiliated with your work that is engaged in similar endeavors.

Simon Levin: Not in the sense you mean, but I'd be happy to discuss it further. Following our work, there have been some developments. For instance, the paper that emerged from a meeting organized by Timothy Geithner at the New York Fed on systemic risk in 2008 led to follow-ups. Additionally, the World Economic Forum organized a special year on systemic risk, where I served as a consultant. This demonstrates some efforts to translate these concepts into more accessible terms.

I've also been consulting regularly for the Boston Consulting Group, which endeavors to translate information on vulnerability and adaptability, drawing from ecological and evolutionary histories, into terms that companies can understand. Consulting groups like this play a role akin to that of an extension service.

Regarding the implications of our work on decision-making in fish schools and consensus formation in human groups for societal problems, there is considerable potential. Some of our research is funded by the military, which is interested in issues such as troop deployments. However, despite the potential societal applications, our work has faced criticism. Senator Coburn from Oklahoma and, more recently, Rand Paul, a prospective presidential candidate, have singled out our work as an example of government spending that could be cut. Their argument revolves around questioning the relevance of studying fish behavior to understanding human societies. This highlights the ignorance regarding the valuable insights that animal models can provide. It underscores the need for an extension service, but I'm unsure of its efficacy in addressing such misunderstandings. Nonetheless, corporations are interested in the lessons that can be derived from our research.

Chris Monterola: Hi, Chris from IHPC. You have discussed a lot of diverse work, involving diverse systems. In these diverse systems, you actually use different techniques to understand the results, ranging from agent-based models to Kuramoto models, do you say...

Simon Levin: When you say "you," you don't mean me, you mean "people do."

Chris Monterola: Yeah, now, eventually in your conclusion, you mentioned that conflict and cooperation are challenges present in all systems, common threads across them. Mathematically, at least in my intuition, conflict could possibly be understood as a diffusion process, as entities strive to maximize resources, while cooperation could resemble an aggregation process. This idea translates to the notion of diffusion-reaction differential equations, as seen in Turing systems, where opposing mechanisms are incorporated, or even in the nonlinear Schrödinger equation, which includes nonlinearity and diffusion terms.

Now, my question is this: Is there a fundamental reason why other ingredients, such as collective phenomena, cannot be captured or incorporated into these differential equations to form a master equation, in an attempt to understand interactions among neighbors?

Simon Levin: Well, I don't think anything I said should be taken to imply that you can't do that, but I want to encourage some caution. Let me give you a simple example: At the beginning, and just in passing, I showed these interacting particle models. Now, there are various ways to model spatial pattern formations. First, you could ignore space entirely and just build a mean-field model. However, if you want to incorporate space, then you can put things into local patches, like a meta-population model, or you can use a reaction-diffusion model, or you could employ an interacting particle model like we had here.

When we set out to understand pattern formation, we began with the interacting particle model. What we found is that it predicted phenomena — I won't delve into the details unless you really want me to. The particular aspects of the patterns that formed couldn't be predicted by the diffusion-reaction models. The main reason for this is that diffusion assumes an infinite speed of propagation, meaning that everything is everywhere almost immediately, albeit not at the same level. Consequently, the sorts of patterns that were forming in our system, which relied on a time lag before species B could displace species A, couldn't be sustained.

So, what we did is similar to what I did with Glenn Flierl and Danny Grunbaum for the fish models: We asked, can we start from the individual-based models and, using formal methods, derive the diffusion limit in the correct way? You can derive the diffusion approximation to it as a stochastic process. That's what we did, and those worked very well. However, they were different from the diffusion approximation that you would get just by adding a diffusion term onto the reaction dynamics, basically changing what had been in the model.

I think the answer is that there should be a goal to develop a master equation or some macroscopic description. A lot of the stuff I showed you was in that spirit, but if you're going to do that, make sure you start from the mechanisms to try to understand what the correct continuum limit to get out of the model is, because what we initially did, what Siegel and I did, what the Turing models did initially, is just take the reaction dynamics between predator and prey and add diffusion at different rates and see what happens. The predictions we got didn't work in that case because we were seeing random diffusion, but we didn't have the correct diffusion limit.

Chris Monterola: Yeah.

Simon Levin: Okay?

Chris Monterola: Yeah. Okay.

Discussant D: Hello, yeah, some aspects of human nature, such as collusion, cheating, and speculation, may lead to certain outcomes. Such behaviors may not be found in animals other than human beings. Could you elaborate more on how we can extend or apply your model to analyze such human behavior?

Simon Levin: First of all, many behaviors that you might think are unique to humans actually apply to other animals as well. For instance, bacteria deal with public goods problems all the time. They produce extracellular polymers that form the matrix for growth, such as the biofilms that support dental health. These polymers also provide signals for quorum sensing, through which one individual can signal others. Understanding why bacteria engage in these behaviors, under what circumstances they should invest in these extracellular polymers, etc., poses a fundamental public goods problem. There's currently a lot of research going on in the field of public goods problems in bacteria. Similarly, many other organisms face public goods problems. For instance, nitrogen fixation by plants is a public goods problem, as is water uptake by plants. All animal groups deal with a variety of public goods problems and evolve cooperation accordingly.

However, this isn't to say that everything is the same as we move up to higher levels of complexity. My approach to modeling is to start from the simplest models and ask how much of the behavior we observe these models can explain. For example, I believe that the herd behavior we see in animal populations can explain a lot of the decision-making behavior we observe in people. There are very few individuals who have the capability or the desire to think through and make decisions themselves, so they attach themselves to others and listen to sources like *Fox News* to determine what they should believe. Therefore, I think it's worthwhile to investigate how much of human behavior is influenced by others. However, we also need to pay attention to the unique problems of human societies and how they make decisions.

I've become particularly interested in social norm formation, which I believe is a crucial factor to some extent. Social norms also form in other animal groups, and individuals can ostracize each other for violating these norms. However, social norms function differently in human groups. For example, there are lots of experiments that drive economists crazy, such as the ultimatum game. In this game, one player offers another player a sum of money, which the second player can accept or decline. If the second player declines, neither player receives anything. Anthropologists have played the ultimatum game in different societies and found that the offers made and accepted vary greatly. Therefore, we need to examine why seemingly irrational behavior arises, how social norms arise and are established, how they're maintained, and how they differ among cultures.

I don't suggest for a minute that by studying fish schools, we can understand human society. However, I do believe that we can learn a lot about human societies and collective behavior. Many of the behaviors observed in human societies are similar to those observed in molecules: Just as with emergent properties and the law of large numbers, lots of individual interactions can give rise to predictable statistical behaviors. The question is this: How much of this behavior can be explained by these mechanisms, and when do we need to introduce new mechanisms?

Gregg Fisher: Thank you. This links quite well with some of the comments you just finished on. In the years before she died, Elinor Ostrom paraphrased, "We lack a coherent overarching framework of collective action in human systems," and I think she's absolutely right. It's critical that we develop that. I believe a major contribution that complexity science can make in social science is to develop some sort of coherent framework for collective action and, metaphorically, have a conversation with political philosophy. Because I deal day-to-day with politicians; I'm up against political ideology, and when I talk about collective action in complex social systems, it doesn't always mix well with certain ideologies.

Is anybody working on this sort of Ostrom-like claim that we need this overarching framework?

Simon Levin: Well, of course, Elinor Ostrom had a whole bunch of disciples who are working, including people like me who had the pleasure of interacting with her. One of the ideas she advocated, an idea that she got from her husband Vincent, was that the steps to international agreements have to begin locally. She talked about polycentric approaches. I don't know if there's any theory that confirms this as the right way to go. When I've talked to Scott Barrett and others, who are leading figures in developing international agreements on climate change and the like, about these approaches, there's a feeling that this approach wouldn't work for Kyoto. In fact, Kyoto failed because it didn't think big enough at the beginning.

So, although it's not yet an area of expertise for me, I find the issue of how to build international consensus fascinating. I'm very influenced by what Lynn had to say about polycentric approaches, local approaches, and building up.

Starting as an evolutionary biologist, where there's been a lot of literature on cooperation, and Murray asked the other day, "Why is it that groups have to fight with each other?" Well, I think, in part, the answer to that is that they have evolved cooperation because they gain nonlinear benefits in the group, largely in competition with other groups. So, this competition with other groups is an intrinsic property. My colleague Anthony Appiah at Princeton talks about the "other," mainly that groups are much more likely to be hostile to other groups that are somewhat similar that they've been in contact with in the past than groups that they've never had contact with because there's a need to distinguish themselves from those other groups.

So, I think we have to find ways to get to the crucial problem you're absolutely right about, and we're not getting through to politicians. I'm certainly interested in talking to anybody who has ideas about that, but you'll not be surprised to know I don't have the answer.

Marcus Karner: Okay, yeah. One of the peculiarities of human systems and collective action is actually that they are, as many have said, usually more complicated and rather simple optimality optimization problems here. Yes, also, well, one of the bigger problems is that people have different rank-ordered preferences, and in your experiments, there's usually one kind of preference or one kind of optimality. So, do you have any experiments or insights done with different rank-ordered preferences?

Simon Levin: Well, I think that's a really fascinating problem. Humans are voting all the time. One of the most interesting aspects, the only one where I've sort of touched on this, is something called hyperbolic discounting. Now, we all know we discount the future at particular rates, and you and I discount at different rates, but we're probably more similar than politicians or CEOs who have much steeper discount rates. Well, what may be surprising to you is that you don't discount the future at a constant rate; you discount the future what's called hyperbolically, and this can be determined experimentally. By this I mean you use a much higher rate for the short term than you do for the long term, and it's easy to do experiments, even with birds, to demonstrate this. Birds have to push a button, they get a reward for pushing the button, but there are two different buttons and they get different rewards, but they're delayed, and so you can see how they trade off delay against the level of the reward, and humans do the same thing. So, this has implications, and it means that every morning — and this is true when I get up — I say, "I know [that] Jan has organized a great reception tonight with all sorts of cakes and things that I shouldn't eat, and when I get there I'm not going to eat them," and then I come there when I say, "Well, I'll start that tomorrow, they really look pretty good," and that's because I'm using a steeper discount rate in the evening than I did in the morning.. So, it leads to what's called intertemporal inconsistency, and therefore, in the morning, I'm saying, "Well, maybe I won't go to the reception, then I won't be tempted." So there's a conflict between the you in the morning and the you in the evening. Now, why might this arise? Well, there are various explanations having to do with uncertainty and things of that sort, but very simply, if you aggregate together different discount rates, you get something that's more or less hyperbolic, not exactly hyperbolic but steeper in the short term when the steepest discount rate comes into effect, and so societies probably discount hyperbolically for that reason, but you probably discount hyperbolically as well because you have different criteria. You're saying on the one hand, "I like the food," on the other hand, "I like to stay healthy," and so you're trading off, and it may be different regions of the brain even.

Marcus Karner: Well, the way I meant it actually was very specifically political, because when you vote for a party, for instance, you don't agree with all their precepts, but you have a certain rank order, and then on average you take this decision, and different individuals have different rank orders for these preferences, and that leads to this…

Simon Levin: I still think it's related to the same thing, because if I agree with you in terms of certain features, [which] I don't, and that model that we looked at is…

Marcus Karner: Yours is a temporal inconsistency, but you also have sort of a synoptic inconsistency in [which] opinions that cannot be aggregated essentially…

Simon Levin: The temporal inconsistency, however, results because I've got different criteria. This is important to me and this is important to me, and I have to vote. So, when I'm dealing with a situation where I agree with you on some of the things you're saying, and this is true, and don't agree with you on others, and [if] I have to balance, I'm going to align with you, I'm voting, and I'm trying to balance these things, so no.

Marcus Karner: I agree with that. But the problem in voting is essentially that, depending on how you design the voting system, you get different results.

Simon Levin: I see what you mean now. So, essentially, when I'm voting, I've got different reasons or different components of myself that are better balancing each other. And indeed, how I aggregate that information is going to affect the outcome. This is not an area of my expertise, but I certainly find it interesting. In that model, I briefly touched on, close to the end, the network model of human decision-making. In that model, every or many of the individuals had intrinsic preferences, and they had to decide: Do I stay a Democrat, do I stay a Republican, or do I bolt and go with what my intrinsic preferences are? So, they're trading it off in that model. I agree that it's a very interesting problem.

Marcus Karner: Okay, thank you.

9.3 Summary of the Talk

Simon Levin's insightful presentation delves into the intricate dynamics of collective behavior in biological systems, offering profound insights into the underlying mechanisms through mathematical modeling. His talk sheds light on different aspects of collective behavior.

Understanding Collective Behavior: Levin delves into the fundamental principles underlying collective behavior in animal groups. He elucidates how local interactions and individual-based rules give rise to emergent group behaviors, exemplified by phenomena like bird flocking and fish schooling. Levin emphasizes the importance of considering both individual behavior and group dynamics to comprehensively understand collective behavior.

One key aspect discussed is the translation of local rules of interaction into a statistical description that accounts for population density. Levin highlights the challenge of closing the system by translating density into probabilities of encountering a certain number of individuals in one's vicinity. Despite these challenges, he demonstrates that mathematical models, when appropriately formulated, can provide a continuum description of the system's behavior.

Furthermore, Levin presents findings from research that incorporates aggregation schemes within the Navier–Stokes-type flow, revealing interesting conclusions. These studies suggest that the efficacy of approximation techniques hinges on the quality of closures employed. Additionally, Levin discusses the impact of turbulence on aggregation dynamics, drawing parallels with concepts like stochastic resonance observed in other systems.

Moving beyond theoretical models, Levin underscores the heterogeneous nature of real aggregations, exemplified by diverse assemblages of individuals like starlings and flying foxes. He highlights the absence of an obvious leader in such groups, where individuals navigate in response to each other's movements.

Collaborating with his postdoc Ian Couzin, Levin delves into the study of fish populations to explore patterns of collective motion and the role of leadership. Using stochastic models, they demonstrate how a small number of leaders can significantly influence group movement, leading to more efficient navigation toward targets.

Overall, in the first part of his talk, he provides a comprehensive overview of the fundamental principles underlying collective behavior in biological systems, highlighting the role of mathematical modeling in elucidating complex phenomena.

Leadership and Decision-Making in Collective Behavior: In the second part of his talk, Levin delves deeper into the role of leadership and decision-making processes in collective behavior. Drawing on research conducted in collaboration with Naomi Leonard, Levin explores models of coupled oscillators to understand how individuals align their movements with those of others in the group.

Levin illustrates how individuals update their direction based on both intrinsic information and social cues from neighboring individuals. By varying parameters such as the weighting of social versus informational cues, Levin demonstrates how different levels of leadership can emerge within a group.

Using simulations, Levin showcases how a small number of informed individuals can effectively lead a larger group toward a target. He emphasizes the importance of uninformed individuals in consensus-building processes, highlighting their crucial role in achieving collective goals.

Moreover, Levin explores the dynamics of decision-making in scenarios where individuals have competing preferences. Through simulations, he demonstrates how groups navigate toward consensus, balancing conflicting interests and adapting their behavior based on contextual cues.

Incorporating insights from game theory, Levin discusses the trade-off between tracking resources and tracking individuals in collective decision-making processes. He underscores the need for cooperative strategies to address common challenges, drawing parallels with real-world scenarios such as environmental resource management.

Furthermore, Levin explores the role of norms and cooperation in resolving conflicts within groups. By analyzing mathematical models, he elucidates how cooperation can emerge through mechanisms such as mutual coercion and norm enforcement.

Overall, in the second part of his talk, he offers a nuanced exploration of leadership, decision-making, and cooperation in collective behavior, highlighting the interdisciplinary approach required to unravel the complexities of biological systems.

Insights and Implications: Simon Levin's talk provides valuable insights into the mechanisms driving collective behavior in biological systems and underscores the importance of interdisciplinary approaches to understanding complex phenomena. By combining mathematical modeling with empirical observations, Levin offers a comprehensive framework for studying collective behavior, with implications ranging from ecological conservation to social dynamics and beyond. Through his work, Levin continues to pave the way for innovative research at the intersection of biology, mathematics, and computational science.

9.4 Relevance of the Talk to the Current Stage of Research

Professor Simon Levin's talk holds significant relevance to the current stage of research in ecology and complex systems science. The following are several points highlighting its relevance:

Interdisciplinary Approach: Levin emphasizes the importance of interdisciplinary collaboration, which remains a cornerstone of modern scientific research. Today, as scientific problems become increasingly complex, collaboration across disciplines such as mathematics, physics, biology, and computer science is essential to developing comprehensive solutions.

Understanding Ecosystem Dynamics: The dynamic nature of ecosystems and the challenges of predicting their behavior are central topics in contemporary ecological research. Levin's insights into emergent properties and collective phenomena provide valuable perspectives for researchers seeking to understand how ecosystems respond to environmental changes.

Sustainability and Conservation: As concerns about environmental sustainability and biodiversity conservation grow, there is a pressing need for effective management strategies. Levin's discussion on the need to relate macroscopic properties to microscopic phenomena offers valuable insights for developing sustainable management practices that consider the intricate interactions within ecosystems.

Advances in Statistical Mechanics: Levin's proposal to develop statistical mechanics frameworks for ecological communities reflects ongoing efforts to apply advanced mathematical and computational techniques to ecological research. These approaches have the potential to revolutionize our understanding of complex systems and inform evidence-based decision-making in conservation and resource management.

Complex Systems Theory: The application of complexity theory to ecology is a burgeoning field of research. Levin's talk highlights the parallels between ecological systems and physical systems, underscoring the importance of studying emergent properties and critical transitions in both domains. This connection fosters cross-fertilization of ideas and methodologies between ecology and other disciplines, leading to innovative approaches to understanding and managing complex systems.

Emergent Patterns and Collective Motion: The study of emergent patterns and collective motion is of particular interest in fields such as biology, physics, and engineering. Levin's exploration of these phenomena, including their underlying mechanisms and modeling approaches, contributes to ongoing research aimed at unraveling the complexities of collective behavior in biological systems.

Overall, Professor Simon Levin's talk addresses fundamental questions in ecology and complex systems science, providing valuable insights and methodologies that are highly relevant to the current stage of research. As scientists continue to grapple with the challenges of understanding and managing complex systems, interdisciplinary approaches and theoretical frameworks inspired by Levin's work will play a crucial role in advancing scientific knowledge and informing policy and practice in environmental conservation and sustainability.

Chapter 10

Resilience for Human Development in the Anthropocene

Johan Rockstrom

YouTube: https://www.youtube.com/watch?v=6qy-j0t8zps

Speaker: Johan Rockström

Moderator: William Clune

Discussants: Sheila Rhonda, Ann Florini, Greg Fisher, John Richardson, Suman Bannerjee, Ryan Chisholm, Simon Levin, Jamie McCaughey

10.1 Talk by Johan Rockström

Thank you, William, and good morning, everyone. What I will do is, in a way, quite simple: I'm trying to add some additional scientific flesh to why Simon's talk is so absolutely critical at this juncture in time. The understanding of how ecosystems connect to the biosphere and how non-linearities and complexities interplay with human development has become absolutely central at this moment in time for human development. We're transitioning from a connected to a hyper-connected state, and from a hyper-connected to an interdependent socially, politically, economically, and ecologically globalized world. I think there is a good argument to say that this may be the most important scientific message to humanity in recent years — namely, the empirical evidence over the past 10–15 years, synthesized in major works from earth system sciences, suggests that we are entering a new geological epoch where humanity constitutes a quasi-geological force, alongside natural geological changes, significantly influencing the Earth's system at a global scale. This notion of the Anthropocene was first suggested by Nobel Prize laureate Paul Crutzen in 2002. It was synthesized in 2004 and has gained prominence since 2007 and 2010. Of course, you're aware of where the decisions regarding geological eras are made, notably the Royal Society in the United Kingdom. The Royal Society has convened a whole commission to decide whether we should change our current epoch from the Holocene to the Anthropocene. So, these are major issues when it comes to changing the face of sustainable development.

The necessity of moving from what could be argued, perhaps provocatively, as an obsolete definition of sustainable development — the kind of Brundtland view of the pillars of sustainability, encompassing social, economic, and environmental sustainability, which has been largely successful in some sectors — needs to shift from merely minimizing environmental impact within a growth model. This model essentially adheres to a linear, incremental, predictable, and optimization-oriented growth paradigm. Instead, we must recognize a very dramatic but simple conclusion: Global sustainability increasingly appears to be, based on evidence, a prerequisite for local human well-being as well. This implies that growth for Singapore or growth for a country like Sweden is interconnected and dependent on sustainability across the entire biosphere. When I lecture my students, I increasingly emphasize that the future they need to navigate is what I call the 3–6–9 world.

It's a world that, according to mainstream climate science, indicates that we are more or less committed to a three-degree warming scenario, have entered the sixth mass extinction of species on planet Earth, and are on track to reach a population of 9 billion people. This 3–6–9 world is essentially what lies ahead of us, within a biosphere shaped by humanity — that is, the Anthropocene. You may have seen the very important synthesis published in *Science* just over a year and a half ago, which presented the evidence behind the drama of the Anthropocene. The synthesis revealed numerous hockey stick patterns, and I can assure you that you can choose essentially any parameter relevant to the economy, and it will exhibit the same exponential rise. While we often focus on population and carbon dioxide, other indicators display similar trends. For instance, in the upper left-hand corner, we see that we have depleted approximately 70% of the fish stocks in the ocean, and there's also the issue of ocean acidification, exacerbated by greenhouse gas emissions, which I'll address further. So, there's a very solid empirical basis for understanding these trends.

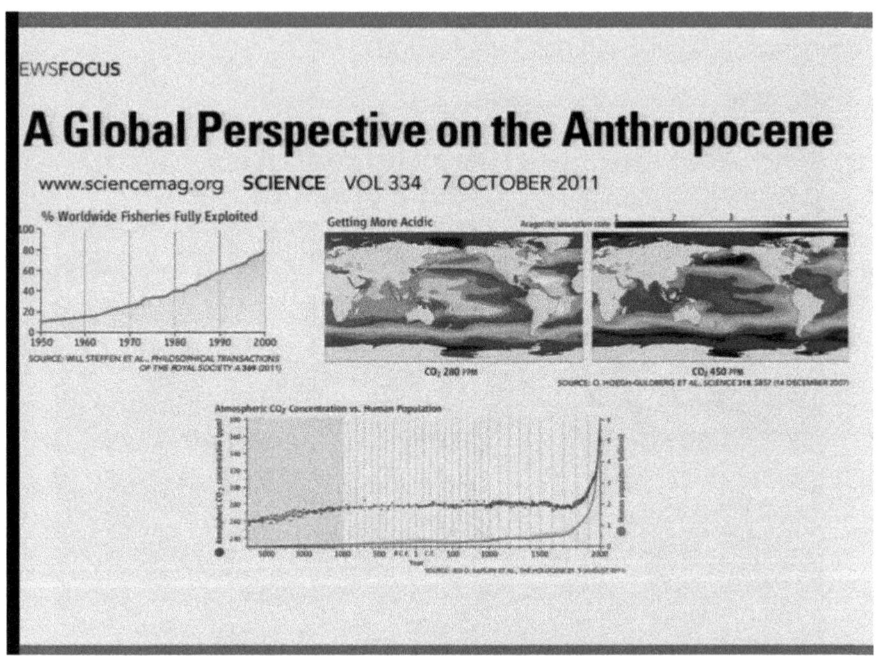

In fact, the magnitude of this challenge is such that we can declare ourselves the first generation to understand that we are beginning to encounter the limits of the hardwired biophysical processes at the Earth system scale. This realization grants us an enormous privilege because it means that we are also the first generation with no excuse for inaction — an advantageous position to be in, particularly considering the 40–50 years of uncertainty we have faced for so long.

The reason why we find ourselves in this predicament, as well understood in this room, is what I refer to as a Quadruple Squeeze. Sometimes, however, we overlook the crude reality of this situation — namely, that while human pressure is indeed a predominant factor, it's not solely the sheer number of people that is the primary issue. It's the 20/80 dilemma: The environmental negative pressures on the planet have largely been caused by the affluent minority of the population. This group, numbering around one billion, embarked on the Industrial Revolution some 150 years ago and has contributed significantly to the current dilemmas. Conversely, the 80% of the population that has had relatively little impact so far is now experiencing positive momentum. We are facing a future where the middle class is projected to grow from 1.5 to 4, 5, or even 6 billion individuals, with purchasing power and aspirations for lifestyles largely equivalent to the unsustainable ones that have caused the problem thus far. This represents a significant social shift, which we are currently witnessing.

Additionally, we have the pressing issue of climate change, which is a crisis in its own right. It's increasingly clear that we need to stabilize greenhouse gas concentrations at 350 ppm for CO_2. However, we currently stand at 390 ppm and are rapidly approaching 400 ppm and beyond. The sheer market value of energy utility companies worldwide, which hold documented reserves of more than five times the amount of carbon that would take us to a safe level, serves as evidence that these companies are not prioritizing climate policy. They continue business as usual, assuming that we will not succeed in curbing emissions.

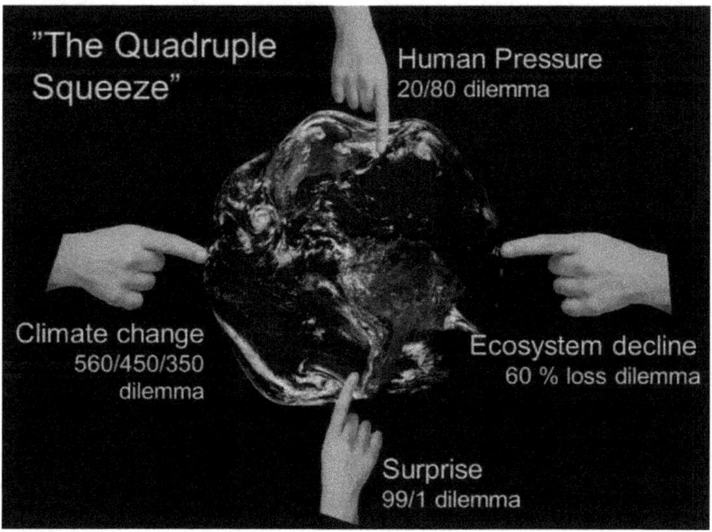

One would have hoped that this perturbation of the entire climate system would have occurred on a resilient planet — a planet capable of buffering such disturbances to the energy system. However, according to the Millennium Ecosystem Assessment, never before, especially in the past 50 years, have we undermined ecosystem services as rapidly as during this period. So, we find ourselves in a dilemma: We are eroding the very foundation of resilience at a time when we need a robust Earth system more than ever.

As if this weren't enough, the scope for humanity seems to be shrinking, with a growing recognition that nature is not akin to a Walmart store, where one can simply walk in and take natural resources off the shelf, replenished seamlessly when depleted. Indeed, long periods of gradual and incremental change can suddenly shift through triggers and feedback mechanisms into state shifts, as eloquently highlighted by Simon. Thus, this is the new reality for human development in the Anthropocene — an entirely new agenda.

It's so new, in fact, that even UN Secretary-General Ban Ki-moon highlighted, for the first time in his Global Sustainability Panel, the equivalent of the Brundtland Commission report for Rio+20, that the current development paradigm is no longer effective. We need to recognize the risk of catastrophic tipping points, as stated by a panel of heads of state informed by science. While it may have little influence in Rio, it represents a significant shift in what we are confronting.

Throughout this conference, there have been numerous examples, including my own, of nonlinear dynamics and state shifts between stable equilibria at the ecosystem scale. This phenomenon appears to be increasing in scale, as evidenced by crucial data from NASA, particularly satellite observations over Greenland. In 2012, which may be remembered as a pivotal moment in the future, feedback directions shifted from predominantly damping negative feedbacks — essentially Earth helping humanity mitigate our impacts of environmental change — to reinforcing and accelerating change.

This is illustrated by albedo measurements over Greenland. The colored lines represent the desired scenario, where roughly 90% of incoming heat from the Sun is reflected back into space — a critical cooling system for Earth and a key feature of the Holocene equilibrium. However, the black line shows how, in July 2012, the albedo plummeted, with less than 50% of heat being reflected back into space.

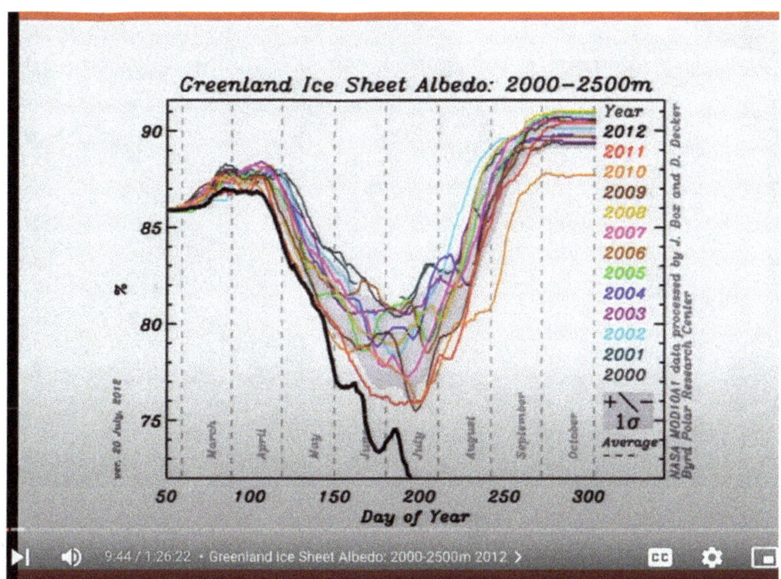

This sudden shift in feedback is due to the entire Greenland continent melting over a period of just a couple of weeks, as demonstrated by Jason Box and colleagues at the Byrd Institute. This change in color, from white to a darker surface, results in the absorption of more than 300 exajoules of energy — equivalent to the annual energy consumption of the United States. Consequently, Denmark surpasses both China and the U.S. to become the largest climate forester on the planet, not because of greenhouse gas emissions — Danish citizens often commute by bicycle — but due to the Earth's system shifting into high gear, transitioning from negative to positive feedback. This response by the Earth's system far surpasses the triggers initiated by humanity in the Anthropocene.

Science has been effectively synthesizing this message to humanity. You may have come across the significant "State of the Planet" declaration, which represents a synthesis of findings from the largest-ever gathering of global environmental change scientists. This declaration, emerging from the Planet Under Pressure conference in the lead-up to Rio, unequivocally states that humanity may be approaching a saturation point.

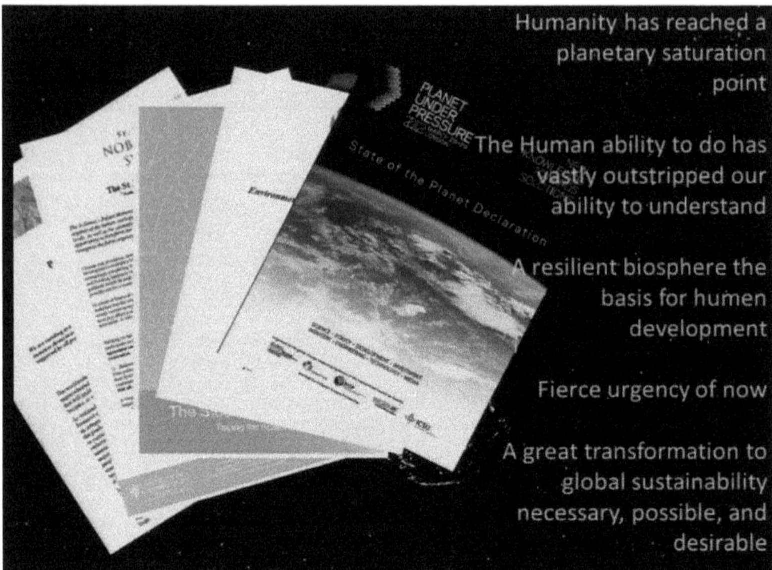

Now, this is incredibly difficult statement to make, but built on empirical evidence that we're starting to see — so much evidence from the oceans, from the land, from the atmosphere, from the stratosphere — the risk of these kinds of nonlinear dynamics. It's not only about the climate, and that we need to start understanding resilience. We need to start to understand the need to look at risks of nonlinear dynamics and state shifts, feedbacks, and how to maintain persistency or robustness in the system.

Interestingly, the conclusion is also that the window seems to still be open for a transition toward a less risky or even safe future. So, it's not as if we're only in a doom and gloom situation. As Brian Arthur and others show through innovation systems and wise economic policy, we can actually trigger a transition, particularly toward a safe energy future which has perfect substitution for the fossil fuel dependency we have today.

Now, this is changing the face also of how the world outside science is starting to understand the complexities facing humanity. This is, as we all know, the way business normally portrays turbulence and complexity: the vortex of Lehman Brothers sucked into the financial crisis in 2008. But even economists welcomed humanity to the Anthropocene in 2011.

So, this is an interesting shift in starting to recognize this interdependent globalized world, and we're starting to see evidence of how these two interact. You may have recently seen Thomas Friedman writing in *The New York Times* referring to the growing empirical science being published of the Arab Spring, of course, being driven largely by a young, well-educated, frustrated, under oppression and socially connected youth, but increasing evidence shows also that[concurrently], you had a 160% rise in phosphorus prices. You had an almost 150% rise in wheat prices because Putin shut the borders in Russia due to forest fires clearly linked to anthropogenic climate change, and Australia after 12 years of drought also reduced its exports of cereals which may be one of the first examples of a regional-scale social-ecological dynamics playing out in big social instability in the world of the Anthropocene. So, this is the kind of turbulence we're starting just to see at the very, let's say, putative level, but it's still growing with increasing evidence. You may have seen this *Economist* issue, and they have a beautiful, in my mind, typical British understatement of how science feels about this situation which goes as follows: "When reality is changing faster than theory suggests it should, a certain amount of nervousness is a reasonable response," and I think that is exactly where we are today.

In fact, the world is changing faster than theories stipulate it should, and precaution must play a role in a situation of this kind of uncertainty. Of course, the pressures of exponential human growth have tremendous social implications. I'll give just one example because I think it's pedagogical. This comes from Olli Varis and colleagues at the University of Helsinki, with whom we've been working for many years. They show here a very simple exponential pressure graph over the last 100 years of water scarcity. On the *y*-axis, you have the number of people suffering from various degrees of water scarcity. The astonishing fact is that 3.5 billion people still suffer today from various degrees of water scarcity, almost half of the world's population. Now, that is a drama in itself, but look at the dotted lines. Those represent the success story of our engineering efforts to try to solve this problem. This is the exponential growth of storing 5,000 cubic kilometers of fresh water in our rivers, leading to 25% of our rivers not reaching the ocean anymore. It's groundwater irrigation, diversion; it's the modern water systems trying to supply water to a growing population.

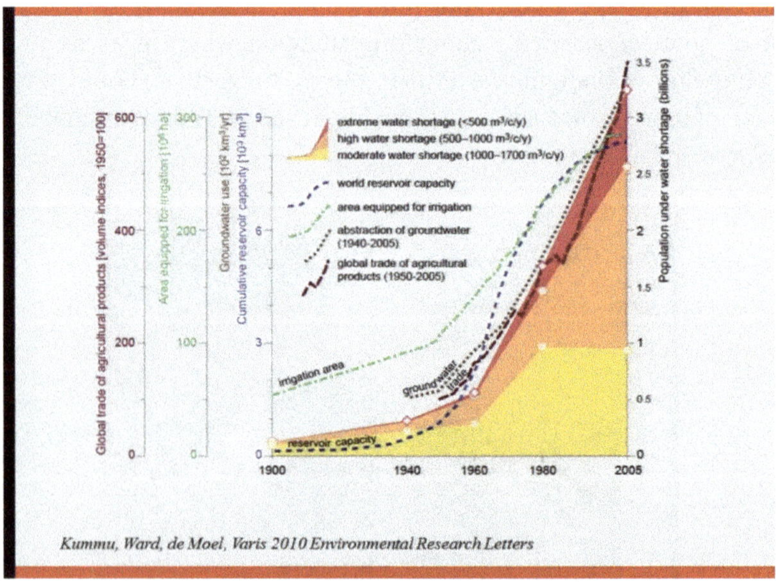

Kummu, Ward, de Moel, Varis 2010 *Environmental Research Letters*

So, we are essentially chasing our own tail. Despite the success story in the Anthropocene, where very high credible evidence indicates that we will have more variability when it comes to water supply, not less, and where there will be more shocks, with water often being the first victim of climate change. These are the kinds of interactions we now need to incorporate into our governance, economics, and development.

This is just a summary of the parameters; you can't read the indicators here, but this basically shows the patterns. If you choose anything from overfishing to deforestation, they all have this general pattern: Up until 1955, humanity had, in fact, at the global scale, very little impact.

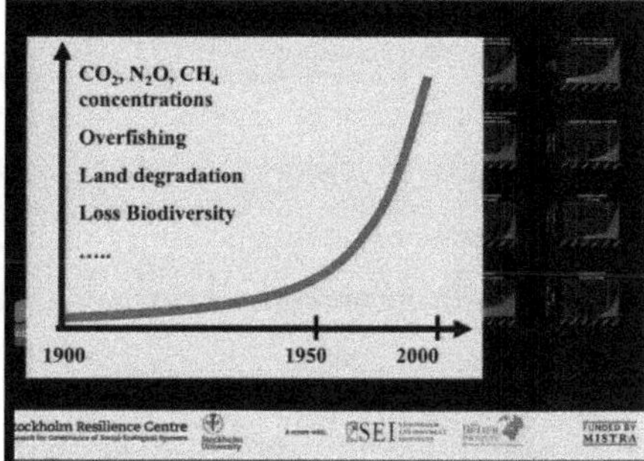

We had localized tipping points and collapses of whole societies, from the Mesopotamian irrigation societies to the Maya irrigation society. But up until around 1955, we had little aggregate impact. That's when we shifted into high gear, marking what is now defined as the Great Acceleration of the human enterprise. Three billion people, just ten years after the Second World War, and off we went into the exponential rise of pressures. The warnings came early, as you know — Rachel Carson's *Silent Spring* warned humanity very early of the risks of this modern trajectory, primarily attributed in her case to chemical pollution of fresh water. Then came the *Limits to Growth*, the Club of Rome report, which was dismissed by conventional economists and policymakers as a kind of Neo-Malthusian future. But if you look at this curve, you could almost say, well, perhaps humanity could be excused. There was so little empirical evidence at that time to say where we were heading, where the world was heading at that point. But today, we're up here.

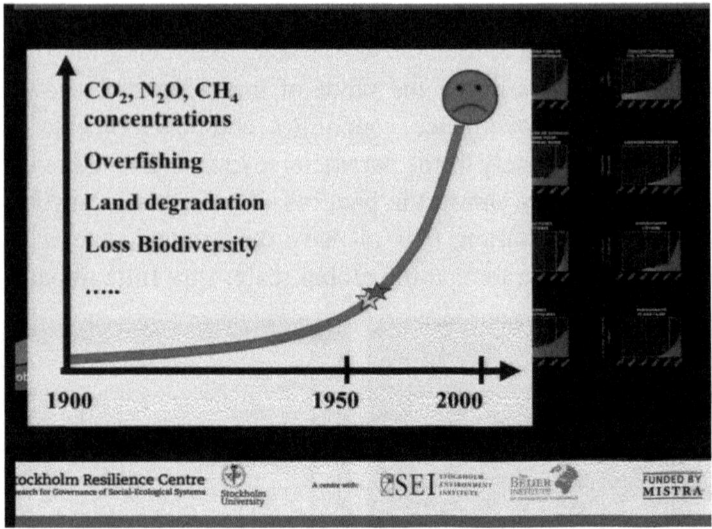

Today, we are standing on a massive amount of empirical evidence at a completely different scale, showing that we are at a very, very top level of exponential pressure. If you delve into the disciplines of oceanography, climatology, ecology, hydrology, there's so much evidence that we need to bend these curves very rapidly in order to avoid nonlinear changes and impacts on the human economy.

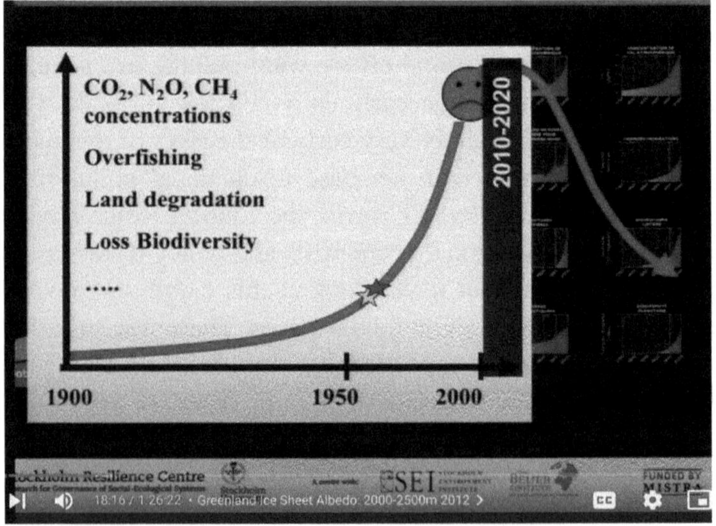

Now, where are we heading today? Well, as you all know, this is just one example of climate. In fact, it seems we are accelerating further in the wrong direction. And this is just one of the observations coming from a recent paper by Glenn Peters and colleagues, showing that... You may all recall when the AR4 came out of the IPCC and they had their scenarios from the B scenarios to the A1 scenarios, and IPCC was very harshly, and I think on good grounds, criticized. You know, why are you presenting these kinds of crazy A1 scenarios? They're totally unrealistic; you're assuming with 11 billion people on Earth, we won't do anything about reducing fossil fuel emissions? It's taking us to 4.2-to-5-degree warmer, it cannot happen. Why are you scaring humanity with these kinds of scenarios? But look at the black dots. In fact, we're following the A1 trajectory. We're following the worst IPCC, the kind of unrealistic scenario which is taking us to a 50% risk of a 4.2-to-5-degree warming this century.

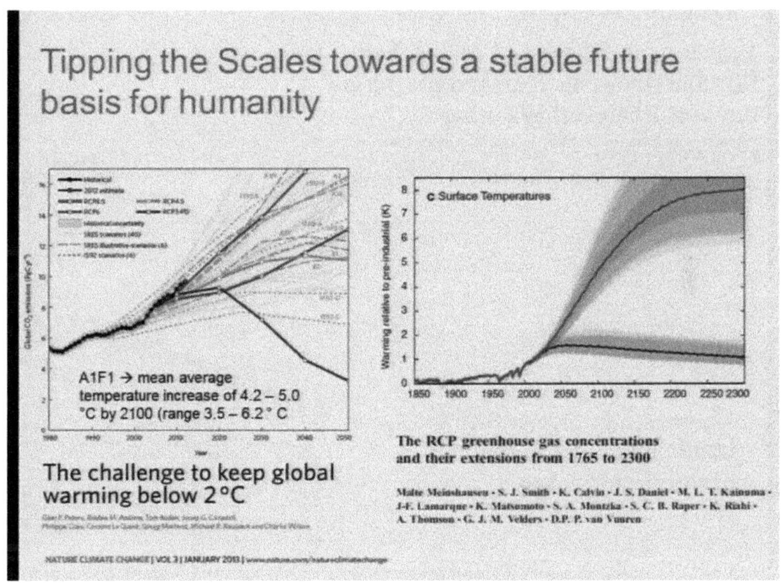

That's why the World Bank recently released a risk report on 4 degrees. And we have more and more research showing that we're approaching a bifurcation point where feedback mechanisms can turn, as in Greenland, from negative to positive, reinforcing a trajectory that could take us way beyond the 4-degree warming.

Now, you may have seen this paper, which is in the latest issue of *Science*, which also, I think, adds a lot of humbleness in terms of how the Earth's system responds to our actions. We've always thought, based on ice core data, that there's a lag time of around 800 years between a rise in temperature and feedback response. So, we know very well that in the past, when we were not in the Anthropocene, it was our position to the Sun and our position in relation to other planets that triggered temperature rise, but it was only through biosphere feedbacks that could take us out of a glacial period. Now, this work by Parrenin and others shows that probably that lag time is in fact 0 to 200 years. But the synchronicity between carbon dioxide and temperature is much tighter than we previously thought, meaning that the biosphere feedbacks may come much earlier, as we're starting to see, for example, in Greenland.

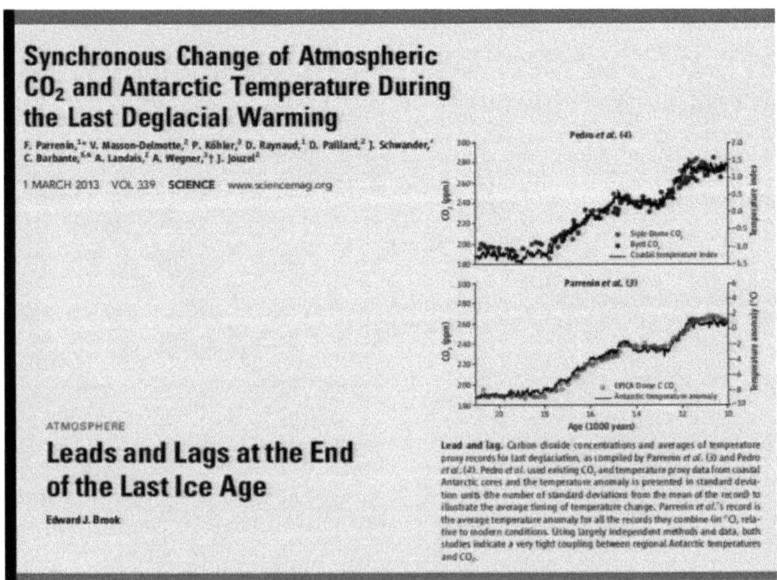

We've been heavily studying recently the canary in the Earth system's gold mine — the Arctic. The Arctic serves as an early red-flag warning to humanity; there's no place on Earth where we're seeing such accelerated change. In 2007, 30% of the sea ice cover was gone in two months. That's one drama, but the other drama was that no science could predict it. It was a total surprise. All models predicted this to be a worst-case scenario, something that could happen in 2060, 2070. Tim Lenton and colleagues have been showing the kind of tipping points we can see in different subsystems in the Arctic. But then comes this report just recently — the six years of fantastic work on ice core dates on Greenland, where Dorthe Jensen and colleagues at the Niels Bohr Institute have been looking at the Eemian, the last interglacial period 200,000 years ago, which is the equivalent of the Holocene 200,000 years back, trying to answer this question: Did Greenland melt? Now, the interesting thing about the Eemian is that all assessments point out that we had a 4,000-year period or so with temperatures 6 to 8 degrees warmer than today. So, it's a perfect example of a really worst-case scenario. What happened then with Greenland, which holds, after all, 7 meters of sea level rise? The drama in this paper is that their conclusion is that Greenland seems to be more resilient than we thought. We've always thought of the Arctic as being more vulnerable of the two poles, but it seems that Greenland only lost in the order of 120 meters of ice mass, which could only explain roughly 2 meters of sea level rise. So, that's the good news; Greenland may, in fact, be able to accelerate but then have some kind of mechanism to stabilize. But the drama is that, well, we know from very strong empirical evidence in the Eemian that the sea levels were in the order of 10 meters higher than today. So, where are the 8 meters? Where did they come from? And if there's only one source, it's Antarctica. So, the paradox here is that it may be that the Arctic is more resilient, but Antarctica is more vulnerable than we previously thought.

So, these are the surprising elements that are, I think, increasingly emerging in the era of the Anthropocene. So, to draw a little line here, and this is just a scientific challenge to us, we're always talking, and we tend to talk, of this period as the massive pressure of humanity on the planet, and now is the time to transition. But what if we've only reached the Double Aperitif? What if we haven't even started the pressures? Because on the social side, it's clear that we haven't put in the social high gear yet. It is a minority that has pushed us to the exponential hockey stick so far. A minority. And now we're becoming a majority. What if the Earth system hasn't even started its feedback mechanisms at the big Earth system scale?

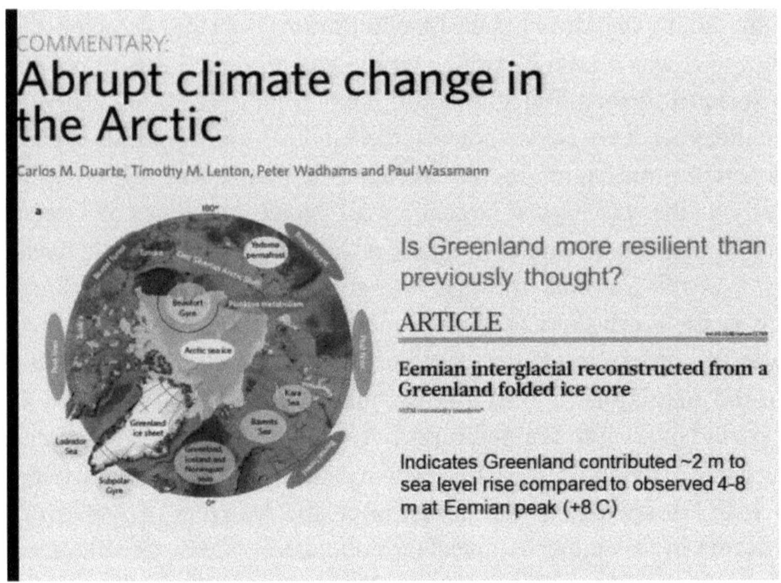

What if there's actually predominantly a dampening effect so far? We know that 50% of our greenhouse gas emissions are actually absorbed by the biosphere, 25% in oceans, and 25% enters into our ecosystems, which are the world's largest free ecosystem services to the global economy. So, what if these two social-ecological accelerators have not even reached the main course? Then we're really in for a very important journey, and this requires new science, and I'll come back to what response science is making there.

That's my little first line, just setting the stage. I know this is not a very attractive way of doing things. I gave this talk once to some business leaders, and Paul Bulcke, who's the head of Nestle, said, "You know, you don't have to be nervous about giving this kind of entry point because my definition of a pessimist is a well-informed optimist." So, bear with me, I'll come to some of the possible solutions in the end. But this really argues scientifically that sustainability cannot only be about resource management. It must also be about understanding the dimensions that Simon has been putting out, and the entire conference here on complexity, namely, resilience thinking, understanding that there are no ecological and social systems... Even Singapore is a social-ecological system and it really needs to be incorporated into the agenda of global sustainability.

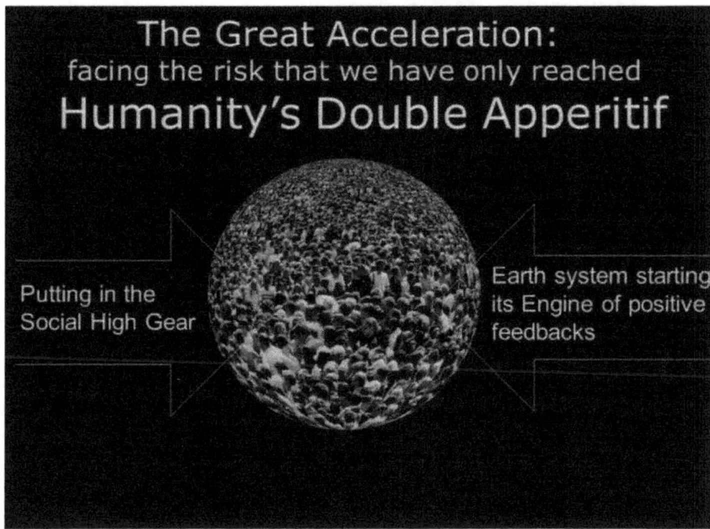

And we've tended to pursue science in this way. Basically, islands of great insights and oceans of ignorance, and we've done great in deepening our understanding in different fields of science.

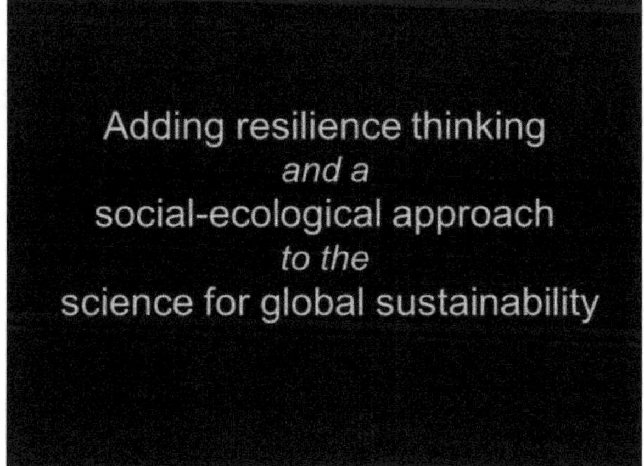

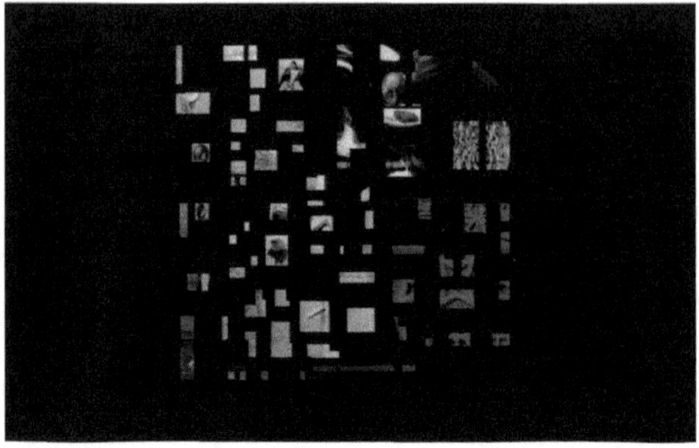

We're starting to understand the crude feeling of the whole but often it comes out as quite blurred pictures like this.

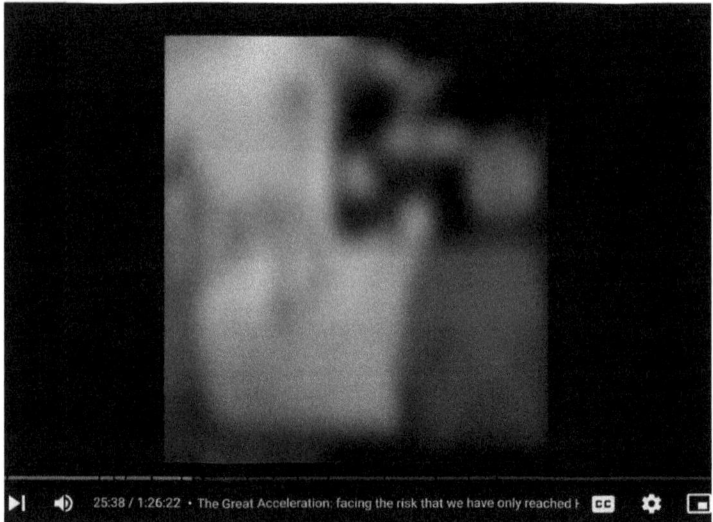

We think we understand the system, but when it clears, we are quite surprised. I think this is one of the great advantages of this kind of gathering: to start recognizing that what actually comes out on the other end may be a surprise and that we need to prepare ourselves for surprises.

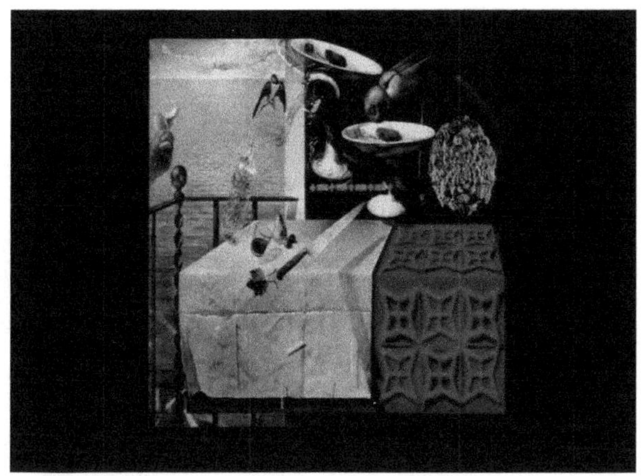

And that, in my mind, is all about resilience thinking, and luckily Simon has covered this much better than I could ever do. But just to summarize, I think what we're moving from is the recognition that we've always thought of systems behaving as in the upper left-hand corner (picture with graph below), that when you put pressure and change the conditions in an ecosystem, it changes incrementally and linearly, to recognizing thresholds and even hysteresis effects, and the fact that we have multiple stable states separated by thresholds. And when, as we illustrate often in this cup and bowl diagram, the depth of your cup is reduced, the system becomes more vulnerable to change. This is the slow return to an equilibrium, as Simon pointed out, and a trigger can push the system across the threshold. That's why we can have such long periods of slow change because the system can actually continue to function even though we are eroding resilience, but then a certain trigger can push it very abruptly over to a new state.

We're looking at this very much from an ecosystem perspective, and this has also been covered very well, recognizing that both functional diversity and response diversity in ecosystems are absolutely critical to building up the functional ecosystem processes — everything from pollination to the quality of water flows, which in turn provides ecosystems and landscapes and seascapes with their ability to self-organize and deal with disturbance, which in turn provides ecosystem services for human

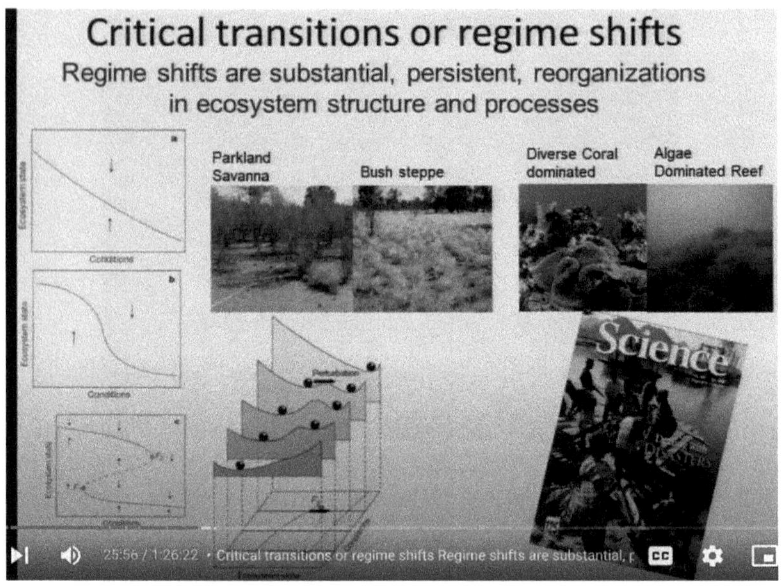

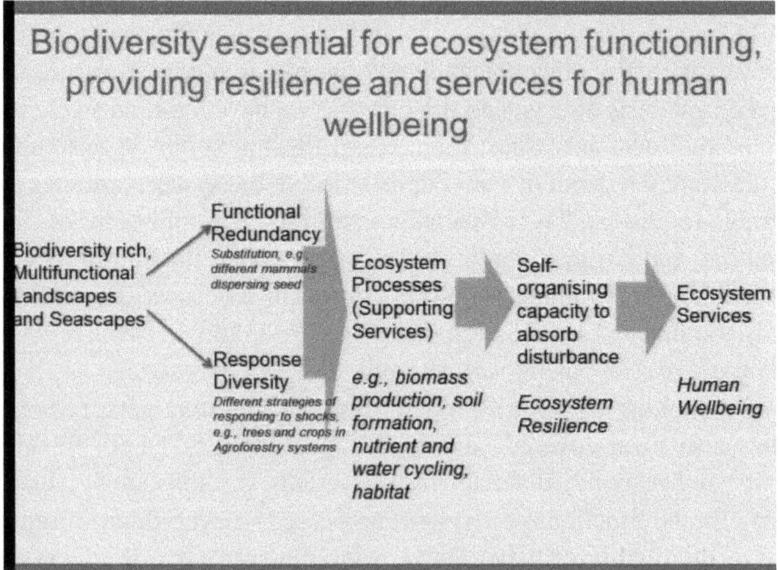

well-being. So, kind of an empirical evidence-based link from the organism, the microorganism, all the way to the economy through ecosystem services and the strength of landscapes and seascapes to deal with change. And I think this is increasingly well anchored empirically.

What we're trying to do is also link this to what are the sources of resilience in complex social-ecological systems, from institutions to sectors like agriculture. There is a paper coming out from something called the Raise the Resilience Alliance young scholars, led by one of our colleagues at the Resilience Center, Oonsie Biggs, and Simon's former Ph.D. student, Maja Schluter, whom we've poached from Princeton quite recently. I would really recommend this paper because it summarizes 30 years of resilience science to see what the attributes of systems are that seem to be building resilience. You have issues like polycentricity from Elinor Ostrom's work, understanding systems as complex adaptive systems, but also a red thread where diversity and redundancy critically play an important role in allowing systems to deal with disturbance. Now, this is at loggerheads with our optimization and efficiency thinking in our conventional economic model because actually investing in redundancy can be quite expensive. It doesn't really fit in a normal cost–benefit analysis. It doesn't fit in a monoculture of any sort, so this is quite an interesting change in paradigm also for economics.

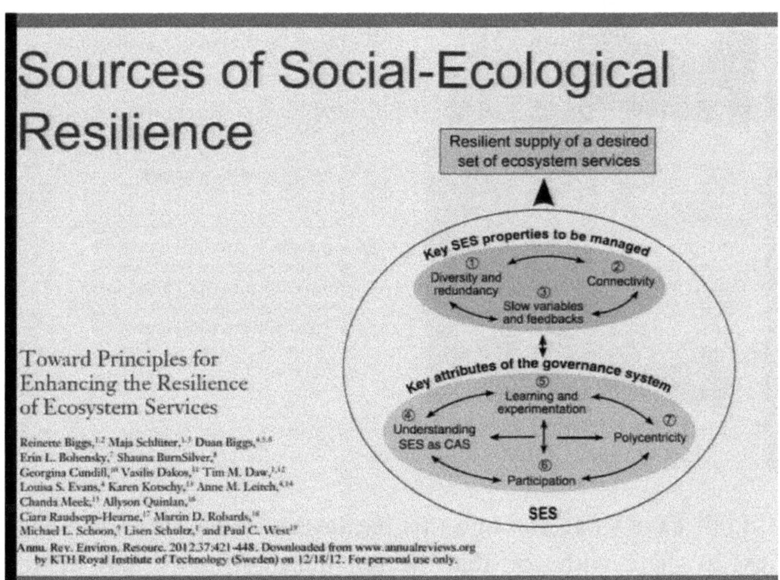

We define resilience in this way, so I think it's fair to say that Simon's talk focused very much on the first core element of resilience: robustness or persistency, the ability of a system to take a shock while remaining in a desired or stable state. It doesn't necessarily have to be desired, by the way; the Mugabe regime in Zimbabwe is very robust and not very desirable. But we also include two other features of resilience, namely, the ability of an ecosystem or society to adapt [and transform]. Basically, you're in a basin of attraction; Singapore, you could argue, is in a desired development basin of attraction. How do you adapt to changing conditions while remaining in that state? And I think an emerging area that we've incorporated within the resilience definition is once you've fallen over a threshold, once you're in a crisis, once you've changed state, what's your ability to move into a completely new development trajectory? What are the social networks, the capitals, the knowledge you need to innovate yourself into a new trajectory, from a fishing system to a business model to an energy utility? So, that's how we define resilience.

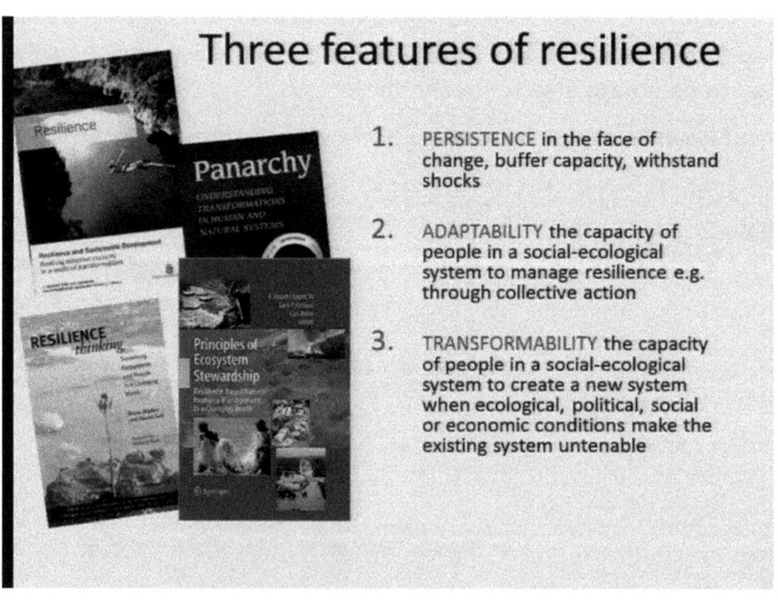

I don't have to go through this because it's been covered very well, but just to share with you that the Resilience Alliance and we've now taken over at the Resilience Center (to be the global repository of) something called the Regime Shifts Database, basically cataloging empirical evidence that we have situations of coral-dominated systems in one basin of attraction, long periods of management which can be more or less sustainable but gradually it's in fact not meeting sustainability criteria, eroding resilience; a trigger like an [inaudible mumbling] event or hurricane pushes the system over threshold and always looking for the feedback mechanism that locks the system in a new state, in this case, the invasion of soft coral systems. And these systems then flip over to a state that cannot, any longer, deliver well-being for the 300 million people or so living along coastal regions depending on tropical coral reefs — from clear water to turbid lakes, from savannas to desertification of steppes. And there are many examples of this, summarized in this Regime Shifts Database, which we think is critical both as a stimulus for science but also as an empirical base for policymaking, to give a strong justification and feeling of certainty that nonlinear changes actually occur.

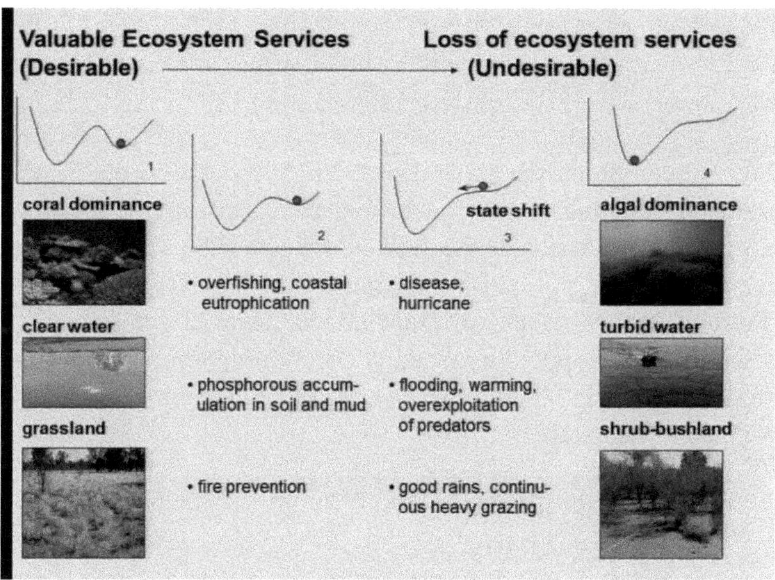

I'll just go through a couple of examples. You, of course, recognize this one, which is the classical collapse of cod fisheries outside of Newfoundland with tremendous growth in exploitation, the success story of efficiency and growing fish fleets, the total collapse, and the surprise of a system that actually changed stability domain and has not recovered since.

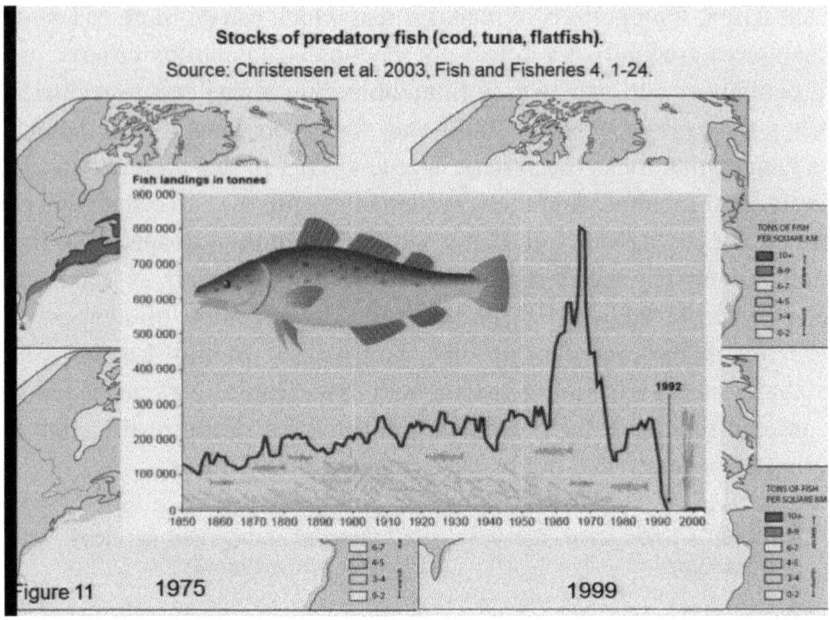

We've been doing the equivalent type of research in the Baltic Sea, unfortunately coming to the conclusion, from very important empirical work in every sub-basin in the Baltics, that in 1989 the environmental change in the Baltic shifted from a largely oxic, high-oxygen, rich in cod, low-nutrient state to an anoxic, cod-free, and algal bloom frequent state, triggered by a likely poor oxygenation event during one year. So, we had a period in the late 1980s that locked oxygen from coming into the Baltic, and we had a state shift.

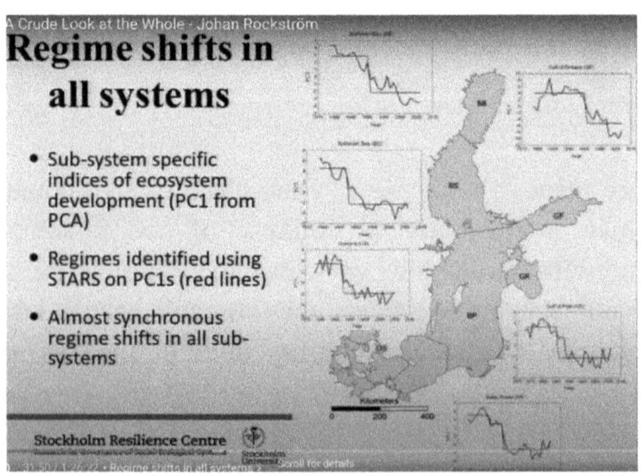

And today, there's a success story here because eutrophication of the Baltic is reducing very fast due to the policies being put in place to really try to save the Baltic. But it's not helping because the system has changed state. Interestingly, the reason for this is not only because of eutrophication from Saint Petersburg and other big cities not having wastewater treatment plants and agriculture but also because the cod has been so overfished that the top predator has disappeared. It means that you get an explosion of sprat, which in turn eats up zooplankton, the grazers, which means that phytoplankton explode. The phytoplankton get even happier when they get more phosphorus and nitrogen, and the whole system goes into a totally eutrophic state. When phytoplankton die, they consume all oxygen, and the whole system, 60% of the Baltic today, is in fact a dead sea.

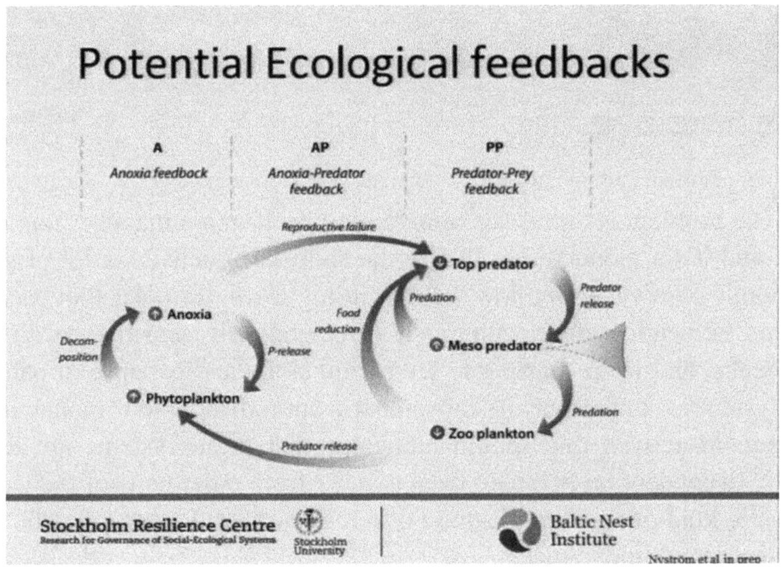

So, biodiversity, climate change, and nutrient fluxes in the Baltic case are interacting to lock the system in an undesired state. I threw these pictures in just before the talk because Simon made a reference to the Maine lobster fisheries. We've been studying that and others as well just to show that this is often perceived as a tremendous success story of a community-based collective action on a tremendously effective industry for lobster production.

The problem is that it has transformed itself into a massive monoculture, and it's a monoculture that in the short term delivers a lot of good economic outcomes. But now we're starting to see examples that because of this tremendous monoculture, it's tremendously sensitive to disease outbreaks, and we're starting to see abrupt shifts and collapse of parts of the system as a result of the movement from a diverse to a monoculture system. Moreover, one should recognize that these systems are today totally dependent on fish feed meal coming from other parts of the world, so we're kind of vacuum-cleaning different parts of the oceans to be able to concentrate in one place.

There are marine regime shifts documented across the world, from the coral reef systems in Australia, as I mentioned, to the Baltic Sea.

There's a very nice paper that came out quite recently by James Estes and colleagues, cataloging the importance of top predators and keystone species in tipping points in ecosystems, for example, as seen in the cod fisheries.

We recently published with colleagues around the world a very important work, I think, showing, as a rule of thumb, it seems to be that when

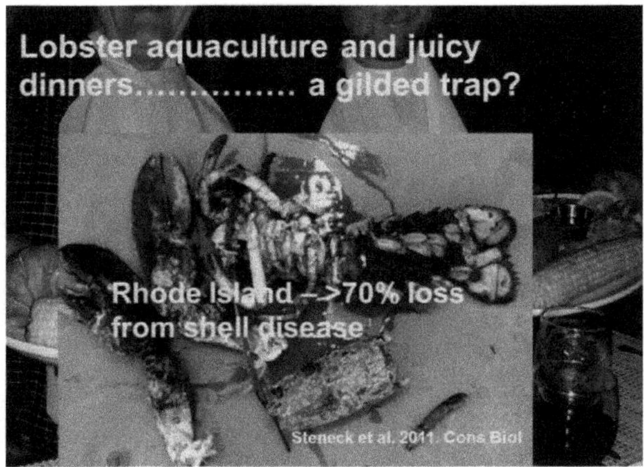

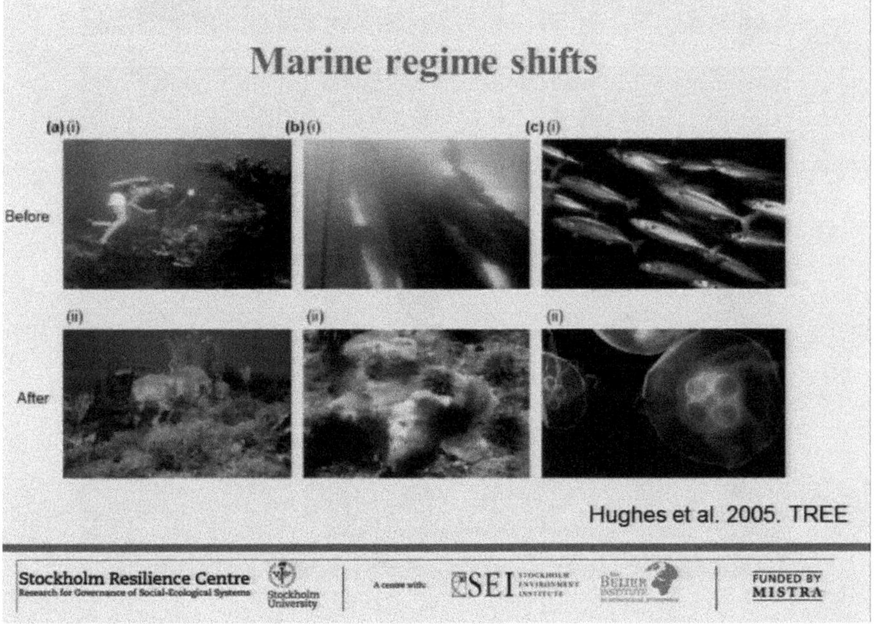

we empty 30% of fish stocks in oceans, seabird populations tend to collapse. So, there seems to be a 30% rule of thumb with regard to resources, which interestingly seems to also apply to terrestrial ecosystems where Bob Scholz and colleagues in South Africa show that when you lose 30% of forest cover, you also seem to risk abrupt changes.

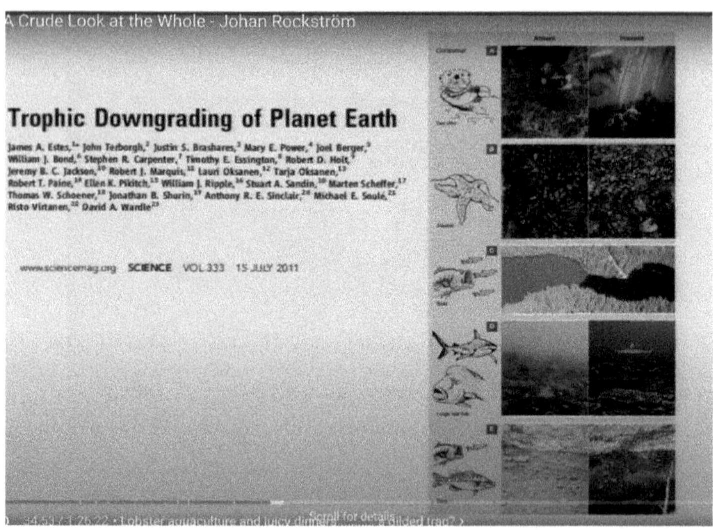

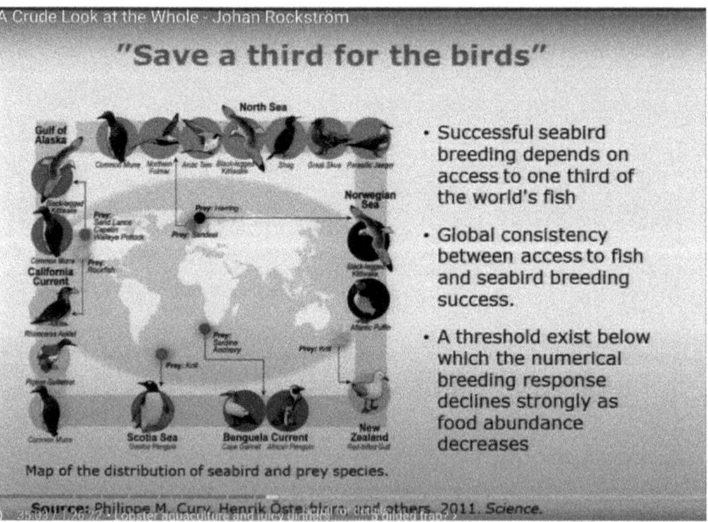

And colleagues show that the Amazon rainforest most likely is more sensitive to a tipping point to savannas than we previously thought. The reason for that, of course, is that climate change interacts with land use change, opening up the system and creating turbulence. The self-regulating feedback of moisture shifts over to become feedback of drying instead. We have a frequency of droughts which is unprecedented in the Amazon.

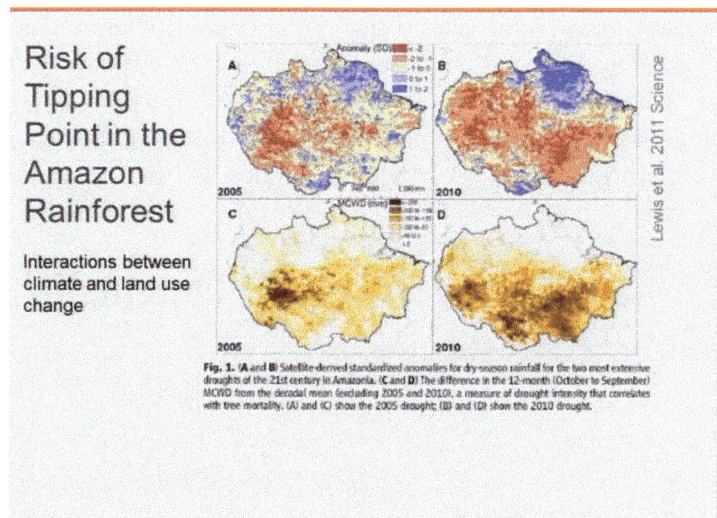

Fig. 1. (**A** and **B**) Satellite-derived standardized anomalies for dry-season rainfall for the two most extensive droughts of the 21st century in Amazonia. (**C** and **D**) The difference in the 12-month (October to September) MCWD from the decadal mean (excluding 2005 and 2010), a measure of drought intensity that correlates with tree mortality. (A) and (C) show the 2005 drought; (B) and (D) show the 2010 drought.

And finally, the work by Anthony Barnosky and colleagues shows that we can no longer exclude the possibility that we're moving toward a point in the mid-century where we could have a global tipping point in terms of loss of genetic diversity in species on Earth. This article is quite debated, as you may be aware, but it's just indicating the direction of understanding.

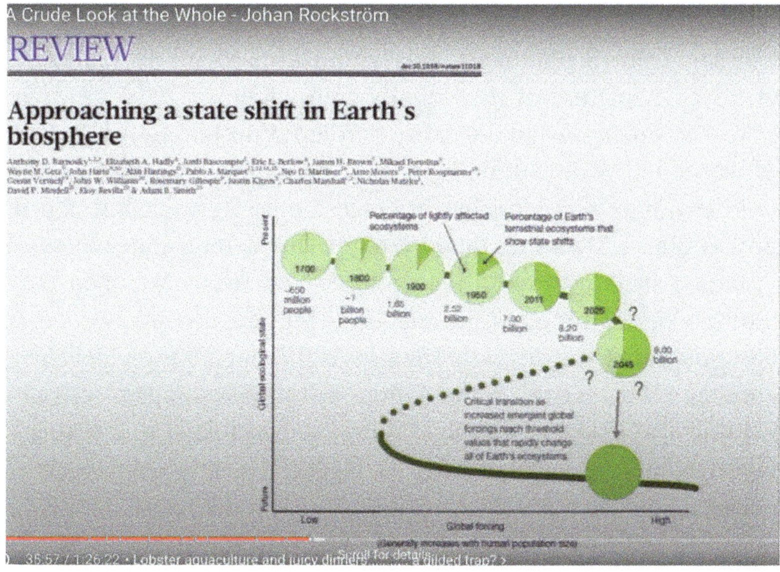

You may have seen this work on climate tipping elements by Joachim Schellnhuber and colleagues, cataloging the systems we need to be nervous about when it comes to climate-induced tipping points — everything from the Indian monsoon, which would affect regions like Singapore, all the way to the ENSO dynamics and the ocean conveyor belts.

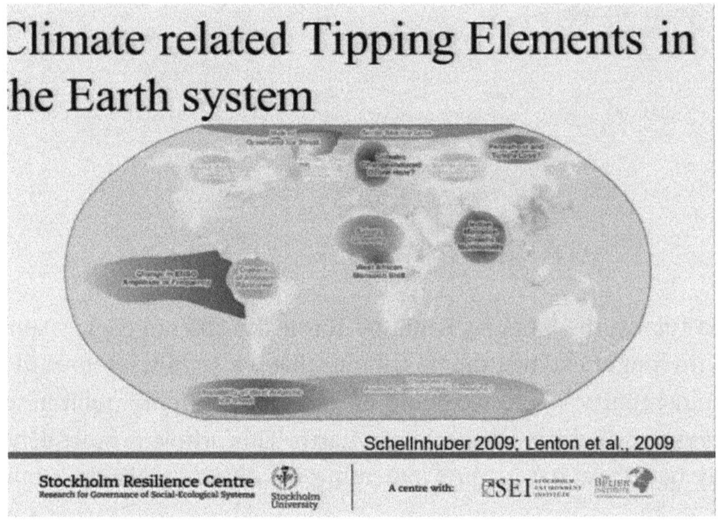

Coming from Scandinavia, this graph is particularly interesting. It shows a NASA picture of the brutally cold winter of 2010 — when climate skeptics emerged and the entire northern Europe said, "Hey, where's global warming? We're sitting here freezing like never before" — shown in the blue minus 4 colder than average temperatures, while the whole planet was plus 1.5 warmer than average. This demonstrates a lock-in of cold weather in the northern hemisphere. You may have seen the very interesting work coming out recently showing that this may be explained by the collapse of the Arctic vortex in 2008 and 2009, which held the high-pressure weather up in the Arctic. With that collapse, cold air was pushed down to Canada, the Nordic regions, and Russia, creating these surprise dynamics in weather patterns. Again, it's early days to say if this is a permanent shift, but it's interesting.

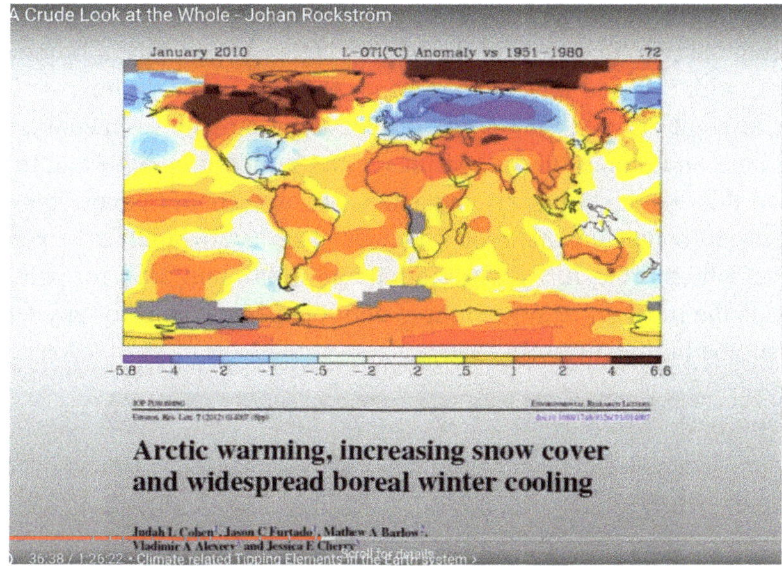

We're now mapping the risk of management-induced tipping points in water systems, so again, interactions between climate and land management are very prominent. And that just shows that we have a very strong base of empirical science to stand on when it comes to nonlinear changes.

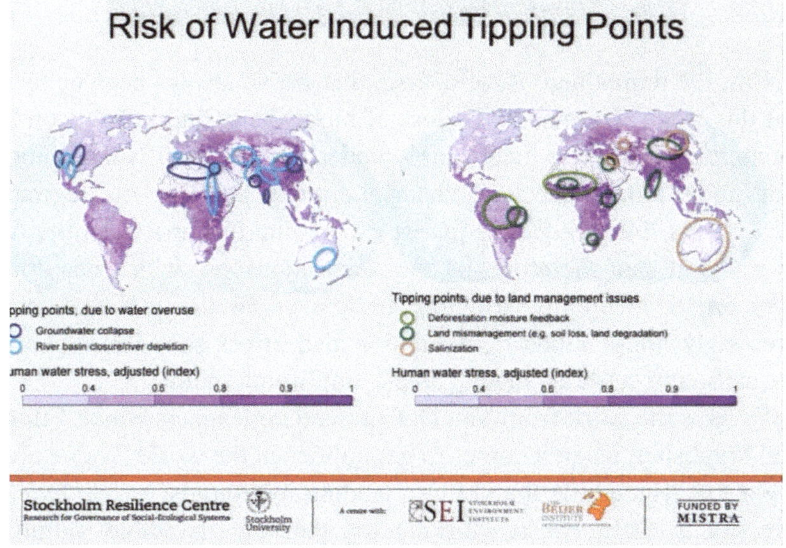

Tele-connections and social-ecological domino effects: Now, before entering the two final parts of the talk — one is to introduce a framework to deal with this and then to have some pathways for solutions — I just want to point out that Simon talked a lot about moving from ecosystems to biomes and biomes to the biosphere, and I think it is important to recognize that we're starting to understand the interconnectedness between systems in a much more profound way. This is so familiar to you, of course, which is the 70% deforestation of the Bali rainforest for palm oil, perhaps the most dramatic and rapid anthropogenic change of any terrestrial biome in direct terms.

Now, the drama here is, of course, that we've always been concerned about this when it comes to the loss of biodiversity and impacts on local communities. But we're increasingly understanding exactly the point that Simon pointed out: that these changes entirely alter the fire regimes in these systems. Fire frequency increases, leading to impacts on the Asian Brown Cloud and, therefore, on the Asian monsoon, which has domino effects on the economy, trickling back even to the global economy. Interestingly, these kinds of dynamics also affect soot fluxes, causing black carbon to settle on glacial areas, amplifying melting.

We have the work from Van Der Ent and colleagues, which I think is a total bombshell when it comes to geopolitics in the world. We've always thought this, [inaudible mumbling], leading hydrologist at the Technical University in Delft; we're working together on [inaudible mumbling] projects, looking at moisture feedback mechanisms in different biomes. And as you're aware, we've always thought that the big rainfall systems essentially originate from evaporation from the oceans. But what this work, indicated in red here, shows is regions in the world where 80% or more of rainfall originates from moisture feedback, from adjacent terrestrial ecosystems. And, for example, in the Chinese case here, with more than 80% of the rainfall coming from local moisture feedback mechanisms, it originates from neighboring countries in the West, so Chinese food supply and stability of the economy depend on how Russia, Kazakhstan, Ukraine, and all the way to the Baltic States manage their forests. So, this is an interdependence which is tremendous. And similarly for West Africa and Latin America. So, these are the kind of interdependencies across big biomes that we need to start considering.

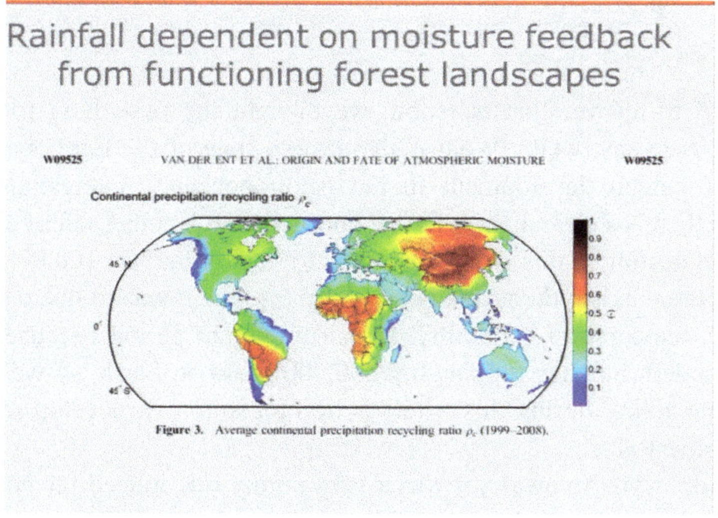

Figure 3. Average continental precipitation recycling ratio ρ_c (1999–2008).

So, that's my line on nonlinear changes. Two sections left here, and we're coming up to the surface, I promise you. Now, what happens scientifically if you connect these two insights? The insight of the Anthropocene, which is largely about pressures, and the insight about the risk of catastrophic tipping points. Well, that was a question we posed to some 30, 40 leading global environmental change scientists some 3 years back, to say, "What does this mean for global sustainability science?"

And to answer that question, we thought the first thing to do, of course, is to say, well, "What is the desired state of the Earth system to support human development in the Anthropocene?" Interestingly, we believe there's a clear answer to that question, and it comes out of ice-core data. Interestingly, this is Greenland, but it's only the last 100,000 years, and on the *y*-axis, you have a good proxy for how it was to live on Earth, namely, temperature variability. It's an important period because we've been modern humans for roughly 200,000 years on Earth, so we've had the same ability during this whole period to, so to say, develop societies as we have today.

And, as you're aware, it was a very jumpy ride indeed for humanity during large parts of this period. We were hunters and gatherers, very few in number. In fact, recent data by Stephen Oppenheimer and others on Paleo DNA-based [inaudible mumbling] shows that at this cold point here, where large parts of freshwater were frozen in the poles, we may have been down to 15,000 fertile adults on Earth. 15,000. So, we were not only virtually [nearly] extinct, [so] we are truly family in this room. And the important thing is that we were stuck in the highlands of Ethiopia. We had a very rough time, and that's the point when we left and walked over the Red Sea and started colonizing all the way down to Papua New Guinea.

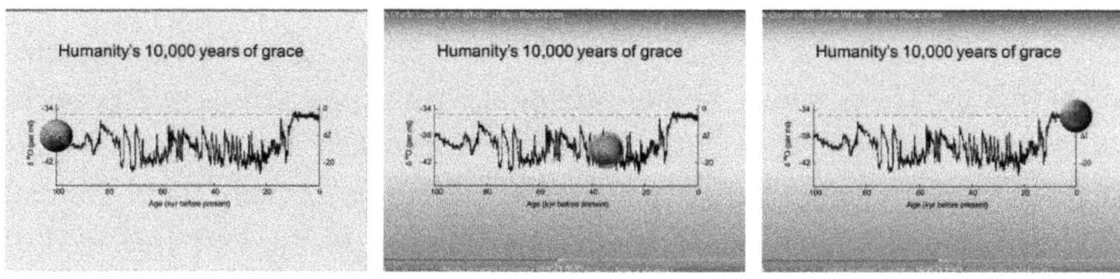

So, this is probably, and deterministically, a result of the tremendous uncertainty we had in our environmental conditions. And then we enter the Eden's garden of the Holocene, the extraordinarily stable interglacial [period] with a temperature variability of plus minus one degree Celsius. This is the period when all the genetic diversity was there, but it's now that things settle in. This is where the rainy seasons become predictable. This is where you know you can rely on four months of temperature above 15 degrees Celsius. This is where the wetlands, the marine ecosystems, the forests, etc., settle in as we know them. And we barely entered this period when we did our most important invention in human history, namely, we invented agriculture. And off we go on the civilizational journey we know. We're three billion people at the great acceleration. We're now 7, committed to 9.

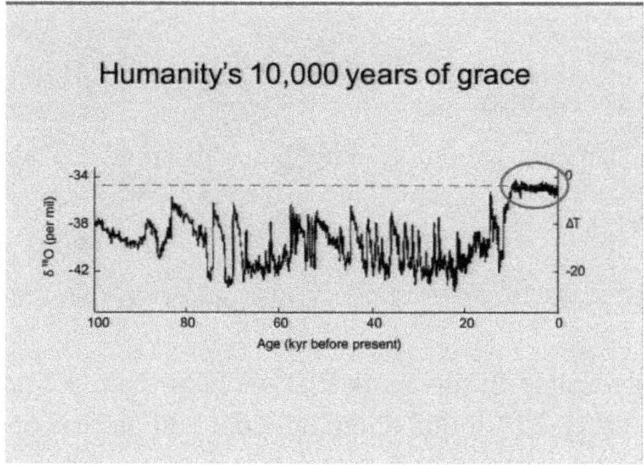

So, our conclusion is as simple as it is dramatic. The Holocene is the only state we know that can support the modern world as we know it. It's not as if we couldn't live here; the question is just what world can operate here. And if you take that as your reference point, as a normative statement that the Holocene is the only state that can support the modern world as we know it, then life becomes much simpler when it comes to navigating the Anthropocene. Because we know the Holocene very well. We can actually set a reference point on what the parameters are that you need to be stewards of to keep Holocene-like conditions. And that's the question we pose: What does Science say in terms of the Earth's system processes we need to be stewards of to have Holocene-like conditions? And could we, for each of those, even identify control variables where we can put a boundary position beyond which you enter a danger zone where you cannot exclude big nonlinear changes that would be undesirable and potentially push ourselves out of the Holocene state?

And that became the planetary boundary framework which led to and was kind of moved through an analysis of looking at, for example, what the importance of maintaining stable polar regions is in order to have them as kind of cooling systems for Holocene state conditioned equilibrium.

And that resulted in this work that we published in *Nature* in 2009, which has resulted in a lot of scientific work and discussions since then. That was the purpose: to trigger science. The important thing, in my mind, about this work is to show that it's not only climate change we have to worry about when it comes to truly securing human prosperity in the Anthropocene. And what we concluded in this work is that we have three, let's say, big systems we need to be preoccupied about: the climate system, the stratospheric ozone system, and the entire stability of the oceans, which have Paleo records of big-scale tipping points in the history of the Earth system.

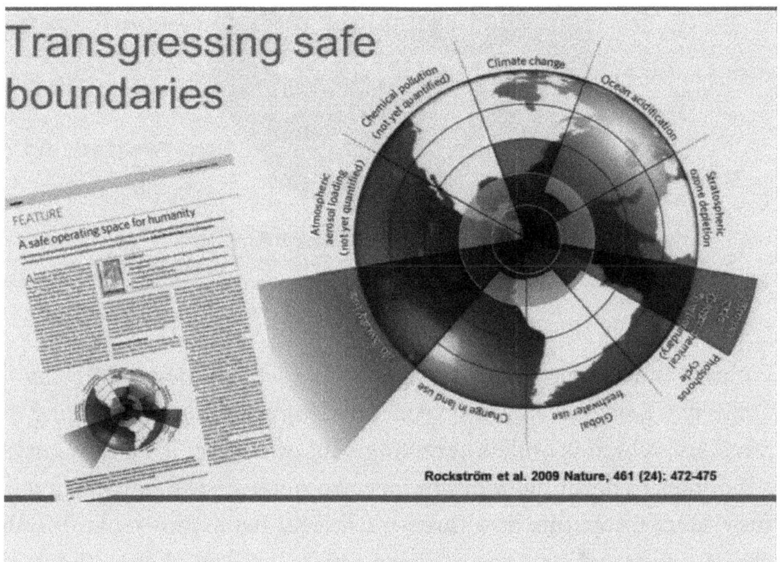

But interestingly, this group of scientists also included four parameters which do not have evidence of planetary-scale tipping points but which have evidence of being the underlying providers of resilience at the Earth's system scale. These are the slow variables working under the hood of the Earth's system machinery: biodiversity, land systems, freshwater, and the big fluxes of nutrients in the Earth system. And there's been a big, very distorted debate coming after this paper, claiming that we said that there were global tipping points also on land and water biodiversity. That was never ever the proposition. The proposition was this: What are the sources of resilience? What are the processes that keep dampening and securing so that the system remains stable? We also included chemical pollution and atmospheric aerosol loading, which we couldn't quantify — we can come back to that in the discussion.

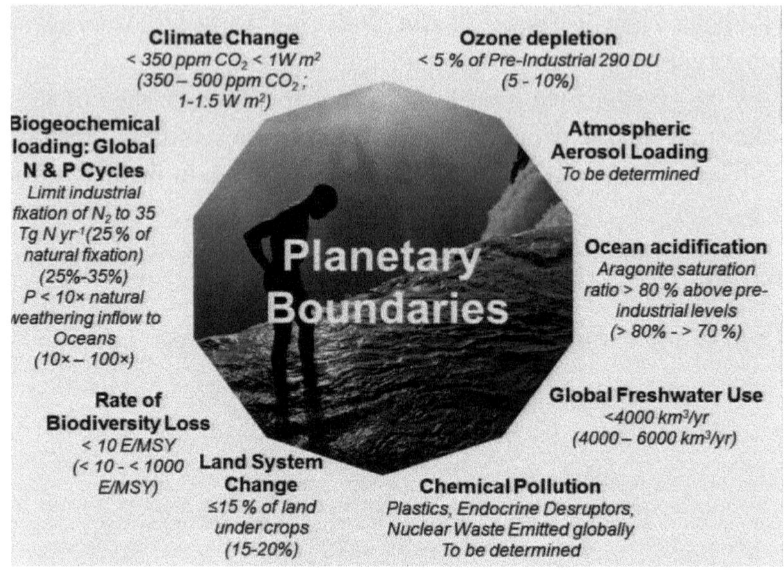

We're now working on a 2.0 paper where we will be quantifying these two, and we have so far not seen any suggestion on changing these 9 parameters, which is quite interesting. But of course, it's a moving target when it comes to defining them. In fact, we suggested in the original paper that they seem to operate as a three-musketeer behavior — one for all, all for one. If you transgress one of these processes, you change the position of others. If you push the climate system too far, you may have to, in fact, secure more land area in order to keep your system resilient, and vice versa if you totally divorce the planet, your climate boundary will move very rapidly in the wrong direction.

We're now working on downscaling these boundaries as they were set at the global scale. I'm just mentioning that, perhaps for your interest, we're, for example, looking at what would be the thresholds for different key biomes on the planet. For example, how much forest cover can you lose in a tropical rainforest before you risk nonlinear changes, which would propel back into the earth system? How far can you lose functional diversity and different keystone species and keystone systems? Because this was a critique coming out of the work.

Earth System Process	Control Variable(s)	Thresholds	Planetary Boundary (zone of uncertainty)	Current Value of Control Variable(s)	State of Knowledge
Land system change	Global: area of forested land Biome: area of forested land	Tropical: Amount of land clearing beyond which self-reinforcing feedbacks lead to land-cover change across a much larger area, with atmospheric circulation teleconnections. Temperate: No known thresholds. Boreal: Possible threshold related to albedo changes associated with land clearing	Global: 75% of early Holocene (pre-agric) forest cover Biome: Tropical: 85% of original forest cover Temperate: 50% Boreal: 85%		Threshold for tropical forests best understood for Amazon, but complex with significant uncertainties. Albedo effect for boreal forest well understood but position of any possible threshold is not known
Biodiversity loss	Genetic diversity (library of life): Extinction rate Functional diversity: Mean species abundance (MSA)	"Soft threshold" proposed somewhere around 50% drop in MSA, beyond which rapid and much larger loss of biodiversity. Threshold known at ecosystem level and proposed for global level	Genetic: no more than 10x background extinction rate but aspirational goal of no loss of genetic diversity. Functional: Maintain MSA at 70% or above (uncertainty range of 70-30%)	Genetic: Current extinction rate is 100-1000x background Functional: Global MSA is currently estimated to be ca. 67%	Aspirational genetic boundary based on first principles. Growing body of evidence for threshold of functional biodiversity loss at multiple levels (ecosystem to global)

There've been a lot of publications suggesting changes in the boundary positions, and I think one of the most important ones was the work of Stephen Carpenter and Elena Bennett, suggesting that we've got it wrong on phosphorus. We suggested that the phosphorus boundary we put on how much phosphorus can we load into the oceans before we get anoxic catastrophic tipping points in the ocean. They said way, way before you knock the oceans over a tipping point, you've actually knocked the whole battery of freshwater systems across tipping points, and therefore you need to couple it with a freshwater boundary, which we have already transgressed.

There's a lot of work advancing on ocean acidification, which we put in as a boundary, (and I think) increasingly showing that this is one of the big, you know, black box important nervous issues when it comes to climate impacts. And the reason for this, as you're certainly aware, is that when oceans take up carbon dioxide, at least for acidification because carbon dioxide plus water becomes carbonic acid, they take and break up calcium carbonates in the ocean. And the calcium carbonate levels here

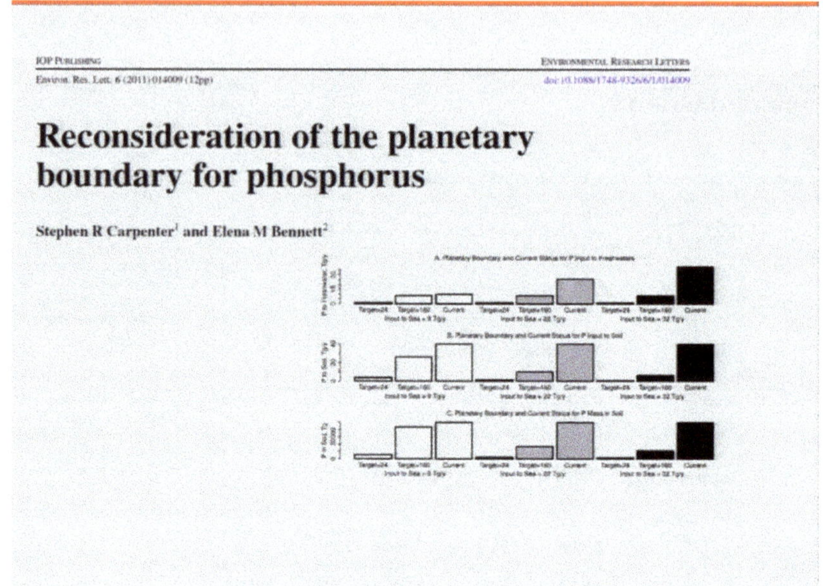

Reconsideration of the planetary boundary for phosphorus

Stephen R Carpenter[1] and Elena M Bennett[2]

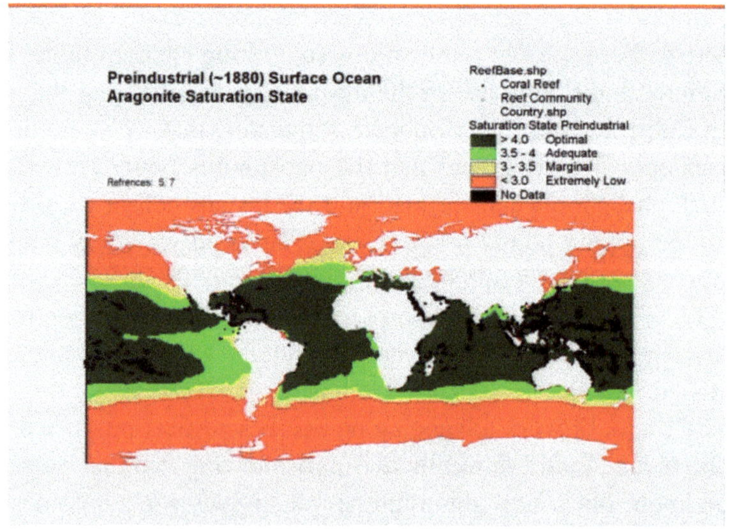

represented in one of them, Aragonite, were high enough in the pre-industrial era, in dark green here, to allow for the development of coral reefs, indicated by black dots here. So, basically, calcium carbonate being the Lego blocks for all marine life, constitutes the backbone for coral reefs and the skeletons and shells of phytoplankton and zooplankton and coral reefs, which led to the establishment of these ecosystems in the Holocene.

And this is the situation today. And these are the concentrations if we do not bend the curves of greenhouse gas emissions.

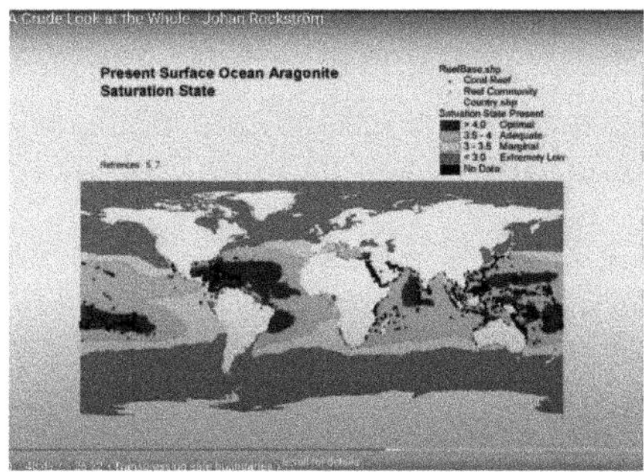

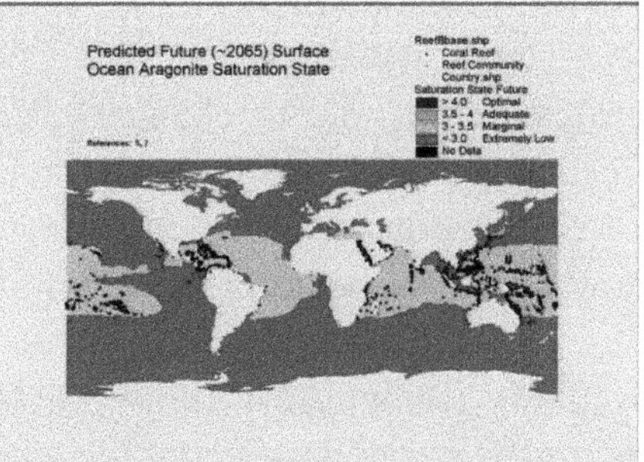

So, this is an example of how we're learning more and more about the interconnectedness and the fact that the oceans are playing an enormous beneficial role in dampening greenhouse gas concentrations and climate forcing but jeopardizing ocean stability.

Steve Running and colleagues have shown, for the first time, that the carbon sink in terrestrial ecosystems, in red here, may be decreasing. This is a drama of tremendous proportion if it is correct, given again that 25%

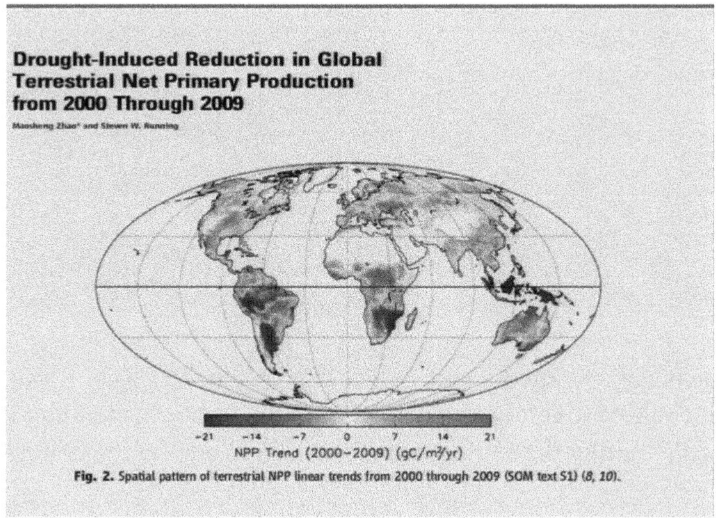

Fig. 2. Spatial pattern of terrestrial NPP linear trends from 2000 through 2009 (SOM text S1) (*8, 10*).

of emissions are taken up by the atmosphere, but also proof that planetary boundaries interact and that the biosphere needs to be managed in a way that can allow for continued dampening, which is the Earth's system response to staying in a Holocene state. Businesses are starting to show a lot of interest in this.

This is where the Business Council is seeking help in science-based influence on how to change business models to incorporate planetary boundary thinking. So, we're forcing ourselves to try and define a set of 2020 "must-haves" when it comes to science-based targets to start moving toward a safe operating space within planetary boundaries. And we're seeing a lot of interest within civil society, particularly to say, "Okay, there might be a biophysical ceiling for human development within a safe operating space, but what about the social floor?"

Planetary "Must Haves" by 2020 In partnership with WBCSD

Planetary Boundary	2020 "Must Have"	Key Links	Key Business Tools
Climate Boundary	Bend Global Emission curve of CO2 by 2020 5-6 %/yr decline thereafter 80-100 % global reduction by 2050	Land, Water, Nutrients, Ocean Acidification	Footprint analysis LCA Target setting
Land Boundary	Sustaining remaining Rainforests on Earth Keep > 70 % forest stand	Water, Climate, Nutrients, Biodiversity	Sustainable Agric and Forest strategy
Water Boundary	Ensure > 30 % river flow	Climate, Land, Biodiversity, N&P	Water productivity indicator Sustainable intensification of production systems Product labelling
Nitrogen & Phosphorus Boundary	>50 % reduction of P leakages in soils > 50 % reduction in N leakage in soils	Land, Water, Climate, Biodiversity	Monitor N and P flows in entire value chain Sustainable Agriculture
Biodiversity Boundary	Absolute global halt of habitat loss Safeguard Critical Biomes (Forests, Marine systems, Polar ecosystems)	Land, Water; Climate	Restoring "hot-spots" Protect critical biomes Economic value of ecosystem services (TEEB)

This is the so-called donut model that Oxfam and Kate Raworth and colleagues have brought forward, to say, "Well, if we're in the Anthropocene and if we're hitting the ceiling and if we need to respect the ceiling, we're actually in the realm of absolutes, not relatives any longer. We simply have to recognize an absolute budget of nitrogen, phosphorus, land, water, and carbon. And then you need to have a fair distribution of that space in a world of 9 billion people." This is, of course, dynamite in the political sphere as well, as you can imagine. But this is where we're moving when it comes to the global sustainability agenda.

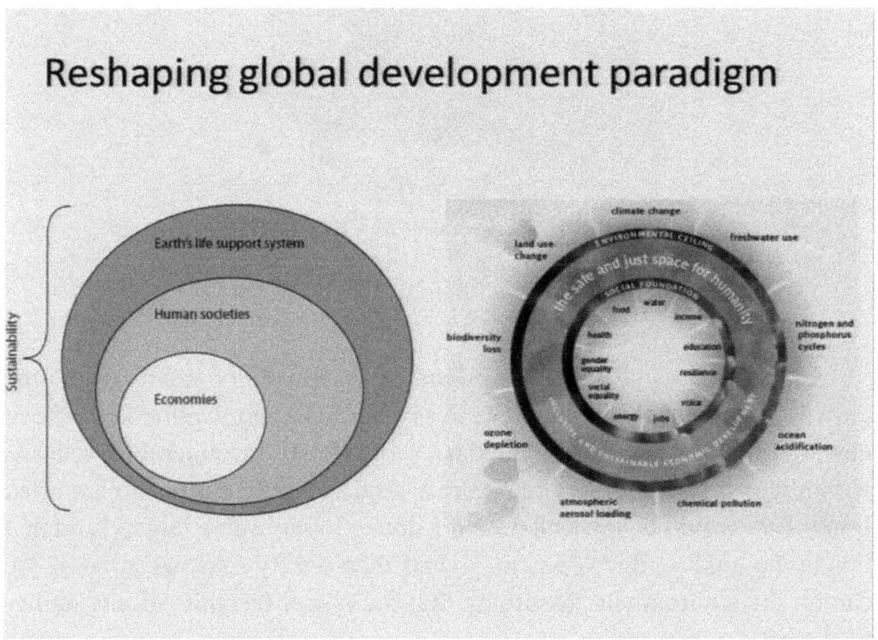

We're just now submitting a paper to *Nature*, and this is — I'm coming to my closure here — to try to inform the United Nations' work moving from millennium development goals to sustainable development goals. What we're saying is that we will probably move to a situation where the economy — the old steady-state economic model that we know so well — is returning into interest today, with the economy to be a subset of society operating within the life support system, and defining a set of sustainable development goals that actually operate in an integrated way with sustainable criteria.

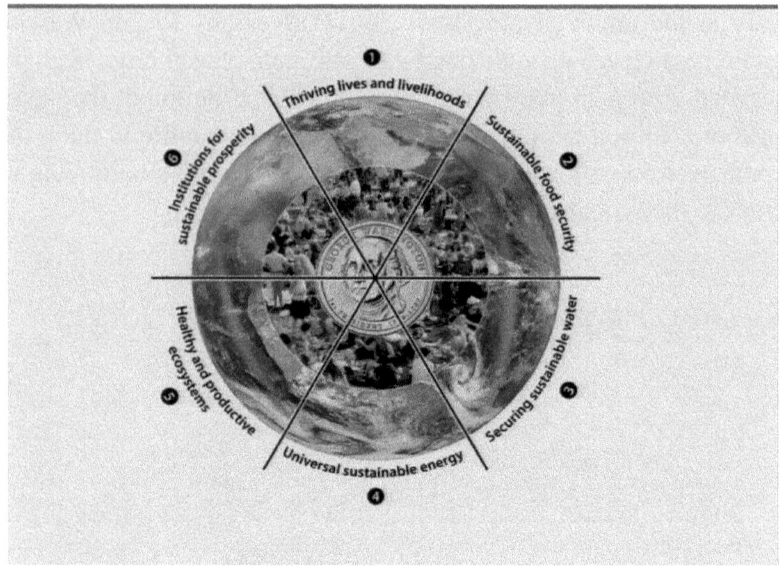

So, moving away from the pillars of sustainability versus economic growth and poverty alleviation, so to say, for lives and livelihoods, we can now define the planetary boundaries of global sustainability criteria, within which social targets can be met. And integrating this into a unified framework would be the first time it's done. So far, we've largely said that this is the goal — development — and then we try to do as good as we can on the environment, assuming that the planet has this infinite ability to absorb our pressures.

The economy is starting to think in these terms. You may have seen Tim Jackson's work on *Prosperity Without Growth*. There's a growing literature, I think, on thinking in terms of prosperity in the Anthropocene, from the perspective of nonlinear dynamics. We're looking into this in terms of planetary stewardship. If we're in the Anthropocene, how can we collaborate? Perhaps learning from bacteria and ants and how populations in ecosystems can collectively act.

We have so much evidence that transitions to a positive trajectory are possible. I think this is really critical; we can adapt, but we can also transform. And we've been looking a lot at, for example, the big transition in

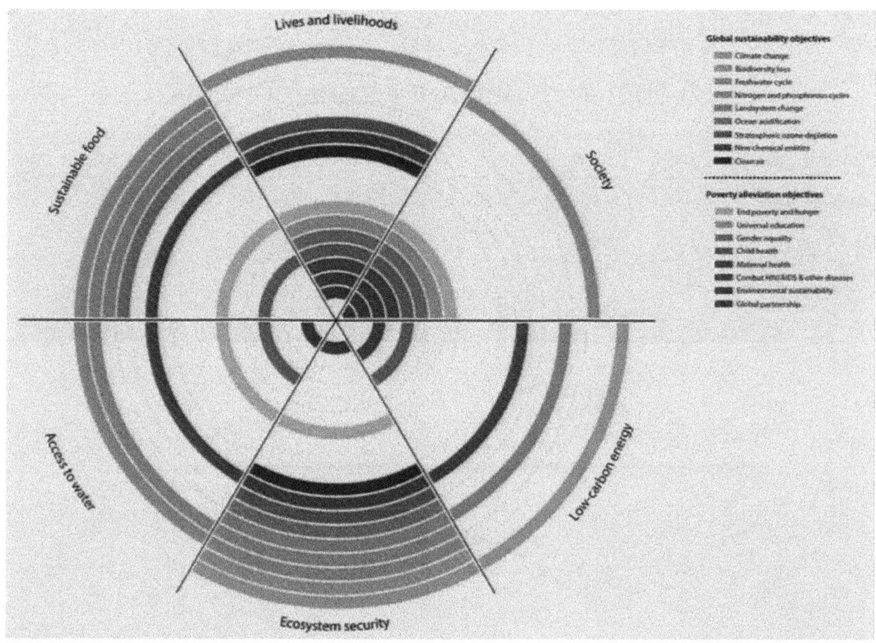

the governance of the Australian Great Barrier Reef, the agricultural revolution in Latin America, moving from plow-based systems to zero-tillage systems, to dealing with the crisis of degradation of an enormously productive and carbon-sequestering trading system, recognizing that you need knowledge to see where you're in a crisis and how to move out of a crisis.

And finally, just how do we deal with this in a world that will need a lot of time to think and rethink our relationship with planet Earth? It's not as if you can just change values overnight and we reconnect societies to the biosphere. What we're landing in more and more is that we probably need to have a kind of a dual-track strategy where there is one fast-track opportunity of really moving within, let's call it a bit sloppily, the obsolete machinery we need to operate in at the current time step, in terms of resource efficiencies, measuring and monitoring, and assessing risks, really pushing for a global energy transition, which I think is showing a lot of promise. You may be aware that on a normal Saturday in Germany, the world's fourth-largest economy today, 50% of their electricity is supplied from solar panels, largely from individual households, thanks to their feed-in tariff system and reciprocity of electricity markets — you can actually sell electricity from your own household energy production. These are huge success stories occurring in real time right now.

And we've recently published this paper showing that you can actually feed the world through sustainable intensified agriculture that meets planetary boundary criteria because we have such enormous untapped potential in yield gaps.

This is the gathering of science to try to address this. You may have heard of Future Earth, which is the effort of the global environmental change research community to move from Earth System science to global sustainability research. It's a tremendous gathering of the International Council for Science, the International Social Science Council, but also the Belmont Forum, which are the big funders of science, to say now we need an integration of social and natural sciences, and we need more solutions-oriented research endeavor. This starts this year as an effort of mobilizing collaborative research.

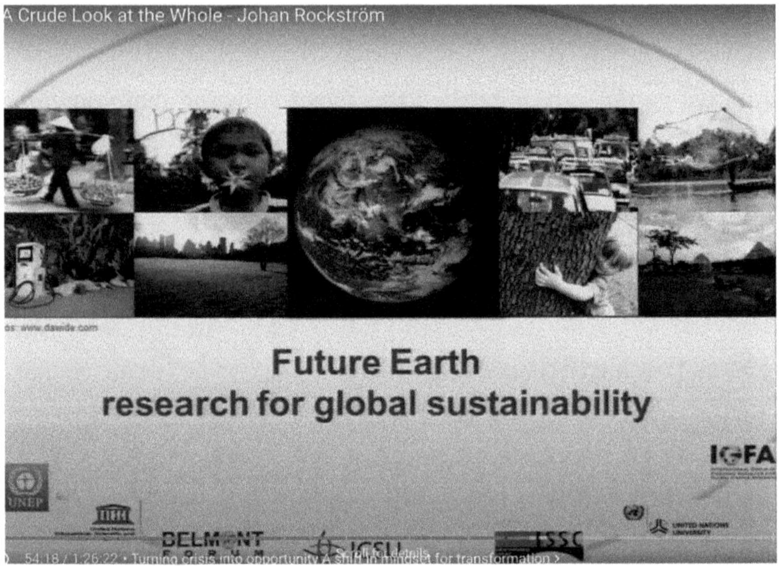

To summarize, we probably need to move along these two areas: First, adopt and support the work from the United Nations, the climate negotiations, and efforts particularly in resource efficiency. It might seem incremental but is clearly a necessity. Second, as we published recently in a book called *The Human Quest*, start really addressing the behavioral change, the value issues, the issues we discussed yesterday on what is it that... How do you couple culture with values, with change?

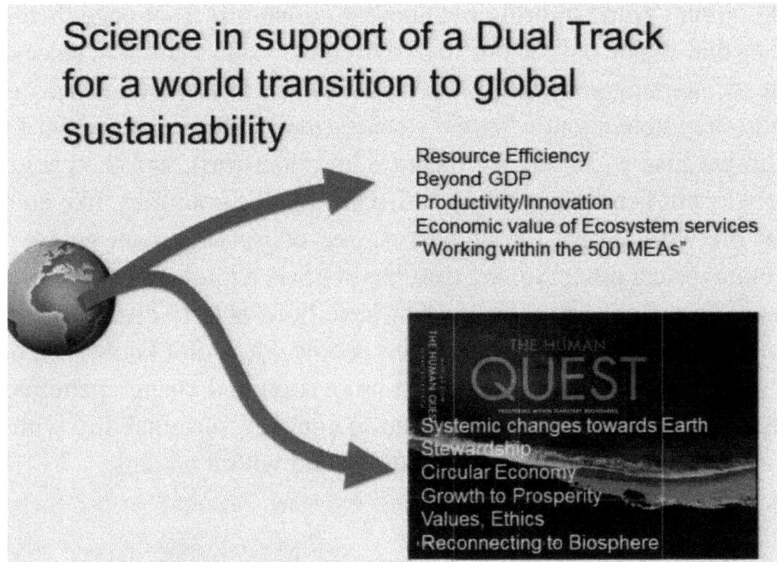

And also, moving [from] that all the way to allowing ourselves to say, "Okay, we live in a world with growth, assuming growth without limits." We had a 40-year period when environmental science talked about limits to growth. Now, I think we need to move into a paradigm of recognizing growth within limits, and that's a much more positive trajectory. We have a safe operating space.

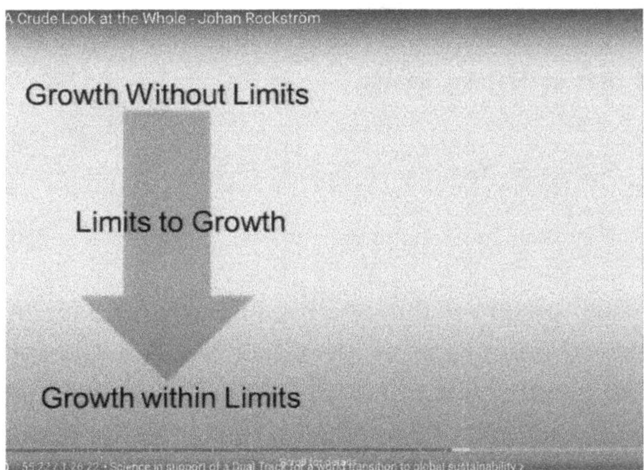

I couldn't help but bring out another *Economist* front page that — of course, this is not only about the environment and natural sciences; it's about recognizing how successful societies actually operate. And I don't want to brag here about what we've succeeded in Scandinavia, but I took this up because yesterday we talked a lot about trust, and the parameter that sticks out here interestingly is that the Nordic countries, like no society in the world, have the highest degree of trust between people who don't know each other. So, the trust base between knowns is high in many societies, but these are societies that have been able to create a sense of security because trust is high between people who don't know each other. And that lowers transaction costs in innovation and change patterns. So, these are the kinds of things we need to connect: sustainability with values, and what builds trust among nations and within nations.

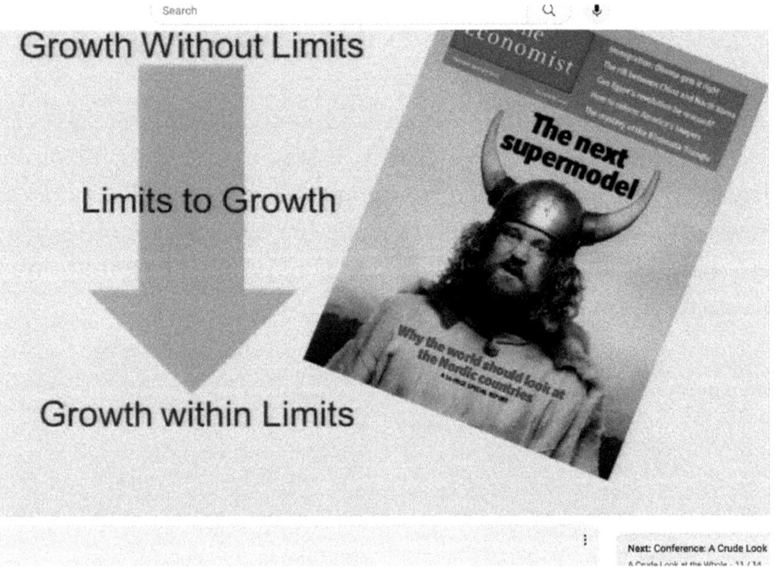

And my final picture is this one because when I gave this talk at the World Economic Forum recently, the Chief Science Advisor to the Dalai Lama sent this picture to me and said, "You know, he really wanted to connect religion and value systems with sustainability research." And he said, "To inspire this, I send you this photograph, which is from a lake up in the Nepalese mountains, 4,600 meters above sea level." I thought at first it was fake, but it's actually a photograph that he's taken, and I just think this encapsulates also the values that are at stake and that all cultures, in fact, have strong connectivity between humanity and nature.

I think nobody really wants to deliberately, let's say, destroy our life support base. Of course, there's optimism also in the universality of understanding the beauty of these kinds of environments. Thank you very much.

10.2 Discussion

Sheila Rhonda: Sheila Rhonda from Welsh College. In a world of greed and avarice, at least in the United States and the world that I've been trying to function in, I'm also seeing a very alarming rate of anti-science thinking and a devaluation of the importance of science. While technology is celebrated for providing entertainment and convenience, it's often disconnected from the scientific principles that underpin it. This is compounded by increasing levels of scientific and mathematical illiteracy among populations. Do you have any suggestions for first-world countries that are regressing in this regard?

Johan Rockström: Yeah, that's a very good question. Well, to begin with, [let's] just say that it's not unique to the United States, even though you might be at the front end of that challenge. Now, I've always been a strong supporter of the fact that we need more science-based decision-making in the world, but frustratingly, we're not seeing much of that happening or at least not enough. What we're experimenting with, I have no clue whether it will work, is to be inspired by scientists like Dan Kahan at Yale University who's done tremendous work showing that, for conservative Americans, the worst messenger is the scientist, because they have a value system that screens off everything that has to do with the environment, per definition, as Big Brother trying to control your freedom. Integrity and freedom are so rooted that in the end, you almost worsen the situation if you try to just have this linear science communication.

The question we're posing to ourselves is this: Should we then, and his conclusion is the same, try to co-design and work more closely in alliances with other stakeholders in society? Have other carriers of the message — business leaders, musicians, cultural representatives — talk about [it], not having Al Gore present *An Inconvenient Truth*, but rather have people like Bruce Springsteen and young rappers, and a lot of that kind of work, and take that very seriously. I think we're trying to have a science–arts focus in a much stronger way to try and communicate in a deeper way.

So, *The Human Quest* book, for example, is written with one of the world's leading photographers, Matthias Klum at *National Geographic*, to try and hit both the stomach and the brain at the same time. But it's a highly frustrating situation. But just to give you a little bit of a surprise (to me), was that we've established a very close collaboration with the World Business Council for Sustainable Development. It's a network of 200 of the largest companies in the world, representing 10% of the world economy. And what they tell us is that, you know, "We, as in our boards, we listen to science. We take science very seriously."

So, sometimes I wonder if on climate, at least, it wasn't [the case] that, you know, up until 2009, everyone was listening to science, and then we all went to Copenhagen and things collapsed. And out of that, the world's political leadership went into some kind of Copenhagen post-trauma, and they're stuck in this trauma still. While [in] business, they never fell over that tipping point, and they had already started to internalize climate thinking, and they're just moving along. So, it seems to me that this was kind of a Tour de France race — that science has the yellow leadership shirt, and business leaders are trying to hang in there and put themselves just behind science, but politicians are way behind. So, what we see in the media may be a lot of the political agenda, while business, in many areas, is working harder I think under the radar screen.

Ann Florini: Hi. Ann Florini from Singapore Management University and Brookings Institution. I have a very specific question for you. I've already been using things like your TED talk and the Oxfam donut in some of my teaching here in Singapore. I find the undergraduates very responsive. I don't find the Asian business community quite so responsive. I think the business trends that you're looking at haven't really registered here to the extent that they have in the West. Here's my very specific question: I am lecturing tomorrow in the advanced management program for two hours. What do I tell them in two hours that is going to get their attention and change the way they think?

Johan Rockström: Well, to begin with, feel free to use any slides you want. So, that's number one. And you know, I find it increasingly effective to turn the whole story around to say that there are no environmental issues left anymore in this world. I mean, let's just drop the environment, let's shut down every ministry for the environment. I don't recognize myself as an environmental scientist, because we've come to a point where it's all about human prosperity. That's the agenda. It's about how we secure growth, development, prosperity, well-being. And because we're in the Anthropocene and because the pressures are so large, you know, I think we've done ourselves a great disfavor by talking about global commons and externalities somewhere out there, you know, this big waste bin that we can dump things in infinitely. Now that things are so saturated and we're so interdependent, it's all about development. And if you turn it around in that way to say if it's all about development and you have a number of capitals to deal with… I mean, Adam Smith started off with having land as prime capital and then we kind of dropped it along the way and focused everything on technology and finance and human capital, and now we're coming back to the notion of recognizing that the biosphere is our most precious capital. And to put that positive story of putting human prosperity upfront and then say how can [we] wisely manageme [these] resources, which is often in my mind, what business people understand first. I mean rare earth metals, land, phosphorus, the [different] kinds of abiotic resources, but then try to lift the living biosphere as much as possible into that picture. So, that's one tactic perhaps. But I think the other important tactic is also, which I haven't talked at all about, just to refer to the great work that is being done increasingly by Pavan Sukhdev and others on the economics of ecosystems, biodiversity, and externalities. Simon and Bryan are here and others. You know you're really trying to show that it's not anymore about ecological economics and some kind of marginal discipline. It's really about mainstream economy. You know, there are four European commissioners today. You've set up something called the European Resource Efficiency Platform. They want to transform the European economy into a circular economy because they're simply recognizing that the momentum in the world is such that if Europe wants to compete in the world, in the Anthropocene, you need to be damn smart on energy efficiency, resource efficiency, and not undermine your ecosystem services. So, you know, somehow change the agenda. We have tended for so long to talk about humans being bad, destroying nature, we need to protect the environment. That is, I think, the absolute dead end of the argument. I'm not sure it helps but I think what you're doing is so important.

Greg Fisher: Thank you. It's more of a statement and see what you think. Because overlaying a lot of what you've said is what I would call the profit maximization algorithm. I think lots of people do have multiple values, but the economy has been run across the world as a single value, which is profit. My background is in economics. I worked for the Bank of England for 9 years, and I've been at the heart of orthodox economics. We need a fundamental paradigm change in economics. I know Brian has been trying to lead in the sort of complexity economics arena, but I think if we're trying to get — if we need — if we're going to get anywhere near the [inaudible mumbling] that you're kind of saying or implying, we need to get right at the heart of economics and have a paradigm change. Economists don't seem to be able to do it, which is just astonishing. So, anyway, any thoughts on that would be appreciated.

Johan Rockström: Yeah, and there are so many others in the room who probably have much wiser thoughts than I can offer on that important question, but I tend to look at this — we, at the Resilience Center, have one very close part of our Center, the [inaudible mumbling] ecological economics that Simon has been deeply involved in over the years. And it arises from, if I may be allowed to say that, let's say the cutting edge of conventional economics, trying to move into and change economic thinking in everything from discount rates to inclusive wealth, and has done so very successfully. But that is really working within the paradigm. It's working with the best economists in the world. The other alternative is to do as the New Economic Foundation in the UK and others do, to basically say the conventional model is obsolete, we need a new model. And it has led, as you know, to tremendous confrontation, at least to kind of an academic confrontation. We tend to adopt the more incremental approach when it comes to economics and try to work with business schools and with the best economists in the world to put economic value on natural capital, to address the fact that discount rates don't work when you have nonlinearities, etc., so, to try to build in that kind of thinking. But I'm a very close colleague and friend to people like Tim Jackson, for example, who I think represents more of the alternative model thinking where we need to totally rethink the paradigm. I think that work that Joe Stiglitz and Amartya Sen and Fitoussi did for the Sarkozy Commission was tremendously important. If we could move beyond GDP, just in terms of what we measure, I mean we only manage what we measure, so in the end, if we just could get a quarterly reporting on our ecological footprint within our national counts, which in fact I should say countries like Sweden are doing already, it would help, you know, it would take us a long way toward at least recognizing that we're building GDP but we're shutting down our ecosystem base. So, I think even here, you need a kind of a dual approach, deep thinking on new paradigms but at the same time work with the economy not against the economy. And I think we need to do what you'll be doing tomorrow; we need to go to the business schools, we need to engage with the best economists, the best economic research training schools in the world, and then work together much more, which, by the way, people like Simon are already doing, so I mean there's a lot of that going on and…

John Richardson: Okay, so John Richardson from the Lee Kuan Yew School. One of the very hopeful cases has been Japan. I wonder if you could just reflect on the new regime of Prime Minister Abe and his re-emphasis not only on growth but the use of nuclear power and so forth. How does this fit into your scheme of things and how do you see the scenario in Japan unfolding?

Johan Rockström: Well, to be honest, I don't know. That's the simple answer. I think both Japan and Germany represent big economies that were hit by a social shock, which has triggered a lot of very important and transformative thinking on an energy transition. In Japan, it seems to be bouncing back toward its previous equilibrium, which is quite interesting. That's exactly what has happened largely in the financial crisis as well. You get a shock to the system, but it tends to be so resilient, it bounces back. So, I think today when it comes to Japan, it's quite uncertain where it's heading. I think the more interesting case is, in fact, whether Angela Merkel will succeed in a complete phasing out of nuclear power and therefore also transitioning to a renewable energy mix in Germany. It would be extremely interesting if Japan persisted and became the second economy in the world to show that you can have a well-advanced society depending to a large degree on renewable energy sources. But I think it's still entirely up in the air. But I should say, I mean, that goes for Japan but certainly also particularly perhaps to South Korea. As far as I can see from the outside, an interesting trend that seems to be ahead of the curve [in South Korea] compared to Europe is to really start integrating growth and sustainability, to start addressing sustainable growth as the integrator rather than have environment and development in two separate parts. And I think that if some of the Asian successful economies can show that you can actually combine good well-being, strong, transparent, democratic societies with a very good mix of sustainability efforts, I think that would be an enormous inspiration and pressure on other countries. I think the most interesting pressure today is, in fact, the discussion in Europe to say, you know, we're not sure about this whole sustainability issue, but we don't want to risk it because what if we're left behind? And I think that's a very good pressure.

Suman Bannerjee: Thank you for this enlightening talk. As an economist, I'm usually the one asking questions, but it seems that economists are given more credit than they may desire. While I'm trained as an economist, it strikes me that there are some micro aspects of economics that people talk about — discount rates, future profits, and the like. But the main idea is that we try to identify natural equilibriums — equilibriums that follow from dominating strategies. Many of the discussions from Simon and others suggest that many of these situations resemble the prisoner's dilemma. If they aren't of that nature, it's just a coordination failure. I don't think we need to dwell too much on it; it will correct itself in due course. However, if we've observed a failure for a long time, it must be that these are indeed prisoner's dilemma scenarios. In such cases, if you are environmentally conscious and I'm not, it's better for me to continue without being environmentally conscious. This kind of prisoner's dilemma scenario might lead to undesirable equilibriums, but there isn't much we can do because these are dominant equilibriums. Essentially, these are situations that are expected to occur. So, we have to carefully separate the ones that are just coordination problems from those that are natural selection models. Natural selection models are harder to change, and I'm not sure if it's even possible. But at least we can focus on addressing the coordination failures that aren't really prisoner's dilemma situations.

Johan Rockström: Yeah, I hope this is constructive, but I recognize that in dialogues with economists, a bit of a provocative statement can be useful. Often, our discussion tends to criticize the economic model. However, I think a potentially more constructive approach, which I believe you're alluding to, is to say that it's not the economic model that is wrong. In fact, it has served us very well. Just look at the past 150 years; during the period when the planet wasn't sending any invoices — which it's starting to do now — it actually served us very well to consider the planet as this huge free service. With seven billion people, we're wealthier than ever; GDP has grown four times just over the past 30 years. So, of course, the question today is whether the model is very effective. Perhaps our error is to think that it can solve all our problems and we apply it everywhere. Instead, we might consider that the conventional economic model can actually only serve us in very few instances. It can only serve us when a resource is very stable, where there is a possibility of perfect information, where there is no conflict between different stakeholders, and where there is substitutability. It's an absolute necessity. So, every time you have bi-stability, every time you have a non-substitutable resource, and every time you have competition between different interests that are not well informed of each other, the model doesn't work. And what you need then is political leadership. You simply need regulatory measures, and every economist recognizes this. A market economy cannot operate without strong regulatory measures, and we're not seeing those regulatory measures play in, which is a bit surprising because, in most other areas, we're cognizant of the need for regulatory measures, even in the financial system. But look at tax policy or social insurance policy or the debates in the US now on the entire medical and education system; when it comes to the environment, we're very weak in putting regulations to the market, to confine and domesticate the market into playing a good role in its confined areas. That might be one way of working as well on the conventional economic area, and I'm sure this stirs up a bit of a debate, but that could be a lunch or dinner issue.

Ryan Chisholm: Hi, I'm Ryan Chisholm from NUS. I'm just wondering about the concept of growth within limits; I'd like a conceptual clarification. Is this idea essentially that you can keep growing as long as you're mindful of those limits? Is this just a new twist on the old idea of internalizing externalities and incorporating environmental factors into our analysis? Is it essentially the same thing, or…

Johan Rockström: No, I would argue it's distinctly different because what "limits to growth" clearly alluded to was that there was a biophysical limit to how far we could generate, let's say, well-being or whatever indicator you use. This was clearly proven wrong because it underestimated our ability to innovate, for example, in the world's energy system, electronics, medicine, or food production. Instead, if you turn it around and say that the Earth's system has these biophysical boundaries, and that they are absolute, we don't, you know, we don't know them exactly because science cannot interview Earth. We cannot sit down and just talk to the planet and have it answer the question, "Where are your boundaries?" There will always be uncertainties, but if you have that safe operating space, then

I think there's a lot of, I mean there's a lot of debate, but there's a lot to support the idea that you can actually have a lot of continued growth, even in GDP terms, within the world within a safe operating space. You can move into, I mean, you could, in fact, have a circular economy operating and generating more and more wealth. It would never be 100% circular, but I mean you could largely confine yourself in a way that stays within, let's say, a biosphere which is not put under too large risks. So, changes in that way are a profound difference, and I think that was the, in a way, mistake of the "limits to growth" analysis. Because it talked about limits to growth, it had to make assumptions about technology, policy, even human needs. What the planetary boundary work and growth within limits does is that it, for a moment, which is a bit awkward for complexity scientists like ourselves, to take out humanity for a while and put it aside, and just ask this question: In the Earth as a system, where are the biophysical boundaries? And once those are defined, you put back humanity and then you allow humanity to innovate or move as she wishes within that playing field. And that becomes a much less threatening approach to development than making assumptions that kind of tie in — you know, so you cannot develop more than this or this. Of course, if we are not successful in finding innovative solutions, you will, in the end, hit the ceiling and then you come to a "limit to growth" situation. So, for climate, for example, we're there. But, you know, I think one of the big paradoxes which we seldom discuss is this: What if we're successful in a transition to a total solar energy future? Will that be sustainable? Well, it could be if we have the regulatory measures to ensure that the rebound effects do not just mean that we propel ourselves into even more accelerated consumption. So, you could actually see a situation where even a totally successful solution to the climate boundary could actually propel negative impacts on other boundaries. So, it's not as if sustainability in itself is the solution. It's really just recognizing that we need to stay within a certain type of playing field. For my students, I try to sometimes just make the analogy — it's like we're running the economy as if you were trying to play football without having the lines on the sides of your field. But in football, you know very well that as soon as the ball is outside of the court, the game stops, and then you play again. But inside the field, you can have both Lionel Messi innovating and you can have a person like myself who can barely hit the ball, but everyone can again fit into that field. So, that's the change I think in terms of analogy.

Simon Levin: Just to follow up on Ryan's question, your last point there, when you talked about what a full solar economy might do, it might change where the limits are, where we can go. And so — which I thought was Ryan's point — there's a feedback, and so maybe we can't take humans out, then define what those boundaries are, and put them back in because the boundaries co-evolved with technology and population growth.

Johan Rockström: That's a very important point. I mean when it comes to the climate system, I think there's a good, let's say, scientific basis to say that it's hardwired in terms of when, for example, the ice sheets melt irrespective of the driver. The driver could be nature, the driver could be human. But essentially, it's a physical system. But you're right — in the Anthropocene, we could geoengineer ourselves to a point where we could shift the position of boundaries, for example, and that kind of geoengineering does not necessarily have to be science fiction; it could be just the way we intensify the use of biomass. You're right. I mean the way we manage oceans could change the position, so you're right, and in the Anthropocene, this is an element that we haven't addressed in full: What happens when you couple and make analyses on different kinds of investments and big schemes of changing management?

Jamie McCaughey: Jamie McCaughey at the Earth Observatory of Singapore here at NTU. When faced with uncertain information about the future, people tend to wait and see. How can we better frame scientific uncertainty, and what kind of decision support might weaken the status quo bias and encourage collective action?

Johan Rockström: You know, that's such an important question. We had a colleague —I think he's somewhere here — talking about this yesterday that… I mean uncertainty needs to be, I think, translated, which we often do also as scientists, into risk, and [we] talk more about risk, and risk in terms of consequences, as far as we can do analytically, which is often difficult. But I think just talking about uncertainty and kind of standard deviations is not very helpful; we need to somehow take it to risk and consequences. One discussion we had yesterday is the fact that perhaps the best industry in the world to do this is the reinsurance industry that basically holds the risk of all natural disasters in the world — companies like Swiss Re and Munich Re. And we're increasingly trying to learn from them on how they deal with uncertainty because they're putting insurance premiums on every risk you can dream of — I mean everything from a drought, a flood, an earthquake, a nuclear power breakdown. And so, I think we as scientists also have a lot to learn from how industry, particularly the insurance business, is dealing with these kinds of uncertainties. So, that's one area. But the second one is, of course, like a donkey shot, just continue to fight — fight the windmills on the need to internalize the precautionary principle. I mean, I did this — the economists' statement here of the need for a certain degree of nervousness as a reasonable response is written in, as you know, into the 1992 UN declaration from the first Rio summit, that the precautionary principle must apply when we have uncertainty. It's not being applied anywhere, but you know it's being applied in so many other areas. I think that… I mean if you look at any business leader or any politician, they're so clever in taking decisions under uncertainty, but when it comes to uncomfortable decisions, we tend to use uncertainty as an excuse for not acting. So, it's not as if we don't act under uncertainty, it's just that we're unwilling to do so in some circumstances. So, I think we need to work on both these areas.

10.3 Summary of the Talk

Johan Rockström's talk delves into the complex interplay between human activities and the Earth's systems, emphasizing the urgent need for a paradigm shift toward sustainable development within planetary boundaries. The following is a detailed summary of his insightful presentation:

Rockström begins by highlighting the critical importance of understanding the Anthropocene, an epoch characterized by human-induced changes to the Earth's systems. He underscores the need to recognize the interconnectedness of various global challenges, from climate change to biodiversity loss, and their potential to trigger catastrophic tipping points.

Central to his discussion is the concept of planetary boundaries, which delineate the safe operating space for humanity within the Earth's environmental limits. These boundaries encompass key biophysical processes such as climate change, biodiversity loss, and freshwater availability, all of which are essential for maintaining a stable and habitable planet.

Rockström emphasizes the significance of resilience in social-ecological systems and identifies diverse and redundant systems as essential attributes for withstanding disturbances. He contrasts this with conventional economic models focused on optimization and efficiency, highlighting the need for a paradigm shift in economic thinking to prioritize resilience-building measures.

The talk explores examples of regime shifts in various ecosystems, from coral reefs to cod fisheries, illustrating how human activities can push these systems beyond critical thresholds, leading to irreversible changes and loss of ecosystem services.

Rockström discusses the planetary boundary framework, which identifies nine critical Earth system processes that must be safeguarded to prevent catastrophic environmental changes. He stresses the need for science-based targets and policy interventions to stay within these boundaries and ensure a sustainable future.

The presentation also addresses the social dimensions of sustainability, including the importance of trust and cooperation in fostering innovation and addressing global challenges. Rockström highlights the Nordic countries' success in building trust among strangers, facilitating societal resilience and adaptation.

Moreover, he underscores the role of cultural values and behavioral change in shaping sustainable development pathways, advocating for a shift toward growth within planetary limits. He emphasizes the need for integrated approaches that consider both environmental sustainability and social equity, citing initiatives such as the "donut model" proposed by Kate Raworth.

Rockström concludes with a reflection on the interconnectedness between humanity and nature, drawing attention to the spiritual and ethical dimensions of sustainability. He shares a photograph of a remote lake in the Nepalese mountains, symbolizing the universal reverence for the natural world and the shared responsibility to protect it.

Overall, Rockström's talk provides a comprehensive overview of the challenges and opportunities facing humanity in the Anthropocene. It calls for bold action and transformative change to ensure a sustainable future for generations to come.

10.4 Relevance of the Talk to the Current Stage of Research

The talk by Johan Rockström remains highly relevant to the current stage of research for several reasons:

Planetary Boundaries' Framework: Rockström's discussion of the planetary boundaries' framework continues to be relevant in guiding research efforts to understand and address global environmental challenges. This framework provides a scientifically grounded approach to identifying key Earth system processes and defining safe operating limits for human activities.

Interconnectedness of Global Challenges: Rockström emphasizes the interconnectedness of various global challenges, such as climate change, biodiversity loss, and freshwater scarcity. This holistic perspective remains crucial for interdisciplinary research efforts aimed at addressing complex environmental issues.

Resilience and Adaptation: The concept of resilience in social-ecological systems, as highlighted by Rockström, remains a central focus of research on climate adaptation and sustainability. Understanding the factors that contribute to resilience, such as diversity and redundancy, informs strategies for building adaptive capacity in the face of environmental change.

Behavioral Change and Values: Rockström's discussion of the importance of behavioral change and cultural values in driving sustainable development pathways underscores the need for research on human behavior and decision-making. This includes studies on public attitudes toward sustainability, as well as interventions aimed at promoting pro-environmental behaviors.

Policy Implications: The talk underscores the importance of science-based targets and policy interventions to stay within planetary boundaries. Research on policy effectiveness, governance mechanisms, and international cooperation remains essential for translating scientific knowledge into actionable policies and strategies.

Integration of Social and Natural Sciences: Rockström advocates for integrated approaches that consider both environmental sustainability and social equity. Research efforts that bridge the gap between social and natural sciences, such as sustainability science and transdisciplinary research, are crucial for addressing complex socio-environmental challenges.

Global Collaboration and Partnerships: Finally, Rockström's emphasis on global collaboration and partnerships reflects the need for collective action to address planetary challenges. Research initiatives that foster collaboration across disciplines, sectors, and geographical boundaries are essential for advancing our understanding of global environmental issues and identifying solutions.

In summary, Rockström's talk remains highly relevant to the current stage of research by providing a comprehensive framework for understanding and addressing the complex environmental challenges facing humanity in the Anthropocene.

Chapter 11

Isotope Chemistry, Climate Change, and the Fate of the Chinese Dynasties: Implications for the Future of Asian Societies

Wang Xian Feng

YouTube: https://www.youtube.com/watch?v=QvZYV9dRECc

Speaker: Wang Xian Feng

Moderator: Kerry Sieh

Discussants: Kerry Sieh, Guaning Su, Peter Schwartz, Tso-Chien Pan, Kevin Xiao, Discussants A, B, and C

11.1 Talk by Wang Xian Feng

Wang Xian Feng: Thank you! Thank you, Kerry. I'm very happy to have this opportunity to talk to you and to share with you a story that blends geology, chemistry, climate, and social science. So, maybe you are also wondering, of course, it is a complex story, but perhaps you are wondering how this guy can do it? Anyway, it's just a simple view from a kid. But I would argue that this is exactly fitting for this workshop — we are looking at a crude view of the whole.

So, in my talk, I want to discuss four things.

Isotope chemistry, climate change
and the fate of the Chinese dynasties

I. Cave calcite records: Dating & interpretation

II. The Chinese cave records
1. Hundreds of thousands of years
2. Last 1,800 years: ties to Chinese culture

III. Ties to climate and culture elsewhere
Guge Kingdom, Greenland and Mesoamerica

IV. Implications for the future of Asian societies

First, I want to outline why we are interested in examining these specific geological rocks found in caves, the types of data we are seeking, and how we interpret these data to construct a narrative about these rocks. Second, I will present two types of records generated from studying these rocks, which we collected from caves in China as well as other locations such as the U.S., Brazil, and India. Additionally, we aim to establish connections between the findings from the rock samples, particularly regarding variations in chemical composition and changes in Chinese culture. Given China's extensive history and well-documented cultural evolution, we are eager to compare our data, interpreted as indicators of climate change, with Chinese historical records. Third, I will discuss how the phenomena we observed in China are not unique to that region but can be observed globally. Lastly, I will address a pressing question: Can we confidently assert that "the past holds the key to the future?"

Isotope chemistry, climate change
and the fate of the Chinese dynasties

I. Cave calcite records: Dating & interpretation

II. The Chinese cave records
1. Hundreds of thousands of years
2. Last 1,800 years: ties to Chinese culture

III. Ties to climate and culture elsewhere
Guge Kingdom, Greenland and Mesoamerica

IV. Implications for the future of Asian societies

How many of you have visited the caves, by the way? Quite a lot, that's great. If you haven't, I would strongly suggest that you also go to see some caves. It's a very different world underground.

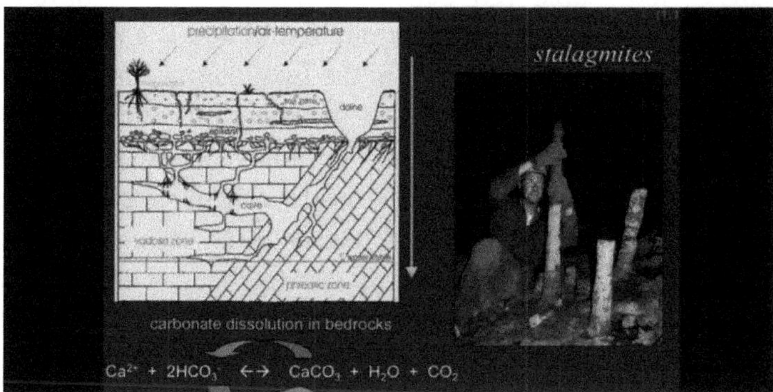

So, if you visit a cave, most likely you will see a variety of chemical formations. Some are hanging from the wall, some are growing from above, some from the ground. For example, like this one, the stalagmite, this is a particular chemical formation we want to work with. Simply speaking, the formation of the stalagmite is due to rainwater. When rainwater reaches the ground, soaking the soil, the soil becomes saturated with carbon dioxide due to organic decay and root respiration. As this water dissolves the carbon dioxide, it becomes acidic. Then, as this water passes through the cracks in the limestone, it dissolves the calcium carbonate from the limestone. This is simply a carbonate dissolution process, where calcium carbonate reacts with water and carbon dioxide to form calcium ions, bicarbonate ions, and a small amount of carbonate ions. When this water enters the cave, because the carbon dioxide concentration or partial pressure in the water is higher than the carbon dioxide partial pressure in the cave air, the CO_2 can degas, moving out from the water into the cave air. This causes the calcium carbonate concentration in the water to become supersaturated, leading to the precipitation of calcite. This is simply a calcite precipitation process.

So, Doug Erwin mentioned yesterday that, as chemists, we can excel at timing. So, this is related to the radioactive isotopes, specifically looking at uranium-238, one of the major isotopes of uranium. This is better explained by my Ph.D. advisor, Larry Edwards, as Kerry mentioned, who developed this during his Ph.D. at Caltech. Uranium-238 has a half-life of around 4.5 billion years. This means that after 4.5 billion years, half of the atoms of uranium-238 will have decayed to form the daughter isotope. Only half of the atoms can persist. After another 4.5 billion years, I believe, only one-quarter of the uranium-238 atoms can remain in the system. But here, we are interested in the two intermediate daughter isotopes of this decay, uranium-234 and thorium-230. Their half-lives are moderately long, around a quarter of a million years. So, we can use this system to date geological samples, mostly calcite or sometimes other minerals. We can date samples as young as a few thousand years up to several hundred thousand years. As you can see, we can also date samples very precisely. Of course, this is possible due to how well we handle the samples in the chemistry lab, as well as the advancement of the mass spectrometer, by the way.

What I'm showing here is the rock slab we cut to obtain a polished surface of a sample. Whenever calcite forms from precipitation from water, uranium tends to enter the rock because, under surface conditions, uranium has very high solubility in water due to its positive charge. However, thorium does not; thorium has very negligible solubility in water. So, this is how we can separate these two elements in the rock. Simply put, by examining this sample, we can count atoms of uranium and count atoms of thorium. We can assume that thorium-230 all comes from or is decayed from uranium-234, and then we can simply conduct measurements on the mass spectrometer, apply the equation, and obtain the age of the sample.

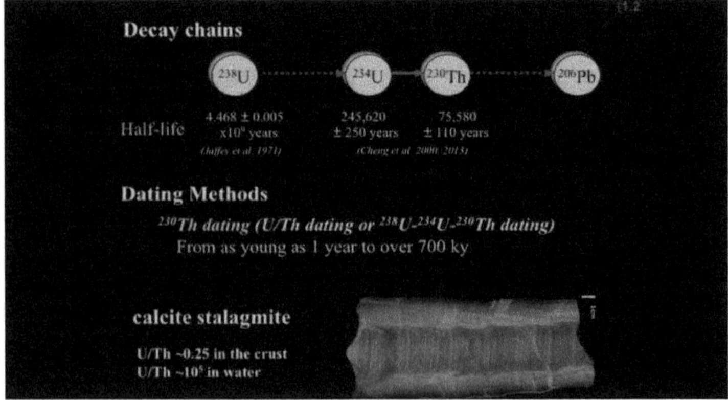

So, in geochemistry, we can not only determine the timing of rocks very accurately but we can also perform many additional analyses. Here, we are examining stable isotopes, specifically oxygen, an element crucial for our survival. If we look at the equations for the calcite dissolution process and the precipitation process, we find that calcite and water share the same element, oxygen. By analyzing the oxygen in calcite carbonate samples, we can trace the oxygen back to the drinking water source. We can further trace the drinking water source to surface water or even to clouds, the water vapor in clouds. Oxygen also exists in isotopic forms, with three stable isotopes, but the most common ones are oxygen-18 and oxygen-16. They have the same number of protons in their nucleus but differ in the number of neutrons. Oxygen-18 has 8 protons and 10 neutrons, while oxygen-16 has 8 neutrons and 8 protons. These two isotopes can undergo fractionation, despite their similar chemical behaviors. During physical processes, they fractionate, albeit with tiny changes. By utilizing a mass spectrometer, we can detect such changes more accurately. These changes are relative to temperature, precipitation conditions, and sometimes the source of the water.

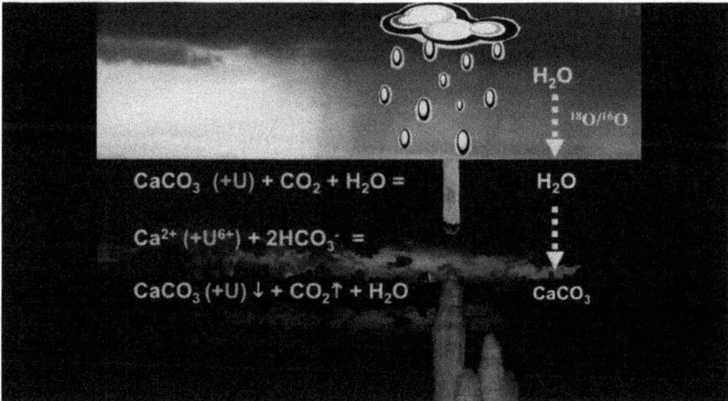

Living in Asia, we are familiar with the wind patterns and rainfall, for example, in Singapore. During summertime, the surface winds are dominated by a large airflow from the ocean to the continent. This is primarily due to the temperature gradient between the continent and the ocean. Many of us have probably heard about the Indian summer monsoon and the East Asia summer monsoon. Meteorologically speaking, they essentially belong to one system. When the wind blows from the Indian Ocean to the Indian subcontinent and moves further to the Indochina Peninsula, the wind also brings potential rainfall to the Chinese coast and further inland.

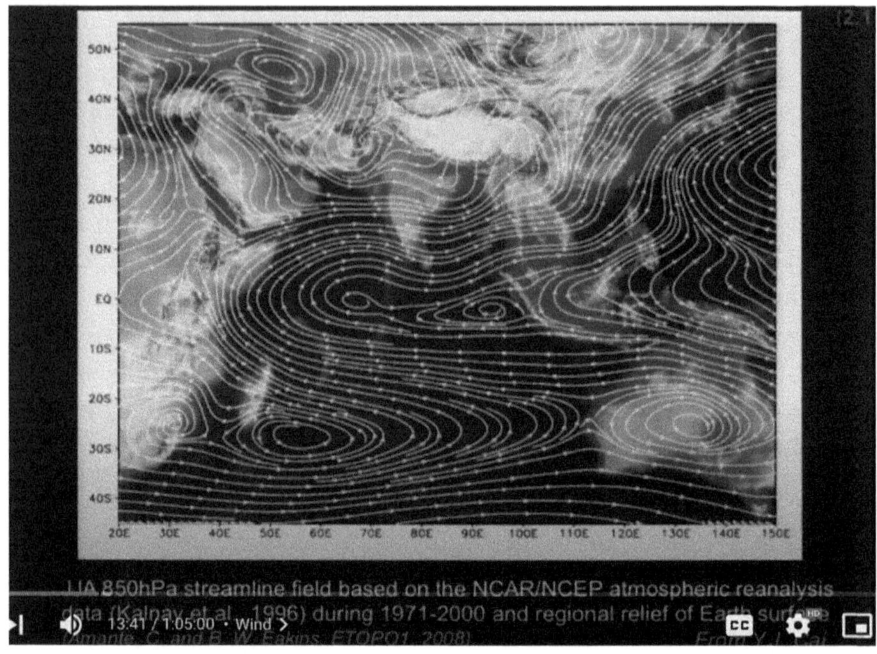

Over the last decade, I have been working very closely with my colleagues in the U.S. and China, along with a couple of colleagues from India. We have studied caves and cave samples from China, as well as a few locations in India. Our aim has been to analyze the cave samples in the lab and conduct chemical studies on them to extract climate information preserved in the rocks. Here, what I am presenting is the result of our collective efforts. Based on the observational data from a dozen meteorological stations, we have observed changes in the isotopic ratio of rainfall during the summer months (June, July, and August) when the monsoon arrives. These changes result in a depletion in the O18/O16 ratio of the rainfall. From this diagram, we can infer that as the monsoon intensifies and brings higher rainfall, the O18/O16 ratio of the rainfall decreases.

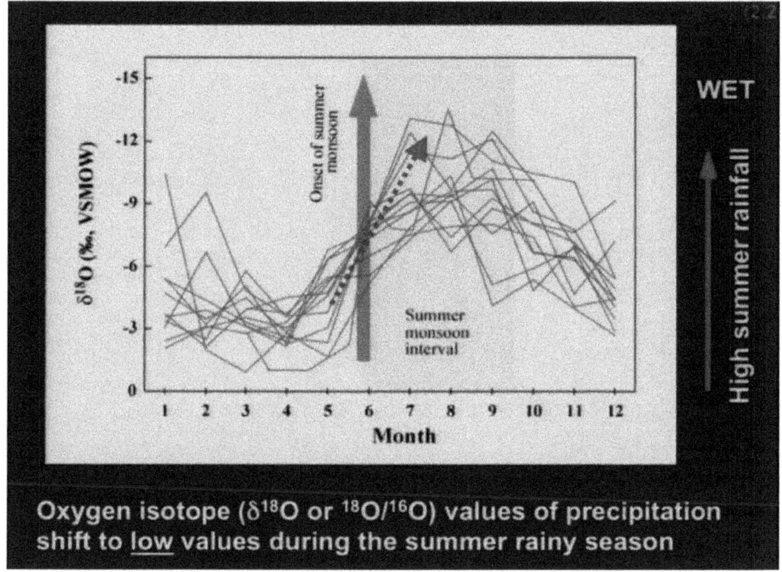

Oxygen isotope ($\delta^{18}O$ or $^{18}O/^{16}O$) values of precipitation shift to <u>low</u> values during the summer rainy season

So, this is a very important observation. Of course, quantitatively, we still face challenges in deriving the link between rainfall and monsoon patterns. But broadly speaking, we can say, the lower the O18 value, the lower the O18/O16 ratio, which suggests stronger rainfall, a stronger monsoon, and higher rainfall. So, please remember, this is essentially the observation we can draw from today's data and apply to what we can extract from geological rocks.

We have been very successful in compiling monsoon records by studying cave samples. What I am showing here is records spanning the last three hundred thousand years that we have assembled by combining different samples and geochemical records. As you can probably see, there are numerous fluctuations, indicating variations in O18 values. The interesting aspect is that we can compare these geological records with insolation. Insolation describes how our planet orbits the Sun, considering that it's not a perfect circle and the Earth is tilted from the vertical axis, leading to seasonal variations. Insolation is a parameter that describes this interaction. It turns out that different parameters can affect how the Earth receives solar radiation. Clearly, we can observe that these two curves match each other beautifully.

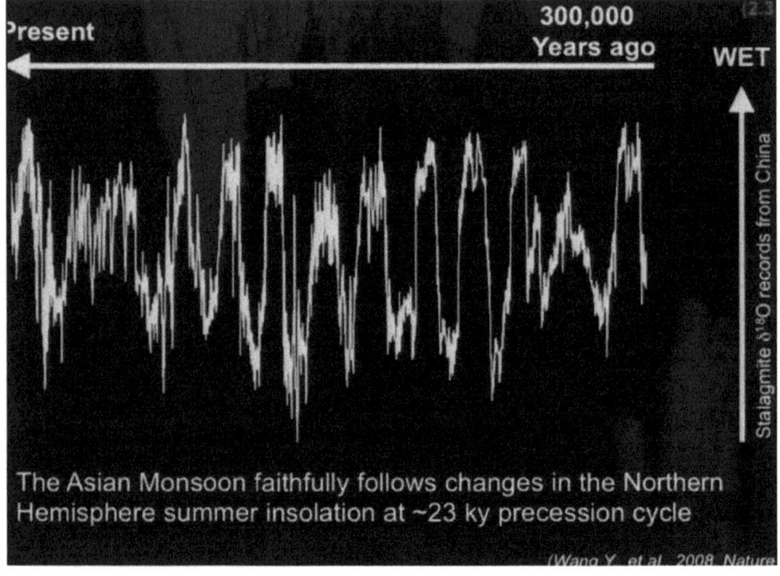

But of course, the main record I want to show you is for the last two thousand years, or in the way that converts the time zone to a more comfortable one, as Johan just mentioned in the previous talk.

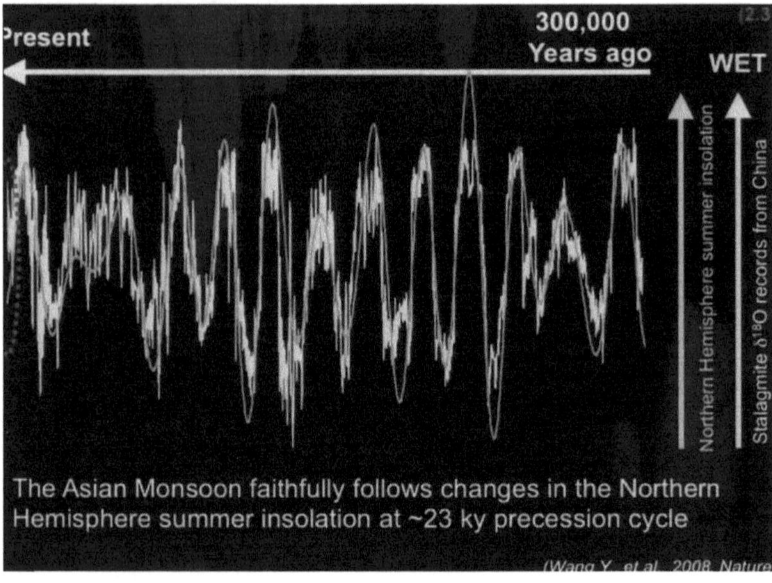

So, this is the record we developed from cave rocks collected from Wan Shang cave. This cave, by the way, is located very close to today's monsoon fringe area, in a place situated between Xi'an and Lanzhou, Jiangsu. As you can see, the light blue area indicates the fringe of the monsoon.

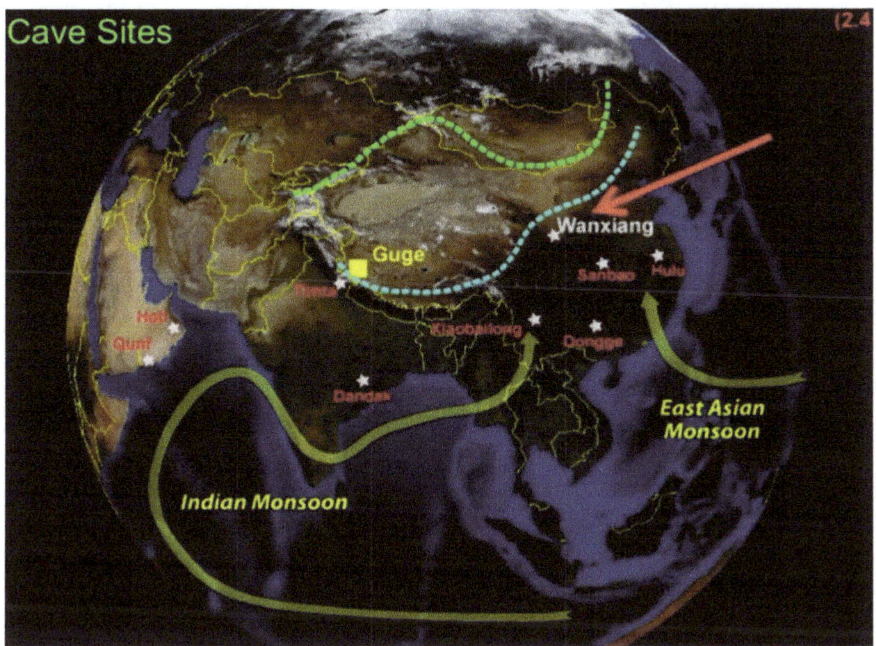

So, here's the record. By the way, if we look at the *x*-axis, it shows the timing, representing a time series. The time spans from 800 AD to the present, 2,000 AD, and we have also included the names of Chinese dynasties from the Han to the Qing Dynasty. Additionally, you'll notice these tiny pink bars, which represent the uranium, thorium dates we obtained from the samples. As you can see, we can obtain very precise dates, with the bars typically spanning only around two years. However, considering other uncertainties, the error bars are typically on the order of five years as well. On the *y*-axis, we have the O18/O16 ratio. It's important to note that when the value is lower, it indicates a stronger monsoon and higher rainfall.

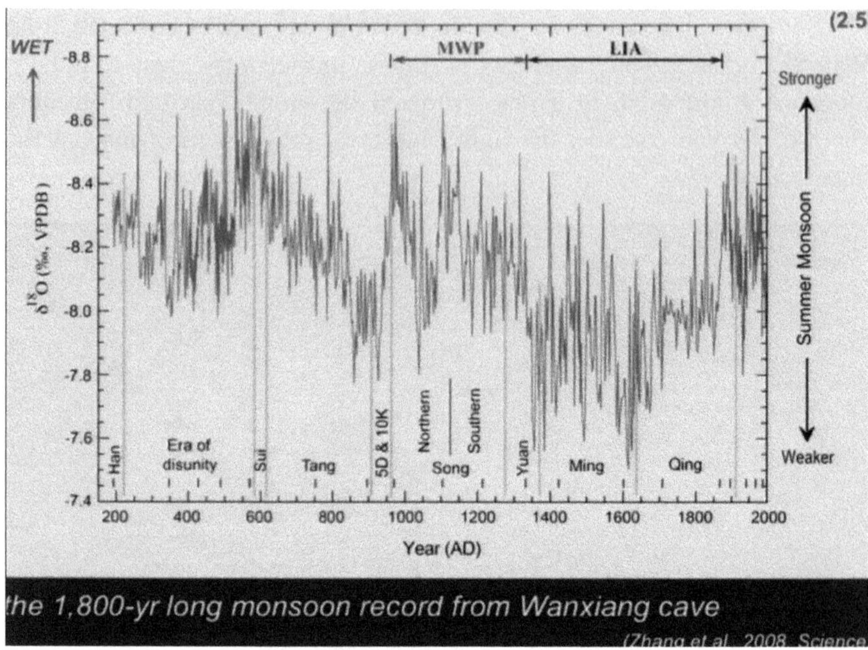

Of course, when we aim to interpret our record as a signal of rainfall, we always question how robust or confident we can be in our findings. Initially, we questioned ourselves, and the approach we took was to compare our findings with historical documents. Our Chinese colleagues compiled regional chronicle documents from their respective counties, which greatly benefited our work. This compilation resulted in what they called the drought and flood index. They quantified this index by simply counting how many years, over a certain time period (e.g., ten years), experienced drought or flooding. On their scale, a value of one indicates severe drought, 0.5 indicates mild drought, zero indicates a neutral climate, and 2.1 indicates severe flooding. Remarkably, you can observe that these two records or curves match each other quite well. The remarkable aspect here is that they represent two completely different sources of information: one derived from geological and geochemical variations developed from rocks, while the other originates from literature documents.

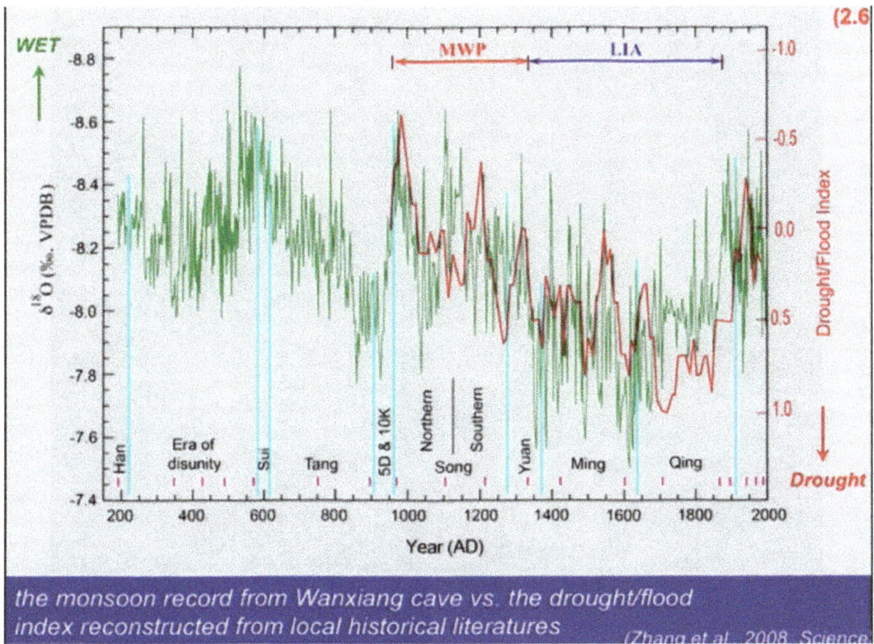

This is truly remarkable; the alignment of these two curves is astonishing. It gives us a strong sense of confidence that our record accurately captures the rainfall changes in this region. However, can we make any assertions about the monsoon by solely examining one cave sample?

To address this question, we sought to compare our record with another cave record developed from the Dandak cave in central east India. These findings were reported by a group from the University of Southern California.

That record from the Dandak cave may not be as extensive in duration, but it does have a period of overlap with the record from the Wanxiang cave, as you can see.

Once again, this shows a beautiful correlation, a very consistent pattern between the two records: One is from India, located upstream of the monsoon, while the other is located further downstream, close to the fringe of the monsoon. This further emphasizes that our record can indeed represent changes in the monsoon.

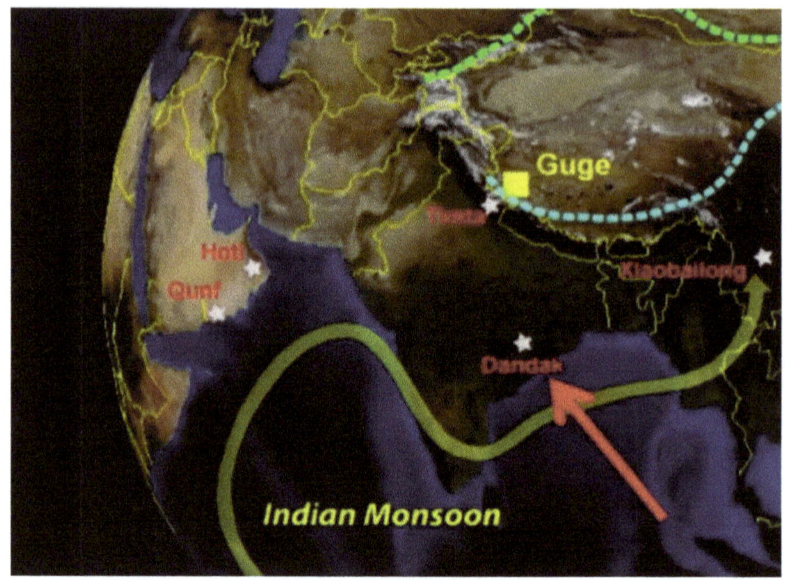

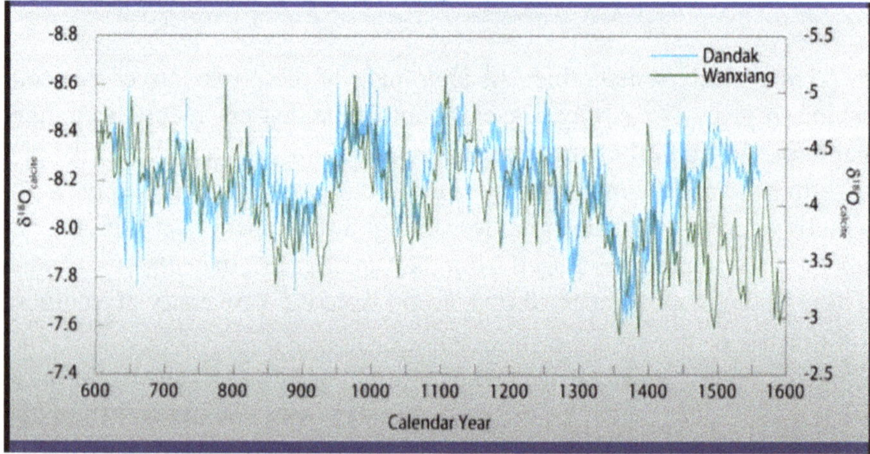

As I mentioned earlier, China is very rich in culture and has very well-documented historical records. Since we aim to interpret our record as a climate signal, as a change in the monsoon, we want to compare it with historical documents. Here, I'll walk you through Chinese history, focusing on five particular time periods. Two of these periods correspond to lower O18 values, indicating a weak climate but abundant rainfall.

The other three periods represent times when the climate wasn't favorable, indicating a dry period.

So, the first snapshot we're looking at corresponds to almost the lowest value of the O18/O16 ratio. We suggest that this probably indicates the highest rainfall over the last 2,000 years. This corresponds to the Northern and Southern Dynasties period in Chinese history. The Northern Qi, one of the kingdoms dominant during that time, controlled the northern part of China.

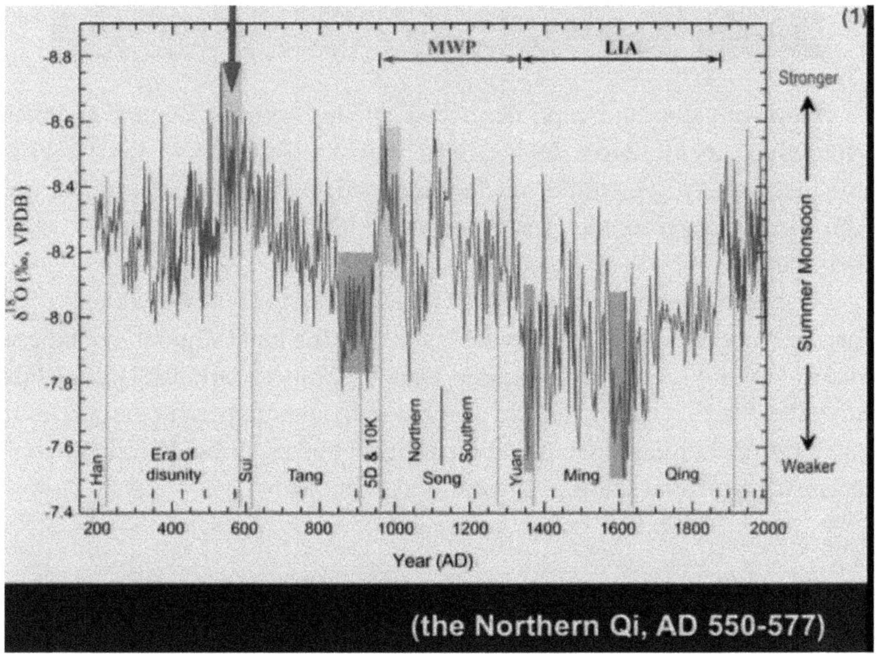

One of the popular songs written during that time is the Chinese Shepherd's song. I still remember learning this song when I was a child, and in fact, it is still taught in Chinese elementary schools. "The boundless grassland lies beneath the boundless skies. When the winds blow and grass bends low, my sheep and cattle will emerge before your eyes." It's a very romantic and beautiful scenery, depicting a romantic life.

However, the landscape described by this song is located in Inner Mongolia, very close to today's desert, the Gobi Desert. If we were to visit this place today, we would not feel inclined to sing this song anymore. This clearly suggests that 1,500 years ago, the climate was much wetter than today.

The second period I want to discuss is the Tang Dynasty, which lasted for a few decades. The Tang Dynasty reached the highest point in Chinese history, marking one of the golden ages of Chinese culture. During this time, the Tang Dynasty was one of the dominant countries in the world. It had a significant influence on surrounding countries through military conquests, commercial trading, and cultural exchanges.

However, if we look at the last few decades of the Tang Dynasty, we find that the O18/O16 ratio in the record was low, suggesting possible wartime conditions during that period. In fact, if we examine Chinese historical documents, we find them well-written, detailing multiple droughts that occurred during that time.

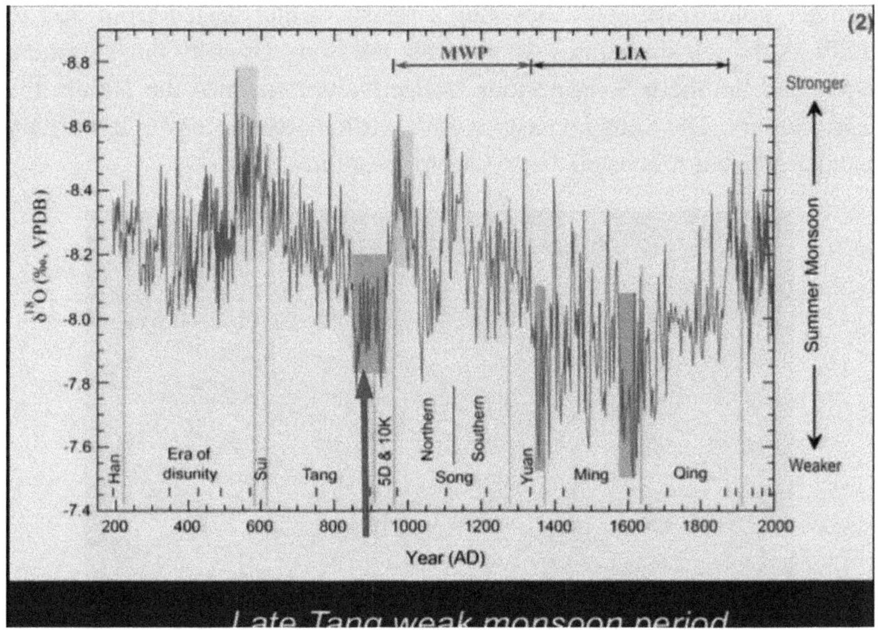

For example, in AD 875, there was a severe drought in China, particularly in the central part of the Tang Kingdom. Locals suffered greatly, and in fact, there was a large-scale peasant uprising. Huang Chao, one of the leaders of the peasant uprising, was a significant figure during this time. Although the Huang Chao uprising didn't immediately bring down the Tang Dynasty, it clearly destabilized it, eventually leading to its downfall.

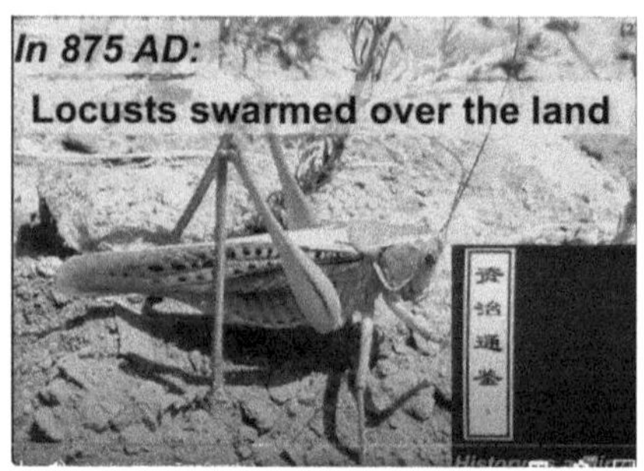

So, 50 years after the collapse of the Tang Dynasty, China entered another golden age with the Song Dynasty, which lasted from 960 to 1279. Although sometimes we separate the Song Dynasty into Northern Song and Southern Song periods, collectively it spanned the 10th to the 13th century. The Song Dynasty wasn't particularly successful in military campaigns, but it boasted a very strong economy.

So, if we look at the records we have obtained from the Chinese chronicles, we find that they correspond to another period of very low values of O18/O16, indicating very wet time periods. Indeed, if we examine Chinese history, during the early decades of the Song Dynasty, the Chinese population dramatically increased, more than doubling within about fifty or sixty years. Rice became the staple of the Chinese diet, and rice cultivation dramatically increased. Rice was even cultivated in northern China, close to where Beijing is located today, which is something unimaginable today.

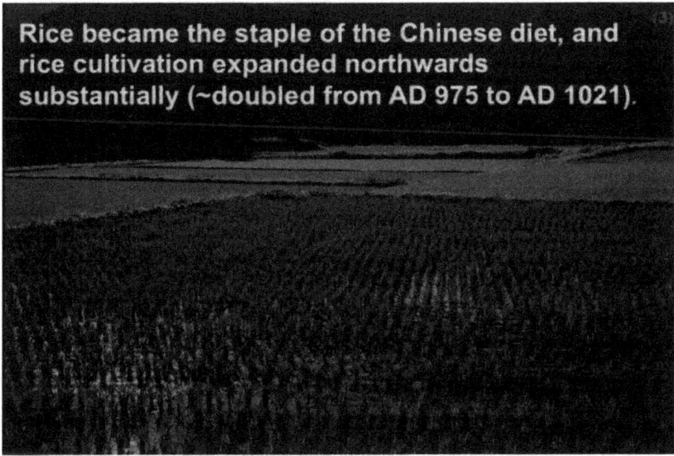

However, the Song Dynasty was eventually conquered by the Mongol Empire. Kublai Khan invaded the Song Dynasty in 1279, leading to the establishment of the Yuan Dynasty of the Mongol Empire.

However, the Yuan Dynasty wasn't successful in maintaining its rule, despite its military prowess. The Mongol Empire couldn't extend its dominion to East Europe or even close to the Mediterranean Sea.

If we examine the records, we find that most of the time during the Yuan Dynasty, the climate wasn't very favorable. Particularly during the end of the Yuan Dynasty, in the few decades leading up to its collapse, our O18/O16 ratio shows quite high values, suggesting possible droughts. We refer to this time period as the "Late Yuan weak monsoon period."

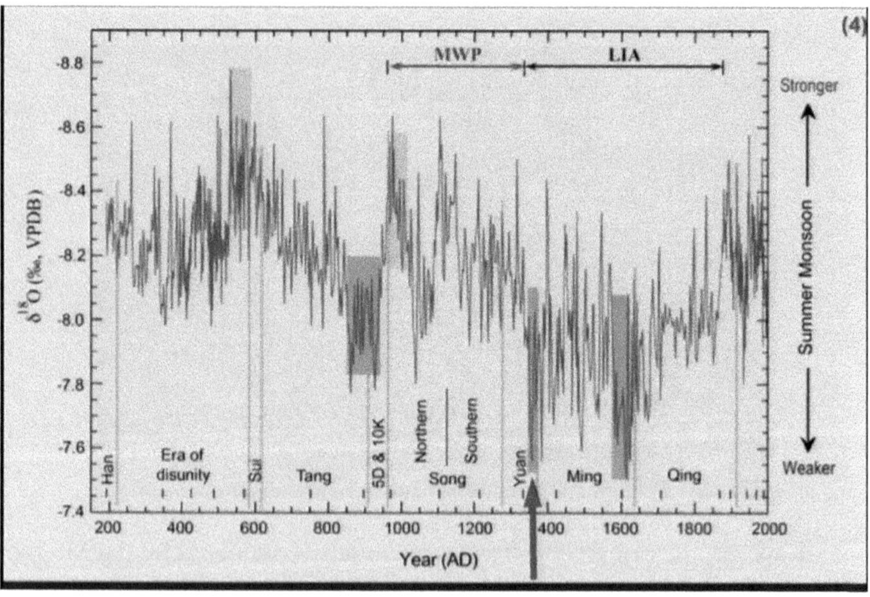

Indeed, multiple droughts occurred during that time, leading to numerous peasant uprisings. One of these uprisings was led by Zhu Yuanzhang. Later, he led his soldiers and captured the capital of the Yuan Dynasty, which is where Beijing is now located. Interestingly, his parents and his son all perished during the droughts.

The Yuan Dynasty itself also faced challenges with the climate. If we look at our records again, we see that for the majority of the Yuan Dynasty, our O18/O16 ratios remained quite high, particularly during the last few decades. The values of the O18/O16 almost reached the highest levels over the last two thousand years. We refer to this time period as the "Late Ming weak monsoon period."

As for the last emperor of the Ming Dynasty, Chongzhen, he wasn't a bad emperor by that time. However, he wasn't lucky enough. As you can see, through most of his reign, there were multiple years of drought, starting from 1637 to 1643. In fact, the drought was so severe that, according to Chinese historical records, [one like] it had never been seen in the past five hundred years. This drought affected both northern and southern China and caused numerous uprisings.

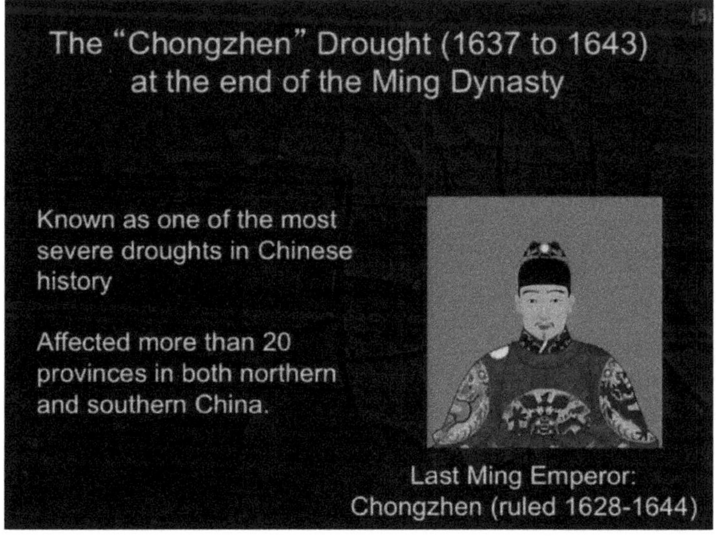

The widespread drought also played a role in aiding the rebel leader Li Zicheng in overthrowing Chongzhen in 1644. Chongzhen, the emperor, ultimately committed suicide in Beijing.

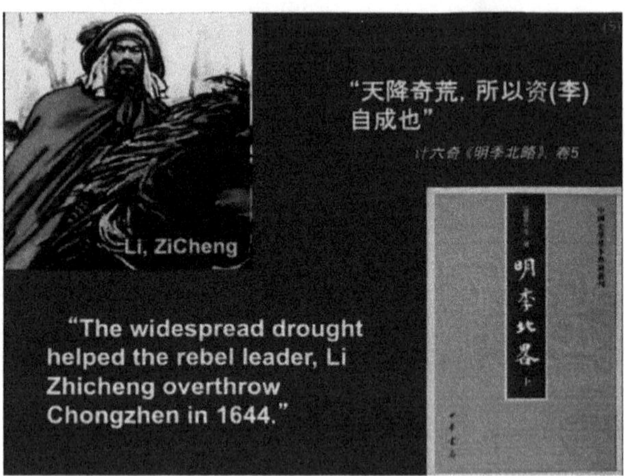

So, I hope that through the last few slides and the examination of the five time windows, I have convinced you that while other factors certainly influenced Chinese culture, the climate indeed played a significant role. Of course, social and political problems such as invasions and civil wars could also easily destabilize kingdoms. However, one could argue that climate played a significant role as well.

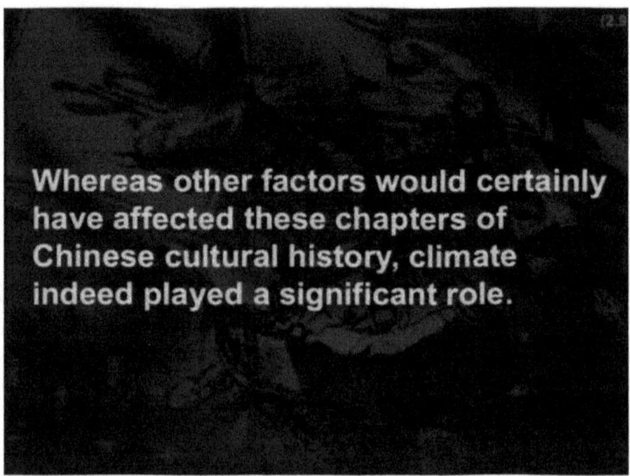

So, in fact, this concept is not something new. One of the ancient Chinese philosophers, from 2,800 years ago, had already made this point very clearly. He said, "Dynasties would collapse if people were poor without harvest from water and land. In the past, the Xia Dynasty collapsed when the Yi and Luo Rivers dried up, and the Shang Dynasty collapsed again when the Yellow River dried up."

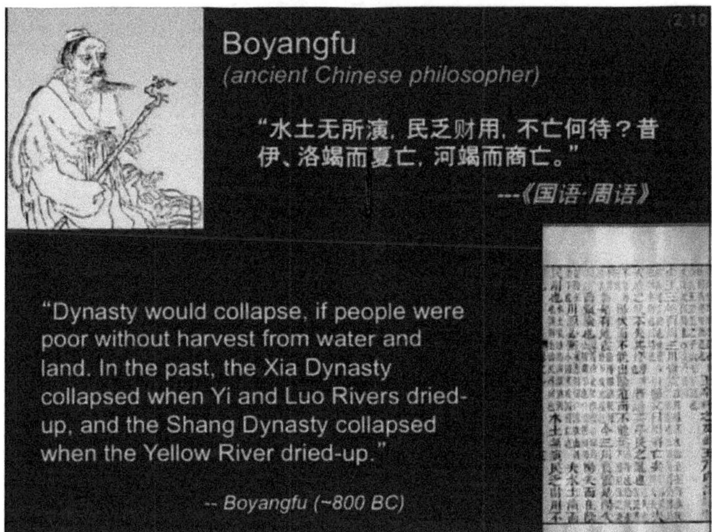

I want to show you an additional example to further argue that this phenomenon is not just limited to ancient China; we can also see similar examples elsewhere. For instance, one of the kingdoms that thrived in the western Tibet region was the Guge Kingdom. The Guge Kingdom began around 950 AD, but around 1630, this kingdom disappeared very suddenly and mysteriously.

When the Guge Kingdom flourished, it played a significant role in transporting culture and fulfilling economic roles in Tibet, particularly in spreading Buddhism to the heart of Tibet. During its peak, there were more than a hundred thousand people living in this region, despite its harsh environment. Interestingly, if we correlate the beginning and end of the Guge Kingdom with the trends of the monsoon, we find a beautiful match: It thrived during times when the monsoon was strong but collapsed when the monsoon weakened.

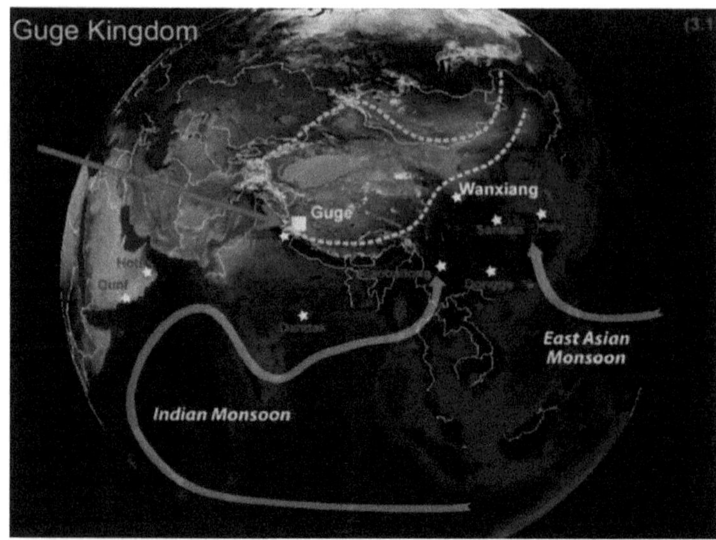

For our European guests, you may have already heard about the well-documented archives regarding the retreat and advance of empires, the Medieval Warm Period and the advances of glaciers during the Ice Age. Interestingly, if we correlate these events, we can find that the Song Dynasty almost matches the Medieval Warm Period, while the Ming and Qing Dynasties correlate with the later Ice Age. Of course, the exact timing of these events may vary from place to place. Those familiar with Viking history may have heard about the colonization of Greenland, which began in 985 AD, coinciding with the start of the Medieval Warm Period. However, the settlements in Greenland were dramatically abandoned during the mid-14th century or as late as the 15th century, around the time of the Little Ice Age.

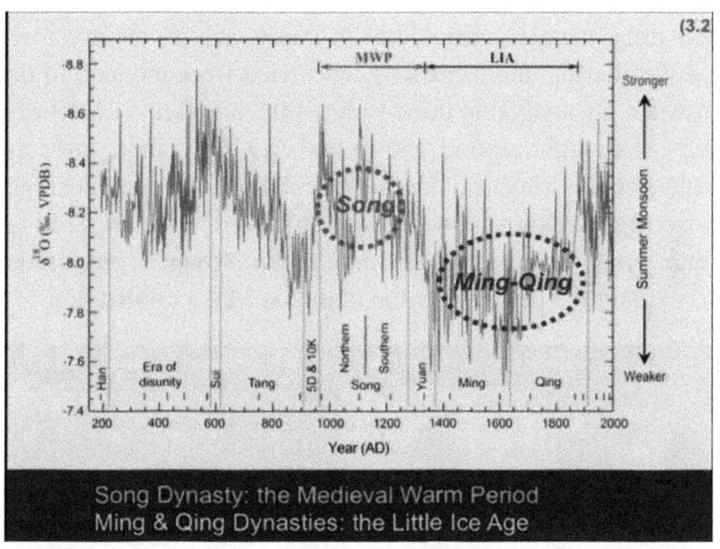

This is another poignant example that unfortunately illustrates how a society was unable to adapt to climate change. We can also observe similar instances of civilization collapse in Central America, particularly with the Maya civilization, which flourished in the region. The Maya built spectacular monuments in the rainforest, but their classical civilization collapsed around the mid-ninth century. Our colleagues have studied marine sediments from the southern part of the Caribbean Sea. They examined trace metals, particularly titanium, which is very resistant to chemical weathering and can easily be carried from the land to the ocean by river water. By analyzing these sediments, they found evidence of multiple years of drought around 850 to 900 AD. Of course, there are other hypotheses for the Maya civilization's collapse, such as tensions within society, overpopulation, and deforestation leading to changes in land use. However, climate change remains one of the strongest proposed mechanisms to explain the collapse of the classical Maya civilization.

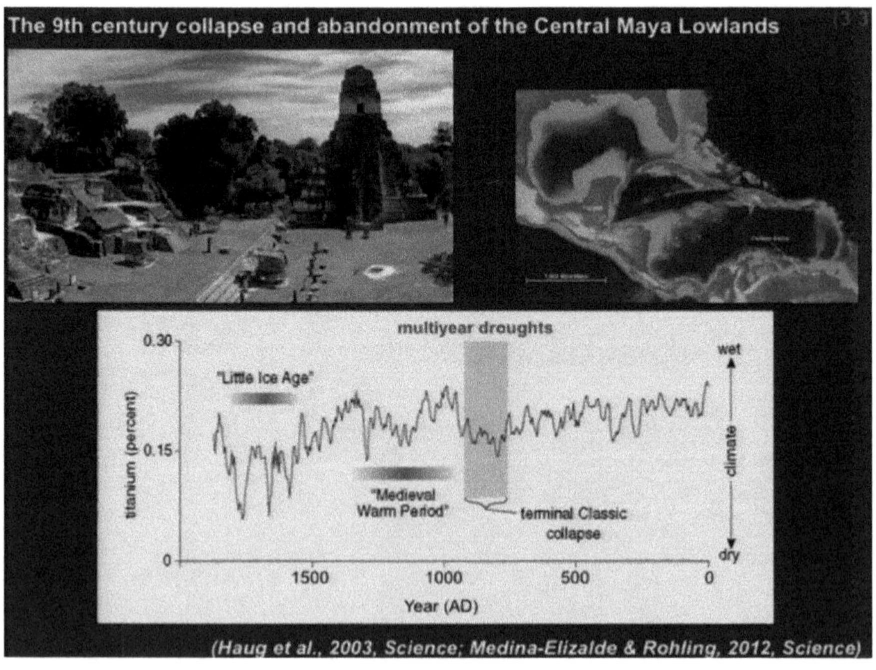

I hope I have presented enough evidence to convince you that there are certainly indicators we can examine to understand how climate change may have affected our civilizations. By studying paleoclimate records, we can investigate how climate change could have impacted our societies. So, what can we learn from this? Can we use the past as a guide for the future?

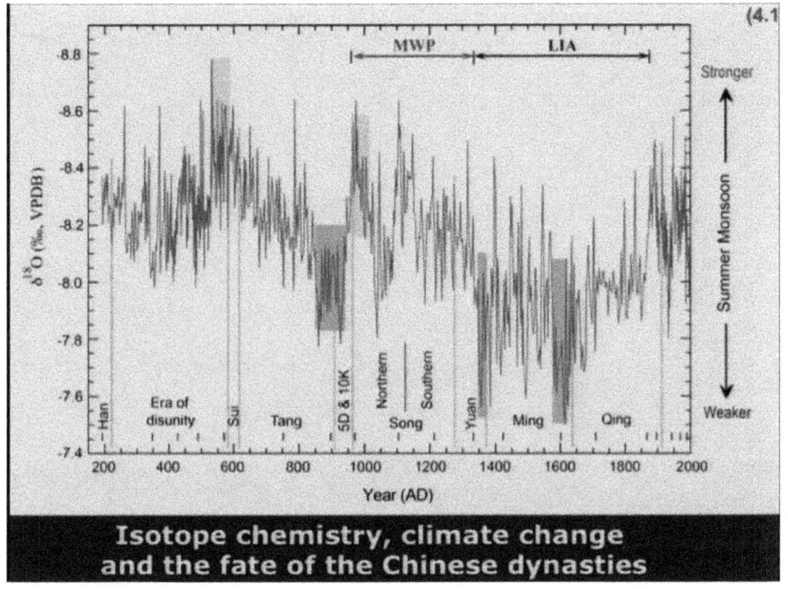

Johan already discussed this with some slides from the ice core. The ice core record is indeed a part of the Holy Grail of the Epica DOME study, representing a truly international effort. A group of European and U.S. scientists drilled the ice core from Antarctica, which is more than 3,000 meters long. What they were able to do was analyze the oxygen and hydrogen isotopic ratios from the ice, which allowed them to infer temperature changes as shown here in red. Of course, the depth of the ice core down here corresponds to the integration time and the ratio time.

Another dramatic finding they made is the change in carbon dioxide levels. They were able to extract carbon dioxide from air bubbles preserved in the ice, providing a record of atmospheric carbon dioxide levels. Just for your information, our modern human civilization flourished during what we call the Holocene, a comfortable time zone for the last ten thousand years.

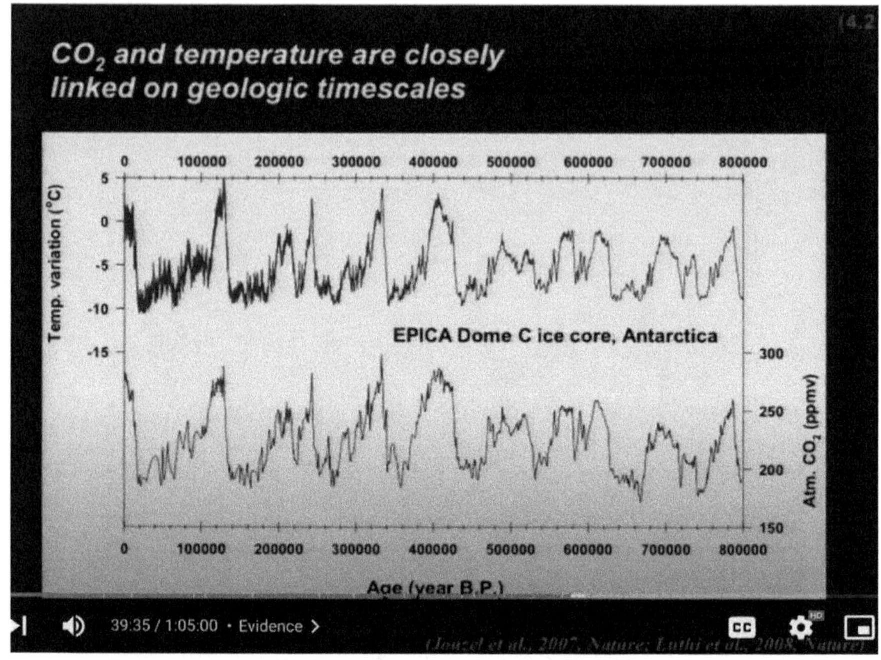

Additionally, I'd like to draw your attention to the changes in atmospheric carbon dioxide levels over the past many years, up to eight hundred thousand years. The concentration of carbon dioxide in the air changed from around 180 parts per million by volume to 280 parts per million by volume.

However, if you look at today, over the last 150 years or so, the concentration of carbon dioxide in the atmosphere has changed dramatically. Today, it is almost 400 parts per million by volume (ppmV). This increase in carbon dioxide levels is consistent with the rise in temperature observed since the Industrial Revolution, which is already around one degree Celsius.

What we are trying to argue here is that by examining paleoclimate records, we can study and characterize the patterns of climate change and understand how the climate system operates. However, it's evident that human influence already exists, surpassing the bounds of natural variability. With a global population of seven billion, we inhabit the only habitable planet in the solar system, and possibly the only one in the entire

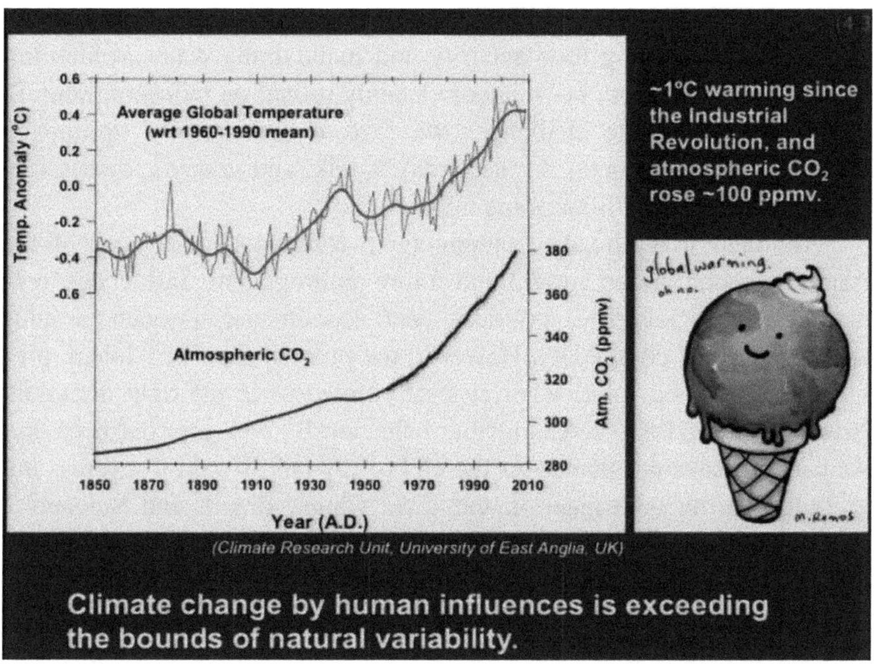

(Climate Research Unit, University of East Anglia, UK)

universe as far as we know. In Asia alone, we have half of the world's population. Sustaining food security and maintaining water availability are critical issues, both of which are heavily reliant on monsoon rainfall. When I consider the challenges we face, I become less optimistic. However, after listening to yesterday's talk and today's discussion, I believe there may still be some hope.

Although it is already challenging to forecast the behavior of the Asian monsoon due to current and future anthropogenic influences, predicting the sociological, political, and economic consequences adds another layer of complexity. However, the previous speaker, Johan, provided a very good solution, or at least pointed us in the right direction. Now is the time for us to change our behavior. By changing our behavior, we can still save our planet. At the end of my talk, I want to express my gratitude to my colleagues in the U.S., China, Brazil, and Singapore. I particularly want to thank the funding agencies SF, NIF, and even private donors. Thank you.

11.2 Discussion

Discussant A: One of the reasons why dynasties change is often correlated with the climate, but there is also the concept of the "Son of Heaven." When the emperor couldn't maintain good relations with heaven, it could lead to floods, killings, or other disasters. Then there would be uprisings, not just because of the floods, but because the Son of Heaven didn't fulfill his duties.

Kerry Sieh: Any other thoughts about this question?

Discussant B: I have a question. Does the rainforest have anything to do with monsoon?

Wang Xian Feng: Rainforest?

Discussant B: Yeah, [does it] have anything to do with monsoon?

Wang Xian Feng: Oh yeah, absolutely.

Discussant B: Oh, so do you mean that, generally speaking, if a country has abundant rainforest, then it is more likely for the country to be prosperous, right?

Wang Xian Feng: There, clearly, the trees cannot survive without water.

Discussant B: I mean for a country, if the country has more rainforests, then the country is more likely to be prosperous.

Kerry Sieh: I think Johan showed that in his last talk; in fact, our rainfall was largely a function of how much vegetation was still around, right?

Wang Xian Feng: Yeah, that's a very good indication. So, if you look at what happened today in Borneo Island and also in the Amazon, if you maintain the rainforest, there is just some hope that we can maintain the amount of rainfall.

Discussant B: Do you mean that countries near the Amazon are very rich countries?

Wang Xian Feng: In terms of the natural resources, yes. Of course, in terms of the GDP per capital, probably not.

Guaning Su: Maybe two things. I want to comment on the exchange just now, and the other one is a question. I think what is really happening is that China has been an agricultural country for a long time, and the prosperity depends on agriculture, which depends on the rain. I don't think that relates very much to rainforests and things like that. The question is, if you look at your chart, the Ming and Qing Dynasties are on the low side of the rain; they are relatively dry. Is there any interpretation you can put on that? First of all, the whole of the Ming Dynasty seems to be low.

Wang Xian Feng: So, of course, I wanted to state one more time. We don't want to say that climate change is the only factor to cause the survival of a dynasty or collapse of a dynasty. But when we look at Chinese history, we did find some indications. For example, as you said, with the Ming and Qing Dynasties, in general, over time, the climate wasn't favorable at last.

In fact, there is also an implication that if the government, if the administrative system worked well, they could shift resources from one place to the other, then they would still be able to maintain the dynasty. In fact, during the early Qing Dynasty, they used the Grand Canal very much, because a lot of parts of China were generally dry due to the monsoon, right? So, they used the Grand Canal to shift a lot of food from the southern part of China, where rice could still be cultivated, to the north. So, in a way, if society can be able to adapt to climate change or if the administrative side can work well, they can share resources, they can. So that's just some hope.

Peter Schwartz: Two questions, one looking back, one looking ahead. You may have answered this question before I came in, so forgive me if you have, you can just skip it. But one argument has been made that when China discovered how to grow rice in rice paddies and let the rice hulls decay in the water, it produced methane approximately 3,500 years ago. We saw a spike in atmospheric methane and the argument has been that it prevented another little ice age or major ice age; do you buy that argument? Number 1. That's the backward-looking question. Number 2. The forward-looking question is, the current momentum of the system seems to be taking us toward at least a peak of somewhere around 550 parts per million, and the likelihood of stopping it much before that is very low. The last time we were there was about 65 million years ago when the whole world was tropical. Is that the likely outcome if we hit 550?

Wang Xian Feng: Very good questions. So, for the first one, you're probably referring to Bill Ruddiman's Anthropocene hypothesis. He proposed that early on, there was anthropogenic forcing that caused greenhouse gases to change. Basically, he argued that as human beings, the most intelligent species on the planet, we changed the climate a long time ago, back in the early Holocene. He's probably right about methane but not CO_2, because he basically had two stories: One is to try to explain the CO_2 change and the other to explain the methane change. He could be right in terms of methane.

For the second question, this actually feels very challenging when looking at the paleoclimate data compared to what we observe today.

If we look at the paleoclimate data, as I showed you and as you mentioned, the last time carbon dioxide was as high as it is today was around 20 million years ago. That was a time when very little ice sheet could exist in Antarctica, and there was no ice sheet at all in Greenland. So, the sea level was much higher than today. Are we ready to head in that direction? We feel very uncomfortable and miserable when we look at the paleoclimate data and compare it to what we observe today.

First, of course, is the CO_2 concentration and second is the rate of increase — the rate at which we are putting carbon dioxide into the atmosphere. This is incredible. As I showed you in the diagram, we — the ice core group — generated data on the timescale of millennia or tens of thousands of years. But today, we are looking at changes over some decades or hundreds of years. So, looking at the future, I am not sure, and this is the big uncertainty. But on the other hand, I want to go back to what Johan said this morning. If we work closely between different governments and we all change our behaviors, it still remains possible.

Kerry Sieh: So, Peter, if we really influenced the Holocene accidentally, maybe we can influence the Anthropocene intentionally — one hopes.

Discussant C: You see, there was a drop in rainfall leading to drought conditions in China around the period from approximately 8,700 to 8,900. Also, dry conditions, abnormally dry conditions, were observed in the Mayan region. In China, correct me if I am wrong, this occurred due to the monsoon, originating from the Indian Ocean. Singapore benefited from this wind, which brought talented people from Yemen and India. When the wind direction changed toward China, it brought about a different scenario. What is the mechanism in the Mayan region that resulted in similar drought conditions? Was it another kind of monsoon? What was the overriding force that, in China, manifested itself as a monsoon carrying moisture, [with] its strength determining the area it covered? Okay, how about in the Mayan region? So, that's my first question. My second question is this: In one of your slides, you showed that there was a rebellion, basically peasant uprising known as the Huang Chao uprising. Now, I noticed that it arose in the northeastern part of China. I would like to have your comments on why this is so. If the pressure due to wind direction is higher in China's northwest, resulting in more suffering and bitterness? People had nothing to lose. Why didn't the rebellion start in the Chinese northwest rather than the Chinese northeast in the form of the Huang Chao? Thank you.

Wang Xian Feng: A very good point. Regarding the first question, we know that the monsoon is simply caused by the temperature gradient between the land and the ocean. This is because they have different heat capacities, resulting in different responses to heating. Most of the caves we study are located inland in China, very close to the monsoon fringe. Therefore, the climate in that region is very sensitive to monsoon changes. Conversely, in maritime regions like Singapore, rainfall may not change significantly during strong monsoon winds because they consistently receive rainfall from the nearby ocean. However, if we consider the entire region, peasant uprisings, famines, and movements of people seeking resources could occur inland, leading to conflicts and migrations.

As for your second question, in China, the monsoon is sometimes referred to as the summer monsoon, which brings rain from the southeast to the inland regions. There's also the northwestern monsoon, which occurs in winter. However, it's the summer monsoon that predominantly brings rain. The Huang Chao uprising was able to develop in the central part of China, close to the east, where agriculture was historically prosperous and sensitive to climate changes.

Tso-Chien Pan: Thank you, Xian Feng, for the engaging talk. I'm from the Institute of Catastrophe Risk Management at NTU. While droughts are certainly problematic for most regions, excessive water from monsoons can also lead to flooding issues, especially when combined with high tides or other factors. I'm interested in exploring potential solutions to mitigate these fluctuations or the data you presented. Looking ahead, in terms of risk management, dealing with fluctuations over periods of tens or hundreds of years can pose significant challenges for societies or governments. So, my question is this: What is the maximum resolution of the technologies or methodologies you are employing or developing? How finely can we see these fluctuations?

Wang Xian Feng: Yeah, this is very good. In fact, over the last decade or so, we've tried to refine the techniques — the technical aspects — to improve timing accuracy and precision. At the same time, we aim to obtain more robust records by collecting samples from different locations to study potential geographical heterogeneity. I think there's still hope with the development of these techniques; we can do better. For example, the record I showed you for the last 2,000 years, the timing precision we can achieve is within a few years. So, we can truly analyze the pattern of monsoon changes at interannual and decadal resolutions. However, we face another challenge: how to quantify it. Remember, what I showed here is the O18/O16 ratio, a static ratio found in archaeological rock samples. Quantifying it in terms of rainfall amounts, like 1,000 millimeters or 500 millimeters, remains a challenge.

Kevin Xiao: Okay, yeah. Actually, I have a comment on the gentleman's question earlier about why there was an uprising in the northeastern part of China rather than the northwestern part. I think one big reason here is that before the Huang Chao uprising, there was already something wrong in the northeastern part, if you still remember. There was the An Shi Zhi Luan, an eighty-year war recorded in Chinese history. I'm sure you know about it, right? That area was totally destroyed and generated so many poor people who simply couldn't survive. For them, it was either uprising or dying. Either way, they had a good chance of dying, so why not try? It's basically like that. I guess in that specific case, maybe the war before the Huang Chao uprising played a bigger role.

Now, let's go back to another case, which I still have some doubts about: the Ming and Qing Dynasties, actually. We briefly discussed this before. According to the historical data we have, the whole Qing Dynasty was relatively dry. However, we also know that the population boomed during the Qing Dynasty. One explanation was that we had this great canal which could deliver food from the south to the north. However, that delivery was mainly for the capital city, not really for the regular people living in the vast areas of the northern part of China. But even inside China, the population size increased quite a lot during the Qing Dynasty.

One argument was that it was mainly because of a policy decided by the country. One of the earlier policies of the Qing Dynasty was, if you have more boys, we won't really increase your tax. So, people may have felt it was better to have more boys, meaning they not only had more labor resources at home but could also live better off, as in they could still feed themselves. And if you have more people to grow crops or whatever, you can actually live better, on average. So, could we say that during the Qing Dynasty, quite a big part of the northern part of China might or might not have been a very serious constraint? Could we say that? I'm not so sure.

Wang Xian Feng: I do agree with what you pointed out. The climate is just one of the factors or one of the straws that brings down the camel. There are certainly other factors, such as how society worked, how politics functioned. We picked these five time windows because we could clearly correlate them with what we found in Chinese historical documents. But what I want to say is that you can't simply understand Chinese history by looking at this simple curve. Clearly, there are more complexities involved.

Kerry Sieh: Let's get to the issue of resistance right, or resilience that Johan was talking about in the previous talk — that the society is more resilient and is more likely to survive some sort of a climate challenge or whatever.

11.3 Summary of the Talk

Wang Xian Feng's talk revolves around the significance of paleoclimate records, particularly focusing on the Asian monsoon system and its impact on civilization. The discussion is grounded in analyzing the O18/O16 ratios in geological samples, primarily from caves, to understand past climate variations. The lower the O18 value, the stronger the rainfall and monsoon, offering insights into historical precipitation patterns.

The presentation starts by emphasizing the challenges in establishing a quantitative link between rainfall and monsoon springs. Despite these challenges, Wang suggests a broad correlation: lower O18/O16 ratios indicating stronger monsoons and higher rainfall. This observation is extended to a comprehensive analysis of monsoon records spanning the last 300,000 years, derived from geological rock samples, particularly cave formations.

Key to Wang's argument is the comparison of these geological records with insolation data, describing the Earth's orbit around the Sun. Strikingly, there is a significant match between the O18/O16 ratios and insolation curves, reinforcing the reliability of the geological records. The focus then narrows to the last 2,000 years, using cave samples from Wan Shang cave, near the monsoon fringe area, to present a detailed record of O18/O16 ratios. This record is correlated with historical documents, specifically a drought and flood index, demonstrating a remarkable alignment between the geological and literature-based records.

Wang further substantiates the reliability of the geological record by comparing it with another cave record from central east India, the Dandak cave. This comparison, spanning overlapping periods, highlights the consistency between two geographically distinct locations, reinforcing the claim that these geological records accurately represent monsoon changes.

The talk explores Chinese history, pinpointing five significant time periods correlating with the O18/O16 ratios. These periods are associated with either strong monsoons and abundant rainfall or weaker monsoons and dry periods. Historical events and societal changes during these periods are discussed, linking climate variations to the rise and fall of dynasties, illustrating the potential impact of climate on civilization.

The Tang Dynasty's collapse was followed by the emergence of the Song Dynasty, a golden age characterized by low O18/O16 ratios, signifying strong monsoons and ample rainfall. The subsequent Yuan Dynasty faced challenges, with higher O18/O16 ratios, indicating weaker monsoons and possible droughts. This pattern was reiterated in the late Yuan weak monsoon period, marked by multiple droughts and peasant uprisings. The Ming Dynasty, especially during the late Ming weak monsoon period, experienced elevated O18/O16 ratios, suggesting challenging climatic conditions and social unrest.

Moving beyond China, Wang draws parallels with climate-induced civilization collapses worldwide. Examples include the Guge Kingdom in Tibet, the Viking colonies in Greenland during the medieval warm period, and the classical Maya civilization in Central America. These instances further emphasize the role of climate change in shaping societies and their resilience or vulnerability to environmental shifts.

The talk concludes with a reflection on the contemporary scenario. Wang underscores the drastic increase in atmospheric carbon dioxide over the last 150 years, coinciding with rising global temperatures after the Industrial Revolution. Despite the ability to study and understand past climate patterns through paleo records, the influence of human activities has now exceeded natural variability. The talk advocates for urgent changes in human behavior to address climate change, recognizing its critical impact on food security, water availability, and the overall sustainability of our planet.

In summary, Wang Xian Feng's talk skillfully intertwines geological data, historical records, and societal events to underscore the intricate relationship between climate variations and the fate of civilizations. It serves as a call to action, emphasizing the urgency of mitigating anthropogenic influences on climate to ensure a sustainable future for our planet.

11.4 Relevance of the Talk to the Current Stage of Research

Wang Xian Feng's talk holds significant relevance to the current stage of research in several ways:

Integration of Multiple Disciplines: The talk seamlessly integrates geological, historical, and climatological data to provide a holistic understanding of past climate variations and their impact on civilizations. This interdisciplinary approach mirrors the current trend in research, where scholars across various fields collaborate to tackle complex issues such as climate change and its societal implications.

Use of Paleoclimate Records: Wang's reliance on paleoclimate records, particularly the O18/O16 ratios derived from geological samples, highlights the importance of studying past climate variations to contextualize present and future climate trends. In the current research landscape, there is a growing emphasis on leveraging paleoclimate data to improve climate models and enhance predictions about future climate scenarios.

Validation of Climate Models: By comparing geological records with insolation data and historical documents, Wang demonstrates the validity of climate models in reconstructing past climate conditions. This validation process is crucial for refining climate models and improving their accuracy in projecting future climate changes, which is a key area of focus in contemporary climate research.

Understanding Societal Vulnerability: The talk underscored the vulnerability of past civilizations to climate variability, highlighting the potential societal impacts of climate change. In the current research paradigm, there is increasing recognition of the need to understand how climate change may exacerbate social inequalities, disrupt food and water systems, and trigger conflicts, driving research efforts to explore these interconnected dynamics.

Call to Action: Finally, Wang's emphasis on the urgency of addressing anthropogenic climate change serves as a rallying cry for action. This aligns with the current stage of research, where there is a growing emphasis on interdisciplinary collaboration, policy interventions, and public engagement to mitigate the impacts of climate change and foster resilience in the face of environmental challenges.

Overall, Wang Xian Feng's talk provides valuable insights that resonate with the current stage of climate research, emphasizing the importance of interdisciplinary collaboration, data validation, and proactive measures to address the challenges posed by climate change.

Chapter 12

A Crude Look at Governance and Complexity by a Former Civil Servant

Peter Ho

YouTube: https://www.youtube.com/watch?v=eV9CnlEq6ks

Speaker: Peter Ho

Moderator: Jan Vasbinder

Discussants: Kerry Sieh, Guangning Su, Robert Axelrod, Kai Xin, Raymond Kwong, Kevin Koh, Greg Fisher, Alec, Peter Sloot, Discussants A, B, and C

12.1 Talk by Peter Ho

Thank you very much, Jan. Okay. Let's start with black swans. We all know now what black swans are. Nassim Taleb made black swans very famous, and I think in Singapore everybody knows about black swans now. Black swans result from complexity. But there's also a whole family of other swans. You've got the white swan, of course; then you've got the dirty white swan — this is an improbable event; but actually, it's not improbable, but everybody ignores it, and then it becomes a dirty white swan. You have a red swan [as mentioned by] Gordon Woo; we've even got the black turkey. A grey swan is highly probable event, predictable, and impact can easily cascade, and these are things like weather and population events. Then we've got the black swan.

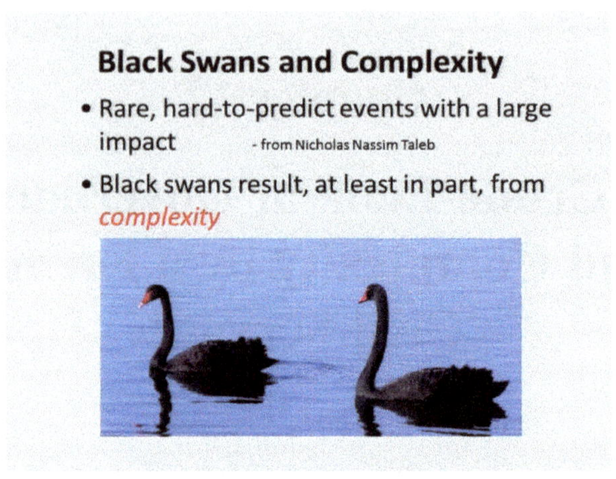

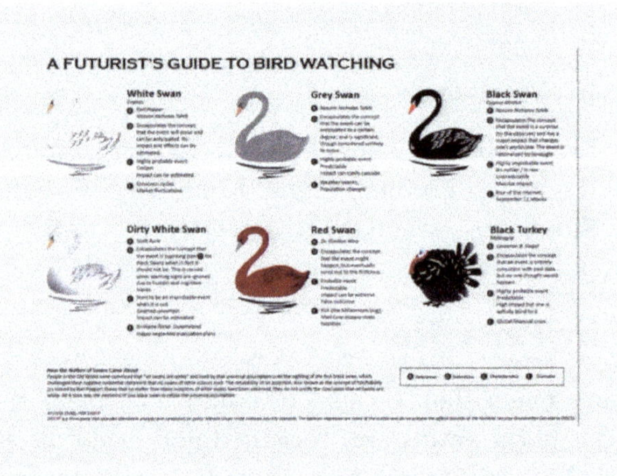

I was introduced to something else because of something that ICRM did just a week ago — Dragon Kings. Supposedly, predictable events in complex systems can often spin out of control leading to extreme events. This is a term coined by Didier Sornette and some of his colleagues — so that's Dragon Kings.

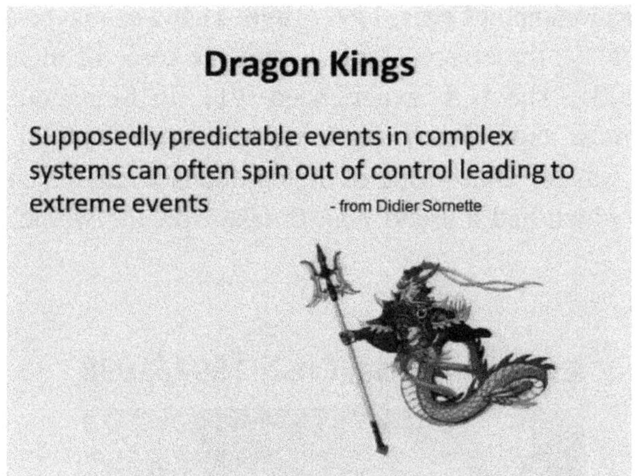

And then, Rumsfeld. There are known knowns, these are things we know that we know. We also know that there are known unknowns, that is to say, there are some things we know that we do not know. But there are also unknown unknowns, the ones we don't know we don't know [as mentioned by] Donald Rumsfeld.

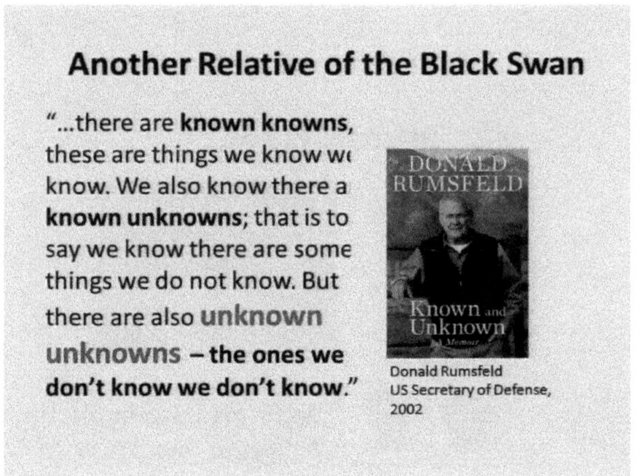

So, we've got black swans, we've got Dragon Kings, we've got unknown unknowns. Now, as far as I know, black swans are real. They exist at least in Perth. Dragon Kings are the stuff of myth and legend. Unknown unknowns are, of course, unknown. Does it really matter?

The Asian Financial Crisis, 1997, 1998: That was maybe a black swan, and it was real — the effects. And we still see some of the effects today. Terrorism, 2001: The U.S. experienced 9/11. In Singapore, we experienced, or almost experienced, our own version of 9/11 on 7 December 2001, which was the uncovering of the infamous Jemaah Islamiyah terrorist network, which had a grand plan to take over the whole of Southeast Asia.

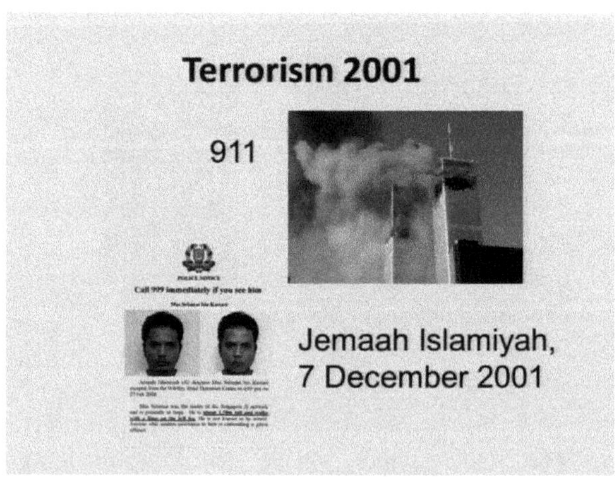

So, that was a real shock to the system for Singapore, which saw itself as a very peaceful, safe, and secure country. In 2003, Singapore faced SARS. That was also a big shock; nobody knew what was happening.

Nobody knew whether we would survive. For some unfortunate reason, we had something called super-spreaders in Singapore. People who were particularly virulent became nodes of infection, and they just spread the infection throughout Singapore. Dozens of people died in this thing; Singapore was one of the most badly hit of the countries affected by SARS. So, this was a big shock to the system.

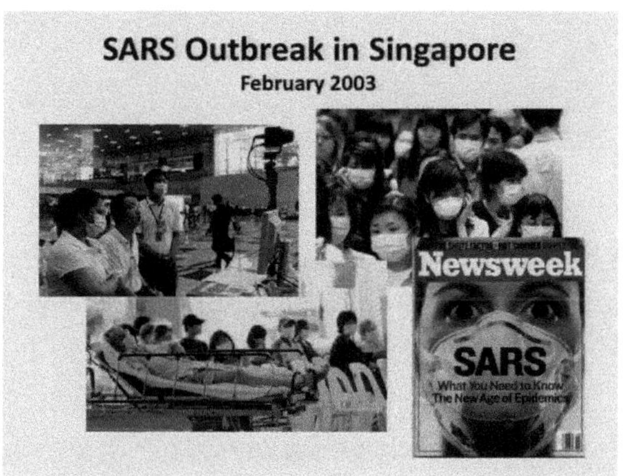

Financial crisis 2008, 2009: We still feel the reverberations of that. Thai floods, 2011: Now, you may ask, what is a black swan — was the Thai Floods of 2011 a big shock? In a way, it wasn't, but in a way, it was. It was a big shock — it wasn't a big shock because in Thailand, it's quite a low-lying country, it floods all the time. But, this time around, the floods were so widespread that they started to flood industrial estates; it caused disruption to global supply chains, so it was a big shock.

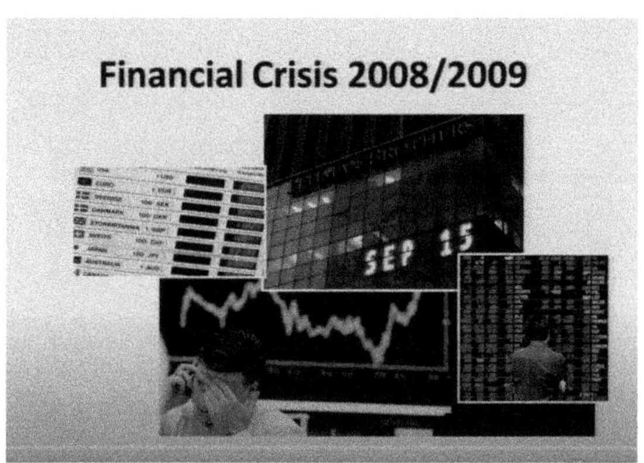

Hurricane Sandy: Those of you who were in the U.S. know this was a big shock, as bad as or even worse than Katrina. Eurozone crisis: I don't think it's over, but I'm not an economist, I'm not a financial expert, but my bet is it's still going to rumble on and create a lot of problems.

And then the Arab Spring: I put Spring in red and in inverted commas because I never believed it was a Spring; I thought it was a winter and I think it is — it is something like a winter. And what happened? It surprised everybody, governments collapsed in 4 countries: there were government changes in Kuwait, Bahrain, Oman, and now there's a civil war in Syria, and no end in sight.

Now, this is a quote I found, which is from a gentleman called H. A. L. Fisher, Harold Herbert Fisher; he was a politician and historian. I think he probably prefers to think — now he's passed away but he probably preferred to think of himself as a historian. And he said, "Men wiser and more learned than I have discerned in history a plot, a rhythm, a

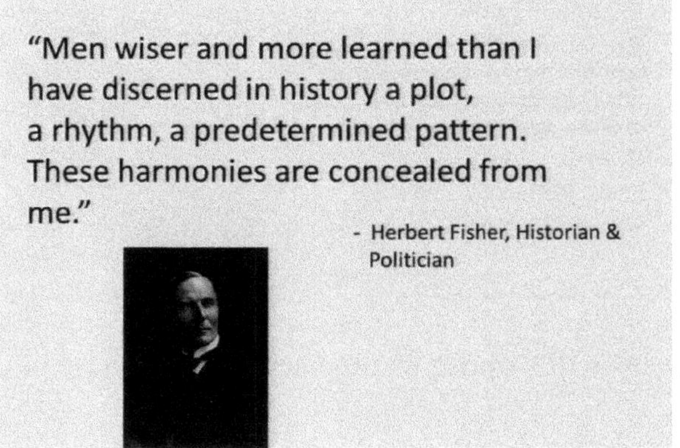

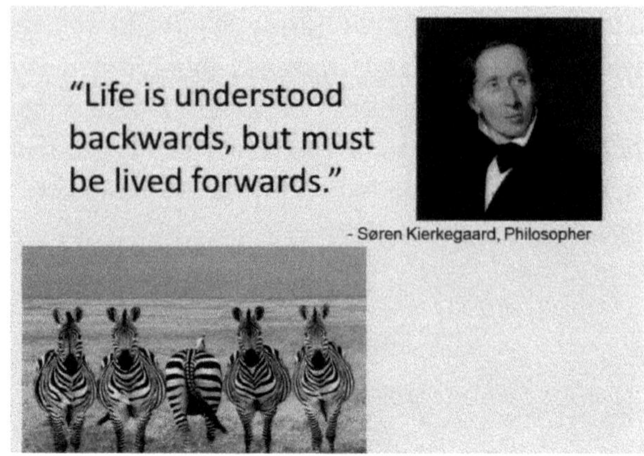

predetermined pattern. But these harmonies are concealed from me." He's being very honest. The philosopher Kierkegaard said, "Life is understood backwards, but must be lived forwards." Now that's very profound.

And this introduces the concept of retrospective coherence. The current state of affairs always makes logical sense but only when we look backward, in hindsight. The current pattern is logical, but it is only one of many patterns that could have been formed, any one of which would have been equally logical. And why does retrospective coherence arise? It arises because of complexity, I think.

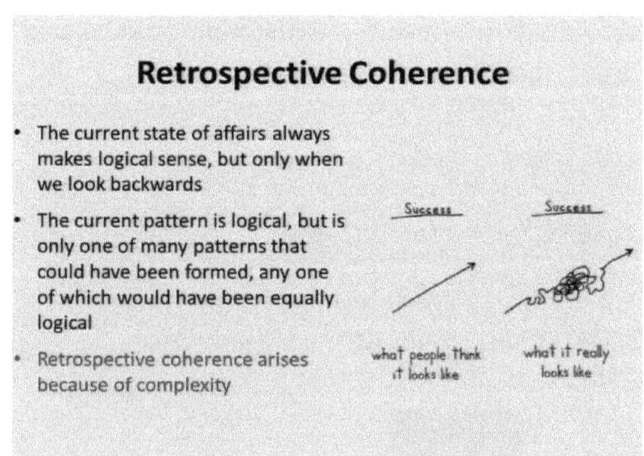

Black swans will surprise us time and again because of complexity.

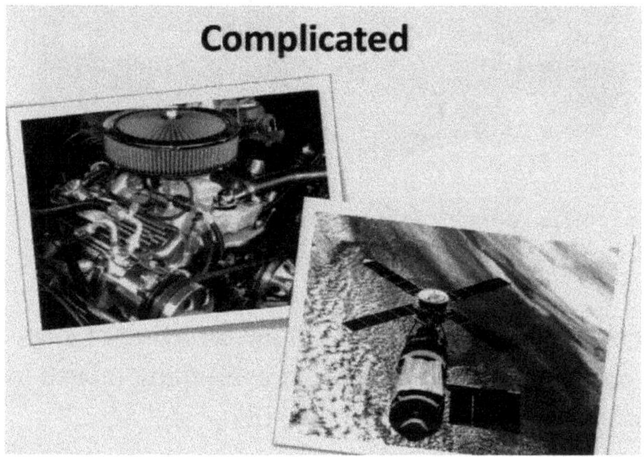

And now, I must say that I always tell people there's a difference between complex and complicated. A complicated system, a complicated environment, well, it can produce outcomes and results that are repeatable. It performs in a predictable way. It may look very complex to you, but actually, it really performs in a repeatable way. So, a satellite in orbit, you know, we say it's very complex, but in fact, it is just a complicated system because it just keeps on repeating itself. The 787 Dreamliner is, I hope, a complicated system; if it's a complex system, we're in big trouble, you know? It's designed to fly; if you steer it right, if you steer it left, it goes to the left; if you bring it down, its nose will point downward. That's a complicated system.

But a complex system, and we heard the presentation just now, human societies are all complex systems, the world is a complex system; there are many things we don't understand. So, we are still trying to understand why civilizations rise, civilizations fall. We're trying to understand, does the monsoon really affect the rise and fall of nations and countries? Maybe it does, we can't be sure. And we're still discovering it and why. Because of complexity.

And this is my only genuflection in the direction of Dao. Just to quote Lao Tzu, "Everything is connected, and everything relates to each other." That's an Eastern description of complexity.

So, what is complexity? You know, these are standard attempts to define it, but the point which I perhaps would like to make is that the collective behaviors of agents within a complex system can deliver discontinuous and nonlinear shocks to the system, which in turn can result in catastrophic failure.

Complexity

- The study of relationships between the system and the agents that act within it.
- These relationships and the actions of the agents can give rise to collective behaviours that are unanticipated and not well understood.
- Such collective behaviours can deliver discontinuous and non-linear shocks to the system, thus resulting in catastrophic failure.
- Complex systems are characterised by "emergent" outcomes that are not always predictable *ex ante*

Paul Ormerod, the English economist, said, "In orthodox economic theory, the agents involved in any particular market are presumed to be able to both gather and process substantial amounts of information efficiently in order to form expectations on the likely costs and benefits associated with different courses of action, and to respond to incentives and disincentives in an appropriate manner." But, this is the important point he's making: "The one thing these hypothetical individuals — who are

"In orthodox economic theory, the agents involved in any particular market are presumed to be able to both gather and process substantial amounts of information efficiently in order to form expectations on the likely costs and benefits associated with different courses of action, and to respond to incentives and disincentives in an appropriate manner. The one thing these hypothetical individuals do not do is to allow their behaviour to be influenced directly by the behaviour of others and their tastes and preferences are assumed to be fixed, regardless of how others behave."

- Paul Ormerod, Economist

agents within a complex system — do not do is to allow their behaviour to be influenced directly by the behaviour of others and their tastes and preferences are assumed to be fixed, regardless of how others behave." So in a way, he's saying that the traditional economists look at a complex economic system as a complicated system rather than a complex system.

And so, this is what happens when you do that.

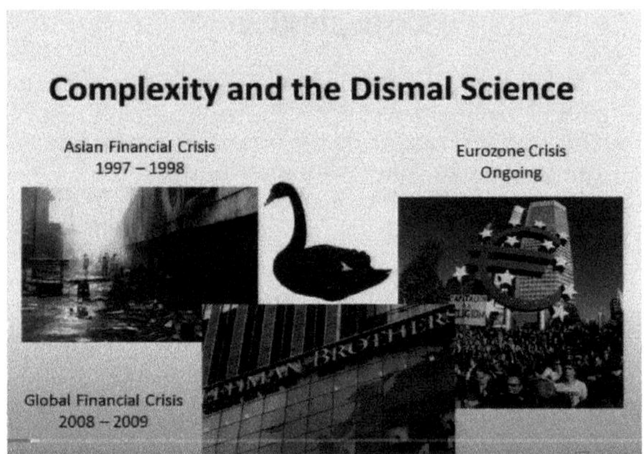

Now, there's another problem with complexity. You know, maybe each event on its own is predictable; so you think about an earthquake: An earthquake may be predictable, maybe. And tsunamis are usually, not always, but usually connected with earthquakes. But the Tohoku Earthquake in 2011, you had one earthquake triggering a massive tsunami, which then flooded the nuclear power station at Fukushima, causing basically a meltdown.

Now, when you combine all of these together, the cumulative effects, is that — can you predict that? It's very, very difficult to combine all these complex events and then say, well you're going to face a catastrophic failure because of that. So, this is one of the challenges of complexity.

Then, we have this other problem, wicked problems. Wicked problem is a term coined by Horst Rittel; these are about highly complex problems, not simple, complicated problems, but complex problems. They have multiple stakeholders. So, you take something like climate change. Everybody has a different view on climate change. If you're sitting in government, you may have a view on climate change; if you're a physical scientist, you might have another view; if you're an economist; you have a third view; if you're part of an NGO, you have a fourth view — everybody has a different view. How do you align these perspectives, and then how do you find a solution that will satisfy all the different stakeholders? Very difficult, that's why it's called a wicked problem.

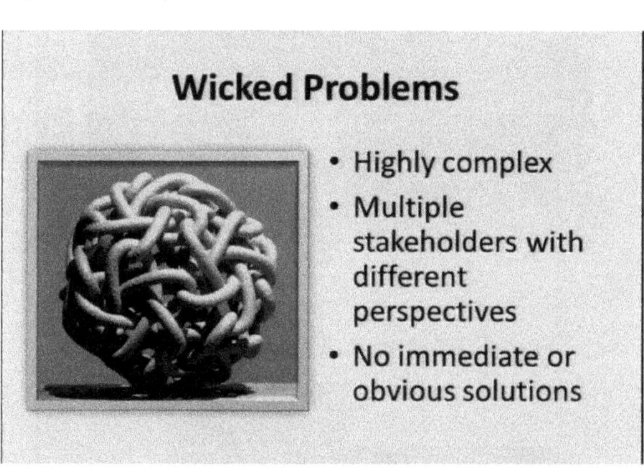

Pandemics, sustainable development, aging are wicked problems. Climate change, food, water, and energy security are wicked problems. You solve one problem, you may create a problem in another space.

So, complexity generates black swans, unknown unknowns, wicked problems, all kinds of things.

And this is my assertion — the frequency of strategic shocks seems to be increasing, and the amplitude of their impact appears to be growing. And the question is, why?

Wicked Problems
Aging
Pandemics
Sustainable
Development

Climate Change

Food, Water & Energy Security

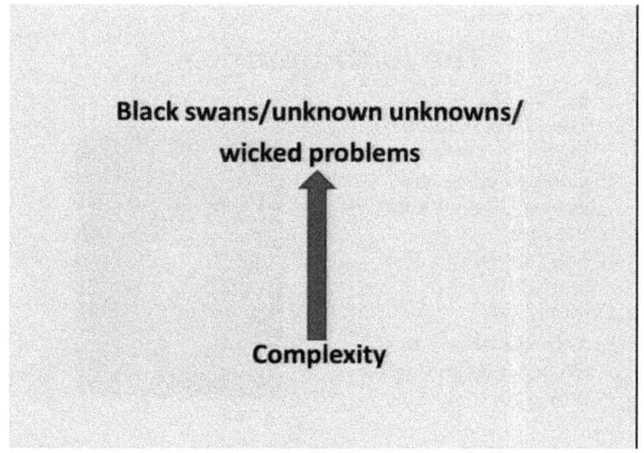

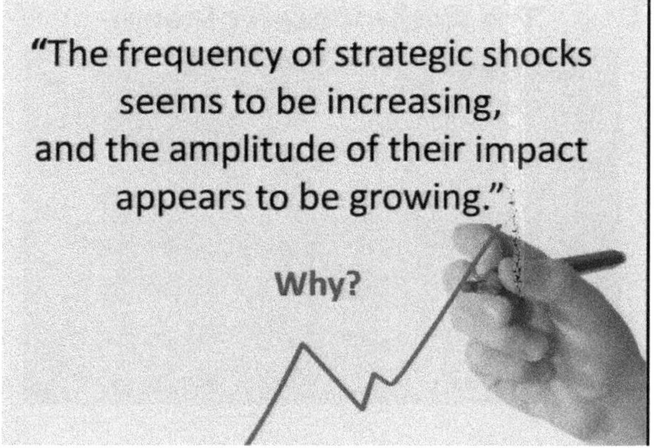

Now, Anthropocene. I think, in fact, in the last complexity conference, there was some presentation on the Anthropocene. This is the period in the Earth's history in which the human impact has become decisive. Before that, it was all geological. Before the Anthropocene, it was the Holocene, which was a geological era.

And, you know, these are just screenshots which show the amount of air networks, urban areas, shipping routes, and so on. Just shows, you know, human beings on the planet, what they do to the planet. This is, look, Dubai. Out of the desert, sheer human will and of course a lot of oil money, you can do that. Is this good, is this bad? I don't know, but this is the Anthropocene.

The Anthropocene

- The epoch marked by humanity's effects on global geology and ecology (beginning with the Industrial Revolution, late 18th Century)
- Before the Anthropocene – the Holocene (a geological era)

The Anthropocenic Planet

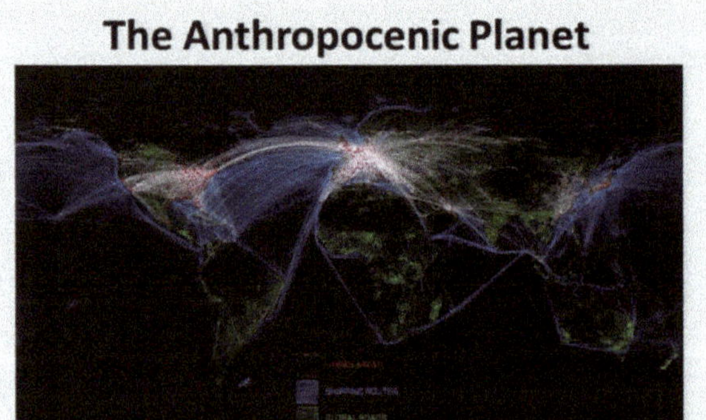

The Human Impact

And there're all kinds of effects — climate change, acidification, ozone depletion, phosphorus cycle, freshwater use, biodiversity and so on, all kinds of things affected.

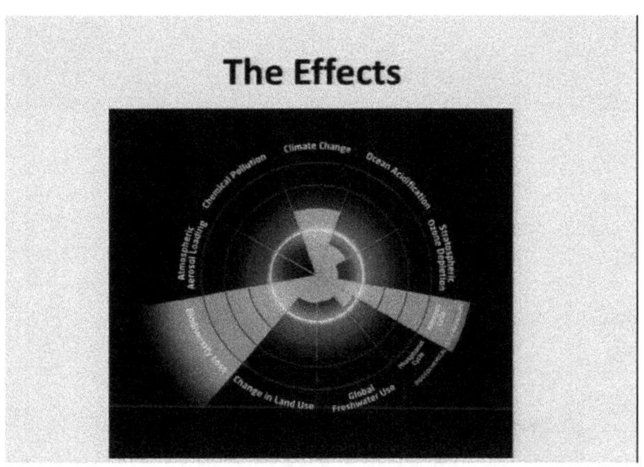

And we were also introduced to this Great Acceleration in the last conference, and it's remarkable. Only in the last 50 years, there's been a surge in all kinds of activities ranging from population, GDP growth, FDI, to even MacDonald's somewhere down there; the number of MacDonald's has shot up since the 1950s, only in the last 50 years. Telephones, motor vehicles. And our colleague, Helga said, "Evolution increases complexity." And Stephen Hawking said, "The 21st century will be the century of complexity."

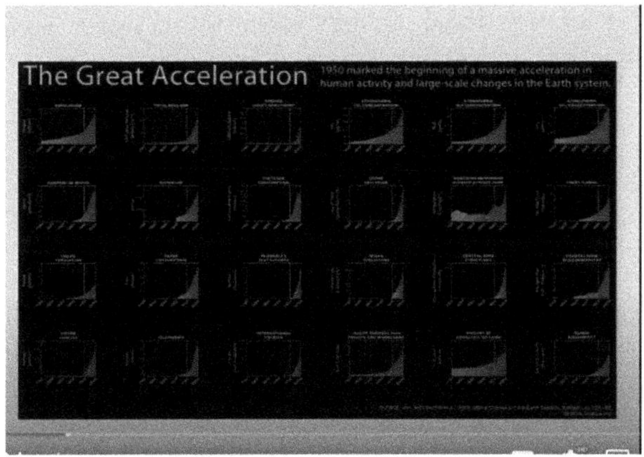

So, we are living in a complex world; unfortunately, it's not just complex, it's getting more and more complex, and we better recognize that we are operating in an unordered, complex environment. We are not operating today in an ordered, merely complicated environment.

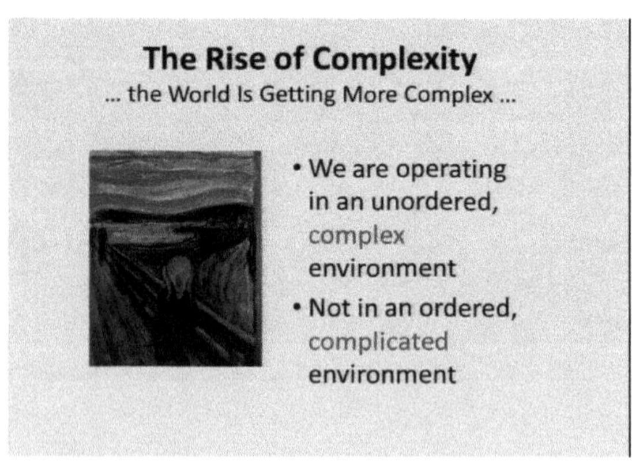

So, the challenges of complexity, and as the connections between the problems and solutions become less clear, that is, it's very hard to connect the cause and the effect, we will sooner or later have to face the unintended consequences of our decisions. So, in other words, you make a decision in a complex system; it may look as if it's working for a while, but after a while, it's no longer valid. It starts to unravel because of all the hidden connections and agents that are operating in a complex system. So, policies and plans can have effects not easily predicted, and policymakers — and that includes all of us — who rely on a linear, mechanistic approach, can get things quite wrong. So, that's everybody. That's a problem. In a complex world, success is not pre-ordained, success is not a permanent condition, and this applies both to companies, organizations as well as to governments.

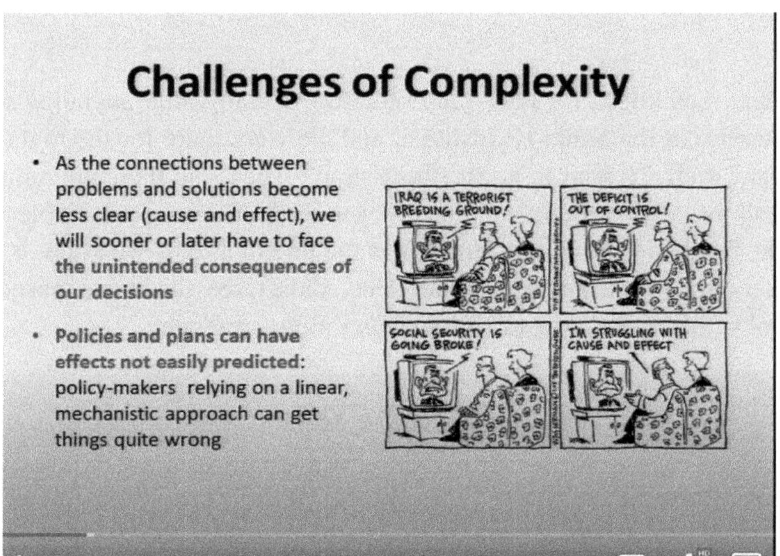

You know, I'm a bit of a camera buff, and I used to take film. Now, Kodak is finished. Its name, I think, is preserved, and I think it made some licensing deal with JK Imaging, but it's finished. They didn't foresee the advent of digital — actually, they foresaw it but didn't realize it was going to be a serious game-changing threat.

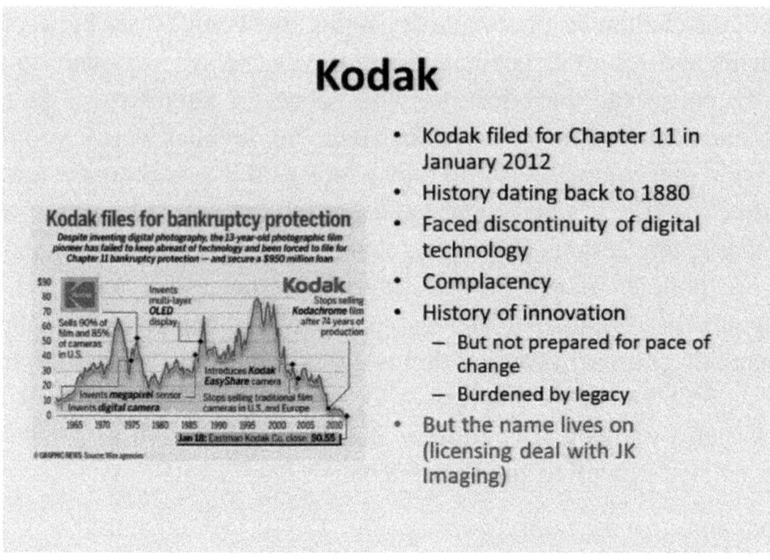

Borders: You know, I was in San Francisco, actually after attending some conference at the Santa Fe Institute, and we were there for the night and we said, well, "I want to go to a bookshop." They said there are no more bookshops left, "Where are the bookshops?" Barnes and Noble shut down, Borders is bankrupt. So, this is the end of life as we know it. But Borders was slow on the uptake. They didn't see the threat posed by e-books; they were slow to establish an online presence.

Nokia: It was first a disrupter; very successful. It established a presence across the entire spectrum, from mass-market telephones to very sophisticated smart telephones. They understood the importance of selling a lifestyle.

Then they got disrupted by Apple and by Samsung. And they had no answer to Android, they had no answer to iPhones. Now, I don't know whether they'll be saved by Windows 8 and Lumia, and I don't know whether they are an innovator anymore, or are they just following the trend. Reader's Digest, another icon, bites the dust. Apple — is Apple next? We don't know. But we have to ask ourselves, will Apple go the way of all its illustrious predecessors? They lost Steve Jobs, but then this is where I'm going to introduce you a bit later on to Clayton Christensen's Innovator's Dilemma.

Dick Foster, who wrote the book *Creative Destruction*, said that 75% of Standard and Poor's 500 in 15 years will be companies whose names we do not know. What it means is that 75% of the companies today will be wiped out; they will be usurped by new innovators, insurgents who come up with new game-changing ideas.

So, this is a constant cycle. Clayton Christensen, and I paraphrase his book, said successful organizations are doomed to fail in the long run because there's tremendous inertia to change a formula that has worked well. This inertia allows the insurgents, the revolutionaries, the start-ups, to sneak in, change the rules of the game, capture market share, and

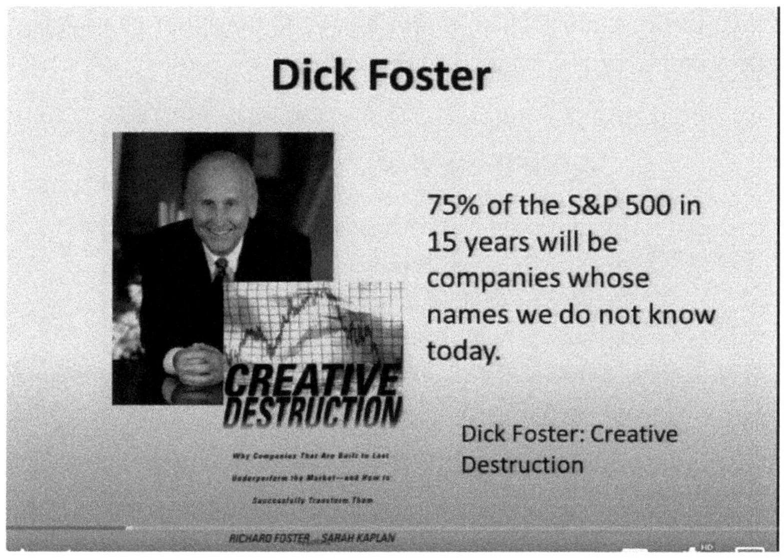

dislodge the incumbent. I asked him about what he thinks about governments; he said he never studied governments. But he said he thinks it applies as much to governments as it does to companies.

Organizations must change and adapt or they die. Companies can fold; they can file for Chapter 11 [Bankruptcy].

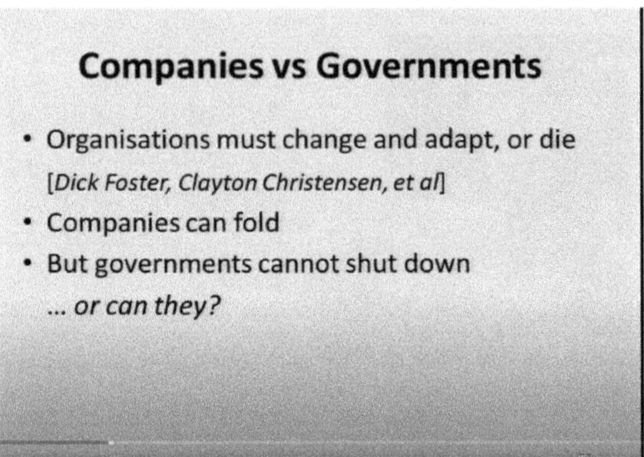

But what happens to governments? If they fail, this is what happens.

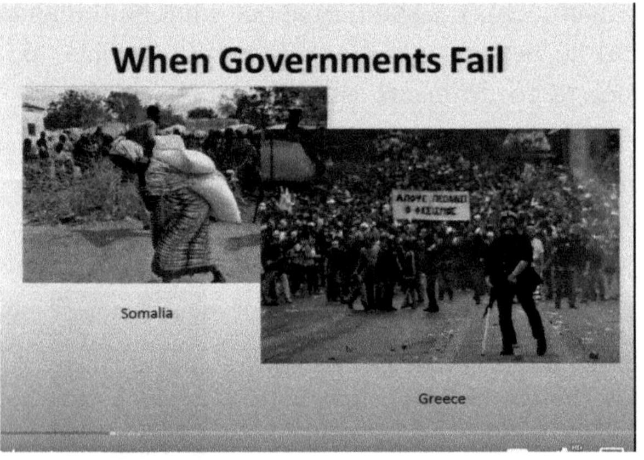

Now, Singapore — a success story; today, the world's second most competitive economy, but this is what the WEF says — world's busiest port, world's best airport, number 2 in home ownership, number 3 in GDP per capita PPP, number 5 in life expectancy. This is a remarkable achievement by any standard — all achieved within the space of about 40, 50 years.

But, this is a Clayton Christensen moment. Are we the Apple of nations, or are we just what we are, a little red dot? This is — you remember that, huh? Maybe we should swap them around. But anyway. The two Peters.

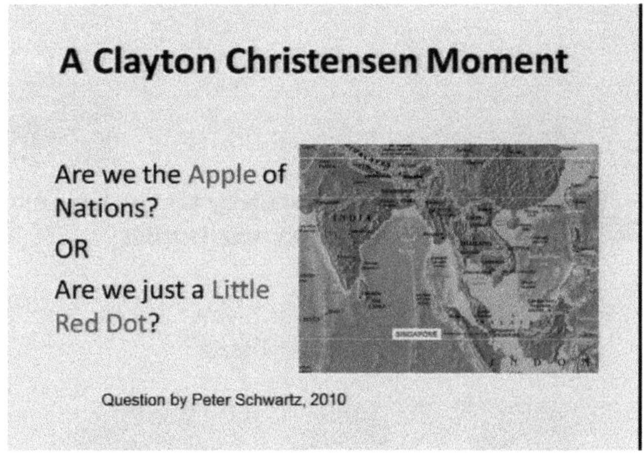

Now, obstacles. Our inability to make sense of complexity.

And I want to introduce you to a special type of failure that all human beings are prone to, and this is accentuated by complexity — cognitive failure.

The little red dot!
The Apple of nations!

Obstacles
Inability to make sense
of complexity

Cognitive Failure

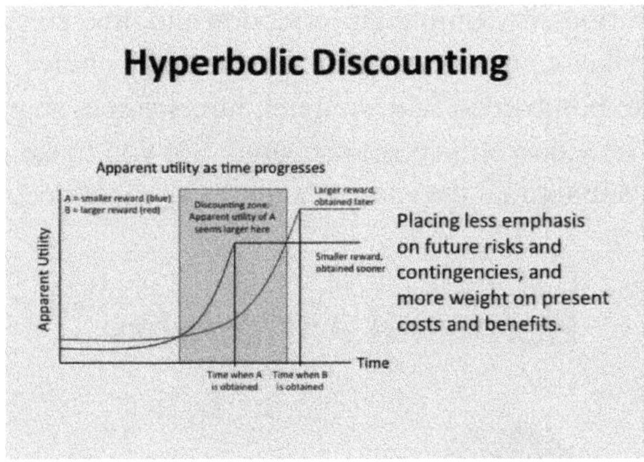

One example is hyperbolic discounting. You place less emphasis on what's going to happen in the future, and give more weight to what happens today. And this is a serious problem. Because governments will tend to invest more in dealing with the fire in front of you, which may be quite cheap to put out — but it's burning, so you put it out — rather than investing long-term, which will perhaps cost money. But why should I pay an insurance premium? Hyperbolic discounting. And for those of you who are interested, just google cognitive biases; you'll have about 100 or so cognitive biases. This is the problem: Human beings have so many cognitive biases. This is just an example of some of them. [This is] Daniel Kahneman's area of expertise.

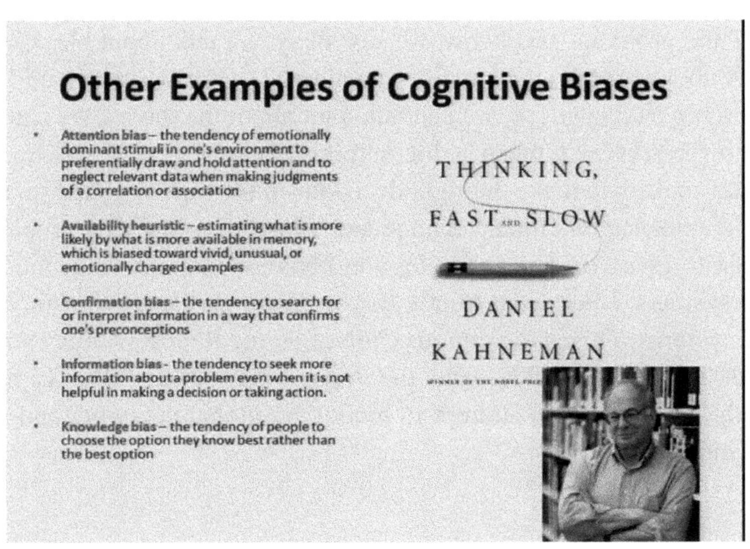

Bounded rationality: Limited information and time pressure mean that the decision-maker cannot possibly make a rational choice. You've got so many — you're bombarded with so much information, so many demands on your time; how do you make a decision? So, you make a decision but it is, at best, a suboptimal decision, it's not an optimal decision.

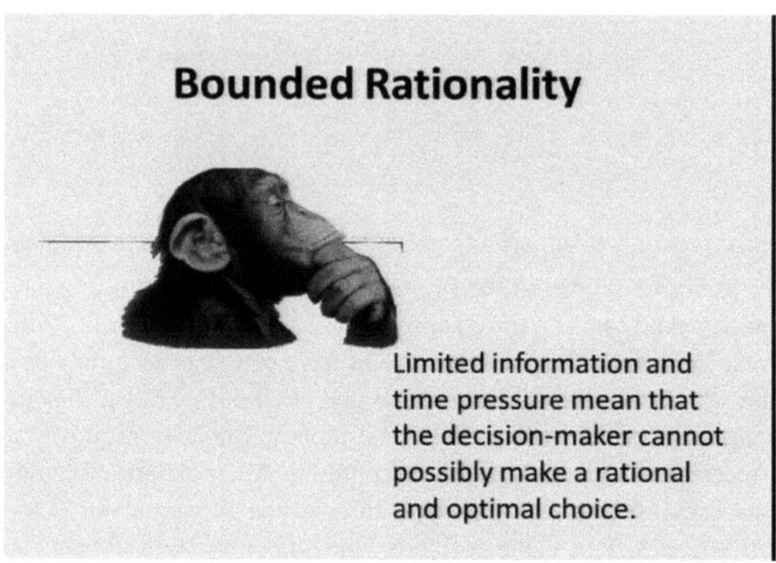

So, human cognitive limitations constrain the ability to deal with complexity and to anticipate black swans and other strategic surprises. This is the problem. You know, we say, okay, we talk about black swans, we talk about complexity, but are we equipped to deal with it? And I think that is a big challenge; so, we can talk a lot about the theory, we can write reams of books and make a lot of presentations, but can we get the decision-makers to make the right decisions in the face of all the incontrovertible evidence on climate change and things like that? Big problem.

Just to give you one example, you know, we talked about modeling world systems. I just want to talk to you about the dangers of modeling world systems. This is the famous Club of Rome Report, which came out I think in 1970 or 1972, and the report was called "The Limits to Growth," and it was an attempt to model the global economy and world population.

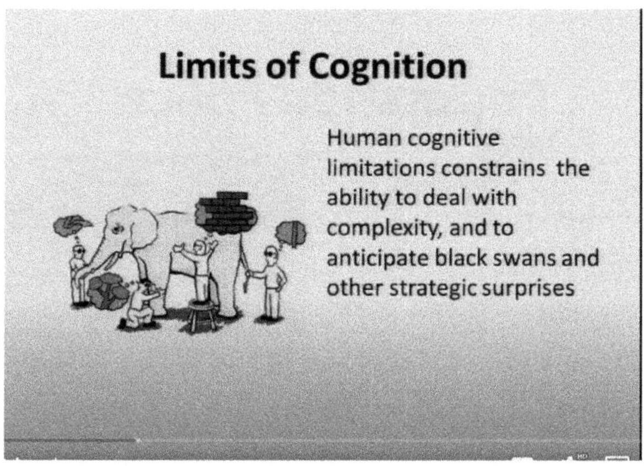

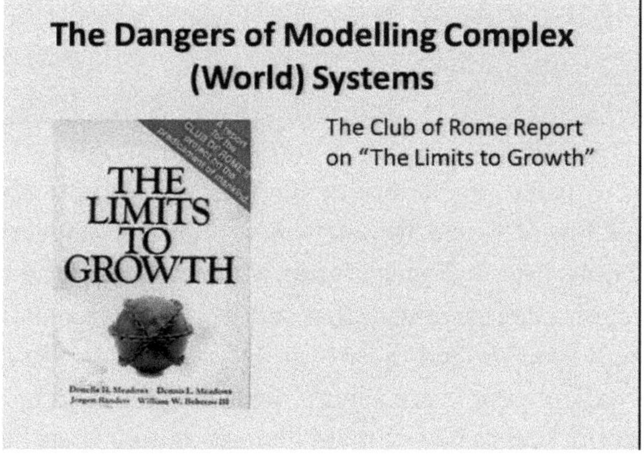

Singapore fell for it hook, line, and sinker; we were persuaded that that model predicted a Malthusian disaster if we did nothing to control the population. So, we introduced a very effective policy called "Stop at 2." So, we didn't just stop at 2, we finally put the brakes on when we were at 1.2. So, we are now at 1.2, our TFR today is 1.2, way below the replacement level of 2.1. So, in other words, in population terms, we became a developed country long before we were really a developed country. And everything the government did after they realized it was a mistake failed to lift the TFR. And what happened? It's very difficult to model the world; it's a complex system.

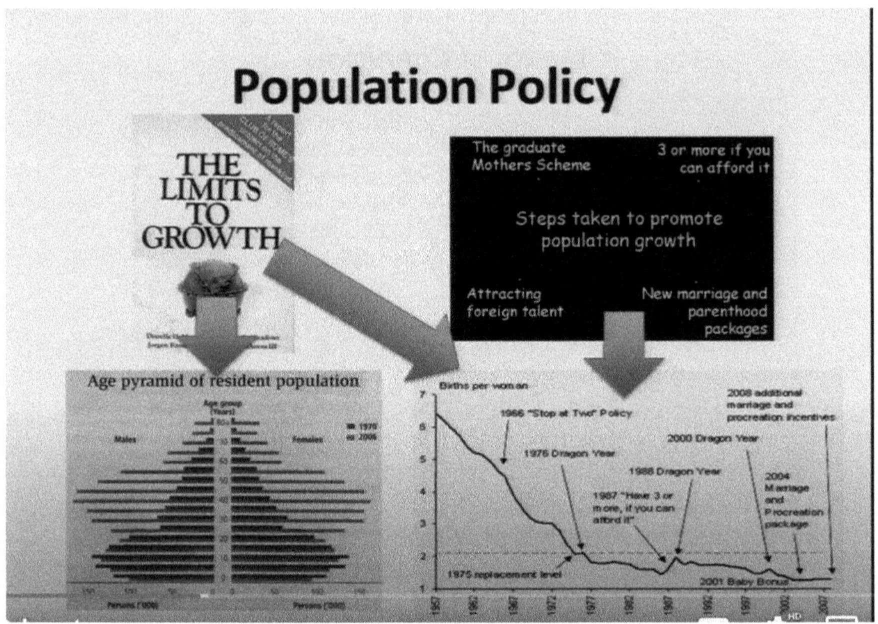

And what was the big failure of the Club of Rome Report? The big failure of the Club of Rome Report was it didn't anticipate the fact that technology would improve agriculture; we've got all these new hybrids, you know, the miracle rice and all that sort of thing, which allowed you to produce enough food to feed the world. Of course, we could be entering another cycle where people are asking, well, maybe we might not be able to feed the world because of climate change and all that, but you know, I'm just giving you the example of dangers of modeling.

So, this has led — this is only for Singaporeans — to problems with the Singapore Population White Paper. And there are winners because there is uncertainty. Without uncertainty, there can be no winners. Instead of seeing uncertainty as a problem, we should start viewing it as a basic source of our future success.

So, the question is, how can we make better decisions in a complex operating environment? Those governments that can learn to manage complexity will gain a strategic competitive advantage.

And I was reading this book, *Ghosts of Empire* by Kwasi Kwarteng. He's British — a British Member of Parliament, but born in Nigeria,

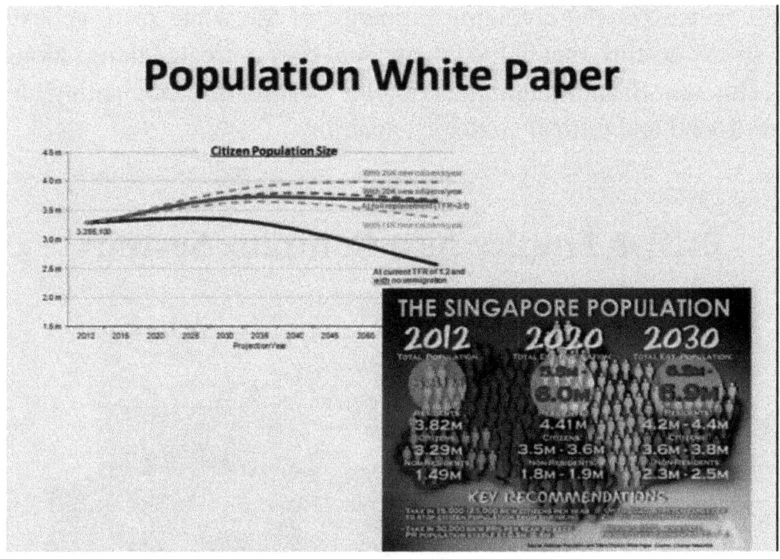

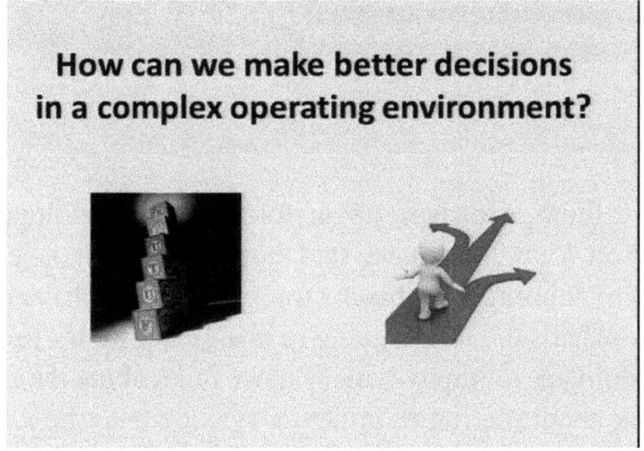

so I guess he's looking at things fairly objectively. The British Empire was a complex system. It established a presence in 6 out of 7 continents, except I think Antarctica, and that's as complex a system as you can imagine. How did they run it so successfully for so many years? They ran it because they learned, maybe not consciously, but they knew — they left it — they learned how to deal with complexity, by not controlling it too much. All they did was a handful of colonial administrators, plus a strong navy and perhaps a few soldiers here and there. So long as they shared the same convictions, the civilizing influence of the white men, believed in their strong destiny, that helped them. And they corrected things along the way. This was Britain running a complex system. So, anticipating change is a profound and critical strategic capability.

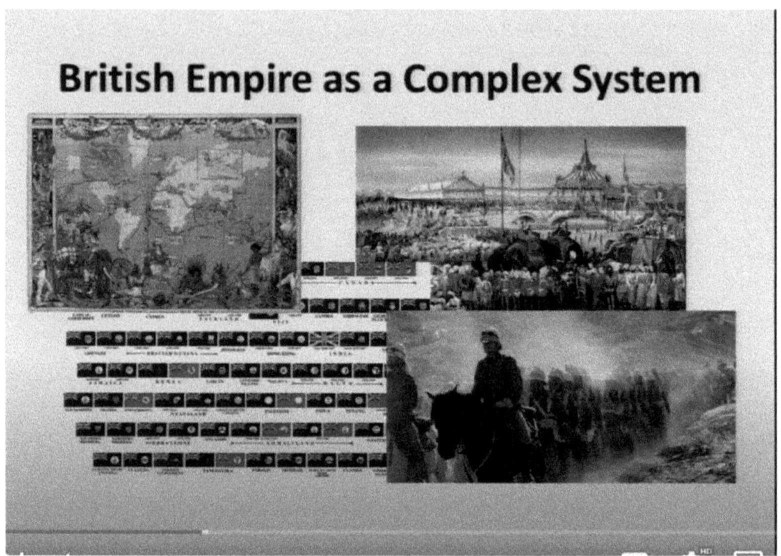

Now, we think a lot here about foresight methodologies — this is important. You know, we have said how difficult it is to predict the future — there're going to be black swans; but we still have to think about the future — what's the future going to bring? So, we try to bring in foresight methodologies to improve the way we think about the future, and we want to think about the future in new ways; it cannot be a linear way of looking at the future. And we also need to use these foresight methodologies to overcome our latent assumptions and biases. The goal is to shape the future, and not to predict what it will be.

Now, scenario planning is one of our basic methodologies in Singapore. And Shell was one of the pioneers; this is one of the key pioneers in that effort, Peter Schwartz I mean. And last year, it commemorated 40 years of scenario planning, and scenario planning helped Shell avoid the oil shock in the '70s.

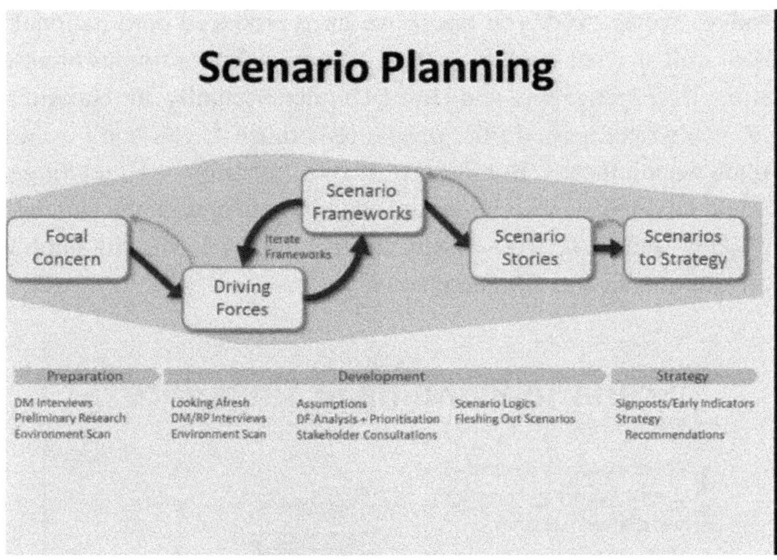

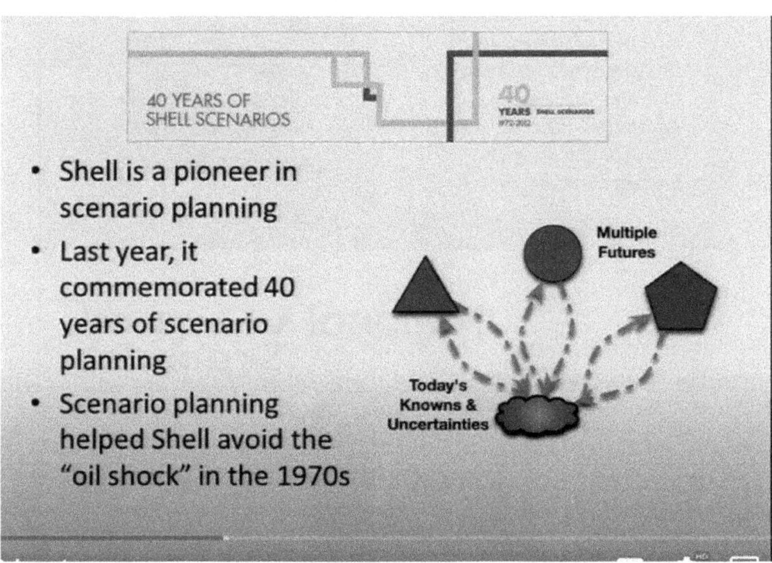

Now, for two decades, the Singapore government has been crafting national scenarios. And it's a key part of the Singapore government's strategic planning process; it's embedded into our annual strategic planning and budget cycles. And, you know, we have produced both national scenarios as well as what we call focused scenarios. So, examples of national scenarios: Hotel Singapore and Home Divided. Actually, in scenario planning, if you've got a good title, people remember it; you don't even have to explain what it means. But these were two scenarios — Hotel Singapore and Home Divided — which I would say kind of foresaw the situation we are in today, and these were produced very early on. We did some work also on new media and climate change.

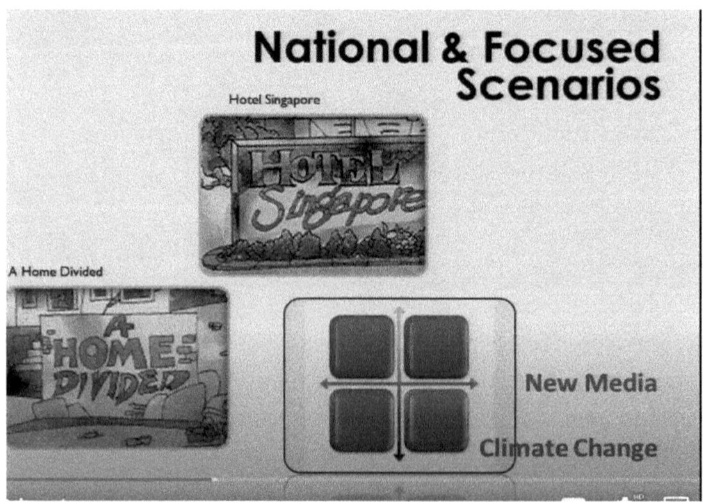

Talking about prostheses for the imagination: So, in Singapore, we use scenario planning partly to surface hidden assumptions and mental models, and — this is quoting Peter Wack — "We aim to design scenarios that lead our decision-makers to question their inner model of reality, and change it as necessary in order to take action."

But scenario planning doesn't quite fully address everything. It's not very good in a nonlinear, complex environment. And it also doesn't work very well in terms of the feedback loops because once you lock into a scenario, it kind of sticks with you for a while and is very difficult to change every few months just because the environment changes.

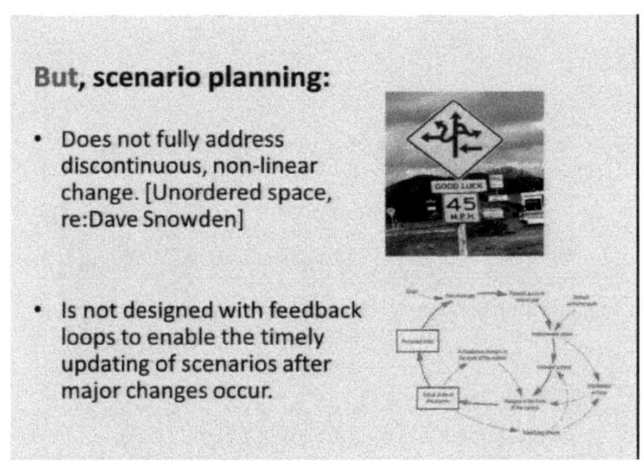

So, we've gone beyond scenario planning. Scenario planning remains our base, but we are doing things like backcasting, policy gaming, which I'll explain in a bit.

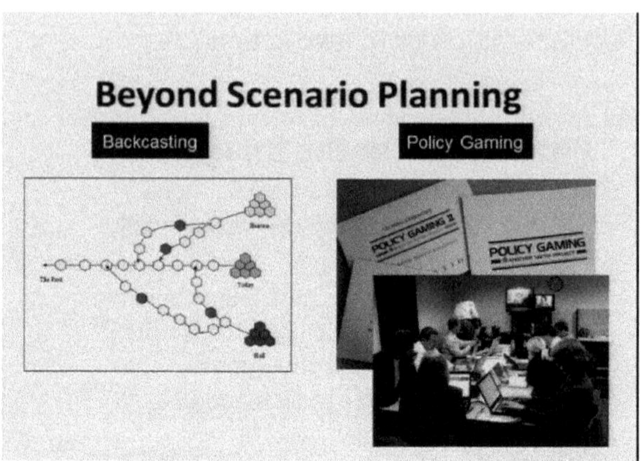

We now have a suite of tools called scenario planning plus, SP+, emphasizing that scenario planning is still there, it's still the base, but on top of that we complement it and supplement it with other methodologies: the Cynefin Framework from Dave Snowden, backcasting, Causal Layered Analysis. This is backcasting, it means you try to imagine the future you want, or your worst case — your worst nightmare — and what could possibly take you from here to there. Then you know what to do because then you're looking for indicators. Siemens calls this Photographs of the Future.

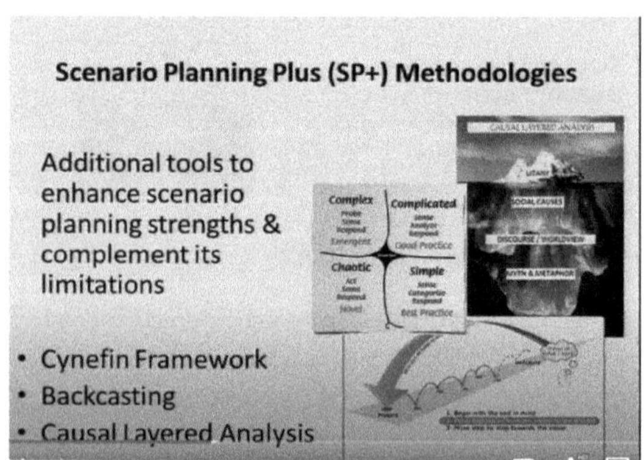

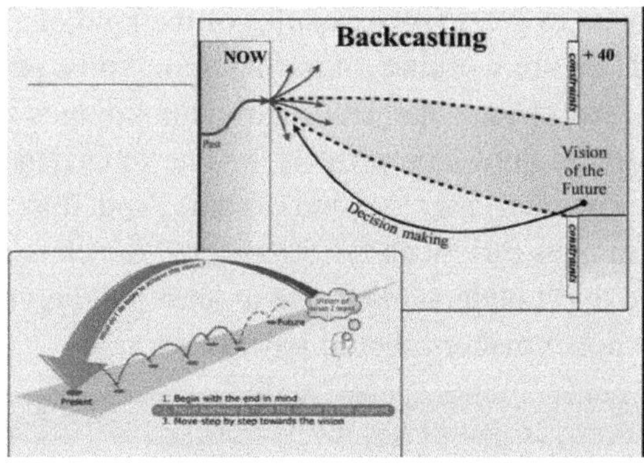

And we also work on emerging strategic issues; that means, what are the issues that are lurking just over the horizon, which could suddenly pop up and then either create challenges or opportunities for you.

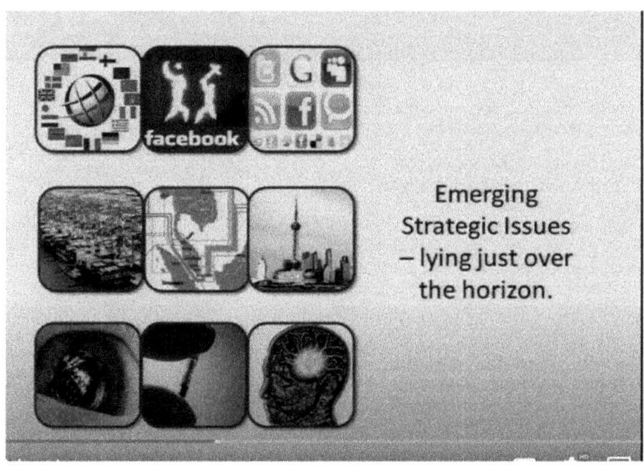

And I managed to get a few examples of the kind of emerging strategic issues which we are working on in Singapore: post-ethnic Singapore, very interesting. Just these data: Transnational marriages where one partner is non-Singaporean now comprise 41% of new marriages in Singapore, while 20% of new marriages are inter-ethnic, and this figure is rising steadily. So, what does this mean for Singapore? It's clearly going to have an impact — we don't understand it, but at least it's flagged as an emerging issue which policymakers should pay attention to.

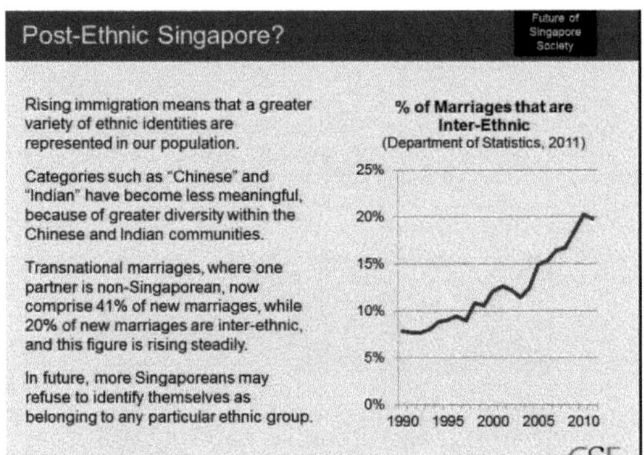

Enhancing our brains — this is just one of these things: Drugs which can improve your cognitive ability. And people will take — apparently, at least, I don't know where they get the data, from which country — but people will take cognition-enhancing drugs despite mild side effects, that's from *Nature*.

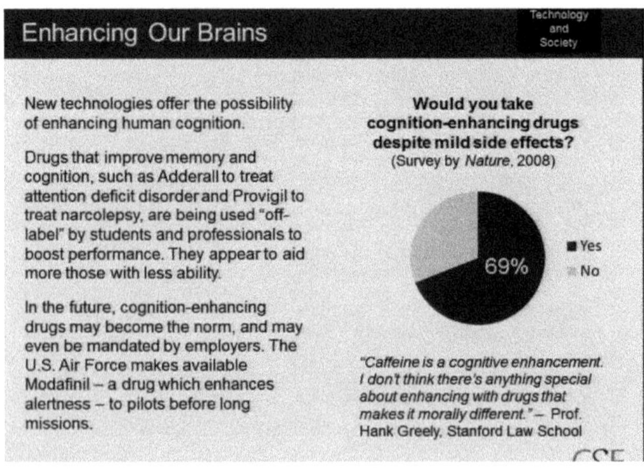

And then, 3D printing: Is there a kind of post-manufacturing economy that we have to start worrying about? You know, what happens when you can make things — customize things to order, and when you can do things in small batches, rather than on the big fort-like industrial scale? So, these are some of the emerging strategic issues we look at.

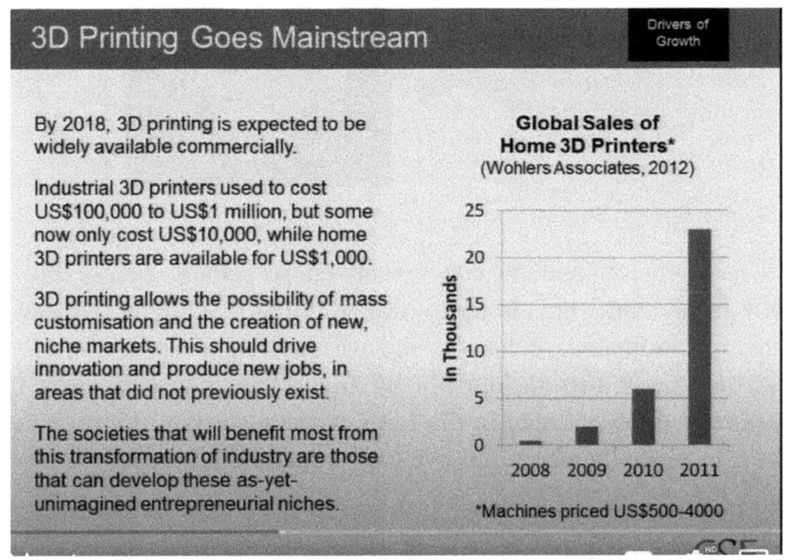

Policy gaming: How do you overcome your cognitive biases? How do you get your people to feel more comfortable making decisions in conditions of uncertainty? And my inspiration for this was the military. Because the military, they know — at least I think the best militaries will know — that the moment the first shot is fired, all their plans have to be thrown out of the window because of uncertainty and complexity. So, for governments also — why can't they behave like militaries? So, instead of calling it war-gaming, we call it policy gaming in Singapore. So, we are beginning to introduce it, and it's a safe environment for contemplating issues outside the comfort zone, and it's also good for helping people to make decisions in conditions of uncertainty and complexity.

Jan spoke about RAHS, and the inspiration for RAHS, by the way, was John Poindexter, who used to head the, he was Reagan's National Security Advisor, involved in the Iran-Contra scandal. But he's an

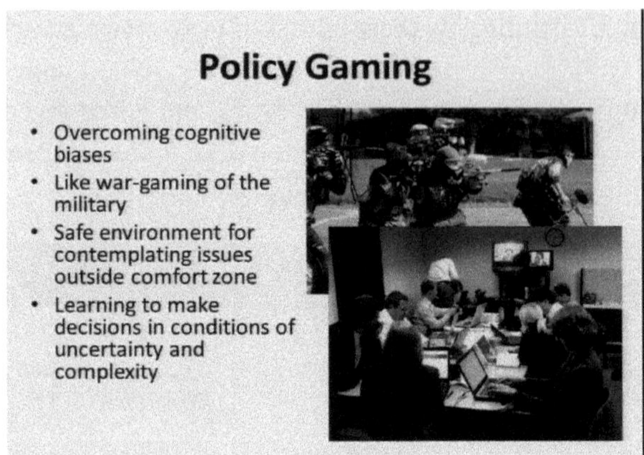

excellent engineer, and he conceptualized something before big data became fashionable: the ability to trawl through mountains of databases, to look for anomalous behavior, and we decided it's possible. He persuaded us that it was technically possible. The only difference is, in Singapore, we have a man in the loop, we don't leave it entirely to the machine to make the decisions. We have a man in the loop.

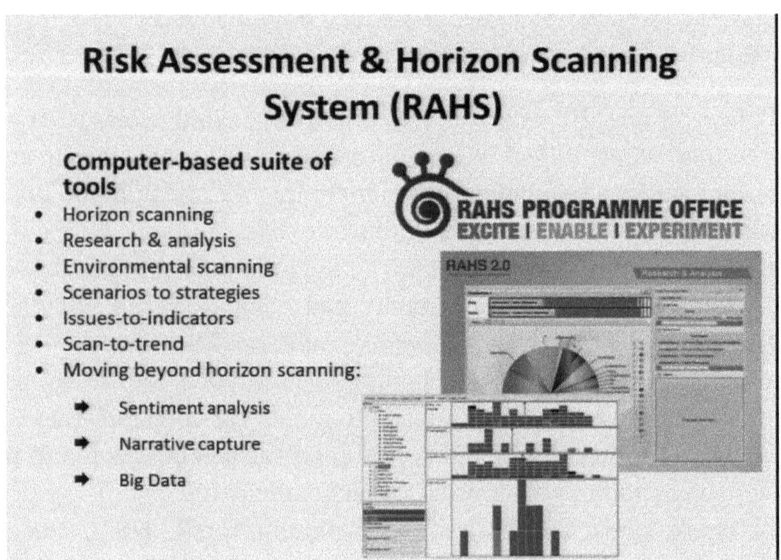

Now, in complex environments, experimentation is more valuable than prediction of analytical models. This is the famous eruption of this Icelandic volcano with a name which I cannot pronounce. Professor, can you pronounce that? [*Discussant A*: Eyja-fjalla-jokull] Eyja-fjalla-jokull. Fjalla…jokull? Jokull? Okay, that is a tongue-twister for me. But, why do we look at Eyjafjallajökull? Alright. Well, I guess in Iceland, you shouldn't be surprised to find glaciers, a lot of glaciers.

Why was this a surprise? Volcanoes erupt all the time. It was a surprise because it threw up a huge dust cloud, which then spread over Europe, forcing the aviation authorities to ground all the aircraft. It disrupted air travel, caused panic among the airlines operating in Europe, and began to affect global supply chains. Then what happened? Lufthansa, feeling the pressure, said, "This cannot go on forever." They proposed a test flight. By that time, the political pressure was so much that they decided to let Lufthansa go ahead. If something happened, it would only involve one aircraft and a couple of pilots. And it flew with no problems. So, after that, they lifted the ban. Remember, the Europeans based it on two data points: One involved British Airways, the other involved Singapore Airlines. Two decades before, there was a big volcanic eruption in Java, Indonesia. British Airways flew through that cloud and almost crashed. Some months later, Singapore Airlines also flew through it and experienced momentary power loss. Based on that, aviation authorities said, "Don't fly through such a dust cloud." But volcanic ash over Java is different from volcanic ash out of Iceland. The ash from Java is very gritty, pitting the turbine blades, while the ash from Iceland is very fine, like talcum powder. So, they let it fly through. Experimentation proved to be a better indicator for action than reliance on models. Exploration and experimentation are often more valuable than predictions of analytical models. We must be prepared to try things out, experiment, and do things in a safe-fail, not a fail-safe, mode. If we had taken a fail-safe approach, the aircraft would have been grounded for weeks more.

In 2008, while still in service, I said, "Rather than plan exhaustively for every contingency before we move, we should adopt a 'search and discover' approach: to act before the window closes and to act boldly in areas where we sense opportunities. We must be prepared to experiment, even if we cannot be entirely certain of the outcome." This approach is to probe, sense patterns, and act, even in the absence of complete information. As I have emphasized before, we must learn to operate not in a fail-safe mode, but instead to operate in a safe-fail mode.

> ## "Safe fail" vs "fail-safe"
> ## "Fail small" and "fail fast"
>
> Rather than plan exhaustively for every contingency before we move, we should adopt a "search and discover" approach: to act before the window closes, and to act boldly in areas where we sense opportunities. We must be prepared to experiment, even if we cannot be entirely certain of the outcome. The approach is to probe, sense patterns, and to act, even in the absence of complete information. As I have emphasised before, we must learn to operate not in a "fail-safe" mode, but instead to operate in a "safe-fail" mode.
>
> Peter Ho
> 4th Strategic Perspectives Conference
> 8 April 2008

So, this is the solution to Clayton Christensen's dilemma. His solution to the Innovator's Dilemma is to create small, self-contained units within the larger organization that have the mandate to experiment with new ideas and new concepts. Mainstream firms establish a timely position in disruptive technologies only when the firm's managers set up an autonomous organization charged with building a new and independent business around that disruptive technology. If the units succeed, then their formulas can be proliferated. If they fail, then the organizational impact will be limited. You know, it's ring-fenced.

And I did this in the Ministry of Defence. This was because I was quite worried about how we keep on improving the Ministry of Defence, the Singapore Armed Forces, so I set up a Future Systems Directorate, whose job is to try out completely new ideas. And the person I chose to head this — we call him the Future Systems Architect — his job is to think about the future. I gave him 1% of the defense budget to experiment — 1%. And I told him, "You have only one KPI — Key Performance Indicator — your KPI is if I don't get any complaints about you, you're not doing your job."

I constructed this family tree of failure — good failure and bad failure. We must learn from failure. Good failure is the lovechild of experimentation and discovery — it's the mother of success. A mother because paternity is a matter of opinion but maternity is a matter of fact. But of course, bad failure is the bastard spawn of ignorance and hubris, and it's a runt of success that's only acknowledged in whispers (never mind).

Managing risk in complex environments means big decisions have to be made. Even if you don't make a decision, you're making a decision. So, you have to make decisions in conditions of incomplete information and uncertain outcomes. In Singapore, because of this, we recently developed something called a whole-of-government Integrated Risk Management system. We're identifying risks at the strategic level, and we use this to mobilize resources, to bring behavioral change to address these risks, and to explore connections among risk factors. You'd be surprised. When you start doing this, at least with the techniques we have used, you can actually start uncovering some of the connections which were previously hidden.

Now, this fellow, Bar Yam, argues that the most basic issue for organizational success is correctly matching the system's complexities to its environment.

What this means is that complex organizations like Jemaah Islamiyah cannot be dismantled by simply setting up a single task force. They are self-organizing and have cells spread across various locations, held together by common beliefs. To address such complexity, we adopted a whole-of-government approach, involving multiple agencies, including those outside the government, to tackle different aspects of the problem. This approach recognizes the need for a comprehensive strategy to deal with a complex network effectively.

Jemaah Islamiyah (JI)

Fighting a Network with a Network

Whole-of-Government
approach to counter
the JI network:
- Greater coordination
between agencies
- Small, active centre
driving the national
counter-terrorism
agenda

Whole-of-Government
aka A Crude Look at the Whole

- A complex approach ➡ Whole-of-Government
vs   vs
- A complex problem ➡ A complex network
(Jemaah Islamiyah)

However, in practice, the biggest challenge lies in the existence of silos within organizations. Bureaucracies tend to operate in silos, with each department optimizing its efforts independently rather than considering the whole-of-government perspective. Yet, effective management of complexity requires the sharing of information horizontally, not just vertically. Breaking down these silos is crucial for discovering hidden connections, emerging challenges, and opportunities.

The whole-of-government approach is particularly effective for addressing wicked problems. Simply creating a separate department or ministry to handle a complex issue like climate change is insufficient, as such problems often involve multiple dimensions beyond the scope of any single ministry. To effectively tackle these challenges, we must adopt a whole-of-government approach. However, achieving this requires a significant shift in mindset and breaking down vertical silos within the government structure.

And this is the big obstacle to the whole-of-government. This is your turf, you don't disturb my turf. So, the whole-of-government is David against the Goliath of "My Turf," and everywhere, Singapore, U.S., Iceland, it's all the same. So, you must nag. Keep on telling people

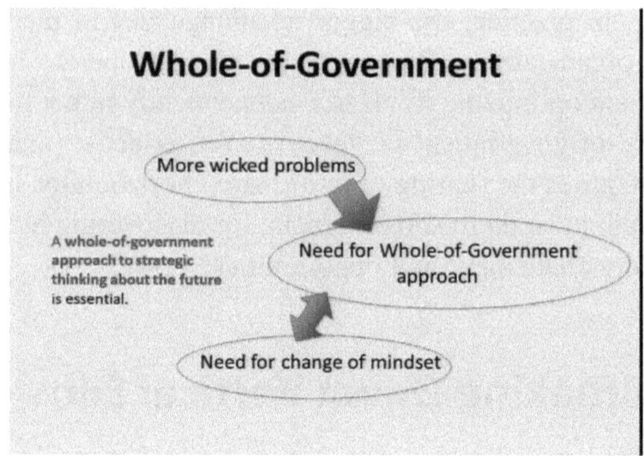

to break down the silos, all the time. Nagging is an inherent skill of leadership.

So, I think in Singapore maybe we are moving toward a kind of 3rd-generation leadership; there'll be many highways and byways along the way. But this means, from government to you, and government for you, we are going to move to government with you. That means we work together: whole-of-nation, whole-of-government. It's not government by agency, which is a silo; it's not even whole-of-government, it's a whole-of-nation coming together.

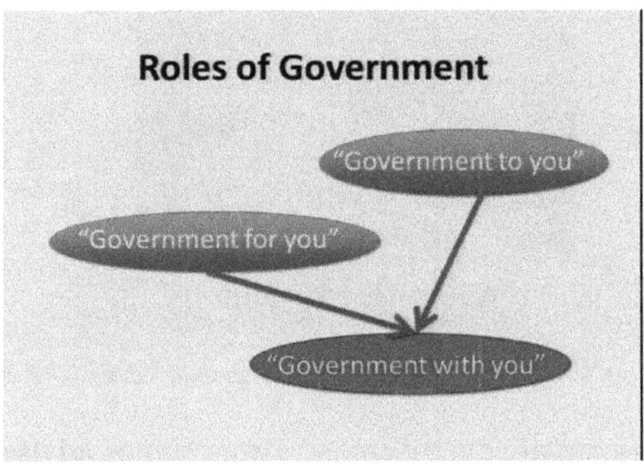

So, we'll talk a bit about resilience. How do you cope with strategic shocks? You must be resilient. But, you know, governments like to talk about lean and efficient. How do you have the extra capacity you need when a shock comes? And it's becoming more of an issue because if I'm right, there'll be more and more shocks. And if you're too lean and too efficient, you don't have the extra capacity to deal with these shocks. So, you must have a resilient system, that hump. Of course, it can also be a horse designed by a committee; you know, there's a resilient system, it's amazingly resilient, if you read the history of the camel. It thrives in Australia. It's not even native to Australia.

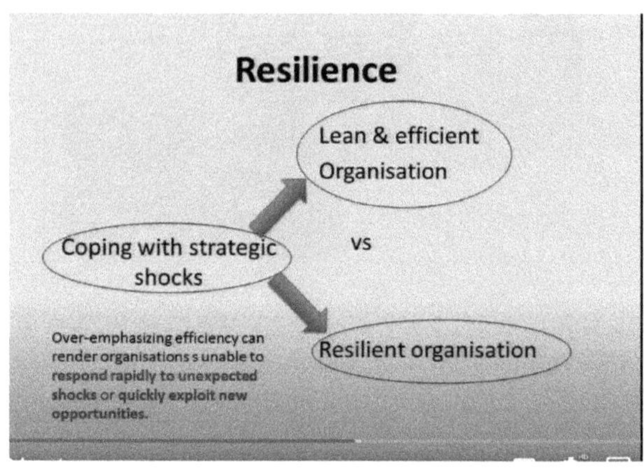

This is just a small advertisement; I'm a senior advisor to the Centre for Strategic Futures — it's one of our efforts to inject some resilience into the system. It's that extra capacity to think about the future that we're injecting into the Singapore Government — the Centre for Strategic Futures.

So, the key question of good governance is this: How can we prepare for a future in which we will be surprised time and again? This includes building deep capacity for strategic planning, employing forward-looking

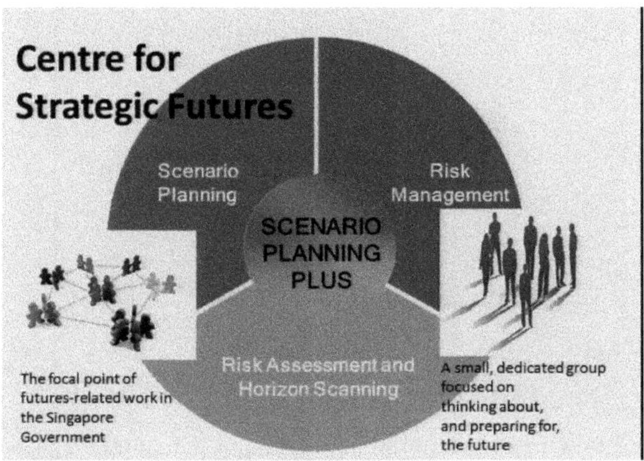

approaches and processes. We must be ready to challenge dominant narratives and paradigms, make decisions in conditions of uncertainty, and have the resources to deal with unexpected shocks. Additionally, we must work together collaboratively and innovatively across departments, adopting a whole-of-government/whole-of-nation approach.

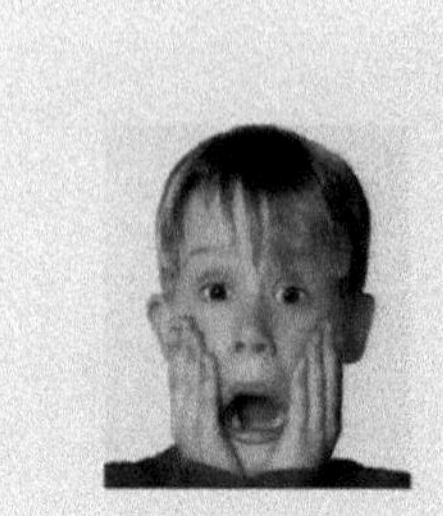
Lesson #1
Recognise that we are operating in an unordered complex environment, not an ordered, complicated environment.

Lesson #2
Acknowledge that human cognition has its limitations in dealing with complexity.

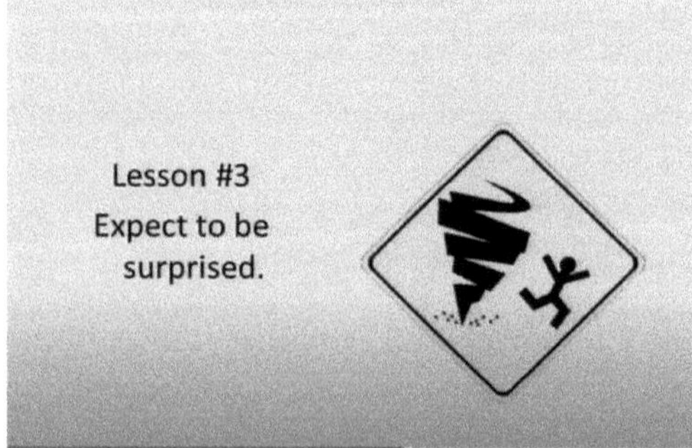
Lesson #3
Expect to be surprised.

The first lesson is to recognize that we are operating in an unordered, complex environment. Human cognition has its limitations, and we must expect to be surprised.

So, this is a final crude look at the whole. Recognize complexity, you must have tools, and the whole-of-government approach.

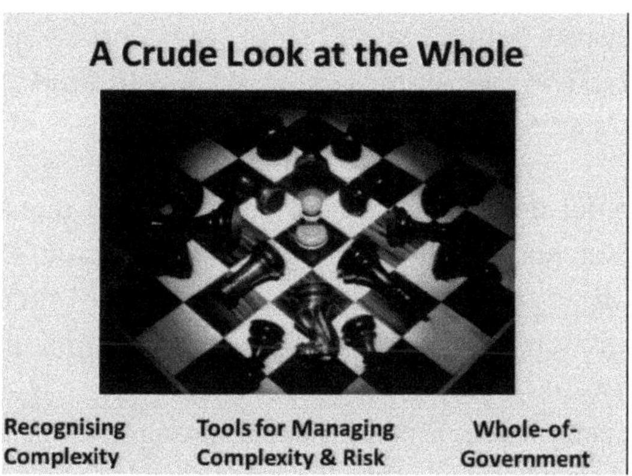

Now, I'd like to finish off with this quote from Winston Churchill about the qualities of a good top official. He said, "It is the ability to foretell what is going to happen tomorrow, next week, next month, and next year, and to have the ability afterwards to explain why it didn't happen."

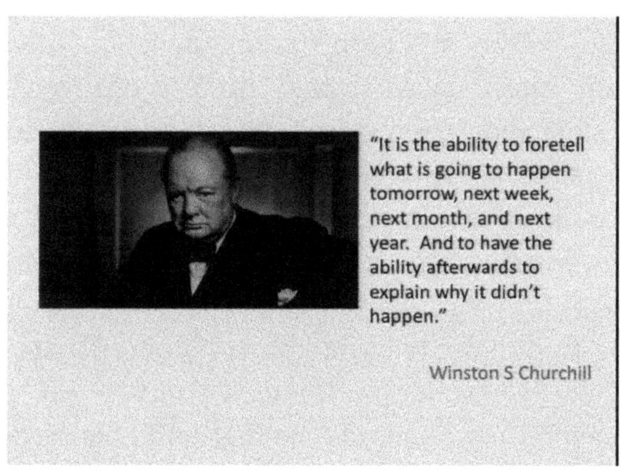

Thank you.

12.2 Discussion

Discussant A: … why do you think they did not want to mention that? Because Singapore tried public entrepreneurship and then decided to discontinue it. The other things that you mentioned about "Stop at 2," we have "Stop at 3," "Stop at 4," and many things; the question is, the government in the past was doing a lot of "stop, stop, stop," where the problem arises thereafter. The question is, is it backward-thinking or forward-thinking? So, it's a foresight problem.

Peter Ho: First, let me attempt to answer your first question by simply saying this was meant to be a crude look; it's not meant to be a comprehensive look. I tried to look at the key issues which I thought were relevant to the discussion. But on your second point, I think governments do try to look forward. They do try to look forward. Where the failure comes, if there is a failure, it is a failure to recognize that no matter what policy you implement, it has a use-by date, because of complexity. So, in other words, a policy that may work for, say, a couple of years may, because of complexity and consequences of complexity, start to become dysfunctional. So, in the case of our population policy, it became clear that it was a dysfunctional policy after about, I think, about 10, 15 years; it became clear that it was a dysfunctional policy. But then, at least you have to give the government credit for acknowledging that it was a dysfunctional policy. And then when they tried to change it, they failed. I showed in the slide they tried everything: incentives, disincentives, all kinds of things, and they're still failing. Now, they have got a new big package, supposed to be very generous, I don't know whether they'll have any impact, because all the evidence of other countries which have this kind of TFR is they will not succeed. There's one thing which I think we might not be able to do, but you know, I won't debate it here, he said France. We generally discourage families from having children out of wedlock, but maybe if you can, maybe that will change things.

Kerry Sieh: The difference probably wasn't the right one to use for your point. In the Java case, there was enough ash that it actually melted on the turbines and shut the engines down. In the case of that test flight, it was already diluted enough that it didn't. The basic problem here is actually complicated, and no manufacturer of engines wants to know what level of concentration of ash will shut their engines down. Singapore Airlines included, does not want to know what those levels are, they don't want to do a test. So actually, here's a case where just simple knowledge and basic experimentation ahead of time would actually have helped them decide when to appropriately shut down or open up the airspace again. But because no one has yet allowed the experiment to be done, nobody knew when to shut down or open up the airspace again. So, the activity you explained, what's the right thing to do? To do an experiment… [*Peter Ho*: Right, it should have been done before.] The experiment should have been done on the ground with engines, by throwing sand into an engine and see when it melts and shuts down. So, there is a case where science can do some good ahead of time if you guys would let us.

Peter Ho: But, I think this is partly ignorance, it's a cognitive bias. Why do I invest in something? The answers may not be entirely pleasant because if you do the experiment, the experiment itself may be cheap, but the medicine you have to take may be extremely expensive, and nobody wants to know. So, let's pretend it's not a problem.

Guaning Su: Two questions, Peter. We've been colleagues for a long time, and I really admire your intellect, and I think that was demonstrated in the talk. I don't know whether this is something you're in a position to comment on, but our Population White Paper has created a lot of uproar and discussion in the population. I'm not sure whether there was a lack of sufficiently good modeling of the system or whether there was too much linear extrapolation as opposed to looking at it as a complex system. I'd like your comment on that. And secondly, we've been talking about overturning conventional economics, and I think Brian Arthur is right here, he's the advocate of that. I wonder whether that would have helped in the formulation of the White Paper, in the sense of the economic growth model, and how economic incentives can work. I would be very keen to hear your comments on that.

Peter Ho: I'll respond to that, not in great detail because I think some of this may not be familiar to non-Singaporean participants. But the Population White Paper is a White Paper, as you know, in the parliamentary tradition, the Westminster parliamentary tradition, and it's supposed to reflect government policy. You table it, it's debated, but it's government policy. So, this was what happened: It was a government position on a projection of how the population was expected to grow. So, from the current 5.3 million, it's projected to grow to 6.9 million by the year 2030, and it created an uproar because there's a question of whether we have the space for this kind of population. But of course, there are a lot of unanswered questions in this policy, which I think is a result of perhaps not very good communication and explanation behind the White Paper. Because even if it grows, hits 6.9 million by 2030, which is quite a lot, what happens after 2030? Nobody has answered this question. And then there's one particular assumption, which I think reflects perhaps too simplistic a view of the situation. That is, there's only one economic model, because it's that economic model that forms the basis to drive the forecast of the population. Is productivity in the conventional sense the only way to go for Singapore? [This] means you must have more warm bodies and if you can't get it through the citizen population growth, then you have to bring it in — foreign population. And so, I think this was not adequately discussed, nor was it adequately explained. You know, we talk a lot about the knowledge economy: Which way is the economy going? How much of it will be a manufacturing economy in 20, 30 years' time? How much of it will be a knowledge economy? How much of it will be a service economy? I don't think that was given sufficient airing. So, I think at a simplistic level, it was not well communicated; at a sophisticated, intellectual level, it was I think also not very well presented. So, I think it caused a lot of confusion and unfortunately caused the government to backtrack on what should have been just a straightforward policy position. My 2 cents' worth.

Robert Axelrod: This is probably a delicate question, but I wanted to ask you about experimentation at the national level. Most democracies believe that there are two institutions that dramatically help in the long run. One is the circulation of elites through elections and competing political parties, and the other is a press that is independent and able to criticize the government. But, from what little I know about Singapore, there is neither one of those. So, do you think that there is a need for major change in the form of government and the press in order to maintain the extremely successful record of top-down management?

Peter Ho: You know, I take a slightly different approach. I'll respond in a slightly different way. I think what has happened is that these changes are going to happen, but it's change that's kind of forced on the government. What should have happened is that some of this should have been anticipated earlier, and the changes ought to have been made so you're not put in this kind of position. But again, this goes back to your Christensen's dilemma: You have been very successful. Look, this is one party that has been in power since 1957; it's got an impeccable track record in terms of delivering the goods, so why would they want to change the formula, even though there are many who say, you know, this is not going to work, this is a new generation, different expectations, the social media is having an impact — all these were said. A lot of the things, as I told you in the scenarios which we did, early in scenarios, Hotel Singapore and House Divided, were foreseen. But the incentive to change is not sufficient to overcome the inertia. So, I think this is a problem. That's why I say, if you are trying to influence decision-makers to make big course changes, you will either have to have a bit more patience and sometimes you don't have that luxury of time. So, I would say, yes, some of these things could have been done earlier, but the fact that they didn't is a reflection more of some of these cognitive problems we have to deal with. So, they are changing but they're changing in response. When you change ahead of the need to change, then you're in control of things. So, that would be my response to your question.

Discussant B: My question is, I completely agree with you and thanks for a very enlightening speech; the experimentation that you mentioned, the best way to learn is to go there and illustrate the failure. But sometimes for resource-crunch countries, this experimentation becomes difficult. Do you have anything, or a third way in mind which maybe, you know, if you're already rich, you can experiment with resources as you like, but if you are in a resource-crunch situation like, say, India and China, this kind of experimental failure might create a backlash.

Peter Ho: No, it's true, I'm not trying to paper over the problems. But don't forget that in Singapore, in the early years, I think we were in a completely experimental mode, and we had no resources, you could say there was a resource crunch, there was no money. Yet the government decided they were prepared to try things out. And when they failed, they failed; when they succeeded, they succeeded. And most of them, because they were clever people, were successful up to a point. The industrialization program, the housing program, education, everything — these were experiments. These were bold experiments, with no guarantee, because they were going against conventional wisdom. So, these were all experiments. And like the population thing, when you realize it's a mistake, well, you either fine-tune or tweak, or like population, you try to reverse, but unfortunately, you can't reverse it, so the damage is done, it can't be helped. But… So, yes, you're right, there are problems, but it's a question of what you want. You're a government, you're in charge, what are you prepared to do? If you say you lay the blame on a lack of resources or the bad reaction, then you just tweak here, tweak there; you won't bring any quantum improvement. You must be prepared to take a risk, but many governments are not prepared to take that risk, so if they're not prepared to take the risk, well, the consequences of that follow.

Kai Xin: Hello I'm Kai Xin from NTU. Could you share with us some of the key predictions for Singapore's future in maybe, like, 20 to 30 years? Like, it's okay if the predictions are wrong because you can always explain that.

Peter Ho: As I've pointed out to you, I hate to try to duck this question, but we're not in the business of making predictions. I wouldn't even dare, I you know, maybe I would have said I won't be alive, but you know, nowadays with the medicine and the healthcare, I might still be alive 30 years from now, so that was the only prediction I was about to say; but then, you know, you see the whole business is not about what the world is going to be. You know, if you take a straight-line prediction — projection — you'll end up at certain end points. But what is going to happen is there're going to be game-changers that will start to insert themselves at different points in time. It could be climate change; it could be a change of political leadership. Supposing suddenly PAP, come 2021, is no longer in power. That is a big game changer. Then, it brings you to a branching point in which you can then go into new trajectories. So, all I'm saying is the processes which we use can help us understand the possibilities, but it's not about assigning probabilities that this or this is going to happen; we don't really know. You know, many of the things which I flashed, whether it's ethnic groups, they're just emerging strategic issues, suggesting to you that there's something you should be looking at, and perhaps we should understand the implications a bit better. For example, manufacturing. Supposing manufacturing is changed dramatically by all this 3D manufacturing — home manufacturing, bespoke manufacturing, all these kinds of things where people do things at home, and then somebody somehow assembles it together somewhere else. What does it mean to your economy? Big changes — but it doesn't necessarily mean it's going to happen. But you'd better understand because if you're behind the curve, you will be in trouble.

Raymond Kwong: Raymond Kwong. Peter, the program is complexities, and my understanding is that nowadays it's getting shorter and shorter, everything is happening in one time. Uh, I'm here. Now the issue is that, do you think that by technologies and science and various other things, are we able to be ahead of the curve, to do what you were talking about, scenario planning plus to be able to [be] sufficient, you know, ahead of the curve, to understand and get a whole suite of deck cards that is able to say, okay, this is actually happening, are we able to, you know, react to this in this way?

Peter Ho: The first step is to have the tools. However, having the tools doesn't guarantee that people will act based on what you've produced. The next step is to get the decision-makers who are making the decisions to recognize the seriousness of the situation and to act accordingly. But that's I think almost a bigger challenge than thinking about the future. In fact, I believe it is the single biggest challenge. And that is why we are beginning to look at things that help us deal with these cognitive biases, like policy gaming and so on. You know, that's a whole big subject.

Kevin Koh: Good afternoon, Mr. Ho, sir. I'm Kevin Koh from NTU, School of Chemical Engineering. So, your 34 years of public service, that's really very illustrious and, as a Singaporean, I'm very appreciative of how you've shaped our country. I would like to ask you a question, wearing your hat as the Chairman of the Urban Renewal Authority. In the past 3 days, actually, our very illustrious guest speakers here have highlighted the impact of climate change on life on earth. In Singapore, and I'm sure many of the visitors here and lecturers here are also very sensitive about this in a very small country, just about a week ago, the past president of the Architect Association of Singapore, after quite a bit of calculations and studies, came up with a proposal in a conference. I think you may recall it; it was reported in the papers, giving a vision of maybe having a New York in the south and a Bali in the north of Singapore to help preserve our greenery, the little greenery that we have, but also to house our growing population. You know, he came up with a plan to house about 2 million more people, actually, in the nice, urbanized southern belt — of course, relocating our harbors and all our ports and all. I'm just wondering if you could comment on this vision, and whether URA sees this as quite a way to go to strike a balance.

Peter Ho: I know which proposal you're talking about. This is Tay Kheng Soon's proposal, and his proposal is to concentrate all the population in a very small enclave, to the south of Singapore where the city currently is. So, this will become a bit like this island country, is it Maldives? Anyway, whatever it is, yeah, these are options, but, don't forget, that kind of density carries certain social costs — if you want to live in that kind of very dense [city], [because], as it is, we're already one of the densest populated countries. So, all I'm saying is, the URA, the Urban Redevelopment Authority, will look at all these options, but in the end, there are very complex trade-offs you have to make. You want some space for your people, and they want the greenery — they're used to the greenery outside their homes. So, if you change their mindset and they're prepared to accept just looking across to their neighbor's window or their neighbor's kitchen, then okay, you can have that kind of thing. All I'm saying is, you can look at all the options, but there are very complex trade-offs because this is a wicked problem. Urban planning in a place like Singapore is a wicked problem because we are running up against very hard limits to our size. You know, you can reclaim a bit more land, assuming we can find the sand, you can reclaim a bit more land, but sooner or later you're going to run into deep waters or our territorial boundaries. So, how much more can you, how much more we can go, and when you start doing that, you have to give up your port. And then you need space for industry, you need space for recreation, so all these require trade-offs. I'm not saying Tay Kheng Soon's proposal is off the table. I'm saying it's only one of many options which are available, and the URA's job is to look at the different options and decide or make a recommendation on which course of action to choose.

Kevin Koh: Just to clarify sir, the idea was to house an extra 2 million. It's not to ship the rest of the population southward, yeah.

Peter Ho: Just housing an extra 2 million carries consequences; that's your city center, you know? There are your businesses, your financial centers, all there. If you start to crowd that [area], you have to kick out… you know, there won't be your nice Marina Bay.

Kevin Koh: Okay, thank you.

Discussant C: Peter, thank you for your talk. Now, since yours happens to be the last presentation, with your permission and the permission of the chairman, may I ask a question that perhaps covers a bit broader ground than Peter's specific talk? Of course, I would appreciate Peter's comments as well. So far, I missed a few talks, so I don't really know if I missed some talks that were on this particular subject. I believe that so far, I have not or it's not obvious to me that anybody has commented on this one thing. There's a lot of talk about the effect of human activities, about global warming, about all these topics. But when we look at these things, isn't it true that a lot of these activities are driven by economic considerations? And currently, when you look at economics, which is the indicator that is widely followed, that has got a lot of weightage, is it not the Gross Domestic Product (GDP), the GDP? Now, if we follow this indicator, then there will be consequences to it. In some of the talks, there was mention briefly of the Happiness Index, etc., but do you feel that perhaps we need to look at an alternative indicator? Because if you keep following one indicator, it has its consequences, and if we see so many consequences, like in the GDP, there is double counting going on. If I understand it correctly, as you up your productivity, there are negative consequences, but you don't count those, you don't take those as negative. So, your GDP goes up, and when you start to clean up again, you cater to your GDPs, so both ways now, the GDP goes up. So clearly, this is not favorable. Can we have some comments on this? I heard about alternative indicators like the GPI, but there are complications, etc. May I have your comments?

Peter Ho: I'm not sure if I'm competent to expand the discussion beyond the scope of this conference, but just a very short comment on GDP. GDP is clearly a very important indicator, but as you correctly point out, it doesn't necessarily describe the whole. It describes one aspect, and the search for alternatives that can describe the whole is also incomplete. For example, Gross National Happiness (GNH), which Bhutan pioneered, has its limitations. Just because you have a low GNH doesn't mean you're necessarily unhappy. I still remember that after a few years at the top of the ranking, Bhutan got displaced by Kiribati, one of these South Pacific islands known for inventing some form of bungee jumping. These are people who jump off rattan towers, and if you're unlucky, you break your neck. People die from jumping off these things. So, there's a search, but I don't think there's any agreement about how to define, in a holistic way, the well-being of a state. There are many factors, and even what is considered well-being today may change in 20 years' time due to changing expectations and technologies. However, I don't think we should expand this discussion to this area.

Greg Fisher: Thanks very much for your talk, Peter. I run a little think tank in London, and coincidentally, my co-director is Paul Ormerod, whom you quoted. I've spent a few years thinking about government policy and complexity, and I've pretty much drawn the same conclusions you have, so it's a kind of relief to see your presentation. But the question I have is about the receptivity, I guess, of complexity, by East and West. One of our esteemed colleagues yesterday talked about the different frameworks of reference and thinking in our heads between the Chinese and the West. And my experience, in general, is that if you speak to a Westerner, especially a politician, and you talk about complexity, they think I'm stupid because they just don't agree with me, okay, whereas if you speak to someone who's Chinese, they think I'm stupid because it's bloody obvious. Okay, so my question to you — that's a massive overgeneralization — do you think the Chinese mindset, which we heard yesterday, is more conducive to understanding complexity theory?

Peter Ho: Well, I think — first, I'll make a general comment. I think human beings, their minds, and their brains are all wired up in the same way. But if you look at a particular study done by Richard Nisbett, and if you've read a book called the *Geography of Thought*, he has studied the cultural differences between people with, say, a Western liberal type of background and those who come from a Confucian, Asian background, particularly the Japanese and the Chinese, and there are big differences in how they look at things. And I'll explain to you the differences. So, I think what it means is that although they've got the same set of cognitive biases, I think your chances of getting an Asian to look at the whole, rather than just look at the part, is higher than with a [inaudible] — and, you know, the difference? If you look at a picture, and the picture is of a horse, the Westerner will remember the horse, but he'll not remember the background, whereas the Asian is more likely to not just remember the horse, but he's going to remember the color of the sky, whether the grass was overgrown, or whether it was properly cropped. So, that's the ability to look at the whole picture rather than just what stands out.

Alex: Hello, my name's Alex, and I'm from the Lee Kuan Yew Centre for Innovative Cities. I like what you said about breaking the silos, and I wonder if you can describe for us what a 21st-century city would look like with an organizational structure that has successfully broken the silos?

Peter Ho: Well, I don't know if there's a real model for this, but I can give you a real example of the trend, at least in the Singapore Government. You know, in the Singapore Government, we're a traditional model of a government: We've got the Ministry of Health, the Ministry of Education, Ministry of Defence, and so on. That's a traditional hierarchical model of government, and we inherited that model from the British. And in normal times, it makes a lot of sense. But then, because of complexity, the demands and the challenges change over time. So, for example, imagine when you're confronted with a problem like Jemaah Islamiyah in 2001. The first important judgment that was made was that this is going to be a problem that's going to stay with us not for a few years, but for maybe generations. It's not going to be an existential problem, but it's going to be a problem we have to deal with. Now, we had two alternatives. One alternative was to do what was done in the United States, and that was to create a Department of Homeland Security, which meshed together something like 24, 25 departments pulled out from all over the place. You know, they took out Customs, they took out Coast Guard, they rammed them together and said, "Now we've got a department to deal with this problem." The irony is, Ash Carter, who's now the Deputy Secretary for Defense, wrote a very interesting piece which influenced us, called "The Architecture of Government in the Era of Transnational Terrorism." That was the piece. And he argued completely against the approach of creating a Department of Homeland Security. He said each of these departments has an existing function, which is not unimportant. So, if you take Customs, what's the function of Customs? It's revenue collection. What's the function of the Coast Guard? It's safety at sea. But if you put them under this Department of Homeland Security, then they will lose their focus on these duties, which are not unimportant, and start focusing on stopping the beast at bay. So, his recommendation was a different one. Go for a matrix organization. And in fact, I think Singapore went that way, and almost every country I know went that way, and created a kind of matrix organization. That means you draw on the strengths of different departments or ministries, and you just have a small coordinating center to coordinate the thing, and in Singapore, we did that by setting up the National Security Coordination Secretariat, and you give it its power by putting it under the Prime Minister. So, if people don't follow, you just crack the whip; the Prime Minister wishes it to be done, and that's it, everything is done. So, that's the... and I think increasingly, because complexity is going to introduce more and more of these kinds of big problems, we can't be setting up more and more departments to deal with... in the end, you'll have 100 departments — what's the point of it? You become dysfunctional, completely dysfunctional. So, you must find instead more of these coordinating mechanisms, but these coordinating mechanisms require the ability of main ministries, the main departments, to also commit to doing work that is outside their main statutory duties. That's whole-of-government, but that requires a whole new change of mindset.

Peter Sloot: Peter, many thanks for your talk. I have one question. You were saying we shouldn't wait for the future to happen; we should act and try to change it. Now, one of the things that you see happening time and again in all these complex problems is the role that the media play. Like, you know, for the Arab Spring, the Arab Winter, and even for pandemics, we know that the role of the media is significant. So, the media, to some extent, is kind of a weapon of mass destruction. Can we turn the media into a weapon of mass peace?

Peter Ho: My own guess is that we are moving through a particular phase in which the mass, or more precisely, the social media, has begun to exert an influence that even the traditional broadsheets and tabloids are not able to do. If you look at Singapore today, the mainstream media is declining; that means the newspapers are losing their readership. But the viewership of news and all that is increasing, the television news is going up. So, these are beginning to change, and my own guess is this will organize itself. Which way, I have no idea. Governments don't know how to deal with the new social media. Which *Facebook* postings do you respond to and which do you ignore? It's a big problem. Furthermore, it's not as if social media is of one voice. In fact, it's worse; it's highly fragmented. There may be groups that agree on one issue and then splinter to disagree on another issue. So, I think this is becoming a real challenge, and I suspect it's going to lead to complexity being an integral part of a government's ability to deal with more fragmented situations because more fragmentation means accentuating the role of individual agents within the complex system. How to deal with it? Maybe the best way is to just let them alone and get on with your business. I don't know, but I think it's not going to be easy to use a traditional model of regulation to make things happen. It may not happen that way. So, that's a long way of saying I really don't know where this is all going to go.

12.3 Summary of the Talk

Peter Ho delivered a comprehensive talk covering various aspects of governance, complexity, resilience, and strategic planning. The following is a summary of his key points:

Complexity Challenges: Ho highlighted the challenges posed by complexity, particularly in decision-making processes. As connections between problems and solutions become less clear, unintended consequences arise. Linear, mechanistic approaches are inadequate in navigating complex systems.

Case Studies: He illustrated his points with case studies of companies like Kodak, Borders, and Nokia, which failed to adapt to changing landscapes due to their inability to foresee disruptive innovations.

Cognitive Failure: Ho discussed cognitive biases and limitations, such as hyperbolic discounting and bounded rationality, which hinder effective decision-making in complex environments.

Scenario Planning: He emphasized the importance of foresight methodologies like scenario planning in navigating uncertain futures. Singapore employs scenario planning as a fundamental tool but supplements it with other methodologies like backcasting and policy gaming.

Policy Gaming: Inspired by military war-gaming, Singapore introduced policy gaming as a method to overcome cognitive biases and enhance decision-making under uncertainty.

Experimentation and Resilience: Ho stressed the value of experimentation and resilience in dealing with strategic shocks. He advocated for a "safe-fail" approach, encouraging proactive experimentation even in the absence of complete information.

Organizational Adaptation: Drawing from Clayton Christensen's work, Ho proposed creating small, autonomous units within larger organizations to foster innovation and adaptation to disruptive technologies.

Whole-of-Government Approach: He advocated for a whole-of-government approach to address complex issues, emphasizing the need to break down silos and foster collaboration across departments.

Resilience and Strategic Planning: Ho underscored the importance of resilience in lean and efficient organizations, particularly in preparing for future shocks. He discussed initiatives like the Centre for Strategic Futures, aimed at enhancing strategic planning and preparedness.

Future Challenges: Finally, Ho emphasized the need for deep capacity building in strategic planning, forward-looking approaches, and resource mobilization to tackle future challenges effectively.

In summary, Peter Ho's talk provided insights into the complexities of governance in an uncertain world, emphasizing the importance of foresight, experimentation, resilience, and collaborative approaches to address emerging challenges and navigate through unpredictable environments effectively.

12.4 Relevance of the Talk to the Current Stage of Research

Peter Ho's talk touches on several themes that are highly relevant to the current stage of research across various fields, including governance, decision-making, complexity theory, and strategic planning. The following points indicate how his insights align with ongoing research:

Complexity Theory: Ho's discussion of complexity theory and its implications for decision-making resonates with current research efforts in understanding complex systems. Researchers are exploring how interconnectedness, emergence, and nonlinearity impact governance structures and policy outcomes.

Cognitive Biases and Decision-Making: His exploration of cognitive biases and their influence on decision-making aligns with ongoing research in behavioral economics and psychology. Researchers continue to study how biases such as hyperbolic discounting and bounded rationality affect individual and organizational decision-making processes.

Foresight Methodologies: Ho's emphasis on foresight methodologies like scenario planning and policy gaming reflects a growing interest in anticipatory governance and strategic foresight. Research in this area focuses on developing tools and frameworks to help policymakers anticipate future challenges and opportunities.

Organizational Adaptation and Innovation: His discussion of creating small, autonomous units within organizations to foster innovation mirrors research on organizational agility and adaptive capacity. Scholars are examining how organizations can adapt to disruptive technologies and changing environments by embracing innovation and experimentation.

Whole-of-Government Approach: Ho's advocacy for a whole-of-government approach underscores the importance of interdisciplinary collaboration and coordination in addressing complex societal challenges. Research in this area explores strategies for breaking down silos and promoting cross-sectoral cooperation in governance.

Resilience and Strategic Planning: His focus on resilience and strategic planning aligns with research on risk management and preparedness. Scholars are investigating how organizations can build resilience to withstand shocks and disruptions while maintaining strategic agility and flexibility.

Overall, Peter Ho's insights provide valuable perspectives that resonate with ongoing research efforts across disciplines. His talk highlights the importance of embracing complexity, fostering innovation, and adopting adaptive strategies to navigate uncertain and rapidly changing environments effectively.

Addendum: Conferences and events organized by Para Limes

The conference "A crude look at the whole" was a special event and, looking back at it after 11 years, it has become an extraordinary event. The other conferences organized by Para Limes since 2012 were special events also, and in due time, taking a retrospective view, have the potential to become extraordinary events as well.

All the conferences shared a common theme: complexity. The purpose of each conference was to explore a special characteristic of complexity or, perhaps more aptly, to examine our world through the lens of complexity. This is evident in themes of the conferences, such as "More is Different", "Emerging Patterns", "Silent Transformations", "Complexities of time", "East of West, West of East", "Disrupted Balance, Society at Risk", "The Illusion of Control" and others.

A full list of the most exciting conferences is provided in the table below.

The list also includes workshops that focus on more specific aspects of complexity, such as "Complexity Methods", "The Complexity Lens" or areas where dealing with complexity is particularly relevant to our daily lives, such as "Governance and Complexity".

Additionally, the list features some discussions and series of lectures.

All the events included in the list were recorded on video, and the videos are accessible at https://www.paralimes.org/past-events/.

In many cases, these events ultimately led to the publication of books, most of which were published by World Scientific Publishers in Singapore as part of their series: Exploring Complexity.

This series now includes 11 volumes, with more to come. The titles of the books directly related to the different events are included in the table below.

Volumes already published under *Exploring Complexity Book Series*

1. *Volume 1: Aha… That is Interesting! John H Holland, 85 Years Young (2014)*
2. *Volume 2: Cultural Patterns and Neurocognitive Circuits: East–West Connections (2016)*
3. *Volume 3: 43 Visions for Complexity (2017)*
4. *Volume 4: Cultural Patterns and Neurocognitive Circuits II: East–West Connections (2018)*
5. *Volume 5: Selected Papers of John H Holland: A Pioneer in Complexity Science (2018)*
6. *Volume 6: Disrupted Balance – Society at Risk (2018)*
7. *Volume 7: Grand Challenges for Science in the 21st Century (2017)*
8. *Volume 8: Buying Time for Climate Action: Exploring Ways around Stumbling Block (2021)*
9. *Volume 9: Fit for Purpose? The Futures of Universities (2022)*
10. *Volume 10: Balanced Sustainability in a Changing World (2024)*
11. *Volume 11: Neuroaesthetics*
12. *Volume 12: A Crude Look at the Whole*

Table: Conferences and other events

Year	Theme	Link	Publication
2011	Workshop Complexity Methods	https://www.paralimes.org/past-events/ workshop-complexity-methods/	
2012	**Conference** More is Different	https://www.paralimes.org/past-events/ conference-more-is-different/	
2013	**Conference** A Crude Look at the Whole	https://www.paralimes.org/past-events/ conference-a-crude-look-at-the-whole/	*A Crude Look at the Whole* World Scientific Series Exploring Complexity, Vol 12
	Workshop Complexity and Governance	https://www.paralimes.org/past-events/ workshop-complexity-governance/	
2014	**Conference** Hidden Connections	https://www.paralimes.org/past-events/ conference-hidden-connections/	
	Workshop East-West Connections — "Cultural Circuits in the Brain"	https://www.paralimes.org/past-events/ workshop-east-west-connections-cultural- circuits-in-the-human-brain/	*Cultural Patterns and Neurocognitive* *Circuits: East–West Connections* World Scientific Series Exploring Complexity, Vol 2
2015	**Conference** Emerging Patterns	https://www.paralimes.org/past-events/ conference-emerging-patterns/	
	Workshop Complexity Lens	https://www.paralimes.org/past-events/ workshop-complexity-lens/	

(Continued)

(*Continued*)

Year	Theme	Link	Publication
2016	**Conference** Disrupted Balance — Society at Risk	https://www.paralimes.org/past-events/workshop-east-west-connections-cultural-patterns-cognitive-patterns-and-a-good-life/	*Disrupted Balance — Society at Risk* World Scientific Series Exploring Complexity: Vol 6
	Conference Silent Transformations	https://www.paralimes.org/past-events/conference-silent-transformations/	
	Conference East of West, West of East	https://www.paralimes.org/past-events/conference-east-west/	
	Conference East-West Connections: Grand Challenges in Brain, Cognition, and Good Life Research	https://www.paralimes.org/past-events/conference-east-west-connections-grand-challenges-in-brain-cognition-and-good-life-research/	*Cultural Patterns and Neurocognitive Circuits II: East–West Connections* World Scientific Series Exploring Complexity, Vol 5
	Discussion Grand challenges of Science in the 21st Century	https://www.paralimes.org/past-events/grand-challenges-for-science-in-the-21st-century/	*Grand Challenges of Science in the 21st Century* World Scientific Series Exploring Complexity: Vol 7
2017	**Conference** Causality – Reality	https://www.paralimes.org/past-events/conference-causality-reality/	
	Series	https://www.paralimes.org/past-events/10-on-10-the-chronicles-of-evolution/	*Sydney Brenner's 10-on-10: The Chronicles of Evolution* Wild-type Books

	Lectures Sydney Brenner's Lectures	https://www.paralimes.org/past-events/ sydney-brenners-lectures/	*In the Spirit of Science* World Scientific
2018	**Conference** Complexities of Time	https://www.paralimes.org/past-events/ conference-complexities-of-time/	
	Lectures W. Brian Arthur Lectures	https://www.paralimes.org/ past-events/w-brian-arthurs-lectures/	
2021	**Webinar** Removing Barriers to Buy Time	https://www.paralimes.org/past-events/ webinar-removing-barriers-to-buy-time-15-17- feb-2021/	*Buying Time for Climate Action,* *Exploring ways around Stumbling* *Blocks* World Scientific Series Exploring Complexity: Vol 8
2023	**Conference** The Illusion of Control	https://www.paralimes.org/past-events/ conference-illusion-of-control/	*The Illusion of Control* University of California Press, Global Perspectives, Special Collection https://online.ucpress.edu/gp/ collection/11457/Special- Collection-The-Illusion-of- Control